DATE DUE

PARKINSON'S DISEASE AND MOVEMENT DISORDERS: DIAGNOSIS AND TREATMENT GUIDELINES FOR THE PRACTICING PHYSICIAN

CURRENT ◊ CLINICAL ◊ PRACTICE

PARKINSON'S DISEASE AND MOVEMENT DISORDERS

DIAGNOSIS AND TREATMENT GUIDELINES FOR THE PRACTICING PHYSICIAN

Edited by

CHARLES H. ADLER, MD, PhD

Consultant and Co-Director, Parkinson's Disease and Movement Disorders Center, Department of Neurology, Mayo Clinic Scottsdale, Scottsdale, AZ; Associate Professor of Neurology, Mayo Medical School, Rochester, MN

and

J. ERIC AHLSKOG, PhD, MD

Chair, Division of Movement Disorders, and Consultant, Department of Neurology, Mayo Clinic Rochester, Rochester, MN; Professor of Neurology, Mayo Medical School, Rochester, MN

 HUMANA PRESS
TOTOWA, NEW JERSEY

PREFACE

The field of movement disorders is relatively broad, encompassing disorders of increased movement, such as tremors, dystonia, and tics, to disorders characterized by a paucity of movement, such as Parkinson's disease. Our understanding of the pathogenic mechanisms and our treatment options are expanding at a rapid pace. This expansion ranges from the medical and surgical advances in treating Parkinson's disease to the flood of genetic abnormalities that have now been found to cause various movement disorders. Although many patients are seen by the movement disorders specialist in neurology clinics around the country, most of these patients receive their followup care from a primary care physician or "general" neurologist who must be versed in the characteristics and treatment plans of this diverse group of disorders.

The major goal of *Parkinson's Disease and Movement Disorders: Diagnosis and Treatment Guidelines for the Practicing Physician* is to distill this immense amount of information and to educate the practitioner about the many facets of the movement disorders field. We believe that this book fills a large void, since most texts on movement disorders are more detailed and geared toward the specialist. We have asked the chapter authors to emphasize the clinical characteristics of each disorder, discuss the differential diagnosis and the diagnostic testing, and then outline the various treatment options, as if they were teaching during a preceptorship in their clinic. To this end, we have not designed the book to be an exhaustive review of each topic; rather, it takes a general approach to each subject. We have avoided referencing each statement; a short list of further recommended reading sources is given at the end of each chapter.

The purpose of this text is to help the practitioner distinguish which disorder is being encountered, to provide a basic understanding of the test and treatment options required, and synthesize any recommendations made by a consulting specialist. As the movement disorders specialist becomes busier and insurance regulations limiting specialty referrals increase, the burden of caring for these patients by the primary care physician will continue to grow. Thus, we hope that this text will offer the reader full confidence in approaching patients with movement disorders.

The text is organized into five sections: basic diagnostic principles, Parkinson's disease, other parkinsonian disorders, hyperkinetic movement disorders, and other movement disorders. In Chapter 1 of Section A: Basic Diagnostic Principles, Dr. J. Eric Ahlskog provides an extensive overview on the neurologic examination. Since movement disorders can involve all parts of the neurologic system, a detailed neurologic examination is often imperative when these patients are evaluated. Changes in speech often occur in many of the disorders, and the speech characteristics may provide important diagnostic clues. In Chapter 2, Dr. Joseph Duffy describes the varieties of motor speech abnormalities that may be encountered and provides a systematic approach to their recognition.

Given the frequency of Parkinson's disease (PD), the tremendous advances made over the past two decades in understanding the disease and its treatment, and the debates on the "best" way to treat the patient, we have devoted 12 chapters to this entity in Section B: Parkinson's Disease. In Chapter 3, Dr. Howard Hurtig discusses the pathophysiology, neurochemistry, and neuropathology of PD. This is followed by Dr. Richard Dewey's description of the clinical characteristics of PD in Chapter 4, outlining not only the typical features, such as rest tremor and slowness of movement, but also clinical signs suggesting an atypical form of parkinsonism.

What causes PD? The intriguing search for etiologic answers has generated volumes of studies and papers, sometimes with conflicting results. This information is distilled in two chapters. The epidemiologic studies of PD, which have generated multiple clues to the etiology, are reviewed by Dr. Demetrius Maraganore in Chapter 5. This chapter covers reported risk factors and the role of genetics (including the defects in the α-synuclein gene) in addition to basic incidence and prevalence data. On the other hand, basic bench research has produced multiple lines of evidence for a variety of possible causal factors. Etiologic hypotheses generated from knowledge of biochemical mechanisms are comprehensively reviewed by Dr. Peter LeWitt in Chapter 6. This chapter includes discussion of the roles of oxidative stress, mitochondrial dysfunction, and neurotoxins and addresses etiologic mechanisms that have received less publicity, such as possible autoimmune and infectious causes.

We have divided the discussion of the medical treatment of PD into three chapters. The issue of whether our current drug arsenal includes medications that will slow the progression of PD is hotly debated, and in Chapter 7, Dr. Ahlskog tackles the theories and practical issues for the practitioner. Chapter 8, also by Dr. Ahlskog, is devoted to the various treatment options available for the patient with newly diagnosed PD. This includes decision-making regarding the use of levodopa, dopamine agonists, and numerous other agents. The complicated issue of how to treat the patient with more advanced PD is covered by Dr. Ryan Uitti in Chapter 9.

Patients with PD also have nonmotor manifestations that can be as disabling as the tremor and bradykinesia. The sleep problems of PD, including insomnia and daytime drowsiness, are addressed by Dr. Cynthia Comella in Chapter 10. Neurogenic bladder and bowel problems and symptomatic orthostatic hypotension are common issues in the PD clinic; these autonomic problems and treatment strategies are covered by Dr. Bradley Hiner in Chapter 11. Dementia, psychosis, and depression can be overwhelming factors in the patient with more advanced PD, and Dr. Erwin Montgomery discusses these in Chapter 12.

Currently, the most visible topic concerning PD is surgical treatment, which has made newspaper and television headlines for the past several years. In Chapter 13, Dr. Kathleen Shannon discusses neurosurgical intervention, including pallidotomy, thalamotomy, deep brain stimulation, and cerebral transplantation, and reviews the prospects for the future. She provides guidelines on which patients may benefit and which of the different procedures may be appropriate for a given patient.

Therapy for patients with PD does not end with medications and surgery; Chapter 14, written by Drs. Padraig O'Suilleabhain and Susan Murphy, addresses adjunctive treatments. They include nutrition and dietary issues, which are especially important in advancing disease. They also address the role of physical therapy in management of parkinsonian motor problems.

All disorders characterized by slowness of movement are not PD, and in Section C we have devoted six chapters to discussing these other disorders. In Chapter 15, Drs. Eric Molho and Stewart Factor cover secondary causes of parkinsonism, such as vascular, toxic, and traumatic etiologies; they also provide a practical strategy for the workup of parkinsonism. Among the more common neurodegenerative disorders sometimes mistaken for PD is progressive supranuclear palsy; the key clinical signs and points that separate this disorder from PD are covered by Dr. Mark Stacy in Chapter 16. When patients have cerebellar signs, prominent autonomic dysfunction, or resistance to dopaminergic therapy, one must consider the multiple system atrophies discussed by Dr. James Bower in Chapter 17. Inherited cerebellar disorders, including the autosomal dominant spinocerebellar ataxias, sometimes resemble sporadic multiple system atrophy and occasionally PD; these familial ataxic syndromes are reviewed by Dr. Bower in Chapter 18. Corticobasal degeneration may resemble PD early in the course. The clinical hallmarks that allow differentiation are covered by Dr. Brad Boeve in Chapter 19. The final chapter in this section, Chapter 20, by Dr. Richard Caselli, describes the various primary dementing disorders, including Alzheimer's disease, that often include components of parkinsonism.

Section D begins the discussion of disorders characterized by too much movement, or hyperkinetic movement disorders. All the chapters in this section address the characteristics of the individual disorders, diagnostic considerations, and treatment options. We begin by a discussion of the most commonly encountered movement disorder, tremor. Chapter 21, by Dr. Joseph Matsumoto, describes the different types of tremor and how to differentiate and treat them. Chapter 22, by Dr. Jean Hubble, goes into further detail about the most commonly seen tremor, essential tremor.

Dystonia is a more common disorder than is often appreciated, occurring in adulthood as torticollis, blepharospasm, writer's cramp, and other focal or segmental dystonias. These along with generalized dystonic conditions, including primary torsion dystonia developing in childhood, are reviewed by Dr. Daniel Tarsy in Chapter 23. Hemifacial spasm is sometimes mistaken for facial dystonia; this disorder, caused by compression of the seventh cranial nerve, is discussed by Dr. Mark Lew in Chapter 24.

The dancing movements of chorea have origins ranging from inherited (Huntington's disease) to infectious (Sydenham's chorea) causes. Clinical characterization and treatment are covered by Dr. John Caviness in Chapter 25. Tardive dyskinesias are sometimes confused with primary choreiform syndromes. These iatrogenically induced conditions are discussed by Dr. Kapil Sethi in Chapter 26.

The lightning-like jerks of myoclonus occasionally cause diagnostic confusion; if repetitive, they may resemble tremor or the phasic movements seen in some dystonic conditions. In Chapter 27, Dr. Caviness discusses diagnostic criteria, categorization, and treatment of myoclonus.

Simple spasm of muscle may have a wide variety of causes, ranging from peripheral to central nervous system origins. In the most elementary sense, the concept of muscle spasm should include the sustained muscle contraction state of dystonia. Primary disorders characterized by muscle spasm, however, have their own distinguishing features, which separate them from primary dystonias. These disorders of muscle spasm, including the prototypical condition, stiff-man syndrome, are discussed by Dr. Michel Harper in Chapter 28.

The most common movement disorder of childhood is that of tics. This problem, however, is not confined to children and occasionally confronts internists with adult practices. The spectrum of motor and other tics, as well as the constellation of symptoms that make up Tourette's syndrome, is the topic of Chapter 29 by Drs. Kathleen Kujawa and Christopher Goetz.

A disorder that has gained much recognition in the past few years is restless legs syndrome, discussed by Drs. Virgilio Evidente and Charles Adler in Chapter 30. Dr.

Adler then covers the various uses of botulinum toxin (Chapter 31), an injectable agent that reduces movement and has application for treating multiple different movement disorders. This drug has revolutionized the treatment of dystonia and certain other hyperkinetic movement disorders.

We conclude this book with Section E, which covers other movement disorders, those that do not fit well into previous sections. Dr. Katrina Gwinn-Hardy, in Chapter 32, discusses the autosomal recessively inherited disorder Wilson's disease, which can present with hyperkinetic or bradykinetic features. This is critical to diagnose, since treatment is available; if unrecognized, it can result in irreversible neurologic damage and even death by hepatopathy.

Abnormal gait is a common component of many neurologic disorders and especially the conditions covered in this text. Recognition of the prototypical types of gaits is critical to diagnosis. Dr. Frank Rubino applies his years of clinical savvy in Chapter 33, which addresses the broad topic of gait disorders.

Commonly, patients attribute their condition to some prior trauma. How often does this occur? Although subject to much debate, the topic of trauma-induced movement disorders is covered in Chapter 34 by Dr. Sotirios Parashos. Possibly the most difficult problem for clinicians is that of a psychogenic movement disorder. In Chapter 35, Drs. David Glosser and Matthew Stern have written a very reader-friendly review of what the practitioner should consider when approaching these patients. We conclude this book with an appendix that lists many of the organizations and foundations devoted to the disorders discussed in the book.

We wish to thank all the authors for their hard work and excellent contributions. We thank the Mayo Clinic Section of Scientific Publications, specifically Roberta Schwartz, Marlené Boyd, Reneé Van Vleet, and John Prickman, and Humana Press, specifically Paul Dolgert, for their diligent effort in publishing this text. We both would especially like to thank our wives, Laura Adler and Faye Ahlskog, as well as our children, Ilyssa and Jennifer Adler and Michael, John, and Matthew Ahlskog, for their support, encouragement, and patience during the long hours it took to complete this book. We hope that our combined efforts have created a readable text for the primary care physician that has distilled the tremendous advances made in the movement disorders field leading up to the millenium.

Charles H. Adler, MD, PhD
J. Eric Ahlskog, PhD, MD

CONTENTS

E. OTHER MOVEMENT DISORDERS   395

CONTRIBUTORS

CHARLES H. ADLER, M.D., PH.D. • *Consultant and Co-Director, Parkinson's Disease and Movement Disorders Center, Department of Neurology, Mayo Clinic Scottsdale, Scottsdale, Arizona; Associate Professor of Neurology, Mayo Medical School, Rochester, Minnesota*

J. ERIC AHLSKOG, PH.D., M.D. • *Chair, Division of Movement Disorders, and Consultant, Department of Neurology, Mayo Clinic Rochester, Rochester, Minnesota; Professor of Neurology, Mayo Medical School, Rochester, Minnesota*

BRADLEY F. BOEVE, M.D. • *Consultant, Division of Behavioral Neurology, Department of Neurology, Mayo Clinic Rochester, Rochester, Minnesota; Assistant Professor of Neurology, Mayo Medical School, Rochester, Minnesota*

JAMES H. BOWER, M.D. • *Consultant, Division of Movement Disorders, Department of Neurology, Mayo Clinic Rochester, Rochester, Minnesota; Assistant Professor of Neurology, Mayo Medical School, Rochester, Minnesota*

RICHARD J. CASELLI, M.D. • *Chair, Division of Behavioral Neurology, and Consultant, Department of Neurology, Mayo Clinic Scottsdale, Scottsdale, Arizona; Associate Professor of Neurology, Mayo Medical School, Rochester, Minnesota*

JOHN N. CAVINESS, M.D. • *Consultant and Co-Director, Parkinson's Disease and Movement Disorders Center, Department of Neurology, Mayo Clinic Scottsdale, Scottsdale, Arizona; Associate Professor of Neurology, Mayo Medical School, Rochester, Minnesota*

CYNTHIA L. COMELLA, M.D., A.B.S.M. • *Associate Professor, Department of Neurological Sciences, Department of Psychology, Rush Medical College, Rush-Presbyterian-St. Luke's Medical Center, Chicago, Illinois*

RICHARD B. DEWEY, JR., M.D. • *Director, Clinical Center for Movement Disorders, Department of Neurology; Assistant Professor, University of Texas Southwestern Medical School, Dallas, Texas*

JOSEPH R. DUFFY, PH.D. • *Chair, Division of Speech Pathology, Mayo Clinic Rochester, Rochester, Minnesota; Professor of Speech Pathology, Mayo Medical School, Rochester, Minnesota*

VIRGILIO GERALD H. EVIDENTE, M.D. • *Director, Movement Disorders Center, St. Luke's Medical Center, Quezon City, Philippines*

STEWART A. FACTOR, D.O. • *Riley Family Chair, Parkinson's Disease and Movement Disorders Center; Professor, Department of Neurology, Albany Medical College, Albany, New York*

DAVID S. GLOSSER, SC.D. • *Clinical Assistant Professor, Department of Neurology, Jefferson Medical College, Philadelphia, Pennsylvania*

CHRISTOPHER G. GOETZ, M.D. • *Professor, Department of Neurology, Rush Medical College, Rush-Presbyterian-St. Luke's Medical Center, Chicago, Illinois*

KATRINA A. GWINN-HARDY, M.D. • *Research Associate, Mayo Clinic Jacksonville, Jacksonville, Florida; Assistant Professor of Neurology, Mayo Medical School, Rochester, Minnesota*

C. MICHEL HARPER, JR., M.D. • *Consultant, Division of Neuroimmunology, Department of Neurology, Mayo Clinic Rochester, Rochester, Minnesota; Associate Professor of Neurology, Mayo Medical School, Rochester, Minnesota*

BRADLEY C. HINER, M.D. • *Staff Neurologist, Department of Neurosciences, Marshfield Clinic, Marshfield, Wisconsin*

JEAN PINTAR HUBBLE, M.D. • *Associate Professor, Department of Neurology, Ohio State University, Columbus, Ohio*

HOWARD HURTIG, M.D. • *Chair, Department of Neurology, Pennsylvania Hospital; Professor of Neurology, University of Pennsylvania School of Medicine, Philadelphia, Pennsylvania*

KATHLEEN A. KUJAWA, M.D., PH.D. • *Instructor, Movement Disorders Section, Department of Neurological Sciences, Rush Medical College, Chicago, Illinois*

MARK F. LEW, M.D. • *Director, Division of Movement Disorders; Associate Professor of Neurology, University of Southern California School of Medicine, Los Angeles, California*

PETER A. LEWITT, M.D. • *Professor, Departments of Neurology, Psychiatry, and Behavioral Neurosciences, Wayne State University School of Medicine, Detroit, Michigan*

DEMETRIUS M. (JIM) MARAGANORE, M.D. • *Consultant, Division of Movement Disorders, Department of Neurology, Mayo Clinic Rochester, Rochester, Minnesota; Associate Professor of Neurology, Mayo Medical School, Rochester, Minnesota*

JOSEPH Y. MATSUMOTO, M.D. • *Consultant, Division of Movement Disorders, Department of Neurology, Mayo Clinic Rochester, Rochester, Minnesota; Assistant Professor of Neurology, Mayo Medical School, Rochester, Minnesota*

ERIC S. MOLHO, M.D. • *Assistant Professor, Department of Neurology, Albany Medical College, Albany, New York*

ERWIN B. MONTGOMERY, JR., M.D. • *Head, Movement Disorders Program, Department of Neurology, Cleveland Clinic Foundation, Cleveland, Ohio*

SUSAN M. MURPHY, M.D. • *Assistant Professor, Department of Physical Medicine and Rehabilitation, University of Texas Southwestern Medical Center, Dallas, Texas*

PADRAIG E. O'SUILLEABHAIN, M.B. • *Assistant Professor, Department of Neurology, University of Texas Southwestern Medical Center, Dallas, Texas*

SOTIRIOS A. PARASHOS, M.D., PH.D. • *Minneapolis Clinic of Neurology Ltd.; Director of Research, Struthers Parkinson's Center; Clinical Instructor of Neurology, University of Minnesota, Minneapolis, Minnesota*

FRANK A. RUBINO, M.D. • *Consultant, Department of Neurology, Mayo Clinic Jacksonville, Jacksonville, Florida; Professor of Neurology, Mayo Medical School, Rochester, Minnesota*

KAPIL D. SETHI, M.D., F.R.C.P. • *Professor, Department of Neurology, Medical College of Georgia, Augusta, Georgia*

KATHLEEN M. SHANNON, M.D. • *Associate Professor, Department of Neurological Sciences, Rush Medical College, Rush-Presbyterian-St. Luke's Medical Center, Chicago, Illinois*

MARK STACY, M.D. • *Director, Muhammad Ali Parkinson Research Center, Barrow Neurological Institute, Phoenix, Arizona*

MATTHEW B. STERN, M.D. • *Director, Parkinson's Disease and Movement Disorders Center; Professor, Department of Neurology, University of Pennsylvania, Philadelphia, Pennsylvania*

DANIEL TARSY, M.D. • *Chief, Movement Disorders Center, Department of Neurology, Beth Israel Deaconess Medical Center; Associate Professor of Neurology, Harvard Medical School, Boston, Massachusetts*

RYAN J. UITTI, M.D. • *Consultant, Division of Movement Disorders, Department of Neurology, Mayo Clinic Jacksonville, Jacksonville, Florida; Associate Professor of Neurology, Mayo Medical School, Rochester, Minnesota*

A

BASIC DIAGNOSTIC PRINCIPLES

1 Approach to the Patient With a Movement Disorder: Basic Principles of Neurologic Diagnosis

J. Eric Ahlskog, PhD, MD

Contents

BACKGROUND

This initial chapter addresses basic principles of neurologic examination and diagnosis and defines some of the broader diagnostic categories as an introduction to the remainder of this text. Since this book is directed to a broad range of clinicians, including primary care physicians, we wanted to provide enough background for non-neurologists to be able to make practical use of the text. Neurology is a field in which the diagnosis is primarily derived by use of old-fashioned methods: the clinical history and examination. Although high-tech imaging and laboratory studies may contribute, an accurate clinical impression is essential for directing the workup and arriving at the correct final diagnosis. In no area of neurology is this more true than that of movement disorders. Assessing and treating patients with movement disorders requires a substantial amount of clinical savvy, for which there is no substitute. In this chapter, we focus on nuances of the neurologic examination as it pertains to movement disorders as well as provide a conceptual framework and definition of terms.

From: *Parkinson's Disease and Movement Disorders:*
Diagnosis and Treatment Guidelines for the Practicing Physician
Edited by: C. H. Adler and J. E. Ahlskog © Mayo Foundation
for Medical Education and Research, Rochester, MN

In a simplistic sense, movement disorders can be divided into conditions in which the problem is primarily reduced movement (hypokinesis) or excessive movement (hyperkinesis). The hypokinetic category is predominantly parkinsonism and its various subgroups, whereas problems characterized by excessive (hyperkinetic) movement include dystonia, chorea, tics, myoclonus, and tardive syndromes. Tremor is obviously a form of excessive movement; however, tremor with the limb at rest is typically associated with hypokinetic signs, and hence resting tremor is usually classified in the hypokinetic (parkinsonian) category.

The distinction between patients who move too much and those who move too little is obvious to even the unseasoned clinician; hence, we have used this distinction in the design of this book. The first half of the book focuses on parkinsonian and other hypokinetic disorders, including conditions in which parkinsonism is only one component. The last half of the book addresses problems characterized by excessive movement. In this introductory chapter, we use this same distinction between parkinsonism and hyperkinetic disorders.

Although most abnormalities of movement are related to disorders of the brain, other levels of the nervous system obviously can also mediate problems of movement. Included are not only lesions of the spinal cord but also conditions that affect the peripheral nervous system (nerve, neuromuscular junction, muscle). This chapter addresses basic clinical distinctions between central (brain and spinal cord) and peripheral disorders and subsequently distinctions between brain and spinal cord (Table 1-1). In addition, recognition of signs of peripheral nervous system or spinal cord dysfunction may be important in multisystem disorders in which the pattern of anatomical involvement may provide clues to the diagnosis.

THE NEUROLOGIC EXAMINATION: OVERVIEW

A comprehensive review of the complete neurologic examination can be found in other texts (e.g., Mayo Clinic Examinations in Neurology, 1998). In this section, we focus on selected components relevant to the diagnosis of movement disorders, recognizing that physicians and students reading this book already have some basic knowledge. Definitions of several basic neuroanatomical terms used in this chapter are found in Table 1-2, with further illustration in Figs. 1-1, 1-2, and 1-3.

Characteristics of Gait

Walking is a fundamental activity, and a discussion of gait is a good prelude to more detailed analysis of neurologic system involvement. Many neurologic disorders affect the gait and do so in a variety of ways and with characteristic patterns. Walking is a very complex motor act, and clues to neurologic disorders may come from observation of each gait component. To appreciate this, consider normal gait. Observation in the office starts with the patient rising from the office couch. This is normally done smoothly, without hesitation and without the need to push off with the arms (as might occur in parkinsonism or with proximal lower extremity weakness). Once the patient is standing, observe the first step. This step should be done without hesitation; in contrast, some patients with parkinsonism may have transiently frozen feet. Observe the feet while the patient walks. Normally with each step, the patient plants the heel with the foot dorsiflexed and then rocks the foot forward to push off with the ball of the foot. Weakness of the feet may

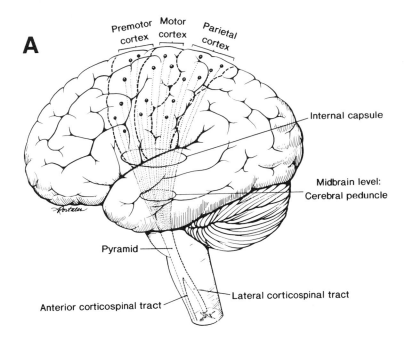

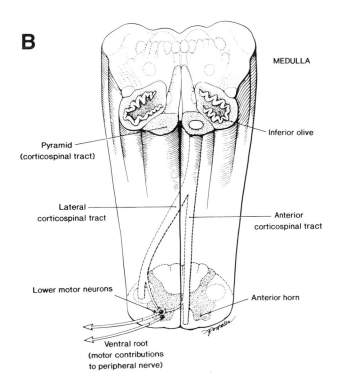

Fig. 1-1. Origins of the corticospinal tract. **A**, Origins. **B**, Pathways to spinal cord anterior horn cells (i.e., to lower motor neurons).

Table 1-1
Motor System Neuroanatomical Distinctions[a]

Primary distinctions	Component circuits
Central nervous system (brain and spinal cord)	• Corticospinal tract • Extrapyramidal system, basal ganglia • Cerebellum • Praxis circuits
Peripheral nervous system	• Anterior horn cells (in spinal cord, but origin of motor nerves) • Nerve • Neuromuscular junction • Muscle

[a]The neurologic examination should provide signs to identify the involved systems.

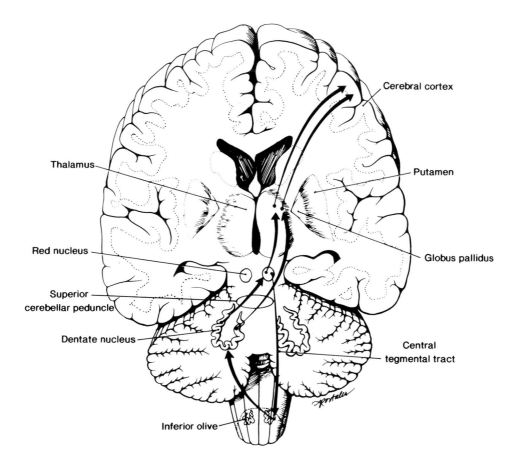

Fig. 1-2. View of cerebellum showing the two hemispheres and selected circuits. The midline cerebellum (vermis), critical for normal gait, is not shown. Lesions of the superior cerebellar peduncle result in severe tremor. Palatal tremor (also known as palatal myoclonus or palatal myorhythmia) is associated with lesions of Mollaret's triangle, which consists of the cerebellar dentate nucleus, the red nucleus, and the inferior olive.

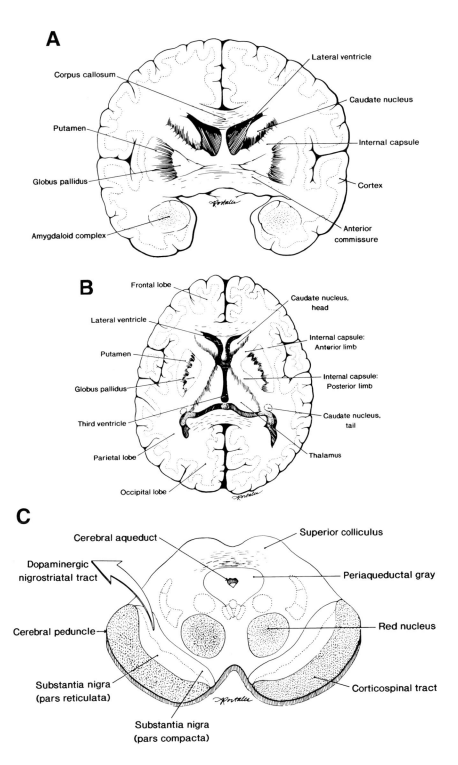

Fig. 1-3. Extrapyramidal motor system. **A**, Section through frontal regions revealing striatum (caudate and putamen) and globus pallidus. **B**, Horizontal section showing these nuclei and the thalamus. **C**, Midbrain section showing the substantia nigra (including the dopaminergic projection to the striatum), the cerebral peduncles (which contain the corticospinal tract), and the red nucleus.

Table 1-2
Neuroanatomical Terms[a]

Term	Definition
Basal ganglia	• Subcortical nuclei contributing to the modulation of movement control. Traditionally, caudate, putamen, globus pallidus, and amygdala were included. Clinicians now use this term to refer not only to these nuclei but also to the closely interconnected substantia nigra and subthalamic nucleus.
Cerebellum	• Posterior brain structure especially concerned with limb coordination and equilibrium. Direct output to the thalamus is via the superior cerebellar peduncle.
Corticobulbar	• The major cortical projection system to the cranial nerve and relay nuclei and to the reticular formation of the brain stem.
Corticospinal tract	• Motor projections to the spinal cord from the cortex. Cortical origins include primary motor, premotor, and anterior parietal cortex. As these projections emanate from the cortex, they congregate to form the internal capsule. As they descend, the compact system forms the cerebral peduncle and ultimately passes through the medullary pyramid en route to the spinal cord. Thus, this system projecting through the pyramids is also termed the "pyramidal tract."
Extrapyramidal	• Implies caudate, putamen, globus pallidus, substantia nigra, subthalamic nucleus, and closely interconnected nuclei. This term was initially introduced to include all motor pathways that influence the lower motor neuron, except those of the pyramidal (corticospinal) tract. More recently, the term has been used somewhat interchangeably with the term "basal ganglia."
Lenticular nuclei	• Refers to the putamen and globus pallidus. The name was chosen because of the macroscopic lens shape.
Lower motor neuron	• Spinal cord anterior horn neurons, which are the source of motor nerves to muscles.
Motor neuron	• Cortical neurons concerned with movement that project to the spinal cord (upper motor neuron), plus anterior horn spinal motor neurons (lower motor neuron).
Neuromuscular junction	• Interface between the lower motor neuron synaptic terminal and the corresponding cholinergic receptor on the muscle cell membrane.
Parkinsonism	• Clinical signs of slowness (bradykinesia), rigidity, resting tremor, imbalance, and attenuated automatic movements that develop after damage to certain components of the extrapyramidal motor system.
Pallidal	• Refers to the globus pallidus of the basal ganglia.
Pyramidal tract	• Corticospinal projections passing through the medullary pyramid.
Striatum	• Caudate and putamen, also known as "neostriatum." In the older literature, the term "corpus striatum" was used to refer to caudate, putamen, and globus pallidus.

[a]Also, see figures.

change this pattern, as does parkinsonism, in which the foot moves parallel to the ground (or slides along). An intorted foot may suggest dystonia. Observe the length and trajectory of stride and spacing of the feet. Normally the feet move in parallel with each other on a narrow base. In parkinsonian disorders, in contrast, stride is shortened. In chorea, there are deviations from the parallel trajectories (e.g., inappropriate steps to the side), whereas ataxic patients have a widened base. Observe the arm swing, which is reduced in parkinsonism and overactive in hyperkinetic movement disorders. Observe the trunk during walking. Stooping beyond that expected for age may suggest a parkinsonian disorder, whereas excessive trunk movements may suggest a hyperkinetic condition, such as chorea or dystonia. Watch the patient turn, a movement that should be a smooth pivot without imbalance. Taking several steps to turn may suggest parkinsonism or extra care due to imbalance. Often, subtle imbalance is manifested only during turns. After walking, the patient can be observed returning to the seated position. Normal persons do this with the feet remaining on the floor. Plopping into the chair with the feet rising off the floor may suggest certain extrapyramidal disorders, such as progressive supranuclear palsy, with truncal instability.

As is obvious from the above, observation of gait requires adequate space. Most clinic offices are too small for all but a cursory examination of walking. When gait characteristics are important, walking in the hallway outside the office should be considered.

The Motor Examination

Patterns of abnormalities on the motor examination allow important distinctions (Table 1-1), such as disorders of nerve or muscle, of the peripheral or central nervous system (CNS), and of the spinal cord or brain, plus identification of specific motor system involvement (corticospinal tract, extrapyramidal, cerebellar, or other). The different components of the motor examination are assessment of tone, reflexes, strength, rapid alternating movements, coordination, and praxis.

TONE

Tone may be assessed in the limbs, neck, and trunk. In the limbs, the major pathologic distinctions are rigidity, spasticity, dystonia, gegenhalten, and hypotonia. The examiner moves the limb both slowly and rapidly across a joint while the patient is encouraged to relax. Tested regions are wrists, elbows, shoulders, knees, and sometimes hips. Assessment of tone in the neck and trunk can be similarly done with nonrapid movement.

- *Rigidity* is characterized by resistance to movement that is relatively uniform across the entire range of motion. If present, this characteristic suggests involvement of extrapyramidal systems within the brain. If tremor is superimposed, the examiner feels a ratchety resistance to passive movement, termed "cogwheel rigidity."
- *Spasticity* is assessed by more rapid limb excursions and is characterized by detection of a catch during this movement. To assess this in the lower limbs, have the patient lie supine, as relaxed as possible, with your hand under the distal thigh, just proximal to the popliteal fossa. Rapid elevation of a spastic lower limb elicits an initial partial flexion at the knee (to gravity) followed by an interruption of the movement (catch). If mild, this catch is overcome, with gravity pulling the leg through the rest of the downward movement. This hitch in the movement of the limb occurs with quick but not slow movements. The similarities to the mechanics of a pocket knife (initial resistance and then relaxation) have led to the term

"clasp-knife reflex." In the upper limbs, quick flexion-extension movements of the wrist or elbow, or pronation-supination of the forearm, may elicit a similar sign.

- *Dystonia* implies an abnormal posture (see below) and is apparent by observation. Some forms of dystonia are present only in certain positions or with certain tasks. A dystonic limb, trunk, or neck is resistant to movement by the examiner.
- *Gegenhalten*, or paratonia, characterized by an inability to relax the limb, is associated with confusional, demented, or frontal lobe states. This effect exceeds the mildly increased tone sometimes detected in a tense but normal person. Characteristically, patients with gegenhalten hamper the assessment by activating the limb while the examiner is moving it.
- *Hypotonia* is the opposite of all the above, with the examiner appreciating flaccidity or floppiness as the patient's limb is moved. This is often difficult to distinguish from normal tone. It may occur in patients with cerebellar syndromes, chorea, or extreme neuromuscular weakness of peripheral origin.

REFLEXES

Reflex assessment is critical to differentiating upper motor neuron from lower motor neuron syndromes.

- *Deep tendon reflexes* are increased in disorders affecting the corticospinal pathways at either brain or spinal cord levels. If the increase is pronounced, clonus is present, reflective of a severely uninhibited deep tendon reflex. Clonus is defined as repetitive muscle reflex responses to a single tendon stretch (e.g., elicited by striking the Achilles tendon or rapidly dorsiflexing the foot).

  Deep tendon reflexes are assessed not only in the limbs but also in the jaw. Gentle pressure is applied by the examiner's index finger on the patient's chin with the jaw relaxed. Tapping the finger with a reflex hammer elicits a brisk jaw jerk in upper motor neuron disorders; if the reflex is very pronounced, clonus is identified.

  A phenomenon also associated with corticospinal tract lesions is diffusion (overflow) of the reflexes to contiguous segments. This is variably present in disorders affecting upper motor neurons. For example, eliciting an ankle reflex may be associated with simultaneous contraction of the hamstring; eliciting the knee jerk may be associated with cocontraction of an ipsilateral or contralateral thigh adductor. The palmomental reflex is also a form of diffusion (overflow). It is elicited by scratching the palm, which provokes chin muscle (mentalis) contraction.

  Disorders affecting peripheral nerves result in reduced or absent deep tendon reflexes. The deep tendon reflexes are decreased in proportion to the weakness in myopathy or neuromuscular junction defects (e.g., myasthenia gravis).
- A *Babinski sign*, with great toe dorsiflexion in response to plantar foot stimulation, reflects an upper motor neuron (corticospinal tract) disorder. Similarly, a Chaddock sign is elicited by scratching the lateral side of the foot from back to front with a slightly sharp object (the same used to test for the Babinski sign). A positive Chaddock sign is also characterized by dorsiflexion of the great toe. The Chaddock maneuver is actually more sensitive to upper motor neuron lesions, and occasional patients have a Chaddock but not a Babinski sign.
- Certain *cranial reflexes* suggest cerebral disorders. Included are the perioral frontal release signs. Gentle tapping with the reflex hammer or tip of the examiner's index finger on the upper lip provokes a brisk contraction in patients with anterior cerebral hemisphere dysfunction. This response does not substantially diminish with repeated

testing in clinically significant cases. With more pronounced disease, a rooting response can also be elicited. This effect is assessed by gently passing a tongue blade or other similar object over one side of the patient's upper lip; a positive response is characterized by contraction of the upper lip toward the test object. The most florid of the perioral frontal release signs is the sucking reflex, in which a puckering response is made to an object placed between the lips. Cognitive impairment typically accompanies these perioral reflexes, especially the rooting and sucking responses.

- The glabellar response suggests extrapyramidal dysfunction and is often elicited in patients with parkinsonism. Blinking that does not stop in response to repeated taps by the examiner in the glabellar region (between the brows) constitutes a positive sign. The examiner's hand should be positioned superiorly and not directly in front of the patient's eyes, using the middle or index finger to tap.

STRENGTH TESTING

Strength assessment of individual muscles provides another means of separating different categories of motor system disorders. General patterns of weakness may allow distinction of peripheral neuropathy, myopathy, and upper motor neuron disorders. Distal weakness, perhaps associated with atrophy, suggests peripheral neuropathy. Proximal limb and axial (e.g., neck) weakness suggests myopathy. In contrast, weakness of CNS origin (brain and spinal cord) may result in weakness that is more pronounced in the extensor muscles of the upper limbs (e.g., triceps, wrist, and finger extensors) and the flexor muscles of the lower extremities (e.g., hip flexors, hamstrings, and foot dorsiflexors). Some diseases of the motor neuron, such as amyotrophic lateral sclerosis and poliomyelitis, tend to be asymmetrical and patchy in distribution. Segmental distribution of weakness often suggests a focal process. For example, isolated weakness of both lower limbs points to a spinal cord lesion (or, rarely, a midline mass in the superior mesial frontal motor or premotor region); diffuse weakness confined to one limb affecting muscles in multiple nerve and root distributions suggests a plexus (brachial or lumbosacral) lesion.

RAPID ALTERNATING MOVEMENTS

Testing of rapid alternating movements (RAMs) often provides diagnostic clues helpful in identifying the specific motor system involved. Such analysis of repetitive voluntary movements can be applied to any moving body part. The tests include repetitive tapping of the finger and thumb, alternating pronation-supination of the hand, and foot tapping. Similar testing is done to analyze the motor components of speech (see Chapter 2).

An approach to the analysis of RAMs is perhaps best illustrated by discussion of the repetitive digit tapping task (Table 1-3). The patient is asked to repetitively tap the tips of the thumb and index finger rapidly and with wide excursions. These repetitive movements should be observed for 5 to 10 seconds, since abnormalities may not be apparent with the first few excursions. Analyze the components of these movements: amplitude, speed, and regularity (rhythm). In patients with parkinsonism, these repetitive movements are slowed, and typically the amplitude is dampened. Sometimes, freezing of the movement occurs (so-called digital impedance), so that the parkinsonian patient's finger and thumb appear to be momentarily stuck together. Often, the amplitude of the movements begins normally but is reduced with continuing repetitions. In parkinsonism, the regular RAM rhythm is maintained except when substantial digital impedance is present, with the appearance of markedly dampened and hesitant movements. Corticospinal tract disorders typically result in slowing but with a regular rhythm and with no substantial loss of amplitude or digital impedance. In contrast, patients with cerebellar disorders perform RAMs with an irregular rhythm that is also slower than normal, but amplitude is preserved. Peripheral disorders (nerve, muscle, neuromuscular junction) result only in slowing in proportion to the degree of weakness. In disorders of praxis (to be discussed in detail below), the patient may have great difficulty performing this simple movement and typically performs it incorrectly, sometimes with a bizarre response, such as flexion of all the digits. Such a patient with apraxia may stare at the hand, mentally trying to will it to do the task.

Alternating pronation and supination of the hand (repetitively slapping the palm and then the dorsal surface of the hand on the thigh) may be similarly analyzed. Similarly, alternating opening and closing of a fist with the arm extended can be assessed.

RAMs are assessed in the lower limbs with foot tapping. This should be done with the patient in a comfortable position, so that the movement of the tapping foot is not impeded by mechanical factors. The patient is instructed to tap the toes of one foot as if listening to music. In parkinsonian patients with lower limb tremor, tapping the heel is preferable; otherwise, it is difficult to separate the voluntary movements from the tremor.

COORDINATION

Coordination testing allows identification of deficits related to the cerebellum or to proprioception. In the limbs, this is assessed by finger-nose testing (patient alternately touches his or her nose and then the examiner's finger at arm length from the patient) and heel-shin testing (patient moves a heel up and down the contralateral outstretched leg). The examiner can observe the finger-nose movements best if the trajectory of the patient's finger movements is approximately aligned with the examiner's line of gaze.

Occasionally, the examiner may be unsure whether the patient has a mild degree of upper limb ataxia, and sometimes this knowledge is critical to the diagnosis (e.g., differentiating multiple system atrophy from Parkinson's disease). Finger-finger testing is a sensitive alternative means of assessing upper limb coordination. The patient is instructed to hold one index finger close to, but not touching, the extended finger of the examiner. With each quick movement of the examiner's finger, the patient's finger should move in parallel, targeted at the new position, but again, close to, but not touching, the examiner's finger. Ataxia is manifested as overshooting or undershooting, with corrective movements as the patient's finger approaches the correct position.

Imprecision due to tremor may make it difficult to interpret whether the patient has true ataxia, and the examiner must mentally discount any tremor component. Sometimes, the overshooting and undershooting movements of a severely ataxic limb give the appearance of tremor (so-called serial dysmetria), and this motion should not be confused with true tremor.

Truncal and lower body ataxia results in a wide-based and unsteady gait, as mentioned above. Assessment of incoordination of eye movements and assessment of speech are discussed below and in Chapter 2, respectively.

PRAXIS

Assessment of praxis usually occurs in the course of the routine motor examination. Praxis implies the ability to appropriately sequence a series of individual muscle movements to make a larger and more complex movement. To illustrate, the simple task of

Table 1-3

Analysis of Rapid Alternating Movements of the Hand (Repetitive Tapping)[a]

Disorder	Speed	Amplitude	Rhythm	Other
Parkinsonism	Decreased	Decreased	Normal (but may appear irregular if progressive dampening or freezing occurs)	Dampening of amplitude may occur only after several excursions; dampening may become severe, and freezing of movement may give appearance of digits momentarily stuck together
Corticospinal tract disorder	Decreased	Normal	Normal	
Cerebellar ataxia	Decreased	Normal	Irregular	
Weakness of peripheral origin (nerve, muscle, neuromuscular junction)	Decreased in proportion to the degree of weakness	Normal	Normal	
Apraxia				Obvious difficulty with the task, which may be performed incorrectly, at least transiently; patient may look at the hand in bewilderment

[a]Patient is instructed to repetitively and rapidly tap the tips of the thumb and index finger together, with wide excursions; observation for 5 to 10 seconds.

waving the hand requires sequential activity in a series of muscles, with the opposing (antagonist) muscles relaxing at the correct times. A computer program simulating this movement would be written with the shoulder abductors activated first and then in sequence the elbow flexors, wrist extensors, finger extensors, wrist flexors, and so on. Patients with apraxia have difficulty using the proper muscle sequences with the proper timing, but they do not have true weakness and are not simply confused. Conceptually, the cerebral computer program for complex movements is dysfunctional in patients with apraxia. This problem is easiest to appreciate in the upper limbs and typically becomes apparent during testing of coordination or RAMs. The patient attempting to close the hand and extend the index finger to make a pointer for finger-nose testing may be unable to easily do this, struggling with what should be an easy maneuver. Such a simple movement as making a pointer-finger may take several attempts, and the patient may stare at the hand, obviously frustrated. Similarly, RAMs may be compromised, with the patient having difficulty performing this simple repetitive act. More formal office testing of praxis includes having the patient imitate simple hand gestures made by the examiner, such as holding up two fingers (or various fingers in different combinations), giving the thumbs-up or -down sign, and waving. Praxis can also be assessed by asking the patient to imagine holding a comb, hammer, saw, or similar item and demonstrate how it is to be used.

Apraxia occurs with parietal cortical lesions, especially those involving the dominant hemisphere. It also is occasionally seen with frontal cortical lesions or with cerebral insults that disconnect the cerebral hemispheres. Apraxia of speech may occur with lesions of Broca's area in the dominant frontal cortex, characterized by dysarthric speech and, if severe, mutism; it is associated with an inability to protrude the tongue or purse the lips on command. More is said about apraxia of speech in Chapter 2.

The term "apraxia" is sometimes applied to parkinsonian gaits with a short-stepped, shuffling appearance, often with freezing of movement (start hesitation). Strictly speaking, this may seem correct, since the patient's motor program for walking has been degraded without weakness or confusion. However, this parkinsonian walking pattern is a stereotyped motor behavior substituting for a more complex gait pattern that is lost with damage to basal ganglia or basal ganglia-frontal connections. Thus, it is debatable whether such "gait apraxia" should be included with general disorders of apraxia.

The Sensory Examination

Examination of sensation is important in the evaluation of certain disorders of movement. In most cases, precise quantitation of sensory deficits is not critical to movement disorder evaluations, whereas documentation of a general pattern is. For example, distal sensory deficits suggest a peripheral neuropathy, abrupt changes in sensation over the trunk (sensory level) are characteristic of a myelopathy, and hemisensory deficits suggest a central (brain or spinal cord) disorder. Assessing pin and light touch sensation with such patterns in mind can usually be done in a short time. Assessing vibratory sense in the toes is typically part of the screening examination for peripheral neuropathy.

Joint position sense should be checked in patients presenting with ataxia. Ataxia of the upper limbs due to any cause is best detected on finger-nose testing. Incoordination on this test can be followed by proprioceptive testing, assessing whether the patient can detect low-amplitude movements of a fingertip (or toe) with eyes closed as the distal phalanx is moved by the examiner. The examiner should use one hand to immobilize all but the distal phalanx, asking the patient to relax the finger. With the other hand, the examiner grasps the tip of the patient's finger, holding it by the sides between the examiner's thumb and index finger and moving the distal phalanx up or down in small, quick excursions. The patient is instructed to report the direction of the movement, up or down (not the position). Persons with normal sensory systems should have no trouble identifying even small excursions of the fingertips. Patients with an ataxic gait should have joint position sense analyzed in the great toes.

Apraxia is often related to parietal lobe dysfunction, and testing for parietal cortical sensory deficits may be appropriate in that situation. This may include assessment of one or more of the following: joint position sense, graphesthesia, stereognosis, and double simultaneous stimulation. Obviously, testing of the last three is uninterpretable if there is major sensory loss to pin and light touch in the tested area of skin.

- *Graphesthesia* is assessed by writing a number with a pointed object on the patient's palm, which is positioned facing toward the patient, whose eyes are closed. The traced number should be oriented as if it were written on paper being held in view of the patient. Other areas of skin may also be tested, as appropriate.
- *Stereognosis* is also assessed with the patient's eyes closed. A small object (e.g., coin or key) placed in the patient's hand can then be felt and identified by the patient.
- *Double simultaneous stimulation* testing assesses whether patients can detect sensory stimuli (usually a touch of the examiner's finger) applied to both sides of the body simultaneously. The patient, with eyes closed, is asked to report which side of the body is touched: right, left, or both sides. The examiner can touch one hand (or foot, side of face), then the other, and then both in various sequences. Patients with a parietal lobe deficit may fail to recognize stimuli on the contralateral side of the body during bilateral presentation (i.e., extinction of the contralateral side).

Eye Movements

BACKGROUND

The same motor disorders that affect limb and trunk function can also affect eye movement, and sometimes the neuro-ophthalmologic examination provides important diagnostic clues.

Normally the eyes are yoked together by brain stem control circuits, which produce conjugate movements in all planes of gaze. Separate, nonyoked movements of each eye (dysconjugate) may reflect disorders of the cranial nerves (CN 3, 4, or 6), ocular muscle disorders, local mechanical factors affecting the globes, or brain stem conditions. Cerebral conditions above the level of the brain stem do not directly result in dysconjugate eye movements.

CONJUGATE GAZE DISORDERS

Eye movements are assessed (1) while the patient is watching a large moving target (usually the examiner's finger or hand) with the head fixed, termed "pursuit," and (2) when the patient is asked to look to the right, to the left, toward the ceiling, to the floor, and so forth, termed "voluntary gaze" or "saccadic eye movements." Some general pathologic categories are as follows.

- *Saccadic pursuit* and *hypometric saccades* may be detected in a variety of CNS disorders, including parkinsonian conditions, encephalopathies, and effects of sedating drugs. Saccadic pursuit is characterized by the absence of smooth ocular tracking of the moving target; the eyes, instead of keeping up, lag behind the target, and catch-up movements (saccades) are made to reestablish fixation. In testing voluntary gaze (e.g., "look to the left"), a related type of abnormality may be observed, with the excursions dampened and coming up short (hypometric); such hypometric saccades necessitate a series of steps, rather than one smooth, single-step movement, to move the eyes fully in the intended direction.

- *Supranuclear gaze paresis* may be associated with one of several neurodegenerative disorders and reflects problems in cerebral conjugate gaze mechanisms. In milder cases, attempts to voluntarily look at or pursue an object in a given direction are marked by hesitancy and incomplete eye movements. If the disorder is severe, no movement occurs. "Supranuclear" implies that the disorder originates at cerebral levels higher than the final common pathways (cranial nerve nuclei) in the brain stem. That the problem is not due to neuromuscular or mechanical disorders is confirmed by having the patient visually fix on an object while the examiner passively moves the head in a plane opposite to the direction of gaze paresis. This head movement with the eyes remaining fixed on the target results in eye movement into the plane of gaze paresis, an argument against a peripheral cause for the paresis. Consider this maneuver in a patient with up-gaze supranuclear paralysis: the examiner passively moves the head down while the patient visually fixes on a target in front, resulting in the eyes moving above the horizontal meridian of the orbits. This passive movement by the examiner producing eye deviation in the plane of paresis confirms that the inability to look up is not due to abnormalities of cranial nerve, ocular muscles, or mechanical factors affecting the globes.

  Up-gaze may be deficient for a variety of reasons, including normal aging. Supranuclear down-gaze paresis, however, has sinister implications; it is typical of certain multisystem cerebral disorders, such as progressive supranuclear palsy, variants of Niemann-Pick disease, and, sometimes, corticobasal degeneration. Supranuclear down-gaze paresis is accompanied by various degrees of gaze paresis in the other directions, as well.

- *Apraxia of eye movements* may be accompanied by apraxia of limb movements, such as in corticobasal degeneration. The disorder is characterized by inability or marked difficulty in moving the eyes in the direction commanded by the examiner and sometimes by similar problems with pursuit.

- *Ataxic eye movements* occur in disorders of the cerebellum. These eye movements are similar to the overshooting and undershooting of an ataxic upper limb attempting to point at a target. Overshooting the target is followed by a corrective movement that may overshoot the target in the opposite direction. This anomaly may best be appreciated by asking the patient to look back and forth between two targets (e.g., between the examiner's right thumb and left index finger).

NYSTAGMUS

These regular periodic movements of the eyes are conventionally subdivided into two basic types, jerk and pendular nystagmus. Most commonly, the back and forth movements of nystagmus are characterized by a slow movement in one direction followed by a fast movement in the opposite direction, which is termed "jerk nystagmus." This is by far the most common type seen in clinical practice. Nystagmus with a sinusoidal rhythm is termed "pendular" and is typical of that occurring congenitally and occasionally with disorders of the brain stem and cerebellum.

Jerk nystagmus may occur with peripheral vestibular disorders as well as with a variety of central (brain stem-cerebellar) disorders. A central origin for the nystagmus is suggested by (1) changing direction of the nystagmus with different eye positions (e.g., fast component to the right on right gaze and to the left on left gaze) or (2) vertical nystagmus (up- or down-beating). Sufficient doses of certain psychoactive drugs (including sedatives and anticonvulsants) can induce a central type of jerk nystagmus. Down-beating nystagmus has specific localizing value, pointing to a process at the craniocervical junction. In normal persons, low-amplitude nystagmus occurs if the eyes are maintained at the extremes of gaze (physiologic end-point nystagmus).

Opsoclonus

Opsoclonus may be thought of as myoclonus of the eyes. Myoclonus is defined as a sudden jerk of a body part (discussed below), and such quick jerks of the eyes, occurring at rest and during eye movements, produce a chaotic pattern. Opsoclonus, myoclonus, and ataxia may occur together as manifestations of a paraneoplastic or parainfectious disorder or as an inborn error of metabolism.

PERIPHERAL NEUROLOGIC SIGNS RELEVANT TO MOVEMENT DISORDERS

Facility in differentiating peripheral (neuropathy, myopathy, neuromuscular junction disorders) from CNS conditions is important in the practice of movement disorders. Peripheral nervous system disorders may result in certain signs that resemble those seen in cerebral disorders. They may also result in weakness and consequently affect motor control. The constellation of findings on the neurologic examination, however, should allow distinction between central and peripheral.

Facility in recognizing signs of peripheral nervous system disease is also important, because polyneuropathy, myopathy, and motor neuron disorders sometimes occur in conjunction with central disorders. In such multisystem conditions, the constellation of peripheral and CNS involvement may be an important clue to the diagnosis. For example, cerebellar signs in isolation raise a question of a structural lesion, whereas in conjunction with peripheral neuropathy, they suggest a more generalized nervous system disorder. Extrapyramidal disorders associated with myopathy may suggest an underlying mitochondrial disorder. Motor neuron findings in conjunction with chorea and tics may point to neuroacanthocytosis. Thus, recognition of peripheral nervous system signs may provide critical evidence toward the diagnosis of multisystem problems.

Peripheral Neuropathy

Characteristic Signs

Peripheral neuropathies affect sensory or motor function, or both. Evidence pointing toward neuropathy is related to the distribution of the symptoms and deficits (weakness, sensory loss), the deep tendon reflexes, and atrophy.

On the motor examination, weakness tends to be distal or predominantly distal, especially affecting muscles of the hands and feet or toes. One exception is the inflammatory polyneuropathies (e.g., Guillain-Barré syndrome and chronic inflammatory polyneuropathies), in which the weakness may be more diffuse or scattered among distal and proximal

Table 1-4
Peripheral Neuromuscular Conditions That Resemble
Movement Disorders of Central Nervous System Origin

Peripheral neuromuscular condition	*Compare with movement disorder of cerebral origin*
Peripheral neuropathy	
Distal weakness with footdrop	• If footdrop is severe, the hip flexion needed to clear the toes off the ground (steppage gait) resembles certain dystonic gaits with dystonic exaggeration of knee and hip flexion
Proprioceptive deficits affecting gait	• Resembles cerebellar ataxic gait
Tremor (occurs with some polyneuropathies, especially demyelinating)	• Similar to essential tremor
Cramps	• Compare with the painful dystonic spasms of parkinsonism or generalized torsion dystonia
Nerve root irritation	
Hemifacial spasm (7th cranial nerve)	• Compare with craniofacial dystonia (bilateral)
Myopathy and neuromuscular junction disorders	
Difficulty rising from seat due to proximal weakness	• Similar to difficulty rising in parkinsonism
Facial weakness, bilateral (especially in myasthenia gravis)	• Resembles the facial masking of parkinsonism
Hanging head due to neck extensor weakness	• Compare with antecollis in cervical dystonia and parkinsonian syndromes
Paraspinal weakness of trunk,with stooping	• Compare with the stooped posture of parkinsonism
Dysphagia	• May also occur in parkinsonism, spinocerebellar degenerations, and dystonia
Dysarthria	• Dysarthria occurs in a variety of central movement disorders, but characteristics distinguish it (see Chapter 2)
Myotonia (impaired muscle relaxation due to repetitive muscle discharge)	• Compare with dystonia
Cramps	• Compare with dystonia

muscles. Focal neuropathies may result in proximal as well as distal weakness if the compressive or inflammatory lesion is more proximal in the nerve or plexus.

Sensory symptoms and signs are also typically distal in the limbs of patients with peripheral neuropathy (the so-called stocking-glove distribution). The symptoms are usually most prominent in the feet and may be described in a variety of ways, including burning, tingling, Novocain-like, and pins and needles. Sensitivity to pressure on the feet is common, for example, manifested as intolerance to bedcovers touching the toes. Often, sensory testing (e.g., pin, light touch) reveals only minor deficits in patients with prominent sensory complaints.

Deep tendon reflexes are typically reduced or absent in disorders of peripheral nerves. General observation may reveal atrophy in severely affected muscles. One caveat, however, is that disuse atrophy can occur with CNS processes if the weakness is severe. Sometimes pes cavus, perhaps with hammer toes, is a clue to a long-standing polyneuropathy. Fasciculations may be present in peripheral neuropathies but typically are not as pronounced as in motor neuron disease. Difficulties performing RAM tasks are in proportion to the weakness.

MOVEMENT PROBLEMS

Gait is affected in peripheral neuropathic disorders if the weakness is pronounced (Table 1-4). In generalized peripheral neuropathies, distal weakness with bilateral foot drop (foot dorsiflexor weakness) may require increased hip flexion to clear the plantar-flexed foot from the floor, producing a high-stepping appearance. There may be superimposed slapping of the feet with each step if severe distal weakness is present.

Peripheral neuropathic conditions producing proprioceptive deficits may result in an ataxic gait. For this to occur, there must be substantial damage to the large fiber components of the peripheral nerves (or the dorsal columns within the spinal cord). Joint position sense should be assessed in patients with ataxia.

Tremor occasionally occurs in conjunction with peripheral neuropathic conditions. The neurophysiologic basis for the tremor may be linked to the neuropathy per se, as occasionally occurs in demyelinating polyneuropathies. In other cases, both may be induced by some pathologic factor, such as certain medications (e.g., amiodarone) or toxic metabolic states (e.g., uremia).

Cramps may occur in a variety of peripheral neuromuscular conditions. These may be confused with dystonia (or vice versa, as in dystonic spasms in parkinsonian patients). By definition, cramps originate in the peripheral nervous system, in contrast to the central origins of dystonia. Cramps are episodic, occurring paroxysmally, and are typically relieved by stretching. Between cramps, there is no abnormality. Dystonia is typically a continuing, chronic problem, although there may be superimposed paroxysms. Cramps are typically acutely painful, whereas the pain of dystonia tends to be more chronic. Cramps typically involve one muscle or a group of contiguous muscles, such as calf or hamstring muscles. Dystonia frequently involves a more widespread distribution of muscles, often spanning a joint. Dystonia typically causes deviation of a body part in a complex vector, such as twisting of the neck or foot; cramps typically pull in a simple plane and then only briefly. Finally, dystonia occurring episodically is often associated with nonparoxysmal extrapyramidal signs, either other forms of dystonia or parkinsonism.

FACIAL NERVE IRRITATION: HEMIFACIAL SPASM

Continuing paroxysms of unilateral facial contractions can be provoked by chronic irritation of the seventh cranial nerve. Typically, such hemifacial spasm is due to compression by a contiguous artery (occasionally a vein) at the point this nerve root exits the brain stem. This spasm resembles facial (cranial) dystonia but can be distinguished by the unilateral distribution (facial dystonia is bilateral, although it may be asymmetrical). Also, synkinesis may be present in hemifacial spasm, in which voluntary closure of the eye elicits contraction of the ipsilateral lower facial muscles (or vice versa, with smiling or grimacing eliciting ipsilateral eye closure). See Chapter 24 for more details.

Motor Neuron Disease

CHARACTERISTIC SIGNS

The prototypical motor neuron disease is amyotrophic lateral sclerosis. In this disorder, weakness, atrophy, and fasciculations are present in an asymmetrical distribution, reflective of lower motor neuron degeneration. Superimposed on this are corticospinal tract signs (hyperreflexia, spastic tone) but with sensation and bowel and bladder function spared. In contrast to most peripheral neuropathies, the weakness is not predominantly distal but rather more haphazardly scattered; it may be more prominent distally, proximally, axially, or in a cranial nerve pattern (i.e., bulbar). Cramps are common in patients with motor neuron disease.

MOVEMENT PROBLEMS

Gait is affected in proportion to the weakness and reflective of the muscles involved. In amyotrophic lateral sclerosis, superimposed spasticity may give the gait a stiff-legged appearance. Differentiation of cramps in motor neuron disease from dystonia is discussed in the section above.

Motor neuron disease is sometimes associated with extrapyramidal disorders. On the island of Guam, amyotrophic lateral sclerosis, parkinsonism, and dementia occur in various combinations, perhaps as manifestations of the same disease process. Outside Guam, the association of motor neuron disease and parkinsonism is rare but described.

Myopathy and Neuromuscular Junction Defects

CHARACTERISTIC SIGNS

Proximal muscle weakness predominates in most myopathic conditions. There are exceptions, however, such as myotonic dystrophy, in which distal weakness appears earlier and is more prominent. In contrast to neuropathic conditions, axial weakness is frequently present in myopathies, most easily appreciated by testing strength in neck muscles. Deep tendon reflexes are reduced in proportion to the degree of weakness, and tone is not substantially affected.

Neuromuscular junction disorders include myasthenia gravis and the myasthenic (Lambert-Eaton) syndrome. In myasthenia gravis, the weakness is due to an autoimmune response directed at the acetylcholine (nicotinic) receptor. In the myasthenic syndrome, an autoimmune response is directed at the presynaptic cholinergic terminal.

The weakness of myasthenia gravis is especially focused in the cranial musculature, particularly the ocular muscles, with resulting diplopia and ptosis. Oromandibular weakness, dysphagia and dysarthria, and neck weakness are frequent problems. In some persons, the weakness is primarily confined to the cranial musculature, whereas in others, it is diffuse, affecting the limbs (typically proximal more than distal). Strength may decline when the examiner repeatedly tests the same muscle. The deep tendon reflexes are preserved, except in severe weakness.

In the myasthenic syndrome, the weakness is typically less pronounced than in myasthenia gravis and is largely confined to the limbs (especially proximal) and trunk. Accompanying clues include reduced deep tendon reflexes, dry mouth, paresthesias, and impotence. Strength may improve with repetitive testing of the same muscle.

MOVEMENT PROBLEMS

Difficulty rising from a chair occurs with proximal lower extremity weakness due to myopathy or neuromuscular junction defects (Table 1-4). This resembles the difficulty rising from the seated position in patients with parkinsonism, in which there is no true neuromuscular weakness.

Facial weakness occurring in bulbar myasthenia gravis may resemble the masked facies of parkinsonism. The distinction should be obvious, however, when cranial strength is tested; also, the ptosis of myasthenia gravis may give the diagnosis away. Similarly, the dysphagia and dysarthria of myasthenia gravis should not be confused with this occurrence in parkinsonism; testing of strength in the cranial musculature should easily differentiate these two disorders, as does the clinical company each keeps.

Severe neck extensor weakness due to myopathy or myasthenia gravis may result in a hanging head appearance. This resembles the antecollis that occurs in certain extrapyramidal disorders, including cervical dystonia, and some parkinsonian syndromes (especially multiple system atrophy). Distinction is by strength testing and by observation while the patient lies supine on a flat surface (antecollis of extrapyramidal origin usually is maintained against gravity). Similarly, weakness of the extensor muscles of the trunk can result in a stooped appearance, to be contrasted with the stooped posture of parkinsonism.

Myotonia occurs in conjunction with some myopathic conditions and is characterized by failure of muscle to relax normally due to inappropriate repetitive muscle membrane depolarization. A voluntary forceful contraction of muscle results in persistent contraction despite the patient's attempt to relax. Forceful eye closure may be involuntarily sustained, bearing some resemblance to dystonia of the eyes (blepharospasm). Myotonia is often most easily recognized in the testing of grip, in which the patient is asked to squeeze the examiner's two fingers and then quickly let go; a couple of seconds' delay in relaxation is typical of myotonia. Confirmation is by electromyography.

Some disorders of muscle ion channels are primarily manifest as cramps. These are paroxysmal and can be differentiated from dystonia by the characteristics described for cramps in the peripheral neuropathy section above.

CENTRAL NERVOUS SYSTEM SIGNS

Brain and Spinal Cord

The CNS includes both brain and spinal cord, and when CNS signs are present, the clinician must make the distinction. This can often be done by the history and examination.

SIGNS AND SYMPTOMS SUGGESTING MYELOPATHY

The anatomical distribution of the symptoms and signs should provide critical clues. Spinal cord lesions sufficient to affect the descending long tracts result in deficits below the level of the lesion, and, hence, usually result in lower extremity symptoms and signs. They spare the upper limbs if the lesion is below the cervical cord and also spare the cranial region except in rare instances when the lesion is at the highest cervical level. A sensory level to pin or light touch over the trunk suggests a myelopathy (i.e., a line on the trunk below which sensation is reduced). With some myelopathic conditions, a distinct sensory level is not as apparent, but the patient is aware of a gradient of sensory loss, diminishing caudally. Similarly, a motor level, with a paraparetic or quadriparetic pattern of weakness, suggests the possibility of myelopathy. Early and prominent bladder dysfunction also raises a question of myelopathy, provided that generalized dysautonomia (occurring in some parkinsonian conditions) and anatomical lesions of the bladder (e.g., outlet obstruction) can be excluded.

With myelopathies, a dermatomal and myotomal level tells the clinician that the lesion is located at that spinal level or above but not lower. For example, a sensory level at T10 (umbilical level) indicates that the spinal lesion is somewhere at or above T10 and could be as high as the upper cervical levels.

SIGNS AND SYMPTOMS SUGGESTING BRAIN ORIGIN

Hemisyndromes (i.e., sensory or motor syndromes affecting half the body) usually point to the brain rather than the spinal cord as the origin of the deficits. This is clearly the case if the face or head is involved. If the hemisyndrome spares the face, a cerebral process is still more likely, although a spinal process cannot be excluded, since hemi-spinal-cord presentations (e.g., so-called Brown-Séquard syndrome) are occasionally encountered.

Motor signs referable to specific brain systems also provide important clues toward localization, such as cerebellar ataxia, frontoparietal apraxia, or extrapyramidal-related parkinsonism, as described below.

Central Nervous System Motor Control Systems

CNS control of movement is mediated by a number of integrated components, conceptually like circuits in a computer. Analysis of a movement disorder requires the ability to appropriately identify the problem as CNS in origin and to determine which CNS motor system or systems are involved. The primary subdivisions of the motor system are corticospinal (pyramidal, or upper motor neuron), cerebellar, extrapyramidal, and frontoparietal praxis circuits (Table 1-5).

Certain neurodegenerative disorders often affect more than one of these motor systems. For example, in patients with atypical parkinsonism (so-called parkinsonism plus syndrome), the clinical diagnosis is based on the additional systems involved. If signs of parkinsonism are accompanied by those of cerebellar ataxia and corticospinal tract deficits, idiopathic Parkinson's disease is unlikely and multiple system atrophy or a spinocerebellar degeneration is suggested. A useful diagnostic strategy in certain unusual cases is to prepare a mental checklist of the major motor systems (cerebellar ataxia, extrapyramidal, corticospinal, frontoparietal praxis) and determine if signs of each of these are present. Obviously, additional diagnostic clues come from symptoms and signs of involvement outside the motor circuits as well. Thus, the hypothetical patient cited above with parkinsonism and cerebellar and corticospinal tract signs probably has multiple system atrophy if there are prominent early signs of dysautonomia.

Hyperkinetic movement disorders have their origins in brain circuits as well. The only major exception is myoclonus, which may be associated with spinal cord pro-

Table 1-5
Central Nervous System Motor Control Systems

System	Signs
Corticospinal	• Hyperreflexia, Babinski and Chaddock signs, spastic gait and tone, upper motor neuron pattern of weakness, slowed rapid alternating movements
Cerebellar	• Ataxia, which may affect gait, limbs, speech, and eye movements; nystagmus; severe postural-action tremor
Extrapyramidal	
Hypokinetic	• Parkinsonism
Hyperkinetic	• Dystonia or chorea; rarely, tics or myoclonus[a]
Frontoparietal cortical praxis circuits	• Apraxia, most easily appreciated in assessment of hand function (but may affect legs, eye movements, or speech)

[a]Myoclonus may originate in a variety of central nervous system locations.

cesses as well as cerebral disorders. Hyperkinetic movement disorders are discussed later in this chapter.

CORTICOSPINAL SYSTEM: SPASTICITY

Damage to the so-called upper motor neurons results in characteristic symptoms and signs familiar to all clinicians. Included are hyperreflexia, spastic tone, Babinski's and Chaddock's signs, a stiff-legged and sometimes circumducting gait, and slowed RAMs with normal amplitude and regular rhythm (Table 1-3). Weakness is often in an upper motor neuron pattern, more pronounced in extensor muscles of the upper limbs and flexor muscles of the lower extremities.

The anatomical origins of the corticospinal system are widespread in frontoparietal cortex, with the efferent projections forming the internal capsule and the descending corticospinal tract (Fig. 1-1).

CEREBELLAR DISEASE: ATAXIA

Damage to cerebellar systems (Fig. 1-2) may result in limited or pervasive neurologic signs. If damage is localized to the midline (vermis), only ataxia of gait may be present (wide-based and unsteady); incoordination on heel-shin testing may also be seen. Lesions of one cerebellar hemisphere may result in a hemiataxic syndrome.

Cerebellar degenerations typically result in a broad spectrum of deficits reflecting pancerebellar involvement, at least later in the course. These include gait and limb (finger-nose, heel-shin) ataxia, ataxic eye movements (perhaps with nystagmus), and ataxia of speech (see Chapter 2).

A severe postural tremor results from damage to the superior cerebellar peduncle, which originates in the dentate nucleus and projects to the thalamus. The wide amplitude of the tremor when the elbows are flexed and the arms abducted produces the appearance of a bird flapping its wings, hence the term "wing-beating."

FRONTOPARIETAL PRAXIS CIRCUITS: APRAXIA

Apraxia implies damage to cortical motor programs that code for patterned movements. As described above, patients with apraxia have difficulty performing motor tasks that should be routine (e.g., hand pronation-supination, imitating simple gestures).

Table 1-6
Signs of Parkinsonism[a]

Location or activity	Manifestation[b]
Face	• Loss of animation (masking) • Decreased blink rate
Speech	• Reduced volume • Dysarthric due to reduced amplitude and precision of the articulators of speech (lips, tongue, and palate)
Automatic movements	• Less gesturing when talking • Reduced arm swing while walking
Gait	• May have difficulty rising from seated position • Stooped • Shortened stride • Shuffling of feet (or feet more parallel to floor in contrast to normal landing on heel and pushing off with toes) • May exhibit freezing in place or hesitancy • Takes several steps to turn • Slowness (bradykinesia)
Balance	• Imbalance typically not an early sign in Parkinson's disease but may be in other parkinsonian disorders • Pull test[c] may detect milder degrees of imbalance (retropulsion)
Tremor	• Hands when in position of repose (e.g., in lap or at sides during walking) • Legs when patient is seated with feet on floor • Chin • Markedly reduced with action (except with concurrent essential tremor)
Tone	• Rigidity of limbs and sometimes neck • If superimposed tremor, examiner appreciates cogwheel pattern
Rapid alternating movements	Slowed, reduced amplitude, and sometimes freezing of movement: • Finger-thumb tapping • Pronation-supination • Opening-closing fist • Foot or heel tapping
Other	• Eye movements may be slowed, and eye movement falls short of target (hypometric) • Meyerson's sign[d]

[a]Traditional cardinal signs: rigidity, bradykinesia, resting tremor, and imbalance.
[b]Limb signs typically asymmetrical.
[c]Pull test: The patient, facing away from the examiner, is instructed to maintain that position while the examiner pulls backward on the patient's shoulders. No movement of the feet to maintain center of gravity and balance is a normal response. Stepping back to avoid a fall is a mildly abnormal response. Frank retropulsion into the examiner signals a more severe abnormality.
[d]Meyerson's sign (glabellar tap): The examiner places a hand above the patient's line of vision and taps the glabellar region. Blinking that does not habituate with repeated taps is a positive (parkinsonian) sign.

The anatomical substrate for apraxia is thought to be primarily in the parietal lobes, especially the dominant hemisphere. Hence, apraxia is a common motor manifestation of certain neurodegenerative disorders that especially localize to parietal regions, most notably corticobasal degeneration (see Chapter 19). The exact underlying neuronal circuitry related to apraxia has not been fully worked out, and other cerebral regions have also been implicated.

Extrapyramidal System: Parkinsonism

Signs of parkinsonism (Table 1-6) may become apparent as the examiner initially interviews the patient. Facial masking with decreased blinking and the soft speech of hypokinetic dysarthria often provide the first clues that the nonspecific-sounding complaints (e.g., fatigue, weakness, stiffness, and slowness) may be attributable to undiagnosed parkinsonism. Early in the interview, the lack of gesturing while talking, which is reflective of the parkinsonian loss of automatic subconscious movements, may also become apparent. A relatively brief examination is typically sufficient to confirm parkinsonism. Observation of gait begins with watching for hesitancy when the patient rises from the office couch. Hesitancy may also be apparent with the first step and if severe may give the appearance of gait freezing. Reduced arm swing, shortened stride, and a stooped posture may be apparent during walking. Instead of the usual heel strike with rocking of the foot forward to push off with the toes, the stepping foot may land more parallel to, and with less clearance from, the floor; if pronounced, this anomaly produces a shuffling appearance. Turning may require several steps instead of the usual pivot.

Prominent imbalance is often a later sign in the course of Parkinson's disease, but subtle evidence may be appreciated with use of the pull test. It is performed by a quick pull backwards on the patient's shoulders as the patient faces away from the examiner and is instructed to resist the pull. Normal persons simply arch the trunk to maintain the center of gravity, whereas parkinsonian patients tend toward retropulsion. This is manifest as a step backwards, if mild, or a fall into the examiner's arms, if profound. The maneuver may need to be repeated if it appears that the patient was surprised by the pull. Parenthetically, severe imbalance with frequent falls early in the course suggests a parkinsonism plus syndrome rather than Parkinson's disease.

The remainder of the parkinsonian examination is performed on the examination table, with tests of tone (rigidity?), RAMs (slowed, dampened amplitude; hesitant?), and eye movements. If resting tremor is apparent, it typically dampens with action on finger-nose testing.

The signs of Parkinson's disease are usually asymmetrical and, early in the course, may be unilateral. Some of the other parkinsonian disorders, however, typically appear with symmetrical signs, such as progressive supranuclear palsy.

Although degeneration of the substantia nigra is responsible for the motor signs of Parkinson's disease, damage to other portions of the extrapyramidal motor system may also give rise to parkinsonism. Thus, parkinsonian motor signs may develop after lesions of the caudate, putamen, or globus pallidus (Fig. 1-3).

HYPERKINETIC DISORDERS

Excessive movements may take on a variety of well-defined forms, including myoclonus, chorea, dystonia, tics, or tremor. Some less well-defined hyperkinetic movements also deserve discussion, including spasms, stereotypies, and myorhythmias. This category also encompasses the tardive disorders, which have a variety of manifestations.

Not all excessive movements are "abnormal." Obviously, we all make a variety of purposeless, non-goal-directed movements as part of our normal behavioral repertoire. Drumming one's fingers on a table while talking or foot tapping when nervous is not viewed as aberrant motor behavior. Three elements characterize the hyperkinetic movements discussed here as pathologic: (1) a general consensus that the hyperkinetic movements are qualitatively or quantitatively outside the limits of what is culturally accepted as normal human behavior, (2) interference of the movements with goal-directed tasks or social relationships, and (3) an inability to consistently inhibit the movements in appropriate situations.

Simple observation is the primary examination strategy for classification of hyperkinetic disorders, and often it helps to observe the patient in different situations, for example, while speaking, walking forward or backward, and standing or lying. In most cases, the formal neurologic examination is primarily used to determine whether the hyperkinetic movement disorder is associated with other signs, such as parkinsonism or ataxia.

Most hyperkinetic disorders arise outside conscious awareness and are entirely involuntary. Some, however, are responses to internal stimuli and thus are partially in conscious awareness and sometimes partly under voluntary control. We first consider movements in response to such internal stimuli.

Hyperkinetic Movements in Response to an Irresistible Internal Urge

This category includes tic disorders and hyperkinetic movements provoked by akathisia (inner restlessness). Certain stereotypies (stereotyped motor behaviors) also belong with this group, sharing features with tics.

TIC DISORDERS

Tics are simple or complex motor acts occurring in response to an inner urge to perform the movement. Usually they are manifested as a single type or a few repeated types; that is, they are stereotyped. These stereotyped movements may be simple or complex. Shoulder shrugging, blinking, grimacing, and grunting are examples of simple tics. More complex tics might include kicking, squatting, or vocalizing words.

Patients with tics are often able to suppress them for a short time; however, such voluntary suppression leads to an escalating urge to release the tic. Thus, after a brief period of suppression, the tics may be released in an explosion of movement. With longer standing disorders, some patients may not be able to suppress their tics.

With rare exceptions, tic disorders begin in childhood or adolescence. An undiagnosed hyperkinetic movement disorder that began well into adulthood without any history of tics usually does not turn out to be a tic disorder. Tic disorders that begin in childhood may persist into adulthood, however.

Some simple tics may be difficult to differentiate from other types of movement disorders, especially focal dystonia and myoclonus. For example, tics manifested as repetitive blinking may be similar to the frequent blinking of ocular dystonia (blepharospasm); tics manifest as jerks of the neck may resemble the phasic movements sometimes seen in cervical dystonia or perhaps the quick jerks characteristic of myoclonus (see below). Clues suggesting that the abnormal movement is more likely a tic are (1) a subjective urge to perform the movement; (2) voluntary suppressibility for at least brief periods; (3) association with other, more characteristic tics, for example, complex tics such as back-and-forth head shaking, spitting, or repetitive touching of a body part; (4) association with abnormal vocalizations, including simple grunts or snorts and words (complex vocalizations); and (5) onset in childhood or adolescence. Further details on tics and Tourette's syndrome can be found in Chapter 29.

MOTOR RESPONSES TO AKATHISIA

Akathisia implies an inner restlessness, which if severe, provokes movement as a means of relief. This may occur as a consequence of administration of dopamine antagonist drugs (e.g., neuroleptic or antiemetic agents) either early, as a reversible adverse event, or later, as a tardive disorder (often irreversible).

The motor responses to akathisia take on a variety of forms, but the usual appearance is an inability to sit still. Repetitive leg crossing and uncrossing, weight shifting, pacing the floor, and similar movements, often done to attenuate the symptoms, form the objective motor component of akathisia.

Akathisia also commonly occurs in parkinsonian syndromes, including idiopathic Parkinson's disease. In contrast to persons with neuroleptic-related akathisia, however, parkinsonian patients typically do not move in response. Hence, it is only subjective in most parkinsonian patients.

Idiopathic restless legs syndrome (see Chapter 30) bears some resemblance to akathisia because of the subjective sensation, which may result in the need to get up and pace. Parenthetically, restless legs syndrome is often linked with periodic leg movements of sleep (recurrent flexion of lower limbs), which originate outside conscious awareness and sometimes occur during waking.

General Categories of Abnormal Hyperkinetic Movements Initiated Outside Conscious Awareness

Myoclonus, tremor, dystonia, and chorea are basic categories of hyperkinetic movements that originate outside conscious awareness. There is no primary subjective sensation that provokes these movements, and in some instances, such as milder cases of chorea, patients may be unaware they are occurring.

MYOCLONUS

Myoclonus is defined as sudden lightning-like jerks of a body part. The jerk, by definition, involves at least one entire muscle and displaces a body part. This is in contrast to fasciculations or myokymia, which occurs in only a segment of muscle and does not result in movements across joints. Myoclonus emanates from the CNS, in contrast to fasciculations, which originate in anterior horn cells or more distally. This is not to say, however, that a peripheral stimulus might not trigger a myoclonic jerk, as is the case in stimulus-sensitive myoclonus.

Myoclonic movements are always simple, in that they are brief, unsustained, and unidirectional. They may take on a slightly more complex appearance, however, if repetitive. Sometimes myoclonus is subtle, manifested as single small-amplitude jerks of fingers or other portions of the limbs. At the other extreme are violent myoclonic jerks of the trunk, resulting in a massive lurch of the body. Myoclonus may be spontaneous at rest or provoked by use of the body part (action myoclonus) or stimulus-sensitive, that is, elicited by tactile, visual, or auditory stimuli.

Myoclonus may originate at any level of the CNS, including cortex, basal ganglia, brain stem, or spinal cord, and hence, may be associated with a broad spectrum of other neurologic signs. Some forms of myoclonus are components of epileptic disorders and in that context are analogous to formes frustes of motor seizures. See Chapter 27 for a comprehensive review of myoclonus.

A brief twitch occurring as a manifestation of a tic disorder may bear close resemblance to myoclonus. However, in tic disorders there are usually other tics with a more complex appearance that allow proper diagnosis.

The brief and repetitive phasic movements sometimes occurring as a component of dystonia may resemble myoclonus (see below). These phasic dystonic movements, however, are typically slightly sustained at the peak of movement, in contrast to myoclonus. Also, phasic dystonic movements are usually associated with more persistent dystonic contractions, which reveal the diagnosis. In rare cases, myoclonus and dystonia occur together (so-called myoclonic dystonia). In such disorders, special electrophysiologic studies may be necessary to properly characterize the disorder.

The term "palatal myoclonus" is well-entrenched in the neurologic literature, but the disorder is not truly a form of myoclonus. It is not lightning-like, in contrast to the accepted definition of myoclonus. Also, it is rhythmical and thus characteristic of tremor (especially the slow tremor sometimes referred to as myorhythmia; see below). The present preferred term for this phenomenon is "palatal tremor."

Asterixis may appear to be similar to myoclonus but is really the opposite. Asterixis is a negative motor phenomenon, with bursts of inhibitory CNS activity momentarily turning off muscle tone. This is manifested as recurrent brief interruptions of attempted sustained limb posture. This phenomenon is most obvious when the patient holds out both arms with wrists fully extended; flapping of the hands downward is seen. A sudden brief relaxation of wrist extensor tone (due to transient muscle tone inhibition) causes the extended wrists to drop by gravity, followed by wrist reextension as the muscle tone is reestablished. Toxic metabolic disorders (including hepatic encephalopathy with so-called liver flap) as well as lesions of the anterior cerebral cortex can cause asterixis.

TREMOR

Rhythmicity is the hallmark of tremor. By definition, tremor implies a regular rhythm with a relatively consistent periodicity. When assessing tremor, it is often a good idea to observe the tremor for more than just a few seconds, attending to the rhythmicity and frequency. In any given body part, organic tremor typically has a fairly constant frequency that does not vary by more than about ± 1 Hz. In contrast, psychogenic tremor often varies widely in frequency.

Repetitive action myoclonus or the phasic movements of dystonia sometimes give the appearance of tremor, but lack of regular periodicity usually allows the distinction to be made. Also, repetitive action myoclonus masquerading as tremor often has an everchanging vector of movement, most obvious when a body part is observed in a sustained posture. This may be easiest to see when the patient attempts to hold an index finger near (but not touching) the extended finger of the examiner; small jerks up, down, left, and so forth, suggest myoclonus. Similar observation of postural tremor typically reveals that the finger deviations do not have chaotic vectors.

Attention to two factors is critical for correctly categorizing tremor: (1) the situations in which it is present, that is, at rest, with posture, with action, and also whether it increases with movements nearing a target (terminal tremor); and (2) the portion of the body affected (e.g., head, chin, hands, or legs). Hand tremor is most commonly the tremor brought to the attention of physicians, and the distinctions among rest, posture, action, and terminal accentuation are easiest to see in the upper limbs; hence, the discussion below focuses on upper limb tremor.

- *Rest tremor* is typical of parkinsonism and may be apparent during the office interview before the examination. It may involve the whole hand or just one or a few digits; for example, thumb tremor alone may occur. Parkinsonian rest tremor is often intermittent but is most consistently seen during walking; it should be specifically looked for when gait is observed. Rest tremor may also be unmasked during testing of rapid alternating hand movements; this test may disinhibit rest tremor of the opposite hand. The usual tremor of parkinsonism is slow (3 to 8 cycles/second) and sometimes so slow that it appears bizarre when of large amplitude. Sometimes patients with postural tremor are mistakenly thought to have rest tremor when the hands are tensely held in the lap.

  Resting tremor may also be observed in other body parts in parkinsonian patients, including the legs and chin. In the legs, it is most typically seen when the patient is seated with the knees flexed and feet resting flatly on the ground; rhythmic bouncing of the knees then is apparent.

  Rest tremor alone does not equate with parkinsonism. Other signs of parkinsonism, such as rigidity and reduced RAMs, are necessary for this diagnosis.
- *Postural tremor* implies that the tremor is present when the body part is held against gravity, such as in the extended upper limbs. Often there are differences in amplitude with different postures. For example, some patients display a mild postural hand tremor when the arms are fully extended at the elbows but a marked tremor when the elbows are flexed and the hands are held in front of the face. Holding an object, such as a cup, to the mouth may reveal a more prominent tremor than simply arm extension. Thus, clinical analysis of postural hand tremor should include observation in at least these positions: with arms extended at the elbows, with the elbows flexed, and with an empty cup held to the mouth.

  Typical head tremor is a postural tremor and not a hallmark of parkinsonism. Because the cervical antigravity muscles must be active to maintain head position in the upright posture, head tremor does not qualify as a rest tremor.
- *Action tremor* implies that the tremor is present when the body part (e.g., hand) is moving. Typically, this tremor is assessed by observation during finger-nose and heel-shin movements. Specific tasks, such as handwriting or pouring water, sometimes bring out an action tremor. Although the usual parkinsonian rest tremor may also be present during sustained posture, it usually is markedly dampened with action.
- *Terminal accentuation* of tremor may be seen during action as the hand (or foot) nears the target. It is important to differentiate this from the serial dysmetria of ataxia, which may have a similar appearance. In serial dysmetria, an ataxic finger nearing a target (e.g., finger-nose testing) imprecisely overshoots the target and then, in an attempt at correction, overshoots the target in the other direction. Such serial overshooting movements may give the appearance of tremor if not closely analyzed.

Which body parts are tremulous is also important to assess. The complete tremor analysis should include attention to all four limbs, head, tongue, and voice. Voice tremor may be apparent during normal conversation but is more easily heard during vowel prolongation, for example, sustaining the "ahhhh" sound.

DYSTONIA

Dystonia implies an abnormal posture of one or more portions of the body with an inappropriate sustained contraction of muscles. For example, a dystonic foot may be involuntarily inverted or a dystonic neck (torticollis) may be manifest as involuntary rotation to one side. Such tonic deviations are usually easy to identify as dystonia. Sometimes, however, diagnostic confusion arises when phasic dystonic movements are superimposed. Such phasic movements may be rapid and repetitive and, thus, confused with tremor. The lack of true rhythmicity, however, allows phasic dystonia to be differentiated from tremor (parenthetically, tremor occasionally accompanies dystonia and is then termed "dystonic tremor").

Phasic dystonic movements are stereotyped (i.e., always in the same direction) and are at least slightly sustained at the peak of movement. Hence, they differ from chorea (see below).

In a dystonic limb (or other body part), voluntary activation of the agonist muscle (e.g., biceps) tends to be inappropriately opposed by involuntary cocontraction of the antagonist muscle (e.g., triceps). This results in a slow, stiff-appearing movement, perhaps sustaining a posture.

Dystonia may be apparent only in certain situations or with certain tasks. For example, foot dystonia during walking may normalize during running or walking backwards. Only very circumscribed tasks may bring out the dystonia, such as so-called writer's cramp (task-specific writing dystonia). In some cases, patients discover that certain tricks (so-called sensory tricks) attenuate the dystonia (e.g., touching the face to reduce the dystonic pull of cervical dystonia [torticollis]). Discussion of dystonic syndromes is found in Chapter 23.

The abnormal sustained postures of dystonia are usually not confused with other movement disorders. A tonically deviated neck or inturned foot is usually recognized as dystonia. If the patient has phasic dystonic movements, however, diagnostic uncertainty can arise. The distinction from chorea is based on the patterned (stereotyped) character of phasic dystonia; that is, the vector of the movements and the form are similar from one movement to the next. In contrast, chorea is characterized by nonstereotyped, free-flowing movements that are somewhat random in space. In some cases, however, chorea and dystonia occur together.

Tics may be identical in appearance to phasic movements of dystonia. Associated motor signs should help make the distinction. If the phasic movements are superimposed on a more tonically sustained posture, dystonia is likely. Conversely, if other tic-like movements involve other portions of the body, a tic disorder is more likely, especially if these other movements are complex (i.e., multivectored or semipurposeful, such as back and forth headshaking or vocalizations). Of note, tics are paroxysmal, with normal motor tone between tics.

CHOREA

Rapid, flowing movements of a part or parts of the body that are random in space and time characterize chorea. Patients may perform simple voluntary movements with a choreiform flourish. For example, voluntary touching of the nose may be preceded by a wide circling movement of the hand before the finger lands precisely on the target. When the patient is observed while seated or standing, various degrees of extra movements are apparent, ranging from a subtle twitch of the face to large flowing movements of the hands. Sometimes the choreiform movement becomes incorporated into a semipurposeful movement; for example, what begins as a choreiform abduction of the arm may end as what appears to be a voluntary smoothing of the hair.

Patients with chorea cannot maintain certain postures more than briefly (motor impersistence). Protrusion of the tongue often cannot be maintained for more than a few seconds (tongue-darting). If the patient is asked to maintain a steady grip on the examiner's fingers, the pressure acquires an undulating pattern after a few seconds (so-called milkmaid's grip).

When first observed, the appearance may simply suggest restlessness. In contrast to the patient with movements in response to akathisia, however, the patient with chorea usually does not experience a sense of substantial restlessness. In fact, patients with milder occurrences may not even be aware that they have chorea. Specific choreiform syndromes are described in Chapter 25.

Unlike tics or dystonia, choreiform movements are not stereotyped in appearance. For example, in chorea of an upper limb, the hand twists and turns without any fixed vectors or pattern of movement. With phasic dystonia, the repetitive movements are simple and in approximately the same plane, one after the other. Patients with tics make the same movement repetitively in contrast to the randomness in space of chorea.

Obviously, some areas of the body, such as the tongue and mouth, are constrained by anatomy and have only limited directions in which movement is possible. In these regions, it may be difficult to differentiate chorea from, for example, phasic dystonic tongue movements. However, the accompanying signs in other portions of the body usually allow the correct diagnosis.

Patients with chorea may have altered deep tendon reflexes. This is most apparent when the knee jerks are tested, giving the appearance of a "hung-up" response.

BALLISMUS

Ballismus is actually on one end of the chorea spectrum. It appears as violent, large-amplitude chaotic movements of a body part, especially the proximal limbs. Typically, occurrence is on only one side of the body and is hence designated "hemiballismus." Most often, a lesion of the subthalamic nucleus is responsible, but lesions of other extrapyramidal structures may occasionally cause a similar syndrome.

ATHETOSIS

Athetosis is a slow, writhing involuntary movement of a distal limb. On close analysis, the appearance is actually that of chorea combined with dystonia. When the choreiform component is prominent, the term "choreoathetosis" may be applied. "Athetosis" and "choreoathetosis" are terms that conventionally have most often been used in cerebral palsy.

Other Categories of Abnormal Hyperkinetic Movements
Not Originating in Conscious Awareness

MYORHYTHMIA; PALATAL TREMOR (PALATAL MYOCLONUS)

The term "myorhythmia" is reserved for slow, pendular, periodic movements; that is, a very slow tremor. The frequency is 1 to 3 cycles/second, which is less than the 4 to 8 cycles/second typical rest tremor of Parkinson's disease. Myorhythmias are typically present at rest and may be exacerbated or attenuated by sustained posture or muscle activation. The term "myorhythmia" is not in favor in some neurologic circles, but it has some specific connotations and is worthwhile to recognize as a subcategory of tremor.

Palatal myoclonus is the prototypical myorhythmia. This slow, repetitive elevation of the soft palate beats with a regular frequency, fulfilling criteria for tremor rather than myoclonus; hence, the new preferred term is "palatal tremor."

Palatal tremor may be associated with similar myorhythmical movements of neighboring branchial musculature, such as the larynx, tongue, jaw, ocular muscles, or diaphragm. Myorhythmias may also be seen in the extremities, with or without association with palatal tremor.

Myorhythmias, including palatal tremor, are associated with disease of the brain stem or cerebellum. The implicated neuronal circuitry (so-called Mollaret's triangle) includes the cerebellar dentate nucleus and its afferent projection to the red nucleus via the superior cerebellar peduncle and the brain stem nucleus, the inferior olive (Fig. 1-2).

The rare syndrome of oculomasticatory myorhythmia is characterized by pendular movements of the eyes that are synchronous with movements of the masticatory muscles. Other craniofacial muscles may also be yoked to these movements, with the same slow periodicity of about 1 to 3 cycles/second. Ophthalmoparesis, especially in the vertical plane, is associated. This anomaly is said to be pathognomonic of cerebral Whipple's disease.

STEREOTYPIES, INCLUDING CERTAIN TARDIVE SYNDROMES

In the context of movement disorders, "stereotypy" implies a recurring and often complex movement that has a relatively fixed character. This is a generic term that might be applied to a recurrent tic seen in Tourette's syndrome, and in fact many tics might be dually classified as stereotypies. The concept of stereotypy also best characterizes some of the adventitious movements of tardive dyskinesia.

"Tardive dyskinesia" is a term that, in its general sense, encompasses a wide variety of abnormal movements that develop in some patients receiving long-term treatment with drugs blocking dopamine receptors (e.g., neuroleptic and antiemetic agents). Although almost any of the classes of hyperkinetic movements may occur as components of tardive disorders (including dystonia and chorea), some of the more classic of these are stereotypies. Included are the hallmark orobuccolingual movements as well as the repetitive body-rocking movements that sometimes are associated with the orobuccolingual movements.

Classic orobuccolingual dyskinesias are the most frequently recognized signs of tardive dyskinesia, and sometimes the terms for these two disorders are used interchangeably. The continuous in and out movements of the tongue, with lip pursing alternating with retraction, fit in the category of stereotypy (even though this point is not emphasized in the neurologic literature). These movements do not fall within the definition of chorea, since they lack motor randomness. The complex pattern of one movement component following the next repetitively suggests that the movement is not simply that of dystonia, even though some components may be somewhat sustained, as in dystonia. However, if the orobuccolingual movements are predominantly sustained, the appropriate classification may be tardive dystonia. For example, neuroleptic-related persistent facial grimacing with sustained retraction of the corners of the mouth might be best classified as a tardive dystonia rather than the classic orobuccolingual tardive dyskinetic syndrome. Such distinctions may have therapeutic significance in some cases, as discussed in Chapter 26.

SPASMS

Spasm is a very general term applied to a sustained hypercontractile state of muscle that consequently reduces mobility of the involved region. The origins may be peripheral, such as in hemifacial spasm (seventh cranial nerve irritation) or lumbar spasms secondary to lumbosacral nerve root compression or even muscle strain. Spasm may also have a CNS origin, such as in stiff-man syndrome (see Chapter 28). In a very broad sense, even the hypercontractile state of dystonia may be thought of as a kind of spasm.

Certain specific types of spasm of peripheral origin require further definition to avoid confusion of terms.

- *Cramps* are painful spasms confined to one muscle or muscle group and are paroxysmal, short-lived, sometimes recurrent, and often provoked by voluntary muscle contraction and relieved by muscle stretching. Cramps may occur in disorders of muscle (e.g., metabolic myopathies), neuromuscular junction (e.g., anticholinesterase toxicity), or motor neurons (e.g., amyotrophic lateral sclerosis). Inborn errors or autoimmune disturbances affecting cellular membrane ion channels may provoke cramps. They may also be induced by metabolic conditions compromising peripheral nerve and muscle function, such as electrolyte disturbances or exercise-induced depletion of muscle energy reserves. Generalized cramp syndromes may be migratory, affecting one muscle and then another. Dystonia differs in that it is of central origin, persists throughout the waking day (albeit to variable degrees), and is nonmigratory. The paroxysmal nature of cramps usually easily differentiates these from dystonia. Rare dystonic syndromes may occur exclusively paroxysmally; however, these involve more widespread regions of the body (e.g., an entire limb), unlike the anatomically limited cramps, which are confined to, for example, calf or hamstring muscles.
- *Tetany* refers to a muscle hypercontractile state due to reduced concentrations of calcium or magnesium or to metabolic alkalosis and manifest by repetitive discharge of motor neurons. Prototypical are the carpopedal (hand or foot) spasms long recognized as a sign of hypocalcemia. In contrast to carpal spasms due to hypocalcemia, writer's dystonia ("writer's cramp") is not temporally random but occurs each time writing is attempted and persists for as long as the writing continues. Tetany can be confirmed by the characteristic pattern of electromyographic discharge reflective of motor neuronal membrane hyperexcitability, as discussed in Chapter 28. Tetany is to be differentiated from tetanus, caused by the toxin from *Clostridium tetani* wound infection and manifested as severe rigidity and painful spasms.
- *Impaired muscle relaxation syndromes* originating in the peripheral nervous system may occasionally resemble movement disorders of CNS origin; these include myotonia, neuromyotonia, and the effect of myxedema on muscle function. Myotonia is manifested as impaired muscle relaxation, as discussed above in the section on myopathies. This is a disorder of muscle membrane instability in which repetitive depolarization precludes normal muscle relaxation after voluntary contraction. Unlike cramps, it is typically painless. Voluntary forceful eye closure or grip may be inappropriately

sustained by the myotonic discharges. Neuromyotonia is a very rare disorder characterized by repetitive depolarization of motor neuron terminals that causes painless stiffness and impaired relaxation. Myxedema may also impair muscle relaxation. Confirmation of these disorders is by electromyography; characteristic firing patterns are seen with insertion of the recording needle into the involved muscles.

FINAL COMMENTS

Nervous system control of movement is complex, with multiple levels of integration culminating in the final common pathway through the peripheral nervous system. Obviously, problems anywhere within this vast array of neuronal circuitry could result in a disorder of movement. The clinician dealing with patients who have movement disorders must be able to recognize the relevant clinical symptoms and signs. This chapter has attempted to provide the diagnostic framework for the more detailed discussions of the spectrum of movement disorders that follow.

SELECTED READING

Brazis PW, Masdeu JC, Biller J. *Localization in Clinical Neurology*, 3rd ed. Little, Brown & Company, Boston, 1996.

Westmoreland BF, Benarroch EE, Daube JR, Reagan TJ, Sandok BA. *Medical Neurosciences: An Approach to Anatomy, Pathology, and Physiology by Systems and Levels, 3rd ed.* Little, Brown & Company, Boston, 1994.

Wiebers DO, Dale AJD, Kokmen E, Swanson JW eds. *Mayo Clinic Examinations in Neurology, 7th ed.* Mosby, St. Louis, 1998.

2

Motor Speech Disorders: Clues to Neurologic Diagnosis

Joseph R. Duffy, PhD

CONTENTS

INTRODUCTION

Speech is the most complex of innately acquired human motor skills, an activity characterized in normal adults by the production of about 14 distinguishable sounds per second through the coordinated actions of about 100 muscles innervated by multiple cranial and spinal nerves. The ease with which we speak belies the complexity of the act, and that complexity may help explain why speech can be exquisitely sensitive to nervous system disease. In fact, changes in speech can be the only evidence of neurologic disease early in its evolution and sometimes the only significant impairment in a progressive or chronic neurologic condition. In such contexts, recognizing the meaning of specific speech signs and symptoms can provide important clues about the underlying pathophysiology and localization of neurologic disease.

Neurologic speech disorders are known as *motor speech disorders*. They can be divided into two major categories, the *dysarthrias* and *apraxia of speech*. There are several types of dysarthria, but each reflects some disturbance of neuromuscular execution or control that can be attributed to weakness, incoordination, a variety of muscle tone abnormalities, or a variety of involuntary movements. The second category, apraxia of speech, is attributable to abnormalities in the programming of movements for speech rather than to their neuromuscular execution. (For readers relatively unfamiliar with concepts of speech production and speech disorders, Table 2-1 defines the basic components of speech production and some of the disturbances within each component that can be caused by neurologic disease.)

From: *Parkinson's Disease and Movement Disorders:
Diagnosis and Treatment Guidelines for the Practicing Physician*
Edited by: C. H. Adler and J. E. Ahlskog © Mayo Foundation
for Medical Education and Research, Rochester, MN

Table 2-1
Definitions of the Basic Components of Speech Production
and the Primary Speech Disorders Caused by Neurologic Disease

Basic components of speech production

Respiration	Exhalatory air provides the transmission medium for speech. The diaphragm and thoracic and abdominal muscles are included. Disturbances at this level generally lead to abnormalities in loudness and in the number of words that can be uttered within a single exhalatory cycle.
Phonation	The production of voice. Laryngeal muscle activity to adduct the vocal cords during exhalation is required. Abnormal voice production, regardless of cause, is called dysphonia, a common problem in dysarthria.
Resonance	The quality of voice determined by the pharyngeal and velopharyngeal muscles, which control the shape of the vocal tract above the larynx, most crucially the degree to which the voice is transmitted through the oral cavity rather than the nasal cavity and vice versa. Excessive nasal transmission leads to a perception of hypernasality, frequently a problem in several types of dysarthria.
Articulation	The production of specific, distinguishable sounds of speech that, when combined, give speech its meaning. This is primarily a function of lingual, facial, and jaw muscle activity, although laryngeal and velopharyngeal movements, which produce many of the distinguishing features among speech sounds (phonemes), also contribute.
Prosody	The variations in pitch, loudness, and duration across syllables that help convey stress, emphasis, and emotion. They are a reflection of the combined activities of respiration, phonation, resonance, and articulation.

Primary speech disorders associated with neurologic disease

Aphasia	A disturbance of language that can affect speech, comprehension of speech, reading, and writing (not discussed herein).
Apraxia of speech	A disturbance in the programming (selection and sequencing of kinematic patterns) of movements for speech production.
Dysarthria	Neuromuscular disturbance of strength, speed, tone, steadiness, or accuracy of the movements that underlie the execution of speech.

Table 2-2
Primary Tasks and Observations
for the Clinical Assessment of Motor Speech Disorders

Task	Observation
Nonspeech	
Jaw, face, tongue, palate	Symmetry at rest
	Symmetry and range of movement during spontaneous, reflexive, and imitated movement
	Involuntary movement at rest and during movement (e.g., fasciculations, tremor, dyskinesias)
Jaw, face, tongue	Strength against resistance
	Ability to smack lips, blow, click tongue (praxis)
Face, tongue	Atrophy
Palate	Movement during "ah"
	Gag reflex
Larynx	Sharpness of cough and grunted "uh"
	Ability to cough volitionally (praxis)
Speech	
Alternating motion rates "puh-puh-puh..." (lips) "tuh-tuh-tuh..." (anterior tongue) "kuh-kuh-kuh..." (mid and back tongue)	Rate, rhythm, range of movement, precision
Sequential motion rates "puh-tuh-kuh, puh-tuh-kuh..."	Sequential accuracy and speed (praxis)
Vowel prolongation ("ah")	Pitch, quality, loudness, duration, and steadiness
Conversation and reading	Rate, phrase length, prosody, voice quality, resonance, and precision

In this chapter, the basic methods for clinical examination of motor speech disorders are described and the distinctions among the disorders highlighted. Each disorder is reviewed in relation to its perceptible deviant speech characteristics, presumed underlying neuropathophysiology, associated signs and symptoms, and localization within the central or peripheral nervous system. Clues to distinguishing among the motor speech disorders also are highlighted. Finally, broad management strategies are reviewed.

CLINICAL ASSESSMENT OF MOTOR SPEECH DISORDERS

The clinical assessment of motor speech disorders often can be accomplished efficiently and simply. It includes (1) an oral mechanism examination, (2) a few examiner-specified speech tasks, and (3) assessment of conversation or reading aloud. A brief description of each task, its purposes, and the basic questions addressed during the task follow. Specific observations derived from these tasks are summarized during subsequent discussion of specific motor speech disorders. The clinical assessment is summarized in Table 2-2.

Oral Mechanism Examination

The oral mechanism is examined to assess the strength, speed, symmetry, range of movement, and steadiness of the jaw, face, tongue, and palate. These structures are observed at rest, during static postures, and during nonspeech movements.

1. *Jaw.* At rest, does the jaw hang excessively open? When opened, does it deviate to one side? Can the patient resist attempts by the examiner to open or close the jaw? Are there any adventitious movements at rest or when attempting to maintain mouth opening?
2. *Lower face.* At rest, are the angles of the mouth symmetrical? Are fasciculations apparent in the perioral area? Do the angles of the mouth elevate equally during an emotional smile or voluntary lip retraction? Is lip rounding symmetrical? Is there any symmetrical or asymmetrical air leakage during attempts to puff up the cheeks?
3. *Tongue.* Is the tongue normal in size bilaterally? Is there any grooving of the tongue indicative of atrophy? Are there any fasciculations? Are there any quick, slow, or sustained adventitious movements of the tongue at rest or on protrusion? Can the tongue be protruded fully, or does it deviate to the right or left on protrusion? Can it be wiggled to the left and right with normal rate, rhythm, and range of movement?
4. *Palate.* Are the arches of the soft palate symmetrical at rest? Can a gag be elicited with equal ease on both sides? Are there any rhythmical or arrhythmical movements of the soft palate? Does the soft palate elevate symmetrically during prolongation of the vowel "ah"?
5. *Larynx.* The integrity of vocal cord adduction can be crudely judged by asking the patient to cough or produce a glottal coup (a gruntlike "uh"). The cough and coup should be sharp, a gross indication of adequate adduction.

Nonverbal Oral Movements

For assessment of the ability to organize and program some fairly reflexive oromotor movements in a volitional way, the patient should be asked to cough, blow, smack the lips, and click the tongue. Any groping, off-target movements or tendency to verbalize the command rather than perform the movement suggests a nonverbal oral apraxia, a problem that tends to co-occur with apraxia of speech and limb apraxia. These apractic difficulties indicate left cerebral hemisphere disease.

Speech Alternating Motion Rates

Following examiner demonstration, the patient is asked to take a breath and repeat "puh-puh-puh-puh-puh…" as fast and as steadily as possible for 3 to 5 seconds. This is followed by similar repetitions of "tuh-tuh-tuh-tuh-tuh…" and "kuh-kuh-kuh-kuh-kuh…." These alternating motion rate (AMR) tasks permit judgments of rate, rhythm, precision, and range of motion of rapid movements of the lips, jaw, and tongue. Normal adults can produce an even rhythm at a rate of about 5 or 6 syllables per second.

Speech Sequential Motion Rates

Following examiner demonstration, the patient is asked to take a breath and produce "puh-tuh-kuh, puh-tuh-kuh, puh-tuh-kuh…" repetitively for 3 to 5 seconds. This task assesses the ability to program a sequence of speech movements rapidly and successively. After brief practice at slow rates, normal speakers accomplish this task rapidly and accurately. The ability to accurately sequence these sounds is the important observation on this task.

Vowel Prolongation

Following examiner demonstration, the patient is asked to take a deep breath and say "ah" for as long and as steadily as possible. Many normal adults can do this for 15 or more seconds, but 5 seconds usually suffices. The purposes of this task are to isolate the respiratory and laryngeal mechanisms and to evaluate their ability to produce a sustained tone at normal pitch, quality, loudness, duration, and steadiness.

Conversational Speech or Reading Aloud

Many important judgments about motor speech ability can be made when patients respond during the medical history and casual conversation or when they read aloud a standard paragraph. Such speech samples permit judgments of speech rate, phrase length, voice quality, resonance, and precision of articulation. Connected speech also permits assessment of the prosodic features of speech (defined below). Prosodic abnormalities often provide valuable clues that distinguish among the various motor speech disorders.

THE DYSARTHRIAS

The dysarthrias reflect neuromuscular disturbances of strength, speed, tone, steadiness, or accuracy of the movements that underlie the execution of speech. By definition, they do not include disorders attributable to anatomical deformities, faulty learning, or psychopathology.

Although dysarthria is manifested as a disorder of movement, it is important to recognize that sensorimotor integration (with tactile, proprioceptive, and auditory feedback representing the crucial sensory components) is essential to speech motor control. From this standpoint, most or all of the dysarthrias localized to the central nervous system should be thought of as sensorimotor rather than simply motor disturbances.

The dysarthrias can reflect disturbances at any one or a combination of the major components of the speech mechanism, including respiration, phonation, resonance, articulation, and prosody (see Table 2-1 for basic definitions). There are several subtypes of dysarthria. Subtype distinctions reflect (1) different auditory perceptual characteristics of speech, (2) different locations of central or peripheral nervous system lesions, and (3) different underlying neuromuscular dysfunctions. The ability to identify the type of dysarthria can be very useful for determining the underlying pathophysiology and lesion localization within the nervous system.

Each of the major categories of dysarthria is described below relative to its lesion locus, pathophysiology, nonspeech oral mechanism findings, and distinguishing speech characteristics. Localization and distinctive neuromuscular deficits are summarized in Table 2-3. Nonspeech oral mechanism findings are summarized in Table 2-4, and distinguishing speech characteristics are summarized in Table 2-5.

Flaccid Dysarthria

LESION LOCI

Lesions associated with flaccid dysarthria lie in the cell bodies, axons, or neuromuscular junction of the lower motor neurons that supply the speech musculature. These most often include cranial nerve V (trigeminal), VII (facial), X (vagus), or XII (hypoglossal) or the cervical and thoracic spinal nerves innervating the diaphragm and other respiratory muscles.

Table 2-3
Types of Dysarthria, Their Associated Lesion Loci, and the Neuromuscular
Deficits That Probably Explain Their Distinctive Speech Characteristics

Dysarthria type	Lesion locus	Distinctive neurologic deficit
Flaccid	Lower motor neurons (cranial and spinal nerves)	Weakness
Spastic	Upper motor neurons (bilateral)	Spasticity
Ataxic	Cerebellum (cerebellar control circuit)	Incoordination
Hypokinetic	Basal ganglia control circuit	Rigidity, reduced range of movement
Hyperkinetic	Basal ganglia control circuit	Involuntary movements
Unilateral upper motor neuron	Upper motor neuron (unilateral)	Weakness, (?) incoordination, (?) spasticity
Mixed	Two or more of the above	Two or more of the above

PATHOPHYSIOLOGY

With the exception of compensatory speech movements, the abnormal speech characteristics associated with flaccid dysarthria are all attributable to underlying weakness.[1] Because neurologic disease can affect the cranial or spinal nerves unilaterally or bilaterally, singly or in combination, the specific speech deficits depend on which lower motor neurons are affected. These effects are reviewed in the next section and are summarized in Table 2-3.

NONSPEECH ORAL MECHANISM FINDINGS AND DISTINGUISHING SPEECH CHARACTERISTICS

TRIGEMINAL NERVE (V)

Unilateral trigeminal lesions generally do not result in significant speech deficits. Bilateral trigeminal weakness, however, can significantly reduce jaw movement, especially closure, and severely affect articulatory contacts among the tongue, lips, and teeth. This can result in significant imprecise articulation of many sounds.

FACIAL NERVE (VII)

Lesions can cause imprecise articulation of sounds requiring facial movement. Facial weakness can affect sounds that require movement or contact of the upper and

---

[1] In persons with flaccid dysarthria associated with myasthenia gravis, speech may be normal if utterances are brief. However, if they are required to read aloud for several minutes without rest, their speech may deteriorate dramatically and any of the characteristics of flaccid dysarthria mentioned here may emerge. Speech may then recover to normal after a brief rest.

Table 2-4
Nonspeech Oral Mechanism Findings Frequently Associated
With Each of the Major Motor Speech Disorders

Physical findings	Dysarthria						Apraxia of speech
	Flaccid	Spastic	Ataxic	Hypokinetic[a]	Hyperkinetic[b]	Unilat UMN[c]	
Hypoactive gag	+	-	-	-	-	-	-
Hypotonia	+	-	+	-	-	-	-
Atrophy	++	-	-	-	-	-	-
Fasciculations	++	-	-	-	-	-	-
Nasal regurgitation	++	-	-	-	-	-	-
Unilateral palatal weakness	++	-	-	-	-	-	-
Weak cough or coup	++	+	-	+	-	-	-
Dysphagia	+	+	-	+	+	+	-
Drooling	+	+	-	+	+	+	-
Pathologic oral reflexes	-	++	-	-	-	-	-
Hyperactive gag	-	++	-	-	-	-	-
Pseudobulbar affect	-	++	-	-	-	-	-
Masked facies	-	-	-	++	-	-	-
Tremulous jaw, lips, tongue	-	-	-	++	+	-	-
Reduced range of movement	-	+	-	++	-	-	-
Head tremor	-	-	+	-	+	-	-
Adventitious orofacial movements	-	-	-	-	++	-	-
Sensory tricks	-	-	-	-	++	-	-
Torticollis	-	-	-	-	++	-	-
Palatal myoclonus	-	-	-	-	++	-	-
Unilateral tongue and lower face weakness	-	-	-	-	-	++	+
Nonverbal oral apraxia	-	-	-	-	-	+	++

+, may be present but not generally distinguishing; ++, distinguishing when present; -, not usually present.
[a]Parkinsonian.
[b]For example, dystonia, chorea, myoclonus, tremor.
[c]Unilateral upper motor neuron.

Table 2-5

Primary Distinguishing Speech Characteristics Associated With Major Motor Speech Disorders

Speech characteristics	Dysarthria						Apraxia of speech
	Flaccid	Spastic	Ataxic	Hypokinetic	Hyperkinetic	Unilat. UMN	
Phonatory or respiratory							
hoarseness	+	-	-	+	-	+	-
Breathiness (continuous)	++	-	-	+	-	-	-
Diplophonia	++	-	-	-	-	-	-
Stridor	++	-	-	-	+	-	-
Short phrases	++	+	-	-	+	-	-
Strained-harsh voice	-	++	-	-	+	-	-
Excess loudness variation	-	-	++	-	++	-	+
Monopitch	+	+	-	++	+	-	+
Monoloudness	+	+	-	++	-	-	-
Reduced loudness	+	-	-	++	-	-	-
Vocal flutter	++	-	-	++	-	-	-
Forced inhalation or exhalation	-	-	-	-	++	-	-
Voice stoppages or arrests	-	-	-	-	++	-	-
Transient breathiness	-	-	-	-	++	-	-
Voice tremor	-	-	-	-	++	-	-
Myoclonic vowel prolongation	-	-	-	-	++	-	-
Intermittent strained or breathy voice breaks	-	-	-	-	++	-	-
Resonatory							
Hypernasality	++	+	-	+	-	-	-
Audible nasal emission	++	-	-	-	-	-	-
Intermittent hypernasality	-	-	-	-	++	-	-

Articulatory or prosodic								
Imprecise articulation	+	+	+	+	+	+	+	+
Slow rate	+	–	+	–	+	++	–	–
Slow and regular AMRs	–	–	–	–	–	++	–	–
Excess and equal stress	+	–	–	–	++	+	–	–
Irregular articulatory breakdowns	+	+	+	–	++	–	–	–
Irregular AMRs	–	–	++	–	++	–	–	–
Distorted vowels	+	+	++	–	++	–	–	–
Prolonged phonemes	+	+	+	–	+	–	–	–
Reduced stress	–	–	–	++	–	–	–	–
Short rushes of speech	–	–	–	++	–	–	–	–
Variable rate	–	–	+	++	–	–	–	–
Rapid "blurred" AMRs	–	–	–	++	–	–	–	–
Repeated phonemes	–	–	–	++	–	–	–	–
Palilalia	–	–	++	++	–	–	–	–
Prolonged intervals	+	+	++	–	–	–	–	–
Poorly sequenced SMRs	++	–	–	–	–	–	–	–
Articulatory groping	++	–	–	–	–	–	–	–
Articulatory substitutions	++	–	–	–	–	–	–	–
Attempts at self-correction	++	–	–	–	–	–	–	–
Articulatory additions	++	–	–	–	–	–	–	–
Automatic > volitional speech	++	–	–	–	–	–	–	–
Inconsistent errors	++	–	+	–	+	–	–	–
Increased errors with increasing length	++	–	–	–	–	–	–	–
Any component								
Rapid deterioration and recovery with rest	–	–	–	–	–	–	–	++

+, may be present but not generally distinguishing; ++, distinguishing when present; –, not usually present.
AMRs, alternating motion rates; SMRs, sequential motion rates; Unilat. UMN, unilateral upper motor neuron.

lower lips (p, b, m, w) or contact between the lower lip and the upper teeth (f, v). Unilateral paresis or mild-moderate bilateral weakness generally results in distortion of these sounds, whereas bilateral paralysis can completely prevent their production.

VAGUS NERVE (X)

Branches of the vagus nerve supply the laryngeal and velopharyngeal muscles for speech. Unilateral lesions of the recurrent laryngeal nerve can produce vocal cord paralysis in the paramedian position, with resultant hoarseness, breathiness, and, sometimes, diplophonia (two perceived pitches rather than one, because the two vocal cords are vibrating at different rates). Bilateral lesions of the recurrent laryngeal nerve do not significantly alter phonation, but the resulting bilateral paramedian positioning of the vocal cords may seriously compromise the airway and cause inhalatory stridor. Unilateral lesions of the superior laryngeal nerve do not produce significant dysphonia, but bilateral lesions can restrict control of vocal pitch. Unilateral lesions of the pharyngeal branch of the vagus nerve can cause mild to moderate hypernasality and nasal air flow during production of consonants requiring pressure build-up in the mouth. Bilateral lesions can result in marked to severe hypernasality, audible nasal emission, and significant weakness of sounds requiring intraoral pressure (i.e., all non-nasalized consonant sounds); nasal regurgitation of liquids may occur during swallowing.

Intramedullary or extramedullary lesions (i.e., above the pharyngeal branch of the vagus nerve) can interfere with both phonation and resonance. The resultant unilateral or bilateral vocal cord paralysis leaves the affected vocal cord in the abductor position, causing short phrases and severe breathiness as well as hypernasality and nasal emission due to velopharyngeal weakness. The cough and glottal coup may be weak with unilateral or bilateral vocal cord paralyses.

HYPOGLOSSAL NERVE (X)

Unilateral or bilateral lesions produce weakness, atrophy, and fasciculations of the tongue. This results in articulatory imprecision of all consonants requiring lingual movements (e.g., as in do, to, no, key, go, sing, them, see, zoo, shoe, measure, chew, jump). Unilateral hypoglossal lesions usually result in only mild imprecision, whereas bilateral lesions can severely affect articulatory precision, sometimes including vowels.

Spastic Dysarthria

LESION LOCI

Spastic dysarthria is the result of central nervous system lesions that affect the upper motor neuron pathways bilaterally. Lesions can occur anywhere along these pathways, from their origin in the right and left cerebral hemispheres to their corticobulbar or corticospinal destinations in the brain stem and spinal cord. When vascular in origin, single lesions that produce the disorder are usually in the brain stem. Offending lesions outside the brain stem are usually multifocal or diffusely located in motor pathways of the cerebral hemispheres.

PATHOPHYSIOLOGY

Upper motor neuron lesions usually produce a combination of weakness and spasticity. Some degree of weakness is usually evident in the speech musculature of persons with spastic dysarthria, and weakness certainly contributes to some of its deviant characteristics. However, it is the resistance to movement and the tendency toward vocal cord hyperadduction generated by spasticity that give spastic dysarthria its distinctive speech characteristics.

NONSPEECH ORAL MECHANISM FINDINGS

Slow rate of orofacial movements, a hyperactive gag reflex, and pathologic oral reflexes (including suck, snout, and palmomental reflexes) are frequently present. Persons with spastic dysarthria often exhibit pseudobulbar affect, or pathologic laughter and crying due to poor reflex control. Drooling and dysphagia are common.

DISTINGUISHING SPEECH CHARACTERISTICS

Spasticity usually affects all components of speech and generates multiple abnormalities in speech, but the diagnosis of spastic dysarthria is often based on a gestalt impression generated by just a few distinctive speech characteristics. Among the most distinctive is a strained-harsh voice quality, often accompanied by reduced variability of pitch (monopitch) and loudness (monoloudness). These features reflect an apparent bias of spasticity toward hyperadduction of the vocal cords during speech as well as a slowing of the rapid muscular adjustments necessary for normal pitch and loudness variability. Also prominent and distinctive is slow rate of speech, often with classically slow and regular speech AMRs.

Ataxic Dysarthria

LESION LOCI

Ataxic dysarthria is associated with abnormalities of the cerebellum or cerebellar control circuit functions that influence the coordination of volitional movement. Cerebellar speech functions are not well localized, but prominent and lasting dysarthria is most often associated with bilateral or generalized cerebellar disease.

PATHOPHYSIOLOGY

Reduced muscle tone and incoordination are associated with ataxic dysarthria. They appear largely responsible for the slowness of movement and inaccuracy in the force, range, timing, and direction of movements that underlie the distinctive speech characteristics of the disorder.

NONSPEECH ORAL MECHANISM FINDINGS

Oral mechanism size, strength, symmetry, and reflexes may be entirely normal. During oromotor AMR tasks, the repetitive movements of the jaw, face, and tongue are performed slowly and irregularly.

DISTINGUISHING SPEECH CHARACTERISTICS

Ataxic dysarthria is primarily a disorder of articulation and prosody. Although many abnormal speech characteristics can be present, only a few give speech its distinctive character. Conversational speech is often characterized by irregular breakdowns in articulation (giving speech an intoxicated or drunken character), sometimes including distorted vowels and excess variation in loudness. Speech rate is often slow, and some patients place excess and equal stress on each syllable produced, giving prosody a scanning character. Vowel prolongation is sometimes unsteady. Speech AMRs are often slow and distinctively irregular.

Hypokinetic Dysarthria

Lesion Loci

Hypokinetic dysarthria is associated with disease of the basal ganglia control circuit, probably bilateral in most instances. It is most frequently encountered in persons with Parkinson's disease or related parkinsonian conditions (see subsequent chapters for more complete discussion).

Pathophysiology

Rigidity, reduced force and range of movement, and slow (and sometimes fast) repetitive movements probably account for many of the distinctive deviant speech characteristics associated with hypokinetic dysarthria. These effects may be evident in the respiratory, laryngeal, velopharyngeal, and articulatory components of speech production.

Nonspeech Oral Mechanism Findings

The face may be masked or expressionless at rest and lack animation during social interaction. Reflexive swallowing may be reduced in frequency, and drooling may be apparent. Tremor or tremulousness may be apparent in the jaw, face, or tongue at rest and may be quite evident when the tongue is protruded. Size, strength, and symmetry of the jaw, face, and tongue may be surprisingly normal, but reduced range of motion may be evident in those structures on AMR tasks. Laryngeal examination may reveal bowing of the vocal cords. Reduced excursion of the chest wall and abdomen may be evident during quiet breathing and speech.

Distinguishing Speech Characteristics

The distinctive speech characteristics are primarily phonatory, articulatory, and prosodic. They often combine to convey a gestalt impression that speech is "flat," attenuated and unemotional, sometimes rapid but slow to start, and lacking in vigor and animation to a degree that is not explainable by weakness.

The very frequent phonatory abnormalities may include a tight breathiness with hoarseness and sometimes a vocal "flutter" (rapid voice tremor or tremulousness) that is most apparent during vowel prolongation. Reduced loudness and reduced pitch and loudness variability (monopitch and monoloudness) are also common, resulting in attenuation of normal prosodic variability. Hypernasality may be present but usually is not prominent.

Imprecise articulation is often evident secondary to reduced range of articulatory movement. Prolonged intervals between phrases reflect problems with initiation of movements for speech. Dysfluency characterized by sound prolongations or rapid sound repetitions sometimes occurs and can be prominent. When these dysfluencies include rapid or accelerating word and phrase repetitions, they are known as palilalia, a disorder rarely encountered in other types of dysarthria. Finally, hypokinetic dysarthria is the only type of dysarthria in which the speech rate may be rapid or accelerated and speech AMRs rapid and "blurred," with associated visually apparent reduced range of articulatory movements of the jaw, face, and tongue. Rapid or accelerating rate combined with prolonged intervals may be perceived as short rushes of speech.

Hyperkinetic Dysarthria

LESION LOCI

Similar to many hyperkinetic movement disorders, most varieties of hyperkinetic dysarthria probably derive from abnormalities in the basal ganglia circuitry (exception: palatal tremor).

PATHOPHYSIOLOGY

Hyperkinetic dysarthrias reflect the influence of quick or slow, rhythmic or arrhythmic involuntary movements that interrupt, distort, or slow intended speech movements. Such movements can affect respiration, phonation, resonance, and articulation, singly or in combination, and often have prominent effects on prosody. Although movements may be present constantly, they may be clinically detectable only during speech, an observation that sometimes leads clinicians to misdiagnose the problem as psychogenic.

Hyperkinetic dysarthrias can be divided into subtypes according to the specific involuntary movements that underlie them. Most of these hyperkinesias can occur elsewhere in the body and sometimes in the bulbar muscles but without affecting speech. The various hyperkinetic dysarthrias are too numerous to review here, but the nonspeech oral mechanism and distinguishing speech characteristics associated with those that occur most commonly are briefly summarized below. They help give a sense of the variety of forms taken by hyperkinetic dysarthria, most of which are clearly distinctive from other types of dysarthria.

NONSPEECH ORAL MECHANISM FINDINGS
AND DISTINGUISHING SPEECH CHARACTERISTICS

HYPERKINETIC DYSARTHRIA OF CHOREA

The adventitious movements of chorea are quick, unsustained, and unpredictable in course. They may occur in the jaw, face, tongue, palate, larynx, or respiratory system. They may be evident at rest, during sustained postures or volitional movement, and during speech. The movements may cause dysphagia, but the oral mechanism may be normal in size, strength, and symmetry and without pathologic oral reflexes. They may range from subtle exaggerations of facial expression to movements so pervasive and prominent that the face never seems to be at rest.

Abnormal speech characteristics depend on the structures affected. Choreiform respiratory movements may lead to sudden forced inspiration and expiration. Quick, involuntary adductor or abductor movements of the vocal cords can produce sudden voice arrests with an intermittent strained quality or transient breathiness. Similar movements in the velopharynx can produce mild intermittent hypernasality and nasal emission. Choreiform articulatory movements can lead to brief speech arrests, irregular articulatory breakdowns, and articulatory imprecision, sometimes including distorted vowels, usually with visually apparent quick adventitious movements such as lip compression, facial retraction, darting of the tongue, and head jerking. Irregular AMRs and unsteadiness of vowel prolongation may be striking because of the unpredictable movements superimposed on the steady vocal tract. Prosodic abnormalities are prominent, and the flow of speech can have a jerky, fits-and-starts character, as if the patient were trying to say as much as possible before the next involuntary movement occurs.

HYPERKINETIC DYSARTHRIA OF DYSTONIA

Dystonic speech movements are relatively slow and sustained, but quick movements are sometimes superimposed. Similar to chorea, dystonia may affect any or all parts of the speech system. Slow adventitious movements of the lips, tongue, or jaw may be evident at rest, during sustained postures and voluntary movements, and during speech. Dystonia in nearby structures that have minimal influence on speech, such as blepharospasm or torticollis, may be present. Patients sometimes develop sensory tricks or postures that help inhibit dystonic movements (e.g., a hand placed under the chin may inhibit a jaw-opening dystonia). It is important to recognize that dystonia can be confined to muscles for speech and swallowing and triggered only by the act of speaking.

When dystonia affects the larynx, it is variably known as spasmodic dysphonia or laryngeal dystonia. It may occur in isolation or be associated with dystonia in other craniofacial or neck structures. Laryngeal dystonia occurs in two forms, the more common of which is adductor spasmodic dysphonia, characterized by intermittent, waxing and waning, or constant strained-harsh voice quality, sometimes sufficient to cause voice stoppages. Less common is abductor spasmodic dysphonia, characterized by intermittent breathy or aphonic segments of speech. Laryngeal dystonia leading to spasmodic dysphonia is the most common focal, speech-induced hyperkinetic dysarthria.

Resonance abnormalities characterized by hypernasality are uncommon as the only manifestation of dystonic dysarthria, but they may be part of the clinical picture in spasmodic dysphonia. Focal face, jaw, and tongue dystonias can devastate articulation and lead to imprecise consonant articulation, distorted vowels, and irregular articulatory breakdowns. Dystonia anywhere within the speech system can have prominent effects on prosody, manifested as monopitch, monoloudness, short phrases, prolonged intervals and phonemes, inappropriate silences, and slow rate, creating a rhythm that reflects both excessive and insufficient prosody.

HYPERKINETIC DYSARTHRIA OF TREMOR

This dysarthria results from tremor, usually essential tremor, that affects speech muscles. The larynx is the most commonly affected speech structure, and organic or essential voice tremor is present in a substantial minority of patients with a neurologic diagnosis of essential tremor elsewhere in the body. Again, it is important to recognize that voice tremor may be present even though tremor is absent elsewhere in the body and may be evident only during speech. Tremor affecting speech can also arise from respiratory muscles and the velopharynx, tongue, and jaw.

Tremor of the arytenoid cartilages of the larynx may be apparent during laryngeal examination, and vertical oscillations of the larynx can sometimes be seen in the external neck during vowel prolongation. Similarly, tremor may be seen in the tongue, lips, or palate during vowel prolongation. It may also be evident during rest or sustained postures, and jaw tremor may be evident at rest or during mouth opening. In general, laryngeal and jaw tremors have the most significant impact on speech production.

Essential voice tremor is most easily detected during vowel prolongation and has a rhythmic, sinusoidal, quavering, waxing and waning character, with fluctuations in the range of 4 to 7 Hz. If the tremor is severe, there may be abrupt staccato voice arrests, sometimes leading to the designation "spasmodic dysphonia of essential voice tremor." When tremor is marked, speech rate may be slow. Prosodic abnormalities may be present, reflected in altered pitch and loudness. Tremor in other speech muscles, especially those of the jaw, may also lead to slow rate and, sometimes, imprecise articulation.

HYPERKINETIC DYSARTHRIA OF PALATOPHARYNGOLARYNGEAL MYOCLONUS

Caused by lesions in the dentatorubroolivary tracts in the brain stem and cerebellum (Guillain-Mollaret triangle), this unusual disorder (so-called palatal tremor) probably results from abnormal activity of a central pacemaker that generates jerks that are time-locked in different muscles. It is characterized by visually apparent, relatively abrupt, semirhythmic, unilateral or bilateral movements of the soft palate, pharynx, and laryngeal muscles at a rate of about 2 to 4 Hz. They are present at rest as well as during speech. Infrequently, the myoclonus includes the tongue, face, or nares. Patients may be unaware of the movements but may complain of a clicking sound or sensation in the ear secondary to myoclonic opening and closing of the eustachian tube.

The effects of the myoclonus on conversational speech often are not detectable, nor do they generally lead the patient to complain of speech difficulty. However, the effects can be very evident during vowel prolongation. They are heard as momentary semirhythmic arrests or tremor-like variations in voice at a rate that matches that of the visually evident palatal myoclonus (2 to 4 Hz). Rarely, intermittent hypernasality is evident. Articulation and prosody are usually normal, but brief silent intervals may occur if myoclonus interrupts inhalation, initiation of exhalation, or phonation. Because of the localization of the offending lesion, it may be accompanied by other types of dysarthria associated with posterior fossa lesions (e.g., flaccid, spastic, ataxic).

Unilateral Upper Motor Neuron Dysarthria

LESION LOCI

The label for this type of dysarthria defines its localization, which is the same as that of spastic dysarthria, except that the lesion is unilateral, not bilateral. Common sites of the lesion are the internal capsule (e.g., as in lacunar stroke) and the brain stem.

PATHOPHYSIOLOGY

The underlying bases for this dysarthria may vary but probably reflect unilateral upper motor neuron weakness in most cases and, in some cases, various degrees of spasticity or ataxia-like incoordination. The variability in underlying substrates probably reflects different effects of lesions at different points along the upper motor neuron pathways between the cerebral cortex and the brain stem and spinal cord.

NONSPEECH ORAL MECHANISM FINDINGS

Most patients have a unilateral "central" facial (i.e., lower face only) weakness that may be evident at rest and during emotional smiling or voluntary lip retraction. Most also have unilateral tongue weakness, most obviously evident as deviation of the tongue on protrusion to the side of the lesion. Clinically obvious jaw, palate, or laryngeal weakness is unusual. Drooling from the weak side of the face may be apparent, and dysphagia can occur, although less frequently and severely than in spastic dysarthria.

DISTINGUISHING SPEECH CHARACTERISTICS

A number of speech abnormalities may be detected, but imprecise articulation is most prominent. Irregular breakdowns in articulation are not infrequent. Harshness, reduced loudness, hypernasality, and slow rate can occur but are usually mild. Speech AMRs can be mildly slow and imprecise and sometimes irregular. In general, the dysarthria is rarely worse than mild to moderate, and when it is, an additional or different speech diagnosis (usually spastic or ataxic dysarthria or apraxia of speech) should be considered. The occasional association of ataxia-like irregular articulatory breakdowns and strained voice quality can sometimes suggest mixed dysarthria (e.g., multiple system involvement) even though the single offending lesion is only in the upper motor neuron pathway. The other results of the neurologic examination, particularly evidence of unilateral limb, face, and tongue weakness and the relative mildness of the dysarthria, often clarify the probable location of the lesion.

Mixed Dysarthrias

Combinations of two or more types of dysarthria occur more commonly than any of the single types just discussed. This reflects the basic fact that many neurologic diseases affect more than one component of the motor system. This does not minimize the value of distinguishing among single types of dysarthria or recognizing combinations when they occur. Recognition of a mixed dysarthria may help confirm expectations for a given disease, call into question a particular diagnosis, or raise questions about the presence of an additional condition. For example, amyotrophic lateral sclerosis can be associated with flaccid or spastic dysarthria and is "classically" associated with mixed flaccid-spastic dysarthria because the disease affects upper and lower motor neurons bilaterally. A type of dysarthria other than flaccid or spastic in amyotrophic lateral sclerosis should raise questions about the diagnosis or the possible existence of an additional disease.

A number of the conditions discussed in this book serve as additional examples. That is, with the exception of medication-related hyperkinesias causing hyperkinetic dysarthria, Parkinson's disease should be associated only with hypokinetic dysarthria. Additional types (e.g., spastic, ataxic, flaccid) in someone with a diagnosis of Parkinson's disease can suggest an alternative disease diagnosis, such as multiple system atrophy or progressive supranuclear palsy.

APRAXIA OF SPEECH

Apraxia of speech represents a disturbance in the selection and sequencing of kinematic patterns necessary to carry out intended speech movements. Its localization, underlying nature, and clinical characteristics differentiate it from the dysarthrias. Although it can affect any component of speech production, it is primarily a disturbance of articulation and prosody.

Lesion Loci

Praxis is a function of the dominant hemisphere, and apraxia of speech is associated with left (dominant) hemisphere disease in the great majority of affected persons. Lesions are nearly always in the distribution of the middle cerebral artery and usually involve the posterior portions of the frontal lobe (Broca's area), the insula, the parietal lobe, or the basal ganglia. Because it represents a left hemisphere pathologic condition, apraxia of speech is very frequently accompanied by aphasia, a disturbance of language that can affect verbal expression. It may also be accompanied by a unilateral upper motor dysarthria.

Pathophysiology

The clinical characteristics of apraxia of speech are not explainable on the basis of weakness, incoordination, disturbances of muscle tone, or involuntary movements. Conceptually, it is a disorder of programming of the temporal and spatial components of movements within and among the many muscles that generate the sequence of speech sounds necessary for intelligible speech.

Nonspeech Oral Mechanism Findings

The size, strength, symmetry, and reflexes of oral mechanism muscles may be normal. Right facial and tongue weakness is often present because of involvement of the nearby corticobulbar pathway, but such weakness does not explain the defining characteristics of the disorder. Frequently, but not invariably, a nonverbal oral apraxia is evident, characterized by groping and off-target efforts on nonspeech oromotor tasks (e.g., cough, blow, click tongue), sometimes with inappropriate accompanying verbalization (e.g., asked to blow, the patient with apraxia may say "blow" while trying to execute the act) (see Table 2-4 for summary).

Distinguishing Speech Characteristics

Articulation is usually imprecise or distorted and often accompanied by perceived substitutions, omissions, or even additions of sounds, with a tendency for more complex sounds and sound sequences to be more frequently in error. False articulatory starts and restarts, repetitive attempts at self-correction, sound and syllable repetitions, and visible and audible trial-and-error groping for correct articulatory postures are often apparent and reflect an acute awareness of errors. Speech rate is usually slow, consonants and vowels may be prolonged, and speech initiation may be delayed. Prosody is further disturbed by a tendency to equalize stress across syllables and words, with restricted alteration of pitch, loudness, and duration within utterances, conveying an impression that speech is being programmed and executed one syllable at a time. Rarely, but sometimes dramatically, prosody is altered in a manner that conveys an impression of a pseudoforeign accent.

In apraxia, unlike in dysarthria, a number of linguistic factors can affect speech accuracy. That is, error rates tend to increase for nonsense or unfamiliar words and as word length or complexity increases. For example, speech AMRs may be normal in apraxia, but sequential motion rates ("puh-tuh-kuh") are very often poorly sequenced. Simple words like "mom," "kick," and "baby" may be articulated without error, whereas more complex utterances like "statistical analysis," "stethoscope," and "we saw several wild animals" may elicit many apractic characteristics. In some patients, automatic speech (e.g., counting, social amenities, singing a familiar tune) may be more normal than more volitional or novel utterances. The patient with severe apraxia may be able to produce only a limited repertoire of sounds or words. Muteness can occur, although after stroke apractic mutism rarely persists for more than a few weeks if there are no accompanying language or other cognitive disturbances (see Table 2-5 for summary).

MANAGEMENT OF MOTOR SPEECH DISORDERS

The effective pharmacologic and surgical management of diseases that cause motor speech disorders can result in improvements in speech. When such treatments are not optimally effective, a number of additional behavioral, prosthetic, and instrumental management approaches may maximize the patient's speech intelligibility or ability to communicate effectively and efficiently. Some of these approaches are common to all motor speech disorders, whereas others are unique to specific motor speech disorders. A few examples are provided here to illustrate the variety of strategies that can have a modest to dramatic effect on speech and communication ability. Identification of the best options usually requires careful speech assessment by a speech-language pathologist, and implementation often requires collaboration among a neurologist, otorhinolaryngologist, plastic surgeon, and prosthodontist.

Pharmacologic Management

Injection of botulinum toxin (Botox) into the vocal cords for the treatment of spasmodic dysphonias or into the jaw, face, or neck muscles for orofacial dystonias and torticollis often results in substantial and sometimes dramatic improvement in speech. Injection of botulinum toxin is often done for the primary or sole purpose of improving speech in persons with hyperkinetic dysarthria of dystonia, especially one of the spasmodic dysphonias, and for many it is the only or most effective treatment for the disorder.

Surgical Management

Persons with flaccid dysarthria and significant hypernasality and nasal emission may benefit from a pharyngeal flap or sphincter pharyngoplasty procedure that provides surgical obturation of the weak velopharyngeal mechanism. Similarly, certain thyroplasty procedures or vocal cord collagen injection may significantly improve weak voice in persons with vocal cord bowing, weakness, or paralysis.

Prosthetic Management

A number of mechanical and electronic prosthetic devices can improve speech or assist communication. A palatal lift prosthesis may reduce hypernasality and nasal emission in persons with flaccid or spastic dysarthria. Voice amplifiers may be very useful for patients in whom reduced loudness is the primary speech deficit, such as in flaccid or hypokinetic (parkinsonian) dysarthria. Pacing boards, metronomes, and delayed auditory feedback devices may be effective in slowing speech rate and reducing dysfluencies in persons with hypokinetic dysarthria. A wide variety of devices for augmentative and alternative communication can dramatically improve communication in severely affected patients, even when the underlying motor speech disorder does not change or worsens. They include picture, letter, and word boards as well as sophisticated computerized devices with a variety of output options, including synthesized speech.

Behavioral Management

Behavioral management can include efforts to improve physiologic support for speech, develop strategies for compensatory speaking or augmentative and alternative communication, and modify the environment and interactions in a way that facilitates communication. These approaches are too numerous to summarize here, but a few examples can illustrate their breadth.

Programs of vigorous vocal exercise may improve vocal loudness and quality for patients with weak voices associated with flaccid or hypokinetic dysarthria. Similar exercise for a weak face or tongue may lead to improved strength for articulation. Postural adjustments and modified breathing strategies may improve respiratory support for speech and result in increased loudness and phrase length per utterance.

Slowing of speech rate may be the most effective strategy for improving speech intelligibility in dysarthria, regardless of type, and a number of behavioral strategies, with and without prosthetic assistance, can help patients accomplish this. Similarly, emphasizing the articulation of each sound or syllable may help slow rate and improve articulatory precision. Patients with apraxia of speech often require intensive and extensive drill work to develop reliable articulatory accuracy.

Behavioral strategies often focus on enhancing communication rather than improving motor speech per se. For example, the speaker may establish the topic of conversation explicitly at the outset of an interaction because it narrows the possible vocabulary and aids predictions about what will come next. When not understood, the patient may learn to point to the first letter of each word on a letter board before saying it. Listeners may learn to confirm their understanding of each word or phrase before moving on with the conversation, so that breakdowns can be repaired immediately.

To summarize, motor speech disorders often can be managed effectively, sometimes beyond what is accomplished through medical treatment of the underlying disease, especially if effectiveness is defined in terms of improving or maintaining the ability to communicate. This can be the case even in patients with chronic or degenerative disease. Some treatments reduce the underlying impairment or otherwise "normalize" speech. Others enhance speech intelligibility or efficiency or improve communication through the use of prosthetic or alternative communication devices or adaptive speaker or listener strategies.

SELECTED READING

Darley FL, Aronson AE, Brown JR. Differential diagnostic patterns of dysarthria. *J Speech Hear Res* 1969; 12: 246-269.

Darley FL, Aronson AE, Brown JR. Clusters of deviant speech dimensions in the dysarthrias. *J Speech Hear Res* 1969; 12: 462-496.

Duffy JR. *Motor Speech Disorders: Substrates, Differential Diagnosis, and Management,* Mosby, St. Louis, 1995.

McNeil MR. *Clinical Management of Sensorimotor Speech Disorders,* Thieme, New York, 1997.

Yorkston KM. Treatment efficacy: dysarthria. *J Speech Hear Res* 1996; 39: S46-S57.

Yorkston KM, Beukelman DR, Bell KR. *Clinical Management of Dysarthric Speakers,* Little, Brown & Company, Boston, 1988.

B

PARKINSON'S DISEASE

3

What Is Parkinson's Disease? Neuropathology, Neurochemistry, and Pathophysiology

Howard Hurtig, MD

CONTENTS

INTRODUCTION

Idiopathic parkinsonism, or Parkinson's disease (PD), is a neurodegenerative disorder of middle and late life with a well-defined morbid anatomy, a specific pattern of biochemical pathology, and an unknown cause. Its name immortalizes Dr. James Parkinson, the British physician whose classic 1817 monograph, *Essay on the Shaking Palsy*, is unrivaled as a lucid and timeless description of the cardinal symptoms and signs of what is now recognized as one of the most common neurologic disorders in the Western world.

The typical parkinsonian patient tells of a subtle decline in motor function, often punctuated at the beginning by the appearance of a rest tremor in one hand. Generalized slowing of movement, stiffness of muscles, shortened stride, easy fatigue, small handwriting, and difficulty with the performance of simple daily tasks, such as brushing teeth or hair, knocking on a door, or turning a doorknob, are common early symptoms. In the beginning, signs of PD commonly are confined to one side of the body, and function on the other side is well preserved. Even late in the course of this progressive disease, at a time when the entire body has become affected, an asymmetry of neurologic deficits persists.

The natural history of the clinical illness is notoriously unpredictable in most patients with PD. Some have symptoms for many years or even several decades before significant disability develops, whereas others, especially the elderly, lose mobility and cognitive function in an accelerated pattern. It is virtually impossible to forecast what course the

From: *Parkinson's Disease and Movement Disorders:*
Diagnosis and Treatment Guidelines for the Practicing Physician
Edited by: C. H. Adler and J. E. Ahlskog © Mayo Foundation
for Medical Education and Research, Rochester, MN

illness will take, particularly within the first year or two after symptoms appear and before treatment is given. A good response to medical therapy can often, but not always, be construed as a good prognostic sign.

The diagnosis of PD is made by recognizing the cardinal features via the history and physical examination and by excluding other, less common neurodegenerative parkinsonian disorders, such as progressive supranuclear palsy, multiple system atrophy, and corticobasal degeneration (collectively called "parkinson-plus syndromes"). The secondary parkinsonisms must also be excluded, such as that which results from multiple small strokes of the basal ganglia, or drugs that impede the neurotransmission of dopamine (e.g., neuroleptics), or other truly rare conditions, such as Wilson's disease and manganese intoxication. These other parkinsonian syndromes tend to be characterized by a greater preponderance of motor slowing and rigidity, a relative absence of tremor, and the early development of gait and balance problems. The rate at which disability progresses tends to be faster in these other conditions than in PD and the pace of deterioration more relentless. Although there is no precise technical way to identify which patients are in each of these diagnostic pigeonholes during life, an experienced neurologist can be accurate 80% to 90% of the time (with pathologic verification at the time of death as the reference standard), provided that the patient is observed long enough for the clinical phenomenology of the disease to evolve. Perfect diagnostic accuracy is usually moot in practice, since virtually all patients are eventually treated at the appropriate time with the same array of antiparkinson medications, particularly levodopa. In fact, levodopa therapy serves as a diagnostic test of sorts: those with idiopathic PD are more likely to have a dramatically positive and sustained response, although the levodopa response may be temporarily substantial in any of the parkinsonian syndromes early in the course. A high degree of diagnostic accuracy is desirable, however, from an investigational point of view. This is necessary for the conduct of clinical trials of new drugs specifically targeted at PD and for the planning of the current and future use of sophisticated and expensive neurosurgical techniques, such as deep brain stimulation, cell implantation, and gene therapy. Experience has taught us that all of these have a greater chance of helping patients with PD than of those with the parkinson-plus disorders.

NEUROPATHOLOGY

The basic neuropathology of PD was well established by a few European investigators in the first 2 decades of the 20th century, but serious debate about the anatomical extent of the brain damage obscured the truth for many years. The early observers localized the abnormalities to a few sites in the upper brain stem, especially a group of darkly pigmented, melanin-containing neurons in the midbrain (mesencephalon) called the "substantia nigra" (SN) (Fig. 3-1). However, controversy soon developed when an accident of nature, the encephalitis pandemic of the 1920s, created a new and different form of parkinsonism in its wake. This peculiar postencephalitic parkinsonism began to appear in thousands of survivors of acute encephalitis and later in those who were exposed to the causative agent (presumably a virus) but did not experience an acute clinical illness. The pathologic features of postencephalitic parkinsonism went beyond the boundaries of the brain stem and so confounded the picture that many careful pathologists found it difficult to separate the two major types of parkinsonism from each other. To make matters worse, cerebral arteriosclerosis was being cited increasingly in the 1920s and

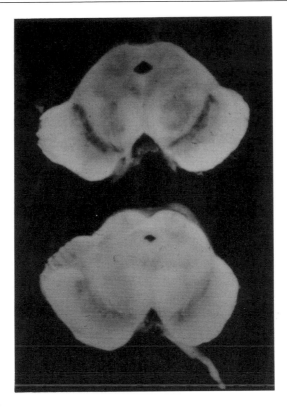

Fig. 3-1. Midbrain sections showing normal brain *(top)* and depletion of darkly pigmented substantia nigra cells in Parkinson's disease *(bottom)*. (From Ahlskog JE, Wilkinson JM. New concepts in the treatment of Parkinson's disease. *Am Fam Physician* 1990; 41: 574-584. By permission of American Academy of family Physicians.)

1930s as a cause of PD, despite the attempts of some thoughtful clinical investigators to differentiate the clinical features of idiopathic parkinsonism from those of parkinsonism caused by multiple strokes. The issue was settled, at least for the time being, in the mid 1950s, when James Greenfield, the leading neuropathologist of the time, concluded in a landmark review paper that the real pathologic location of PD was not only just where the early investigators said it was—mainly in the SN of the upper brain stem—but also in the few other sites where pigmented neurons cluster (e.g., the locus ceruleus of the pons and the dorsal vagal nuclei of the medulla). Shortly afterward, the pigmented cells in the SN were found to synthesize dopamine. Pathology attributed by some authorities to other anatomical locations, particularly the subdivisions of the basal ganglia (see below), was judged by Greenfield too inconsistent to be valid. More recent observations have added other dopamine-producing regions to the list of affected neuronal populations, specifically the ventral tegmental region of the midbrain (near the SN). This locus of cells projects its axons to the basal forebrain and frontal cerebral cortex and is thought to contribute to normal cognitive and behavioral function. Mild-to-moderate depletion of dopamine in the ventral tegmental-basal forebrain-cortical circuit occurs consistently in PD and may contribute to the cognitive-behavioral problems, especially the slowed

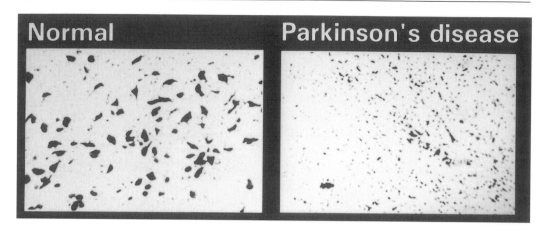

Fig. 3-2. Photomicrographs of substantia nigra. *Left,* In normal subject, there are large darkly staining cells. *Right,* In a patient with Parkinson's disease, small, reactive glial cells have replaced the large cells. (Courtesy, University of Pennsylvania School of Medicine, Department of Pathology.)

mental processing experienced by many nondemented patients with PD. Autonomic neurons in the hypothalamus, thoracic spinal cord, and even the myenteric plexus of the esophagus and colon have also been shown to degenerate, but severity varies. New technology in immunostaining and histologic identification is likely to show still more pathologic sites as more brains are studied.

For still unknown reasons, the pigmented, melanin-containing, dopamine-producing neurons of the SN undergo selective and progressive cell death during the course of the illness at rates that roughly parallel the clinical course of the patient's progressive clinical disability. Microscopic analysis of the brain in patients with PD shows severe loss of neurons and proliferation of reactive glial cells in the anteriormost layer of neurons of the SN (the zona compacta) (Fig. 3-2) and severe secondary depletion of dopamine stores in the caudate and putamen (neostriatum). Moreover, data derived from the study of postmortem brain tissue and from in vivo isotopic radiographic brain imaging in very early PD strongly suggest that the pathologic process begins years before parkinsonian symptoms are actually recognized by patient and family. It is estimated that depletion of total stores of dopamine in the basal ganglia must exceed 50% to 75% in this preclinical or subclinical phase of the degenerative process before symptoms appear.

Another unsolved mystery is the true significance of the classic intracellular Lewy inclusion body typically found in the cytoplasm of the nigral neurons that survive to be visualized at postmortem examination (Fig. 3-3). The Lewy body is also found abundantly in neurons of the cerebral cortex in the approximately 30% of patients in whom dementia develops as a *late* component of the illness. The pathologic finding in these cases is often a mixture of Alzheimer's disease and widespread cortical Lewy bodies. Dementia can be an *early* feature of one of the less common atypical parkinsonian syndromes, known as diffuse Lewy body disease, a combination of relatively mild parkinsonism and early-onset, progressively severe dementia.

Speculation about the cause of the selective nigral cell death that typifies PD has been broad, ranging from intoxication by an environmental toxin (as yet unidentified) to a genetically mediated vulnerability of melanized neurons to the killing effect of uncon-

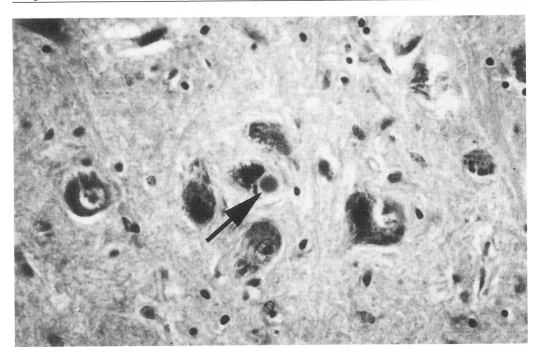

Fig. 3-3. Photomicrograph of intracytoplasmic Lewy body inclusion (at *arrow*) adjacent to clump of melanin granules. Note halo surrounding the eosinophilic Lewy body. (Hematoxylin and eosin.) (Courtesy of the University of Pennsylvania School of Medicine, Department of Pathology.)

trolled oxygen free radicals (the oxidant stress hypothesis—see Chapter 6). The nigral toxin MPTP (1-methyl-4-phenyl-1,2,3,6-tetrahydropyridine) was identified as the cause of severe parkinsonism among users of illicit drugs who self-injected this narcotic analogue in the early 1980s. The recognition that MPTP was responsible led to animal experiments in an effort to create a model of the human illness. Despite the unexpected success of MPTP as an agent of progress in understanding experimental parkinsonism, it has not brought us closer to identifying the cause of PD except by the creation of viable parallel hypotheses that are still at the center of the research spotlight. For example, the oxidant stress hypothesis has gained favor in recent years, as more evidence has accumulated to support the general concept of cell damage caused by abnormal oxidation. The oxidative mechanism by which MPTP damages the mitochondria of nigral cells has been closely investigated and held up as a model for how nigral cells die in PD. Increasing information suggests that the brain's natural defenses against oxidation are somehow impaired in the environment of the SN in PD. Factors cited in favor of oxidative toxicity include the presence of dopamine and melanin (which facilitates the generation of toxic oxygen free radicals) and an unusually high concentration of iron (another promoter of oxidation). The natural attrition of cells attributable to aging may add to the risk of disease, although there is no direct evidence to firmly support its role in pathogenesis.

Notwithstanding the evidence suggesting a role for oxidative toxicity, nothing is known of the primary initiation of this process. Moreover, there is no plausible explanation for what sustains the active degenerative pathologic condition over many years.

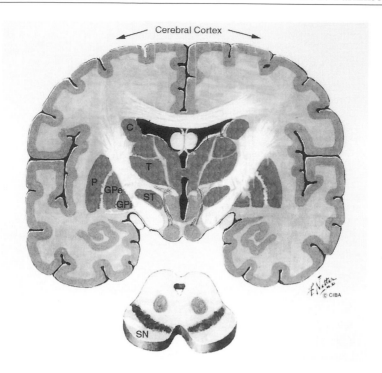

Fig. 3-4. Schematic coronal section of the brain showing the anatomy of the basal ganglia in addition to the midbrain at the level of the substantia nigra. C, caudate; GPe, external globus pallidus; GPi, internal globus pallidus; P, putamen; SN, substantia nigra; ST, subthalamic nucleus; T, thalamus. (Modified from Duvoisin R. Parkinsonism. *Clin Symp* 1976; 28: 2-29. By permission of CIBA Pharmaceutical Company.)

NEUROCHEMISTRY

The discovery of the biochemical pathology of PD in the late 1950s was preceded by a fertile period of research into the biology of catecholamines (epinephrine, norepinephrine, and dopamine) and indolamines (serotonin). Anatomists during this time were using new staining techniques to identify previously suspected but unproved neuronal projections connecting the SN to the basal ganglia (Fig. 3-4). It is now recognized that the basal ganglia, including the striatum (caudate and putamen) and globus pallidus, are highly integrated with cerebral cortex as well as thalamus, cerebellum, and brain stem centers. Cortical visual, perceptual, cognitive, sensory, and motor inputs modulate the basal ganglia's motor programs (Fig. 3-5). Understanding of the chemical neurotransmitters contained within these circuits is evolving.

The molecular activity of dopamine and other neurotransmitters is highly conserved in the mammalian brain. The neurotransmitter is packaged into subcellular containers called "vesicles" and is released by an energy-dependent mechanism from the axon terminals across the tiny synaptic gap. Once the vesicles empty their contents into the synapse, the transmitter migrates to the postsynaptic receptor, where it attaches by a sophisticated lock and key mechanism and activates the receptor (Fig. 3-6). At termination of the transmitter's brief physiologic action, the "used" molecules are taken up by recently discovered proteins known as transporters that penetrate the cell's membrane with transmitter in tow and reenter

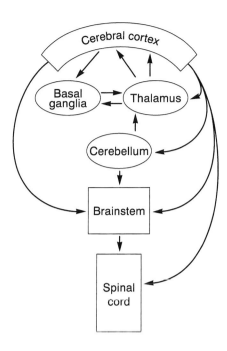

Fig. 3-5. Diagram of the major components of the motor system and their relationships. The basal ganglia and cerebellum are depicted as two parallel circuits that influence the motor areas of the cerebral cortex through discrete and segregated portions of the thalamus. The diagram also shows the parallel projections of the cerebral cortical motor areas to the various deeper structures.

the cytoplasm for repackaging of transmitter into vesicles and recycling. Excess neurotransmitter is enzymatically catabolized intracellularly and extracellularly. For dopamine, catabolism is by the enzymes monoamine oxidase (MAO) and catechol *O*-methyltransferase (COMT).

The dopamine transporter has become the target for isotopic labeling and computerized imaging, with use of the techniques of positron emission tomography (PET) and single-photon emission computed tomography (SPECT), for diagnosis and research. The dopamine transporter has also become the focus of investigation into the variability of genetic susceptibility to PD because of its critical role in transporting MPTP into the cell's cytoplasm, where it disrupts mitochondrial respiratory function. One leading researcher hypothesizes that susceptibility to the development of PD could vary directly with the genetically determined quantity of dopamine transporter in the basal ganglia. High levels of dopamine transporter would increase susceptibility because of the greater likelihood that an etiologic agent would be transported inside the cell to disrupt its basic biologic machinery.

In the late 1950s, the Austrian pharmacologists Hornykiewicz and Ehringer noted that the neurotransmitter dopamine is severely depleted from the caudate and especially the putamen of patients dying of PD and other parkinsonian syndromes (with more modest reductions in other amines, such as norepinephrine and serotonin). As pieces of the puzzle fell into place, it became clear that dopamine was almost certainly the chemical messenger responsible for mediating transmission of neural impulses from nigra to striatum

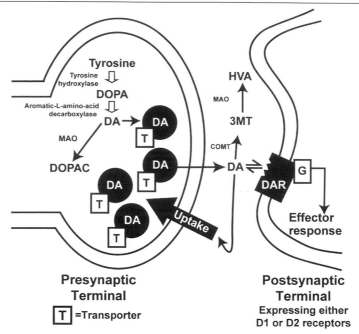

Fig. 3-6. Diagram of a dopaminergic synapse indicating sites of pharmacologic actions. Dopamine (DA) is synthesized within neuronal terminals from the precursor tyrosine by the sequential actions of tyrosine hydroxylase (conversion of tyrosine to dopa) and L-aromatic acid decarboxylase (conversion of dopa to dopamine). Inside the presynaptic terminal, DA is transported into storage vesicles (black circles) by a transporter protein (T) in the vesicular membrane. Release of DA into the synapse allows the DA molecule to activate the dopamine receptor (DAR) and the transduction protein (G) to create the effector response. The actions of DA are terminated by the sequential actions of the enzymes catechol O-methyltransferase (COMT) and monoamine oxidase (MAO) or by reuptake of DA from synapse back into the presynaptic terminal, as indicated in the diagram by the largest arrow. Metabolites: 3 MT, 3-methoxytyramine; HVA, homovanillic acid; DOPAC, 3,4-dihydroxyphenylacetic acid. (From Standaert DG, Young AB. Treatment of central nervous system degenerative disorders, in *Goodman & Gilman's The Pharmacological Basis of Therapeutics* [Hardman JG, Gilman AG, Limbird LF eds], 9th ed, McGraw-Hill, New York, 1996, pp. 503-519. By permission of The McGraw-Hill Companies.)

(especially in the putamen, which, more so than the caudate, receives the bulk of the motor input from the SN). These results provided the necessary platform for the stepwise and successful development of levodopa therapy. Levodopa is the precursor of dopamine, and oral levodopa administration replenishes depleted brain dopamine stores.

PATHOPHYSIOLOGY

The Basal Ganglia: General Considerations

Defining the basal ganglia's exact role in producing normal movement has eluded clinical and basic neurophysiologists for centuries. For many years, few could fathom the apparent complexities. One early 20th century neurologist concluded that "the ganglia situated in the base of the brain still, to a large extent, retain the characteristic of basements, namely darkness." Our current understanding, based on clinical and experimental evidence, holds that the basal ganglia are responsible for the automatic execution of learned motor plans, such as the sequential actions (the motor programs) that together constitute habitual, everyday activity (e.g., brushing one's teeth, drinking from a glass, signing one's name). The specifics of how this broadly conceived set of motor activities is carried out in the normal brain remain largely a mystery, although numerous hypothetical models exist. The central problem confronting investigators in this field is that most conclusions about normal basal ganglionic function have been derived from animal lesioning experiments or inferred from clinical abnormalities in patients with basal ganglia disease. To make matters more complicated, some insults to the basal ganglia cause no motor deficits or only transient loss of function that later resolves (perhaps because other brain motor centers compensate by reorganization of circuitry or by some unknown mechanism).

The pathophysiology of disturbed brain function in PD is necessarily linked to the slow and cumulative degeneration of nigral neurons with the associated progressive depletion of dopamine in the basal ganglia. Much of the basal ganglia's outflow is directed at motor centers of the frontal lobe that are segregated from the classic primary motor strip of the cortex (area 4). These adjacent secondary motor regions, specifically the supplementary motor and premotor areas, are thought to participate in the cognitive or anticipatory aspects of movement—the planning phase before actual kinetic displacement of muscle takes place.

A large body of research, accumulated over the past several decades, shows consistent deficits among parkinsonian patients who are asked to perform tasks that require quick decisions and follow-on motor responses. Patients with PD do better in these experimental tasks when given external cues to guide them, especially visual ones. Without these cues, reaction time is slower than in age-matched controls. It has also been noted that the performance of sequential and simultaneous movements is faulty, so that a patient with PD or another parkinsonian syndrome characteristically does well at the beginning of a series of actions, such as finger tapping, speaking, or writing, only to lose amplitude and force of movement with successive repetitions of the same action. These simple examples are often cited to illustrate how each motor program (e.g., a single finger tap) in the total motor plan (the series of finger taps) is subject to decay when implemented—a deficiency that occurs despite normal strength and coordination. Therefore, the basal ganglia's principal role in the heirarchy of movement, according to this model (proposed by Marsden), is to plan the *next* movement rather than to make any contribution to *this* movement.

A corresponding loss of pure cognitive function can often be documented on neuropsychologic tests that reveal deficits in "executive" function, or the ability to shift quickly from one conceptual pattern to another. This type of abnormality is probably at the root of the classic "bradyphrenia" (slowed processing) that characterizes the thinking of many parkinsonian patients who lose their train of thought in conversation or complain of not being able to follow a set of simple instructions (such as how to start and increment the doses of an antiparkinson medication). These patients are not truly demented, but mild deficits of this type are distinctive. Perhaps these cognitive problems are the result of the domino effect of declining levels of dopamine in the SN and ventral tegmental nuclei directly affecting secondary connections to cognitive centers in the frontal lobe and basal forebrain.

Today's conventional wisdom holds that the pathways linking cerebral cortex and basal ganglia conduct physiologic "business" by way of parallel processing. It is postulated that at least five sets of parallel and mostly segregated circuits are operational; these

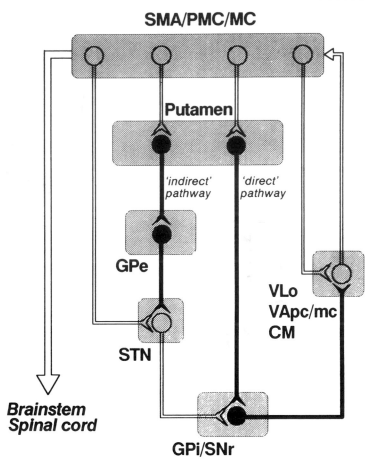

Fig. 3-7. Diagram of the "motor" circuit with its cortical and basal ganglionic components, including the designated indirect and direct pathways. Inhibitory neurons are filled; excitatory neurons are open. The indirect pathway is sidetracked through the external segment of the globus pallidus (GPe) and subthalamic nucleus (STN), with the ultimate output of both pathways to the major outflow nuclei of the basal ganglia: the internal segment of the globus pallidus (GPi) and the nondopaminergic substantia nigra pars reticulata (SNr). Note that the SNr is structurally and functionally distinct from the dopaminergic substantia nigra pars compacta. CM, centromedian nucleus of the thalamus; mc and MC, motor cortex; PMC, premotor cortex; SMA, supplementary motor area; STN, subthalamic nucleus; VApc/mc and VLo, motor nuclei of the thalamus. (From Alexander GE, Crutcher MD. Functional architecture of basal ganglia circuits: neural substrates of parallel processing. *Trends Neurosci* 1990; 13: 266-271. By permission of Elsevier Science Publishers.)

are locally active within the basal ganglia and also transcend regional boundaries. This line of thinking represents a major departure from earlier theoretical models of brain circuitry, which, before the 1980s, supported the concept that neural transmission was "funneled" through the cerebellum and basal ganglia and converged at specific sites in the motor thalamus before projecting to the primary motor cortex. Research has proved that there is very little overlap and that segregation is the rule rather than the exception (Fig. 3-5). The most important of the parallel circuits are the direct and indirect motor pathways that connect striatum (caudate and putamen), globus pallidus, subthalamic nucleus, and the motor centers of the thalamus, as proposed by DeLong and associates

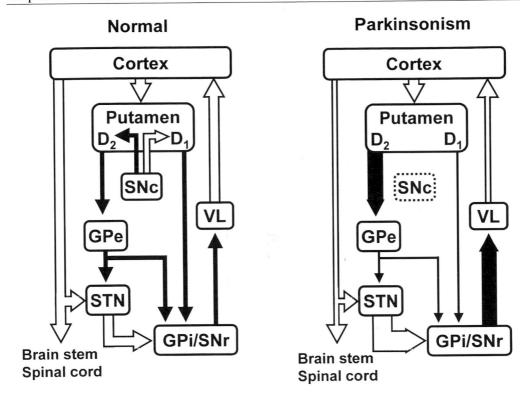

Fig. 3-8. Diagram of direct and indirect motor pathways (not labeled; see Fig. 3-7) of the basal ganglia in normal and parkinsonian brains. Degeneration of the nigrostriatal pathway leads to differential changes in the two striatopallidal projections; increased activity is indicated by increased thickness of the connecting arrows. Two types of dopamine receptors are contained within the putamen: D_1 and D_2. GPe, globus pallidus externa; GPi, globus pallidus interna; SNr, substantia nigra reticulata; SNc, substantia nigra compacta; STN, subthalamic nucleus; VL, ventral lateral nucleus of the thalamus. (From Wichmann T, DeLong MR. *Movement Disorders: Neurologic Principles and Practice.* [Watts RL, Koller WC eds], McGraw-Hill Company, New York, 1997, pp. 87-97. By permission of The McGraw-Hill Companies.)

(Fig. 3-7). In this schema, the final common pathway from the basal ganglia is the output from the internal segment of the globus pallidus and the nondopaminergic cells of the SN, termed "substantia nigra pars reticulata." The pars reticulata is functionally very separate from the dopaminergic substantia nigra pars compacta. Output from the basal ganglia via these pathways normally tends to inhibit the motor thalamus and, hence, the thalamocortical system. In PD, the net effect of dopamine loss is to increase the basal ganglia inhibition of the thalamocortical system (Fig. 3-8). Conversely, disorders characterized by excessive movement (e.g., chorea in Huntington's disease) are associated with reduced basal ganglia inhibition of the thalamocortical system. The normal equilibrium of these two pathways is upset in PD by the degeneration of the nigrostriatal pathway and the secondary depletion of dopamine in the striatum. Although this physiologic model has numerous imperfections, it currently provides the most coherent theoretical framework for comprehending the basic functional disturbances of the motor system in PD and serves as a guide to understanding why patients have a favorable response to levodopa replacement pharmacotherapy and to the newer neurosurgical manipulations of the globus pallidus and the subthalamic nuclei (see Chapter 13).

Clinical Phenomenology

The pathophysiology of the various objective neurologic deficits in PD and the other parkinsonian syndromes is still unclear, even after almost a century of investigation. Each of the four cardinal signs—rest tremor, cogwheel rigidity, bradykinesia, and postural instability—has been studied or speculated on with various levels of elucidation. Taken together, they are classically called extrapyramidal signs to differentiate them from the grouping of weakness, spasticity, and hyperreflexia commonly associated with disorders of the pyramidal (corticospinal) tract. Tremor, rigidity, and bradykinesia are common early signs, but each may occur in isolation, at least for a while, during the course of the disease.

Tremor frequently is prominent in the beginning when the other signs are absent, and it can be restricted to an arm or a leg (or both, usually on the same side of the body) for many years before appearing on the other side. As the other cardinal signs emerge with the passage of time, tremor may gradually subside and even cease to be a prominent part of the clinical picture, irrespective of treatment.

Tremor in PD classically occurs when the affected body part is in repose and is typically in the range of 5 to 7 cycles per second. It is more likely to be a prominent symptom in PD than in the other parkinsonian syndromes, such as progressive supranuclear palsy or multiple system atrophy.

For many years, researchers have sought to identify the nervous system's tremor generator, but no definite source has been identified. Current evidence supports the notion that active "tremor" cells in the ventral nuclei of the thalamus are important, firing at the same frequency as the clinical tremor. However, recent intraoperative recordings during pallidotomy from cells in the internal segment of the globus pallidus have revealed "tremor" cells firing in register with the clinical tremor and at the same frequency as similar cells in the thalamus. Thus, these globus pallidus cells are also a part of the tremor circuitry, but the exact relationships of the various anatomical tremor connections have not been fully elucidated. PET imaging has also implicated the cerebellum in these tremor circuits. Finally, circuits in the spinal cord involved in what physiologists call hyperactive long-loop reflex arcs may also add a physiologic element to the generation of tremor, but the evidence on this is conflicting. In fact, interruption of the normal reflex arc of the spinal cord has no effect on tremor.

Certain drugs can effectively suppress the tremor of PD by influencing one of two neurotransmitters—dopamine and acetylcholine. Both are critical to basal ganglionic function but have polar opposite pharmacologic effects. In the normal brain, dopamine and acetylcholine are counterbalanced in an equilibrium that promotes proper motor function. In the parkinsonian brain, this equilibrium is upset in favor of a relative excess of acetylcholine as the supply of dopamine gradually declines. Thus, tremor can be reduced by the administration of drugs that enhance dopamine neurotransmission (e.g., levodopa) or block acetylcholine (e.g., benztropine).

Lesions in one of several brain areas can abolish parkinsonian tremor. They include surgical lesions in the pallidum (pallidotomy) and ventral thalamus (thalamotomy). Furthermore, pathologic lesions (e.g., strokes) in the interconnected primary motor cortex or efferent corticospinal (pyramidal) tract may also secondarily reduce tremor.

Rigidity of the limbs in PD has the feeling of sustained, plastic resistance throughout the entire range of passive flexion and extension movements when the doctor examines the patient. It can be differentiated from the clasp-knife spasticity associated with disorders of the pyramidal tract, such as stroke. The muscle stretch reflexes are normal in PD, an extrapyramidal disorder, and hyperactive in pyramidal disorders. A hyperactive long-loop reflex connecting brain and reflex arcs in the spinal cord has (as with tremor) been implicated in the generation of rigidity, but concrete proof is lacking.

Akinesia-bradykinesia is arguably the most disabling and least understood of the cardinal signs of PD. This peculiar characteristic is just as much a visible hallmark of PD as rest tremor and the signature postural deformities. The words "akinesia" and "bradykinesia" designate absent and slow movement, respectively. This phenomenon underlies the masked face, the loss of manual dexterity, the difficulty parkinsonian patients have in getting out of chairs, and the loss of spontaneous animation and "associated" movements (e.g., decreased arm swing when walking). It is separate from rigidity, which may be minimal or absent in patients with the most severe akinesia.

Parkinsonian patients also have trouble initiating or carrying out two actions simultaneously, such as the patient whose walking is only mildly impaired when unencumbered but who freezes or falls when carrying a large object. In other words, the parkinsonian brain is easily "jammed" when multiple circuits are activated. It is as though there were a pathologic lowering of the threshold at which the system of parallel processing is overloaded and shut down. Evidence that something of this sort takes place in the parkinsonian brain has been provided by PET scanning during the time patients are engaged in motor and mental tasks. Metabolic activity in the premotor, supplementary motor, and association cortex of the frontal lobe in this situation is significantly diminished in comparison with that in age-matched controls. Conversely, it is significantly more robust after levodopa or after pallidal stimulation, both of which increase the speed of processing through the pathologically jammed circuits.

Posture and its control mechanisms are disordered in all the parkinsonian syndromes, although postural abnormalities may be minimal early in the course of PD. The severe shuffling, freezing, and, particularly, falling that cause such disability in advanced disease are the most difficult symptoms to treat; the response to dopaminergic therapy is almost always inadequate. Not only is postural instability the single most disabling and demoralizing problem in advanced PD but also it almost always heralds a permanent constriction in the quality of life and in lifestyle. Many experts in movement disorders have concluded that the failure of levodopa and dopamine agonists to alleviate postural instability must mean that the pathophysiology lies outside the dopamine system and is caused by a disturbance in another transmitter system. There is no direct evidence, however, implicating another specific neuronal circuit.

Parkinsonian patients often lose their balance and fall for no reason, frequently "retropulsing" backward uncontrollably. They may fall during turning, because the lower body does not adjust to changes in the center of gravity as readily as the upper body. It is assumed that the normal basal ganglia automatically make corrective adjustments of the body's center of gravity to prevent disequilibrium and falls. Environmental cueing from vestibular, neuromuscular, spinal proprioceptive, and visual systems (all unaffected by dopamine deficiency, except possibly vision) enables the basal ganglia control center to keep the body in the most favorable upright position to suit the requirements of a particular motor activity.

Other features of postural dyscontrol in PD are notably prominent when the patient walks. These include involuntary flexion of the knees, a narrow standing base, and a poverty of "associated" movements, such as the normal sway of the body and swing of the arms.

SUMMARY

The clinical, pathologic, neurochemical, and pathophysiologic aspects of PD have been briefly reviewed and integrated to give the reader a broad overview of the current state of knowledge. Much progress has been made in understanding the nature of PD since James Parkinson's *Essay on the Shaking Palsy* in 1817, especially in the past four decades. Much is left to be discovered and much to be learned about better forms of treatment. The near horizon is crowded with promising therapeutic opportunities that have grown out of the partnership between clinical researchers and basic scientists.

SELECTED READING

Gibb WRG, Lees AJ. Pathological clues to the cause of Parkinson's disease, in *Movement Disorders* (Marsden CD, Fahn S eds), 3rd ed, Butterworth-Heinemann, Oxford, 1994, pp. 147-166.

Marsden CD. Neurophysiology, in *Parkinson's Disease* (Stern GM ed), Johns Hopkins University Press, Baltimore, 1990, pp. 57-98.

Marsden CD. The mysterious motor function of the basal ganglia: the Robert Wartenberg Lecture. *Neurology* 1982; 32: 514-539.

Stern MB. Contemporary approaches to the pharmacotherapeutic management of Parkinson's disease: an overview. *Neurology* 1997; 49 Suppl 1: S2-S9.

Uhl GR. Hypothesis: the role of dopaminergic transporters in selective vulnerability of cells in Parkinson's disease. *Ann Neurol* 1998; 43: 555-560.

Wichmann T, DeLong MR. Physiology of the basal ganglia and pathophysiology of movement disorders of basal ganglia origin, in *Movement Disorders: Neurologic Principles and Practice* (Watts RL, Koller WC eds), McGraw-Hill, New York, 1997, pp. 87-97.

4 Clinical Features of Parkinson's Disease

Richard B. Dewey, Jr., MD

CONTENTS

Involuntary tremulous motion . . . in parts not in action and even when supported; with a propensity to bend the trunk forward, and to pass from a walking to a running pace . . . .

—James Parkinson, 1817

Since James Parkinson's original description of the shaking palsy in 1817, the classic clinical features of moderate-stage Parkinson's disease (PD) have become widely known to physicians and even to the general public. Statistically speaking, most patients who present with resting tremor, slowing of movement, gait difficulty, and limb rigidity have idiopathic PD. When a typical resting tremor appears in the elderly, patients and their relatives may begin to suspect the diagnosis before a formal evaluation by a physician. Because treatments for the symptoms of PD abound today and putative "neuroprotective" agents are entering clinical trials, early diagnosis is essential.

In this chapter, the presenting features of early PD are discussed first, followed by a brief discussion of the major motor and nonmotor symptoms and signs of PD, the natural history of the disorder, clinical clues that help to differentiate PD from several mimicking conditions, and a recommended strategy for correct diagnosis.

From: *Parkinson's Disease and Movement Disorders:*
Diagnosis and Treatment Guidelines for the Practicing Physician
Edited by: C. H. Adler and J. E. Ahlskog © Mayo Foundation
for Medical Education and Research, Rochester, MN

Table 4-1
Clinical Features of Parkinson's Disease

Primary motor	*Psychiatric*
Tremor	Depression
Rigidity	Anxiety
Bradykinesia	Panic attacks
Postural instability	Drug-induced hallucinations
Gait difficulty	*Autonomic*
Other motor	Sweating abnormalities
Hypokinetic speech	Orthostatic hypotension
Facial masking	Impotence
Decreased blinking	Constipation
Foot dystonia	Urinary frequency
Drooling	*Sensory*
Micrographia	Pain
Cognitive	Numbness or tingling
Bradyphrenia	Decreased smell
Tip-of-the-tongue anomia	
Dementia (late)	

PRESENTING FEATURES OF EARLY PARKINSON'S DISEASE

Since PD is a gradually progressive disorder, the first symptoms and signs are often subtle and vague. Often, patients and their families reflect on years past and realize that several personal characteristics they had attributed to normal aging were actually early features of PD. This is commonly the case for such minor motor manifestations as stooped posture, loss of facial expression, and "mumbling" speech quality that often presage the onset of PD. Although the list of possible presenting features is long (Table 4-1), the most common one is unilateral resting tremor, often associated with a reduction of spontaneous arm swing on the affected side. The tremor invariably begins gradually and may be first noticed by a spouse or relative. A curious feature of parkinsonian rest tremor is that it usually disappears instantly on willed movement of the affected limb, which explains why relatives are sometimes the first to observe the affliction. Patients with early PD, whether or not they have tremor, tend not to swing the affected arm when walking, and many carry the arm slightly flexed at the elbow. This posture occasionally prompts acquaintances to ask whether the person has suffered a stroke. Diminution of arm swing is often the first evidence for the slowing of movement (bradykinesia), arguably the most significant feature of PD. At the same time or shortly thereafter, other signs of bradykinesia often emerge, including subtle loss of manual dexterity in the affected hand and a trailing off of the script near the end of a written sentence (micrographia). Rigidity of the affected arm, characterized by a velocity-independent increase in passive tone, is also common at presentation. Limb rigidity may cause pain in the upper arm or shoulder, which is occasionally the patient's chief complaint at presentation, often leading to a fruitless search for diseases of the bone, joint, or bursa.

Although the actual mix of symptoms and signs may vary at presentation, a key aspect of early PD is its often unilateral onset. Most frequently, the symptoms appear in the upper extremity first, but unilateral onset of lower extremity resting tremor or "dragging" of one leg when walking is by no means a rare presenting feature. Younger patients may even present with unilateral foot dystonia, which most often consists of involuntary internal rotation of the foot, especially during walking. Accordingly, focal action dystonia in one foot in an adult should suggest PD as the most likely cause. So characteristic of PD at onset is unilaterality of symptoms and signs that this feature is incorporated as the key element in the clinical staging of early PD (discussed below). Naturally, patients examined later in the course of the disease very likely have bilateral features of parkinsonism, though even then, symptoms and signs are typically worse on one side. Postural instability and falls are unusual presenting features of Parkinson's disease and, except in patients who present for the first time with advanced disease, should be regarded as "red flags" suggesting an alternative diagnosis.

MAJOR MOTOR SYMPTOMS AND SIGNS

The major motor features of Parkinson's disease consist of the three classic features—tremor, rigidity, and bradykinesia—and a fourth, designated "postural instability." Gait difficulty also is discussed as a separate major motor symptom of PD, although the origin of gait difficulty lies in the other cardinal features. For most clinicians, at least two of the classic triad are required for a diagnosis of PD.

Tremor

Tremor is the most obvious and well-recognized symptom of PD. Parkinsonian tremor is characterized as a "resting tremor," which means that it is present chiefly when the patient is not using the limb in directed activities. Thus, the tremor seen in the hand when the patient is walking has the same significance as a tremor appearing when the hand is at rest in the lap. Parkinsonian rest tremor typically attenuates or resolves with action, differentiating it from the action tremors seen in disorders such as hereditary essential tremor and multiple sclerosis (action tremor due to cerebellar lesions). At onset, the resting tremor may be mild and highly intermittent, becoming obvious only when the patient is upset or anxious, for example. With progression, however, the tremor typically increases in amplitude and becomes a consistent feature of the resting state.

For eliciting a nonspontaneous resting tremor in the office, mental activation maneuvers are useful. Having the patient recite the months of the year backward or count back from 100 by sixes increases psychologic tension and overcomes volitional suppression of tremor. In this way, an otherwise subtle or intermittent resting tremor can be easily observed.

Parkinsonian tremor occurs most often in the extremities and is typically unilateral at onset. With time, the tremor may spread to the other side of the body while often remaining most severe on the side initially affected. In some patients, the lip, chin, and jaw may exhibit tremor, but tremor almost never involves the head. Since head tremor is common in essential tremor, some authors have argued that head tremor in a patient with PD indicates coexisting essential tremor.

The tremor of PD, at 3 to 5 Hz, is slower than most postural tremors, which are 8 to 10 Hz. Some authors have also emphasized that parkinsonian rest tremor is usually related to alternating contractions of agonist and antagonist muscles, whereas the faster postural tremors are most often due to simultaneous co-contraction of agonists and antagonists. However, these features are less helpful in differentiating tremor for diagnostic purposes, in part because they are difficult to measure and in part because exceptions occur to both rules. Additionally, the diagnostic importance of differentiating resting and postural tremors is diluted somewhat by the circumstance that some patients with PD have both types of tremor. The significance of this finding has been debated by experts, with some suggesting that parkinsonian rest tremor may have a postural component and others arguing that when both types are seen, two diagnoses (such as PD and essential tremor) should be given. Although it is true that patients with PD can have postural as well as resting tremor, the converse is much less often true, so that any resting component the patient has suggests parkinsonism. Although a mixed resting-postural tremor is seen in some patients with PD, the tremor typically dampens in amplitude with action (e.g., the action of a finger-nose maneuver).

Although the most obvious motor symptom of PD, tremor is often the least disabling owing chiefly to its tendency to diminish or disappear during volitional movement. In some patients with more advanced disease, however, tremor may markedly increase in amplitude during trough levels of medication and lead to marked distress and disability. During severe exacerbations of parkinsonian tremor, this symptom may be misdiagnosed by emergency department personnel as an epileptic phenomenon. As PD progresses to the later stages, however, tremor usually becomes less and less significant, because severe bradykinesia and postural instability begin to dominate the clinical picture.

Rigidity

Rigidity is an important clinical sign of PD, correlating with limb stiffness. Rigidity is defined as a velocity-*in*dependent increase in passive tone in a limb. This is differentiated from spasticity, which is a length- and velocity-dependent form of hypertonia. The term "lead-pipe rigidity" was coined to connote the key feature of this sign, which is a constant resistance to passive movement that varies little with the extent or speed of limb excursion. By contrast, a spastic limb can be moved with little resistance for short distances and at slow speeds, but when it is rapidly moved throughout the range of motion, resistance increases significantly. Spastic limbs may also demonstrate the clasp-knife phenomenon, whereby resistance is high on initiation of movement but yields with continued force, a feature not seen in parkinsonian rigidity.

Many descriptions of rigidity in PD use the term "cogwheeling" to signify the jerky quality of the hypertonia felt when moving the limb. Cogwheeling is due to the superimposition of tremor (which may not be visible) on the increased tone of the limb. Since cogwheeling is strongly indicative of parkinsonism, it is a very helpful sign when present. However, many patients with PD have rigidity without cogwheeling, an indication that the significance of this feature may have been overemphasized in the literature.

Rigidity can be very difficult to assess in some patients with PD because of poor relaxation. The tendency of some patients, particularly those with cognitive impairment, to help the examiner by moving the limb when rigidity is tested is termed "gegenhalten" or "paratonia." When present, this phenomenon may lead to the perception of increased or decreased tone, depending on whether the volitional movements are opposing or augmenting the examiner's effort. Paratonia should be recognized if the limb continues moving after the examiner stops applying pressure. In many cases, repeatedly entreating the patient to relax is of limited use, and clinical estimates of the severity of rigidity are therefore compromised.

In addition to being a useful sign of PD, rigidity is a common cause of the limb pain some patients experience. Rigid limbs are due to simultaneous contraction of agonist and antagonist muscles, which can lead to muscle fatigue and myalgia. In chronic limb pain due to rigidity, nonsteroidal anti-inflammatory drugs and range-of-motion exercises may be useful adjuncts to typical antiparkinsonian drugs. In some cases, severe rigidity can be manifested by sudden painful cramps occurring as the medications for PD wear off, and the result may be admission to a hospital emergency room. This painful off-period muscle contraction is interpreted as dystonia (rather than rigidity) by some authors. Pain medications are sometimes required, but a more effective treatment is rapid correction of the parkinsonian off-state by an oral load of levodopa.

Bradykinesia

The defining element of all parkinsonian syndromes is slowing of movement, or bradykinesia. Bradykinesia may appear with a variety of vague manifestations, including loss of arm swing, difficulty with manual dexterity, micrographia, facial masking, mumbling speech, and slowing of gait. Even the drooling seen commonly in more advanced disease is primarily caused by bradykinesia of the swallowing musculature.

Generalized bradykinesia in its simplest manifestation represents the loss of normal automatic movements. Normal people rarely sit perfectly still in the doctor's office; mild adventitious movements, shifting of position on the seat, and tapping the foot or crossing the legs are such universal signs of normality that they are scarcely noticed. Yet patients with parkinsonism are quite still when seated at rest, an important diagnostic clue to the observant physician. Bradykinesia is also manifested by the decreased blink rate that is so common in parkinsonian patients and results in a staring facial appearance.

Clinical testing for bradykinesia in the office involves observing the patient stand up from a chair and walk in the hall; in PD, both tasks are executed more slowly than normal. It is also useful to have patients rapidly tap the index finger to the thumb in each hand separately. This act reveals slowing and a remarkable tendency for the amplitude of movement to decrease as the task proceeds. Although a patient with significant bradykinesia may start out tapping well, within several seconds the amplitude drops greatly, so that the rate of tapping may still be adequate but at the expense of reduced amplitude.

Of the three classic features of PD, bradykinesia is, in most cases, the most disabling. Patients often use the term "weakness" to describe this symptom, by which they mean that they cannot get their muscles to obey the commands of their mind. When pronounced, slowing of movement can become akinesia, the total loss of voluntary willed movement. Few things are more terrifying to patients with PD than true akinesia. In this state, they are helpless to do anything for themselves and may beg their relatives to take them to a hospital for emergency care. Bradykinesia or akinesia is often the key feature of parkinsonian off-states, periods when brain levels of dopamine are inadequate. Not infrequently, akinetic off-states are accompanied by nearly panic attack reactions, complete with cardiac palpitations and hyperventilation. Sometimes, crushing substernal chest pain accompanies these reactions, leading to urgent evaluation for myocardial infarction. Fear of having an akinetic spell in an unfamiliar setting may cause some patients to refuse to leave home (agoraphobia). Such patients become increasingly dependent on others for continual supervision and ultimately require nursing home placement unless the off-states can be eliminated by adjustment of medications.

Postural Instability

Progressive loss of postural reflexes, rare as a presenting feature, eventually develops in most patients with PD, leading first to gait difficulty and balance problems and ultimately, in some advanced cases, to frequent falls and wheelchair confinement. Early in the course, postural instability may be subtle and evident only on specific testing. The pull test was developed specifically for this purpose. While standing securely behind the patient, the examiner quickly pulls the patient backward by the shoulders. A normal person does not need to step backward on a second pull; knowing what to expect from the test pull, this person simply sways backward at the waist momentarily, and effortless righting of the trunk follows. In contrast, the first sign of postural instability in the patient with PD is the need to step backward with one foot to prevent a fall, and this maneuver recurs on subsequent pulls. With greater impairment of postural reflexes, the patient would fall backward if not caught by the examiner. With severe loss of these reflexes, patients are unable even to stand without support, toppling backward or to the side without perturbation.

Unlike the other cardinal manifestations of PD, postural instability has a neuroanatomical basis that appears to be primarily in neuronal circuits other than the dopaminergic substantia nigra. Thus, postural instability is the clinical feature most refractory to treatment with dopaminergic medications. Therefore, progression of this symptom leads to greater and greater disability, which can be controlled only through supportive means such as physical therapy and the educated use of assistive devices.

Gait Difficulty

The gait problems in PD are beautifully described in the original essay by James Parkinson: the patient is "irresistibly impelled to take much quicker and shorter steps, and thereby to adopt unwillingly a running pace." Gait difficulty can be the presenting feature of the disease, especially for more elderly patients, in whom the disease may be heralded by the onset of stooped posture and slowed gait. A characteristic feature of the parkinsonian gait is a shortening of the stride length that gives rise to the appearance of "shuffling." Shortening of the stride length is analogous to the loss of amplitude of finger tap movements and as such is an example of bradykinesia. Accompanying the shortened stride length is a tendency toward forward leaning, which results in propulsion, or the involuntary acceleration of gait in the forward direction. Because patients have a limited ability to increase stride length as they pick up speed, a fall may result. In most cases, this so-called festinating gait disorder is responsive to levodopa and with proper drug therapy can be substantially ameliorated.

A similar but distinct gait problem is termed "freezing gait," which is an unexpected freezing in place when gait is initiated, doorways or narrow halls are entered, or turns are executed. Freezing usually appears as a sudden halt in forward movement. It is typically followed by futile attempts to resume walking, characterized by the taking of tiny, ineffective steps. Sometimes spells of gait freezing can last for more than a minute, and they are often exceedingly frustrating for patients. Freezing phenomena, sometimes accompanying levodopa off-states, may also appear during states of proper drug levels and even overdose states. Frequently, patients learn tricks that help them reinitiate gait, including counting a walking cadence out loud, visualizing marching in their mind, or stepping over a nearby object on the floor. The use of visual cues for gait reignition is often effective, and a specially modified walking stick has been developed for this purpose (Fig. 4-1).

Fig. 4-1. A modified walking stick provides a visual cue for gait reignition in patients with freezing phenomena.

Patients often find that by placing the platter of the stick on the floor in front of one foot, they can step over the platter, thereby reigniting a more normal gait.

MINOR MOTOR AND NONMOTOR SYMPTOMS AND SIGNS

Several minor motor features and nonmotor features are also associated with PD.

Speech Impairment

Speech is commonly affected in PD, and impairment may be among the most obvious and disabling symptoms in some patients. The speech disorder of PD is characterized as a hypokinetic dysarthria manifested by a decrement in speech volume, sometimes with a paradoxical increase in speech rate; the result may be a "mumbling" quality. The drop in volume and acceleration of rate are manifestations of bradykinesia affecting the bulbar musculature. Analogous to the typical parkinsonian finger tapping response, the amplitude of the contraction of the articulatory muscles (lips, tongue, and palate) decreases while the rate of contraction remains fast or increases. See Chapter 2 for a more detailed discussion.

Cognitive Impairment

Although PD is not considered to be a dementing illness, most patients have some reduced scores on a complete battery of neuropsychometric tests. In addition, significant and disabling dementia develops in about 10% to 30% of patients during the course of the disorder. When dementia occurs in PD, it typically is a late finding.

Dementia in PD is usually "subcortical," meaning that cortical features such as apraxia and aphasia are relatively rare. In most cases, the dementia begins with short-term memory loss and difficulty retrieving recently stored memories rather than encoding new ones. This problem is frequently recognized as the "tip-of-the-tongue syndrome" in patients with PD who seem to know what information they are looking for but cannot bring it forth. Other types of cognitive deficiency are bradyphrenia (slowed thinking) and diminution of executive functions (passivity, loss of planning and goal-directed thinking). At the time of autopsy, PD dementia is found to be related to superimposed Alzheimer's disease or to the more widespread Lewy body disease in many cases. See Chapters 12 and 20 for further details.

Psychologic Disorders

Depression is very common in PD, occurring in 40% to 60% of patients. It is difficult to distinguish the extent to which the depression in PD is a reaction to the disabling aspects of the disease versus a primary manifestation of the neuronal degeneration. Anxiety is common, particularly in patients who are prone to unpredictable off-states, and in the treatment of this problem, care must be taken to avoid benzodiazepine dependency. Hallucinations are very uncommon in early PD, and if they are prominent as initial manifestations, an alternative diagnosis such as diffuse Lewy body disease should be considered. In advanced PD, and especially with the administration of high doses of dopaminergic agents, hallucinations and sometimes full-blown psychotic reactions with delusions and paranoid ideation occur. Further discussion is found in Chapters 12 and 20.

Autonomic Involvement

The autonomic system is affected in the PD neurodegenerative process, although symptoms and signs of autonomic involvement are less pronounced in this disorder than in multiple system atrophy (e.g., Shy-Drager syndrome). Patients who have battled hypertension for years may enjoy resolution of this problem when PD develops. Orthostatic hypotension, although often causing mild positional lightheadedness, is rarely severe in untreated PD. However, dopaminergic therapy with levodopa tends to exacerbate orthostatism, resulting in frank syncope in some cases. Salt supplementation and treatment with mineralocorticoids are usually effective. Constipation is a nearly universal problem in advancing PD, with more than 70% of patients complaining of defecatory dysfunction. Other clinical features of autonomic involvement are intermittent profuse sweating, urinary hesitancy with nocturia, and sexual dysfunction. Autonomic involvement is discussed in detail in Chapter 11.

NATURAL HISTORY OF PARKINSON'S DISEASE

The natural history of Parkinson's disease in the era preceding levodopa was perhaps best characterized by Hoehn and Yahr (1967), who followed a cohort of 802 patients with parkinsonism between 1949 and 1964. Eighty-four percent were believed to have idio-

Table 4-2
The Hoehn and Yahr Scale for Staging Parkinson's Disease

Stage 1	Unilateral involvement only, usually with minimal or no functional impairment
Stage 2	Bilateral or midline involvement without impairment of balance
Stage 3	First sign of impaired righting reflexes; patients are physically capable of leading independent lives, and their disability is mild to moderate
Stage 4	Fully developed, severely disabling disease; the patient is still able to walk and stand unassisted but is markedly incapacitated
Stage 5	Confinement to bed or wheelchair unless aided

From Hoehn MM, Yahr MD. *Neurology* 1967; 17: 427-442.

pathic PD (the remaining patients had postencephalitic parkinsonism or secondary parkinsonism due to various insults such as stroke, trauma, and drug intoxication). The mean age at onset of PD was 58, and the mortality rate among parkinsonian patients was three times that of the general population matched for age and sex.

They recognized that in most cases, PD begins with unilateral symptoms and signs, later progressing to bilateral findings. They also stressed the importance of postural instability as a marker of more advanced and disabling PD. Using these key clinical features, they developed the clinical staging scale that now bears their names (Table 4-2). The median times to reach stages 1, 2, 3, 4, and 5 from the time of symptom onset were 3, 6, 7, 9, and 14 years, respectively. In this group of patients not treated with levodopa, 80% were either severely disabled or dead within 10 to 14 years of the onset of symptoms. Hoehn and Yahr noted, however, that this exact progression does not occur in all patients. Some have bilateral features at onset, thus bypassing stage 1, and others may reach a clinical plateau at the intermediate stages that continues for years.

More detailed clinical scoring tools have since been developed to measure severity and disability in PD, although the Hoehn and Yahr scale is still used. The primary scale now applied in most clinical trials is the Unified Parkinson's Disease Rating Scale (contained in the Langston et al. article in the reading list at the end of this chapter). This instrument consists of 42 items divided into four major sections, including mental function, activities of daily living, motor examination, and complications of therapy. These rating scales are primarily research tools, although they are sometimes adapted to the clinic for routine use.

The advent in 1969 of levodopa to treat the symptoms of PD dramatically altered the natural history of the disease. Whereas in the era before levodopa mortality in PD was three times that of age-matched controls, the introduction of levodopa reduced mortality rates to levels just slightly higher than those in persons of similar age and sex without PD. As a result, modern treatment strategies have shifted the times to reaching later stages of PD significantly to the right, so that it is common to see patients with PD, particularly those who were relatively young at onset, still at stage 3 after more than 15 years. The use of dopaminergic therapy has been far short of curative, however, and with more advanced PD, the response to levodopa therapy is often uneven, inconsistent, and associated with motor complications.

In the modern era, the diagnosis of PD is usually followed quickly by the institution of drug therapy. Effective dopaminergic treatment (usually with levodopa or dopamine agonists, or both) typically produces marked improvement and often nearly complete resolution of parkinsonian symptoms for a period lasting between 4 and 7 years. During this period, patients usually enjoy a sustained, smooth, beneficial response to the drugs. This phase, often called the "honeymoon period," eventually ends as the duration of effect of individual levodopa doses shortens. This results in the wearing-off effect, whereby symptoms of parkinsonism return as brain levels of levodopa decrease rapidly. This phenomenon ushers in the second major phase of PD, the period of motor fluctuations. In many patients, levodopa-induced choreiform involuntary movements (dyskinesias) also become prominent at this stage of the disease (occasionally sooner). During this time, patients typically must be seen more frequently for adjustment of drug dosage while adverse motor reactions to levodopa become increasingly common. Late in the clinical course, many patients who formerly responded well to levodopa suffer increasing loss of the beneficial effect, resulting in increased immobility and rigidity.

CLINICAL CLUES IN THE DIFFERENTIAL DIAGNOSIS OF PARKINSON'S DISEASE

Although idiopathic PD is by far the most common cause of parkinsonism, a number of less common disorders may mimic PD and require a different diagnostic and therapeutic approach (Table 4-3).

Medication-Induced Parkinsonism

All patients presenting with parkinsonism should be questioned about current and previous medication use, since dopamine-blocking drugs are an important cause of treatable parkinsonism. The most common offenders are the classic neuroleptic agents, such as chlorpromazine and haloperidol, but even low-potency agents (molindone, thioridazine), dopamine-depleting agents (reserpine), and some newer antipsychotic drugs (risperidone, olanzapine) can induce or worsen parkinsonism. Antiemetics (metoclopramide) and certain anxiolytic drugs (most notably, buspirone) have also been associated with drug-induced parkinsonism. Discontinuation of the offending agent is obviously the appropriate treatment strategy, but with some of these drugs, many weeks and, occasionally, many months may elapse before drug-induced parkinsonism resolves completely.

Vascular Parkinsonism

Parkinsonism due to small strokes typically presents as a gait disorder with bradykinesia in elderly patients, especially in those with a history of hypertension or diabetes mellitus, conditions that predispose to small vessel arterial disease. Sudden onset of deficits or stepwise deterioration may mark the course in some cases. Often, the signs occur primarily in the lower limbs, producing a shuffling, parkinsonian gait. In this condition, brain magnetic resonance imaging (MRI) scans typically show diffuse, periventricular increased T2 signal, suggestive of microvascular disease. Such scans should be interpreted with caution, however, since many older patients with typical PD also have periventricular T2 signal abnormalities. For this reason, brain imaging is indicated chiefly for parkinsonian patients who have atypical features or no response to treatment.

Table 4-3
Differential Diagnosis of Parkinsonism

Primary
    Idiopathic Parkinson's disease
Secondary[1]
    Drug-induced (neuroleptics, antiemetics)
    Toxin-induced (MPTP, carbon monoxide, manganese)
    Vascular parkinsonism
    Postencephalitic parkinsonism
    Post-traumatic parkinsonism
    Normal pressure hydrocephalus
    Wilson's disease
Parkinsonism-plus disorders[2]
    Progressive supranuclear palsy
    Multiple system atrophy
        Shy-Drager syndrome
        Striatonigral degeneration
        Olivopontocerebellar atrophy
    Parkinsonism-dementia-ALS complex of Guam
    Corticobasal degeneration
    Diffuse Lewy body disease
    Alzheimer's disease
    Huntington's disease
    Rare inborn errors of metabolism appearing early in life with parkinsonism and
        other features

ALS, amyotrophic lateral sclerosis; MPTP, 1-methyl-4-phenyl-1,2,3,6-tetrahydropyridine.
[1]See Chapter 15.
[2]See Chapters 16 through 20.

Normal Pressure Hydrocephalus

This condition is rare but important to remember because of the implications for treatment. This disorder typically is manifested as a gait disorder in the elderly characterized by a slow, shuffling gait with few other findings typical of PD. Often, the gait is wide-based, unlike the narrow base of PD, and the arm swing is relatively preserved. Frequently, patients give a history of urinary incontinence marked by bladder evacuations of large volume that catch the patient by surprise ("unwitting wetting"). This disorder is also typically associated with at least some degree of dementia, although severe dementia is a late finding. Brain imaging reveals enlarged lateral ventricles, out of proportion to cortical atrophy, and the definitive treatment is ventriculoperitoneal shunting, which, if done early in the course, may result in complete remission of symptoms.

Postencephalitic Parkinsonism

Although of historic importance, postencephalitic parkinsonism is almost never seen today. Similarly, toxin exposure is an uncommon cause, requiring investigation only when the history suggests possible exposure (see Chapter 15).

Parkinsonism-Plus Syndromes

The parkinsonism-plus syndromes are a group of idiopathic neurodegenerative disorders that are commonly mistaken for PD, particularly early in the course. Within this category are such disorders as progressive supranuclear palsy, multiple system atrophy, and corticobasal degeneration. Absolute differentiation of these disorders from PD can sometimes only be made at autopsy; in one autopsy study of 100 patients with a clinical diagnosis of PD during life, only 76 were found to have Lewy body PD. However, a number of useful clinical clues to the parkinsonism-plus conditions can greatly increase the likelihood of a correct diagnosis. These "red flags" alerting one to a parkinsonism-plus disorder are briefly discussed here, and these conditions are covered in detail in Chapters 16 through 20.

Progressive Supranuclear Palsy

Postural instability is a prominent feature of progressive supranuclear palsy. Patients experience falls at the onset, which become increasingly frequent as the disease progresses. Speech develops a spastic quality, and many patients experience a "pseudobulbar affect" wherein sudden spells of inappropriate emotionalism (laughing or crying) occur unpredictably. Rigidity occurs, affecting axial musculature (neck and spine) more prominently than the limbs. Inability of the patient to look down associated with supranuclear paresis in the other directions of gaze is an important clinical hallmark of the disorder. However, this finding may not appear until later in the course. Although upgaze limitation is common in PD and even in normal aging, downgaze impairment is so rare in PD that this finding is considered strong evidence in favor of a diagnosis of progressive supranuclear palsy. In the early stages of this disorder, the eye findings are often not prominent, and for a time patients may have a misdiagnosis of PD.

Multiple System Atrophy

Three major clinical syndromes can be identified within this group of diseases, although clinical features and pathologic changes overlap considerably among syndromes.

SHY-DRAGER SYNDROME

When parkinsonism is associated with early and profound autonomic failure, Shy-Drager syndrome is recognized clinically. The most obvious and disabling symptom of autonomic involvement is orthostatic hypotension, which may be so severe that presyncope and often syncope occur whenever the patient assumes an upright posture. Caution is necessary when assessing the significance of orthostatic hypotension in parkinsonism, because this can be a feature of more advanced idiopathic PD, especially with the added hypotensive effect of antiparkinsonian drugs. However, when severe orthostatism occurs early in the course and is not primarily attributable to antiparkinsonian agents, Shy-Drager syndrome should be suspected. Patients with this disorder often have prominent corticospinal and cerebellar signs in addition to parkinsonism.

STRIATONIGRAL DEGENERATION

When adequate doses of levodopa seem to have no effect on the features of parkinsonism, striatonigral degeneration is suspected on clinical grounds. Because PD is primarily a disorder of dopamine deficiency with intact dopamine receptors, levodopa produces a robust clinical response. A sustained beneficial response is typically not seen in striatonigral degeneration because postsynaptic striatal degeneration parallels the loss of substantia nigra neurons.

OLIVOPONTOCEREBELLAR ATROPHY

Only the sporadic form of olivopontocerebellar atrophy is considered to be a subtype of multiple system atrophy; hereditary cases are classified as autosomal dominant spinocerebellar ataxias. Clinically, olivopontocerebellar atrophy is suspected in patients who have cerebellar deficits (gait ataxia, limb dysmetria) associated with corticospinal tract signs and parkinsonism at presentation. In contrast to Shy-Drager syndrome, dysautonomia is less symptomatically prominent.

Other Parkinsonism-Plus Syndromes

A number of other conditions can masquerade as PD. Corticobasal degeneration begins much like PD, with unilateral onset of rigidity and bradykinesia. The disorder progresses more rapidly than PD, however, and remains strikingly unilateral or very asymmetrical until late in the disease. Motor apraxia is a common feature, which is manifested by an inability of patients to execute simple motor tasks (such as brushing teeth or combing hair). An inability to use the affected limb for meaningful tasks develops in patients with corticobasal degeneration, and the limb may move spontaneously (the alien limb syndrome).

Advanced Alzheimer's disease is often associated with parkinsonian features, albeit usually not very prominent. Patients who have significant dementia when signs of bradykinesia and rigidity appear most likely do not have PD but more likely have Alzheimer's or diffuse Lewy body disease (see Chapter 20).

DIAGNOSTIC APPROACH IN PARKINSONISM

Although many conditions mimic idiopathic PD, most are rare in comparison to PD itself and can be suspected on the basis of a careful history and neurologic examination. Computed tomography or MRI imaging of the head is typically unrevealing in patients older than 50 who have the usual presentation. A robust response to antiparkinsonian drug therapy with nearly complete remission of symptoms is an adequate confirmatory test, so that brain imaging can be reserved for atypical cases. Patients presenting with parkinsonism who are younger than 55 should be tested for Wilson's disease, an autosomal recessive disorder leading to deposition of copper in the liver and brain; this is fatal if undetected. Serum ceruloplasmin determination, 24-hour urine copper analysis, and a slit-lamp examination for Kayser-Fleischer rings (a brownish-green discoloration of the cornea) are useful screening tests. Patients with Wilson's disease are treated with copper chelating agents (see Chapter 32).

Brain imaging in cases of atypical parkinsonism may reveal diagnostic clues. As mentioned, hydrocephalus or signs of vascular parkinsonism may be present in certain cases. Cerebellar and brain stem atrophy raises the possibility of multiple system atrophy. Decreased intensity of basal ganglia signals on MRI is consistent with mineral deposition and suggests one of the parkinsonism-plus syndromes rather than idiopathic PD.

Routine blood tests are rarely helpful in the diagnosis of parkinsonism, although thyroid testing may be indicated in certain patients. If hydrocephalus is supported by findings on brain imaging, patients should have a trial of cerebrospinal fluid drainage, because gait improvement after removal of a large volume of cerebrospinal fluid correlates with long-term benefit from ventriculoperitoneal shunting.

Positron emission tomography (PET) has been used to confirm the diagnosis of PD for research purposes. This technique involves intravenous administration of a radiolabeled isotope of levodopa, ^{18}F-dopa, which is avidly taken up by nigral neurons and stored in the neuron terminals as ^{18}F-dopamine. The rate of accumulation of ^{18}F within the striatum correlates with the number of presynaptic dopaminergic neurons. Thus, in PD, marked decreases occur in the rate of accumulation of ^{18}F, more pronounced on the side of the brain contralateral to the more severe clinical symptoms. Sequential fluorodopa PET scans performed over time show progressive loss of ^{18}F accumulation that roughly parallels the progression of clinical symptoms. Because ^{18}F-dopa has a very short half-life (less than 2 hours), the compound must be synthesized on-site, a requirement that limits this technique to only a few research centers.

A cocaine analogue, ^{123}I-β-CIT, binds strongly to the dopamine transporter, a protein embedded in the presynaptic membrane of nigrostriatal nerve terminals. Consequently, this agent can be used to mark dopaminergic striatal terminals. It is diminished in proportion to the degree of loss of substantia nigra neurons among patients with PD. Unlike ^{18}F-dopa, this radiotracer is relatively stable and can be prepared as a radiopharmaceutical product for widespread distribution. Single-photon emission computed tomography (SPECT) is used to construct an image of striatal ^{123}I-β-CIT activity, which affords another advantage to this technique, since SPECT scanning equipment is much more widely available than PET. To date, the Food and Drug Administration has not approved ^{123}I-β-CIT for use in the United States, but this radioligand, as well as analogues, may become available for routine clinical use in the future. Currently, however, the clinical history, examination, and response to antiparkinsonian agents are the key elements of accurate diagnosis.

SELECTED READING

Calne DB, Snow BJ, Lee C. Criteria for diagnosing Parkinson's disease. *Ann Neurol* 1992; 32 Suppl: S125-S127.

Dietz MA, Goetz CG, Stebbins GT. Evaluation of a modified inverted walking stick as a treatment for parkinsonian freezing episodes. *Mov Disord* 1990; 5: 243-247.

Hoehn MM, Yahr MD. Parkinsonism: onset, progression, and mortality. *Neurology* 1967; 17: 427-442.

Hughes AJ, Daniel SE, Kilford L, Lees AJ. Accuracy of clinical diagnosis of idiopathic Parkinson's disease: a clinico-pathological study of 100 cases. *J Neurol Neurosurg Psychiatry* 1992; 55: 181-184.

Langston JW, Widner H, Goetz CG, Brooks D, Fahn S, Freeman T, Watts R. Core assessment program for intracerebral transplantations (CAPIT). *Mov Disord* 1992; 7: 2-13.

Parkinson J. *An Essay on the Shaking Palsy.* Printed by Whittingham and Rowland for Sherwood, Neely, and Jones, London, 1817.

Stacy M, Jankovic J. Differential diagnosis of Parkinson's disease and the parkinsonism plus syndromes. *Neurol Clin* 1992; 10: 341-359.

Uitti RJ, Ahlskog JE, Maraganore DM, Muenter MD, Atkinson EJ, Cha RH, O'Brien PC. Levodopa therapy and survival in idiopathic Parkinson's disease: Olmsted County project. *Neurology* 1993; 43: 1918-1926.

Ward CD, Gibb WR. Research diagnostic criteria for Parkinson's disease. *Adv Neurol* 1990; 53: 245-249.

5

Epidemiology and Genetics of Parkinson's Disease

Demetrius M. (Jim) Maraganore, MD

CONTENTS

The purposes of this chapter are to establish the public health significance of Parkinson's disease (PD) and to provide epidemiologic insights into its causes and possible prevention.

HOW COMMON IS PARKINSON'S DISEASE, AND WHAT IS ITS PUBLIC HEALTH SIGNIFICANCE?

Incidence

Incidence is the number of new cases of a disease with onset during a given period for the total population at risk. In a recent study of the incidence of parkinsonism in Olmsted County, Minn., for the 15 years 1976 through 1990, 363 patients with parkinsonism were found. Of these, 154 (42%) met the criteria for PD (the most common cause of parkinsonism in the population). The incidence rate (per 100,000 person-years) of PD was 10.9 in the general population and 49.7 in persons older than 50. The cumulative risk of PD developing up to age 95 years was 2.7%.

From: *Parkinson's Disease and Movement Disorders:*
Diagnosis and Treatment Guidelines for the Practicing Physician
Edited by: C. H. Adler and J. E. Ahlskog © Mayo Foundation
for Medical Education and Research, Rochester, MN

Prevalence

Prevalence is the number of existing cases of a disease for the total population at a given time. In a door-to-door survey performed in Sicily, Italy, the crude prevalence of PD was approximately 3 cases per 1,000 in the total population, and it increased exponentially from approximately 1 case per 1,000 at ages 50 to 59 years to 31 cases per 1,000 at ages 80 to 89 years. PD is widely distributed throughout the world, without convincing racial disparity.

Survival

For the Olmsted County patients with PD, median survival from symptom onset was 8.8 years. This finding in part reflects the advanced age at onset for the patient population. By comparison, patients with other causes of parkinsonism combined (including progressive supranuclear palsy and multiple system atrophy) survived a median of 6.7 years. Women with PD, patients with early onset, and patients with rest tremor tended to live longer. In an earlier study from Olmsted County, although patients with PD, treated or untreated, had shorter survival than the general population, levodopa therapy was associated with a marked survival advantage.

Utilization

Also for Olmsted County, utilization of medical services by PD patients was compared with that by age- and sex-matched reference subjects. Patients with PD had significantly more physician encounters and emergency room visits per year, and the relative risk for nursing home placement was significantly increased (relative risk, 6.7). Older age at onset, lower education, and dementia were risk factors for early nursing home placement. Patients with PD were also more likely to use antidepressant and neuroleptic drugs. Most patients with PD had computed tomography or magnetic resonance head imaging, although results of these studies rarely influenced diagnosis or management. The increased use of neuroleptic agents, which may aggravate parkinsonism, and the frequent use of computed tomography or magnetic resonance head imaging are practices in PD demanding further scrutiny.

Significance

The total U.S. population aged 65 and older is projected to increase to 80,109,000 by the year 2050. Most rapidly growing, however, is the 85 and older age group (oldest old), doubling in size from 1990 to 2020, and increasing sixfold by 2050. The increasing number of elderly people, particularly the oldest old, implies a large growth in the number of persons with age-associated neurologic disorders such as PD. On the basis of available age-specific prevalence figures and official population projections, the total number of Americans affected by PD is estimated to have been approximately 700,000 in 1996, to be 900,000 in 2010, and to reach 1.8 million by 2050. Therefore, the effect of PD on health services and the cost to society are continuously expanding.

WHAT ARE THE RISK FACTORS FOR PARKINSON'S DISEASE?

Age

Olmsted County data demonstrate that PD with onset before age 40 is rare. The incidence of PD increases steeply after age 50 and continues to rise after age 85.

Sex

Also for Olmsted County, across all ages, men have higher incidence rates for PD than women, a suggestion that men are at increased risk.

Family History

Case-control studies have revealed that persons who have a first-degree relative (mother, father, brother, sister, or child) with PD are at three times greater risk for PD. One such study further found that persons with two or more affected first-degree relatives are at 10 times greater risk. Since the lifetime risk for PD in the general population is about 3%, 1 in 10 persons with a single affected first-degree relative and nearly one-third of persons with two or more affected first-degree relatives are predicted to acquire PD during their lifetimes!

Herbicide and Pesticide Use

Case-control studies report that PD is about three times more likely to develop in persons exposed to herbicides and pesticides than in those unexposed. However, only 10% of the occurrences of PD in the population can be attributed to such exposures. No single herbicide or pesticide has been linked to PD. Although well water consumption and rural living have been similarly linked to PD, multivariate analyses suggest that these are merely surrogates for herbicide and pesticide exposure.

Drug-Induced Parkinsonism

A historical cohort study of drug-induced parkinsonism in Olmsted County, Minn., found that persons who had a reversible episode of drug-induced parkinsonism were more likely to have subsequent development of PD. Clarification of the mechanism of this association may have implications for prevention.

WHAT ARE THE PROTECTIVE FACTORS AGAINST PARKINSON'S DISEASE?

Smoking

Case-control and cohort studies have noted an inverse association between tobacco smoking and PD. It has recently been suggested that this risk is strongly modified by age; smoking may be protective in younger patients but not in older ones. The risks of tobacco smoking, however, would seem to far outweigh this potential benefit from a public health standpoint. Further, less tobacco smoking among patients with PD may simply reflect a premorbid personality type. In support of this conjecture, preliminary data from Olmsted County show that PD incidence patients are also less likely to consume other stimulants, such as coffee. Might this premorbid personality type and these inverse associations reflect genetic determinants in PD? Recent studies have linked smoking addiction to specific genotypes, including some associated with PD!

Estrogen

Preliminary data from Olmsted County show that PD patients are younger at menopause and have received less estrogen replacement therapy than controls. Similar observations have been made for Alzheimer's disease. These observations are provocative and are in keeping with the finding of higher incidence rates for PD in men (who generally are not exposed to estrogen) than in women.

WHAT CAUSES PARKINSON'S DISEASE (CLUES FROM EPIDEMIOLOGY)?

Time Trends

It was recently reported that for Olmsted County, Minn., the incidence of PD did not consistently change over 15 years. Although the time frame of the study was relatively short, the findings argued against environmental factors for PD having been introduced or removed from the study population over the study period. These results are compatible with a stable environmental cause, a genetic cause, or gene-environment interaction as the cause of PD.

Environmental Toxins

The consistent association in case-control studies of PD with herbicide and pesticide use suggests that PD is caused by an unidentified environmental toxin or toxins. Also in support of a toxic cause are the observations that (1) a PD-like syndrome develops in intravenous drug users exposed to the selective neurotoxin MPTP (1-methyl-4-phenyl -1,2,3,6-tetrahydropyridine), (2) consumption of cycad flour, which contains the excitatory neurotoxin β-MAA (beta-N-methylamino-L-alanine), may be associated with the amyotrophic lateral sclerosis, parkinsonism, and dementia complex of the western Pacific, and (3) poor metabolizer genotypes of the detoxification enzyme debrisoquine hydroxylase predispose to PD. Against a toxic cause for PD, in addition to the low population-attributable risk of herbicide or pesticide exposure, are the failure to identify a specific toxin, the insidious onset and progression of PD, and the lack of precedent for neurodegeneration on the basis of a defined toxin.

Genetics

As noted above, case-control studies consistently report that a positive family history is a risk factor for PD. Family studies also indicate that about 20% of patients with PD recall a similarly affected family member. Occurrences of PD in these families are often clinically indistinguishable from sporadic cases of PD. However, familial clusters can occur on the basis of shared environmental or genetic factors. The low concordance rates for PD among twin pairs, both monozygotic and dizygotic, largely discounted a genetic basis for PD. However, these studies failed to account for preclinical PD, variability in ages at onset, and possible formes frustes of PD. Indeed, recent ^{18}F-dopa positron emission tomography (PET) imaging studies of twin pairs suggest that the concordance rate for PD in monozygotic, but not dizygotic, twin pairs is increased. This finding supports a genetic cause for PD, because although monozygotic and dizygotic twin pairs are equally likely to share environmental exposures, the former have completely overlapping nuclear DNA.

Rare autosomal dominant kindreds with levodopa-responsive, Lewy body parkinsonism have also been reported. It was subsequently observed that genetic anticipation, or the progressively earlier age at onset for patients across generations, was common in such families, in keeping with a trinucleotide repeat gene defect. Alternatively, anticipation in such families was explained on the basis of selection bias. At least two distinct gene loci (on chromosomes 4q and 2p) have since been identified for some of these families, and for the 4q locus, two separate mutations of the α-synuclein gene (but not an unstable trinucleotide repeat gene defect) have been cloned. The occurrence of PD in two families with separate mutations of the same gene and the pronounced and largely specific immunostaining of α-synuclein in Lewy bodies strongly suggest that the α-synuclein gene mutations are pathogenic (although a transgenic animal model has not as yet been developed). Nevertheless, only a minority of PD kindreds are associated with these chromosome loci or genes, suggesting that at best PD is genetically heterogeneous and that the population-attributable risk of these known mutations is very low.

Gene-Environment Interaction

Case-control studies provide support for both environmental and genetic risk factors for PD. Several years ago, a Mayo familial aggregation study concluded that PD has a multifactorial causation, the result of gene-environment interaction. This unifying hypothesis has been the subject of surprisingly little study during the past 25 years.

Etiologic Heterogeneity

Ultimately, PD may have multiple causes. For example, sporadic cases may be secondary to an environmental toxin. Rare autosomal dominant cases may be due to single gene defects transmitted in a classic mendelian fashion. The 20% or so of patients with PD who can recall one or two affected relatives but who do not have a convincing autosomal dominant pattern of transmission may share susceptibility genes, and occurrence is on the basis of gene-gene or gene-environment interaction.

CAN PARKINSON'S DISEASE BE PREVENTED?

Preclinical Parkinson's Disease

Pathologic studies indicate that incidental Lewy body disease may be 10 times more common than PD among elderly persons. It is assumed that incidental Lewy body disease represents preclinical PD.

Screening Tools

Longitudinal ^{18}F-dopa PET and β-CIT (2β-carbomethyoxy-3β-[4-iodophenyl]-tropane) single-photon emission computed tomography (SPECT) imaging studies suggest that PD has a preclinical period of roughly 5 years. Results of studies with these imaging techniques of twins or other family members at risk for PD suggest that preclinical abnormalities in nigrostriatal function can be identified. PD has subsequently developed in some persons with abnormal findings on imaging studies.

Prevention

As neuroprotectant therapies for PD emerge, it will become a public health priority to select persons in the population who are at highest risk. Use of ^{18}F-dopa PET and β-CIT SPECT may prove helpful in identifying subjects at risk who have preclinical PD. See Chapter 7 for a discussion of neuroprotective strategies.

WHAT CAN BE CONCLUDED FROM EPIDEMIOLOGIC STUDIES OF PARKINSON'S DISEASE?

Significance

PD, a common and disabling disorder of the elderly, is of considerable public health significance.

Etiology

Although environmental and genetic risk factors have been defined, further consideration of potential gene-environment interactions may provide significant inroads into our understanding of the causes of PD.

Prevention

Clarification of PD risk factors and development of screening tools may yield strategies for disease prevention.

SELECTED READING

Eldridge R, Rocca W A. *Parkinson's disease,* in *The Genetic Basis of Common Diseases* (King RA, Rotter JI, Motulsky AG eds), Oxford University Press, New York, 1992, pp. 775-791.

Schoenberg BS. *Epidemiology of movement disorders,* in *Movement Disorders 2* (Marsden CD, Fahn S eds), Butterworth-Heinemann, London, 1987, pp. 17-32.

6

Parkinson's Disease:
Etiologic Considerations

Peter A. LeWitt, MD

CONTENTS

PATHOLOGIC UNDERPINNINGS
BACKGROUND AND GENERAL COMMENTS
ETIOLOGIC HYPOTHESES
SELECTED READING

PATHOLOGIC UNDERPINNINGS

Parkinson's disease (PD) is associated with distinctive and highly localized pathologic changes in the brain. Degeneration of substantia nigra pars compacta neurons within the midbrain is pronounced in patients with PD. These dopaminergic neurons project to the striatum, and loss of striatal dopamine is responsible for most of the motor deficits of PD.

A distinctive intracellular inclusion, the Lewy body, is present in surviving nigral neurons. Researchers have regarded the process generating Lewy bodies as closely linked to the cause of the disease. The clinical picture of PD is not entirely due to nigrostriatal degeneration, however, and the degenerative process is not confined to this brain system. Lewy bodies are abundant in other catecholaminergic, melanin-bearing neurons, such as those of the locus ceruleus. They are also found in non–catecholamine cell groups, such as the cholinergic substantia innominata. Certain limbic regions, notably the amygdala, contain Lewy bodies, and they are also found outside the brain in the sympathetic ganglia. With modern histochemical techniques, Lewy bodies have been detected in the cerebral cortex, a region previously thought to be spared from the parkinsonian neurodegenerative process. Most likely, cortical Lewy bodies contribute to the cognitive impairment that occurs in some patients with PD.

Many aspects of PD pathology remain mysteries despite intensive study over the past century. Although the substantia nigra is a major focus of the parkinsonian degenerative

From: *Parkinson's Disease and Movement Disorders:
Diagnosis and Treatment Guidelines for the Practicing Physician*
Edited by: C. H. Adler and J. E. Ahlskog © Mayo Foundation
for Medical Education and Research, Rochester, MN

process, nigral cell loss is not uniform. Even in advanced cases, normal-appearing dopaminergic neurons survive in certain portions of the substantia nigra pars compacta. Why some nigral regions are consistently spared is unknown, but the answer may be critical to the understanding of the underlying pathophysiologic process.

The pathologic condition provides no clues about when the disorder begins, and it is unknown if the degenerative process develops slowly or rapidly. Loss of substantia nigra neurons does not result in clinical symptoms until depletion is about 60% to 80%. Thus, the degenerative process could conceivably smolder over years or even decades until nigral cell loss is sufficient to result in the clinical manifestations. This inability to date the onset of the underlying degenerative process makes it difficult to link it to potential causative mechanisms. Parenthetically, occasional brains from persons lacking clinical features of parkinsonism contain small numbers of nigral Lewy bodies (so-called incidental Lewy body disease). Such cases might represent an early stage or arrested development of PD. If so, they may provide an opportunity to study the initial developments in neuronal degeneration.

BACKGROUND AND GENERAL COMMENTS

Currently, there is no consensus on the probable cause of PD. Extensive research has led to a number of intriguing hypotheses. The comments that follow provide background for considering the various etiologic proposals.

- PD is not simply an inevitable consequence of normal aging. Positron emission tomography with ^{18}F-fluorodopa provides an index of dopaminergic nigrostriatal integrity. Over time, there is a substantial and progressive decline of striatal ^{18}F-fluorodopa uptake among patients with PD; in contrast, age-related declines are relatively minimal among neurologically normal persons. However, this is not to say that age-related factors may not contribute to the development of PD, since most studies suggest that the incidence of PD increases steadily with age.

- Any etiologic hypothesis must account for the diversity of the clinical manifestations of PD. Initial motor impairment is often limited to one limb or one side of the body (hemiparkinsonism), whereas it may develop symmetrically in others. Not all the "classic" features of parkinsonism are present in every patient, and those present vary in severity from patient to patient. For example, as many as one-third of patients lack the characteristic resting tremor that can be so prominent in others. Although PD is generally progressive, it can stabilize after the first few years in some fortunate patients. Dystonia, postural instability, bulbar involvement, depression, and dementia are additional clinical features that can evolve for some but not all patients with PD.

- The difficulty in identifying a single cause of PD may be related to the possibility that multiple factors contribute. For example, an inherited predisposition may not result in clinical disease unless some environmental factor comes into play, perhaps at a specific time in life. In addition, if the risk from a given factor were relatively low, its relationship to PD might be obscured.

- PD appears to be predominantly a sporadic disorder. Although this pattern is in keeping with an environmental cause, instances of obvious clustering in neighborhoods, workplaces, and similar areas have not been encountered. In the clinic, one frequently encounters patients with PD who have one or two family members with a parkinsonian disorder. Does this occurrence reflect an inherited disorder with low penetrance or recessive inheritance or some common environmental exposure? Families with a clear dominant inheritance pattern, with involvement of successive generations, have been identified, some with genetic linkage confirmed; these families, however, are rare and appear to be etiologically distinct from the usual patients with idiopathic PD.

- Epidemiologic study of PD has provided few clues. Although increasing age is clearly correlated with increasing risk for PD, no evidence exists for any other risk factor of this magnitude. A slightly greater incidence of PD in males has been observed in a number of studies, but whether this gender effect is a biologically determined propensity or the result of a specific lifestyle factor needs further investigation. Race is unlikely to contribute much to the risk for PD, as indicated by recent studies showing similar prevalence of PD in comparisons of Asian, African, and European-white heritages. Because PD is a disorder of sporadic occurrence without clustering, it would seem that if environmental or other nonconstitutional factors contribute risk for PD, these influences are relatively subtle. A meta-analysis of incidence surveys worldwide provides little evidence that geography by itself contributes to the risk for PD.

- Certain environmental exposures or lifestyle influences on the development of PD have been suggested in epidemiologic surveys. So far, the strongest factors to emerge are rural residence (including drinking well water and farm work) and not smoking cigarettes; each of these adds only minimally to the age-related risk for PD. Certain industrial activities, such as mining of metals or use of chemicals in factories, may add regionally to the risk for PD. However, such influences are small and not clearly proven by the types of prevalence studies conducted. The lack of strong evidence for environmental, employment, dietary, or other etiologic factors is the outcome of extensive research from several large-scale analyses. These negative results are not the final word on the possibility that various epidemiologic factors might contribute somewhat to causation of PD. An environmental cause might consist of repeated insults, each of relatively small consequence, toward the damage of substantia nigra neurons. There might be a long latency between an exposure and the loss of neurons (making identification of the earlier toxic exposure difficult).

- The mechanism or mechanisms causing PD might not be the same as those causing the disorder to progress. For example, once the substantia nigra cell population has been diminished by an initial insult, the compensatory increased activity of the remaining dopaminergic neurons may place increased metabolic demands on these cells or the oxidative stress from this increased dopamine turnover may be damaging to neurons. Alternatively, an initial neuronal insult might trigger a local inflammatory reaction in response to the products of cell death, which might perpetuate the process of damaging neurons and glia.

ETIOLOGIC HYPOTHESES

Discussed below are the major theories currently under consideration for the cause of PD. Although each is unproven, these speculations have guided laboratory research and, in some instances, have prompted clinical trials.

Neurotoxic Mechanisms

For many years, the notion of a neurotoxic cause for PD has been popular. Support for this concept has come from several directions. In the laboratory, several substances are known to cause selective damage to dopaminergic neurons in the nigrostriatal pathway. Since parkinsonian features are the outcomes of toxicity produced in animal models, similar compounds in the environment or of endogenous creation have been sought.

Foremost among the neurotoxins capable of producing striatal dopamine depletion is the meperidine analogue 1-methyl-4-phenyl-1,2,3,6-tetrahydropyridine (MPTP). This was first recognized as a nigral toxin after it was discovered to be the cause of irreversible, levodopa-responsive parkinsonism among a number of Northern Californian users of illicit drugs who had self-injected a "synthetic heroin." This discovery in the 1980s led to animal models that identified the mechanism of the selective substantia nigra destruction. MPTP is a substrate for monoamine oxidase type B, which converts it into a deaminated species, MPP+. This charged molecule is actively taken up by dopaminergic nerve terminals, and once in the cytoplasmic environment, it is toxic to mitochondria by inhibiting mitochondrial respiration at its initial electron transfer step (complex I).

MPTP and MPP+ are not present in the environment. Several compounds with structural similarity to these compounds, such as the insecticide paraquat, have been investigated; however, these also do not contribute to the development of PD.

With the recognition that MPP+ toxicity is mediated by inhibition of mitochondrial respiratory chain (complex I) activity, other potential mitochondrial toxins were investigated. Of particular interest have been the endogenous isoquinoline compounds, such as tetrahydroisoquinoline (THIQ). This substance is a condensation product generated in very low concentrations from the phenylalanine metabolite phenylethylamine. THIQ and its N-methylated derivative are produced in the brain and also can be transported into the central nervous system after systemic production or from dietary sources. Like MPP+, THIQ is an inhibitor of mitochondrial electron transport at complex I and can kill dopaminergic mesencephalic neurons in culture by this or possibly other mechanisms. The toxicity of THIQ is blocked by its metabolite N-methyl-THIQ. A decreased concentration of the N-methyl derivative, as has been observed in the PD striatum, could conceivably facilitate the toxicity of THIQ in physiologic concentrations. In experiments with nonhuman primates, continued administration of THIQ produced a reduction in striatal dopamine and behavioral changes resembling parkinsonism. It is not clear, however, that brain THIQ ever reaches sufficiently high concentrations to pose a threat to nigral neurons.

Two other classes of endogenous neurochemicals could confer toxicity on dopaminergic substantia nigra neurons. Beta-carbolines, derived from indolamine compounds, are present in the human brain and share structural similarity with MPP+. Like MPP+, some of these compounds can diminish mitochondrial respiration and have other interactions with dopamine metabolism. Salsolinol, a condensation product of dopamine, is also found in the human brain. Together with its N-methyl derivative, salsolinol interacts with neuromelanin to promote ferric ion–catalyzed formation of oxyradicals. These compounds also inhibit tyrosine hydroxylase.

6-Hydroxydopamine, a derivative of dopamine, is used as an experimental agent for selectively destroying the nigrostriatal dopaminergic pathways by direct brain injection in animal models. Although this toxin can be generated in vivo by nonenzymatic mechanisms, there is no convincing evidence for 6-hydroxydopamine production in the brain.

Excitotoxic mechanisms have also been proposed to contribute to the development of PD. The concept of excitotoxicity came from recognition that excessive stimulation of certain excitatory neuronal receptors could initiate a cascade of events damaging to neurons. This phenomenon has been most closely studied with glutamate, a major excitatory neurotransmitter within the striatum. Experimentally, excessive stimulation of glutamate receptors (especially the N-methyl-D-aspartate receptor) damages neurons by prolonged opening of calcium channels; excessive calcium influx can then trigger cytodestructive mechanisms. Excitotoxicity has been regarded as a possible disease mechanism in several neurodegenerative disorders besides PD, including Huntington's disease and amyotrophic lateral sclerosis. Some have proposed that although glutamate excitotoxicity may not be the initial cause of PD, it could contribute to the continuing degenerative process. Currently, clinical trials with glutamate antagonists are being carried out to assess their neuroprotective potential.

Exposure to heavy metals has also been considered a causative factor in PD. Iron is known to promote oxidative stress (see below), and local iron injection into rodent substantia nigra can induce neuronal damage. However, high environmental exposure to iron is not associated with an increased prevalence of PD, and hemochromatosis, an iron storage disorder, does not induce PD. Excessive environmental exposure to manganese, as rarely occurs in certain unsafe industrial environments, can result in parkinsonism, but with a clinical and pathologic appearance quite different from PD. Parkinsonism may be a clinical component of Wilson's disease, the copper deposition disorder, but this condition also differs greatly from idiopathic PD.

Oxidative Stress

One of the more compelling directions of research into the cause of PD has been the notion that oxidative stress may have a critical role in substantia nigra damage. The brain operates at a high metabolic rate and is potentially vulnerable to injury arising from failure in antioxidant defenses or by excessive production of free radicals. These oxyradicals can result in damage to cellular lipid membranes, various intracellular organelles (especially mitochondria), enzymes, and nucleic acids.

Dopamine metabolism may impose toxicity on substantia nigra neurons in several ways. First, this neurotransmitter is subject to spontaneous auto-oxidation into quinoid forms, which can yield oxyradicals as by-products. Enzymatic degradation of dopamine by monoamine oxidase (types A and B) produces hydrogen peroxide, which then can convert spontaneously into the hydroxyl and superoxide radicals. Certain evidence indicates that mitochondrial electron transport and oxidative phosphorylation are impaired in PD (see below), and dysfunctional mitochondria are another source of free radicals. Finally, excitotoxic mechanisms may also generate free radicals by way of cytotoxic by-products of nitric oxide; nitric oxide reacts with the superoxide anion to generate a highly reactive ion, peroxynitrite.

Oxidative stress could result in acute damage if its effects reached a threshold for producing irreversible toxicity. Alternatively, oxyradicals could produce cumulative impairment over years of exposure, slowly degrading the viability of substantia nigra neurons. One example of the latter mechanism might be chronic oxyradical damage to DNA, impairing its ability to regulate cell function. Support for this mechanism comes from measurements in PD substantia nigra of 8-hydroxyguanosine, a by-product of free radical damage to guanosine in DNA base pairs. The increased tissue concentrations of 8-hydroxyguanosine in PD support other evidence of either deficient antioxidative mechanisms or an excessive amount of oxidative stress. The cumulative damage to DNA beyond the capabilities of repair mechanisms could result in a gradual loss of control in cellular functions. Another indication of increased oxidative stress localized to the substantia nigra has been the finding that PD specimens from this region of the brain contain greater concentrations of lipid peroxidation products than do age-matched controls.

The potential toxicity of oxyradicals may be amplified by other factors, most notably iron and neuromelanin. The substantia nigra accumulates iron to a greater extent in patients with PD than in age-matched controls. Some of this iron is bound to the neuromelanin, which also accumulates in nigral neurons. Neuromelanin provides a chemical environment for converting iron to its reactive ferric species, which then can catalyze hydrogen peroxide into the toxic peroxyl radical. The combined presence of neuromelanin and increased tissue concentrations of iron in the PD substantia nigra has been regarded as a biologic "Achilles' heel" for accelerating the formation of oxyradicals.

Free radical scavenging and other antioxidant mechanisms in neurons of the substantia nigra appear to be diminished in PD. The major tissue antioxidant, glutathione, is actively synthesized by glia and neurons and is biochemically recycled via a reduction reaction. Measurements of reduced glutathione concentrations in PD substantia nigra specimens have shown decreases by as much as half. These prominent changes in glutathione content of the brain in PD are found despite an intact capacity for synthesizing and reducing this antioxidant molecule. Although other small molecules and proteins contribute to a neuron's antioxidant defenses, they cannot make up for the loss of glutathione effect. Interestingly, no decrease of reduced glutathione in substantia nigra has been found with two other disorders involving loss of dopaminergic substantia nigra neurons, progressive supranuclear palsy and multiple system atrophy. The apparent specificity of diminished substantia nigra glutathione content in PD suggests an intriguing target for exploring neurodegenerative mechanisms.

The theme of oxidative stress has led to consideration of trials with antioxidative strategies. Among these are investigations using the monoamine oxidase B inhibitors selegiline (deprenyl), lazabemide, and rasagiline. These drugs can lessen the rate of peroxyl radical formation from oxidative deamination of dopamine. A recent study of spinal fluid from selegiline-treated patients demonstrated, however, that a large fraction of dopamine is metabolized by central nervous system monoamine oxidase A. Hence, these clinical trials with the selective inhibitor of monoamine oxidase B have not fully tested the hypothesis that free radicals generated from dopamine catabolism are a possible cause of PD. Another antioxidant tested for possible neuroprotective effect was alpha-tocopherol (vitamin E), which did not slow progression of PD at a daily intake of 2,000 IU.

Immune System Activation

Circumstantial evidence has raised the possibility that the immune system could have a role in the progression of PD. It has been suggested that perhaps another distinct process initiates nigrostriatal damage and then secondary activation of the immune system perpetuates the destructive process. In support of this thought has been the observation that years after symptom-onset, reactive microglia (with macrophagic properties) are found in the substantia nigra of patients with chronic PD. Also, biologic markers of immune system activation are contained within Lewy bodies. Recently, cerebrospinal fluid from some patients with PD was reported to be toxic to dopaminergic cells in culture but not to nondopaminergic cell lines; specimens from control subjects were not toxic. However, since the immune system acts as a scavenger to clear the debris from any cytodestructive process, such findings could represent epiphenomena rather than a primary disease mechanism.

Programmed Cell Death (Apoptosis)

The dropout of neurons in PD is thought to be a long-standing, gradual process. Typical findings of acute necrosis have not appeared in PD brain specimens. The other well-studied process of cell death, apoptosis (programmed cell death), has been suspected as a mechanism of substantia nigra loss in PD. Supporting observations have been made in PD brain tissue, but a consensus that apoptosis is present and meaningful is lacking. Furthermore, the typical apoptotic process usually evolves over no more than a few days, in contrast to the more chronic rate of PD clinical progression. Apoptotic cell death can be modeled experimentally by several interventions, including inadequate supply of neurotrophic factors (see below).

Deficiency of Neurotrophic Factors

Neurotrophic factors have been recognized as essential for the viability of neuronal populations in the central nervous system, including those of the substantia nigra. Removing these substances experimentally is injurious to dopaminergic neurons. In experiments with the neurotoxin MPTP, administration of either glial- or brain-derived neurotrophic factors can reverse the toxin-induced damage (even after exposure to MPTP). One hypothesis for the cause of PD is that there might be a deficiency or malfunction of these trophic factors. Although there is no evidence directly supporting this speculation, a clinical trial of intracerebral administration of neurotrophic factors is under way.

Mitochondrial Defects

In patients with PD, reduced activity has been consistently found in mitochondrial respiratory complex I (NADH-ubiquinone oxidoreductase, the first step in the mitochondrial electron transport chain). This has been demonstrated not only in substantia nigra but also systemically, in both platelets and skeletal muscle. It has been found even in the initial clinical stages of PD, before administration of medication. Apart from mitochondrial complex I defects, less striking and inconsistent defects in other mitochondrial complexes have also been reported. Whether such mitochondrial dysfunction reflects a primary and early factor in the neurodegenerative process or evolves later is uncertain.

At least two mechanisms associated with mitochondrial dysfunction could have a pathophysiologic role. First, diminished respiratory chain activity could compromise cellular metabolism; this alone could lead to cell death or could render the neuron more susceptible to other metabolic or toxic insults. Second, dysfunctional respiratory chain complexes may fail to adequately quench oxyradical intermediates, allowing them to be released into the cytoplasm, with cytodestructive consequences.

Genetic Factors

Most cases of clinical PD do not appear to be familial. However, recent studies have confirmed an estimate by Gowers one century ago that approximately 15% of patients have at least one first-order relative also affected with PD. Some have speculated that this estimate may represent the "tip of the iceberg," with undetected subclinical PD being far more common. Pathologic studies have suggested that the prevalence of Lewy bodies detected during postmortem examination is perhaps 10 times more frequent than is the occurrence of clinically apparent PD.

The clinical epidemiologic search for hereditary influences on PD is subject to ascertainment bias and many other methodologic challenges. For example, patients with PD are more likely to have considered whether others in the family had any similar disorder; in contrast, neurologically normal persons probably have not given this much thought. Also, descriptions of elderly or deceased relatives can be misleading about whether they were actually affected by PD. Essential tremor is a common diagnosis that is occasionally confused with PD. Postencephalitic parkinsonism in older relatives might also be confused with PD. It occurred during the first half of the 1900s and was sometimes indistinguishable from PD.

Studies of twins argue against a major genetic basis for PD. Concordance for parkinsonism has been found to be similar among monozygotic and dizygotic twin pairs. The most extensive investigation of twin concordance is a recent study of male twins who had served in the United States Army during World War II and for whom follow-up was extended beyond age 70. The concordance rate, 16% among monozygotic twins and 11% among nonidentical twins, was not significant. There was a suggestion of a possible genetic basis contributing to those cases in which parkinsonism began before age 50. Of course, concordance in instances of PD in both members of these younger twin pairs is not proof of a genetic basis, since environmental causes and shared lifestyle could be the key factors.

Unequivocal cases of familial, dominantly inherited PD have been described and confirmed; however, these represent much less than 1% of all cases of PD. In some of these families, the clinical or pathologic picture differs from that of idiopathic PD.

Recently described were six families in Italy and Greece with multigenerational instances of PD in whom the same responsible gene was identified. The family initially studied was from southern Italy (the Contursi kindred), with PD occurring in four generations with a dominant inheritance pattern and affecting 45 family members. Lewy body disease was confirmed in two autopsied cases. Linkage analysis using this family revealed that the gene was a missense mutation at chromosome 4q21. This gene codes for the synaptic terminal protein alpha-synuclein. Subsequently, multiple laboratories screened for this genetic mutation among hundreds of cases of sporadic PD, and none were identified. Thus, it appears that the genetic abnormality linked to PD among these Mediterranean families is not a cause of idiopathic PD.

An etiologic link between alpha-synuclein and idiopathic PD may exist, however, despite the failure to identify the genetic mutation in sporadic PD cases. Recently, it was recognized that alpha-synuclein is concentrated in Lewy bodies, which are the pathologic hallmark of idiopathic PD.

Alpha-synuclein also is a precursor protein for the nonbeta amyloid component of senile plaques in Alzheimer's disease. Whether this common feature among these two neurodegenerative conditions is of any significance has been debated.

Another gene, associated with juvenile parkinsonism in Japan, has been localized to a different chromosome (the so-called parkin gene). However, the pathologic picture differs from that of typical PD.

Currently, many investigators are discounting the possibility that idiopathic, sporadic PD has an inherited basis as the primary and sole cause. As a contributory factor, however, inheritance may have a key function in conjunction with other elements (multifactorial hypothesis).

Central Nervous System Infection

Most research into the possibility of an infective cause for idiopathic PD has been inconclusive or unrewarding. Several decades ago, an infectious cause was proposed for many or perhaps all cases of PD. This was based on the experience from a pandemic of influenza in the early 1900s in which parkinsonism was an early or later manifestation, following clinical encephalitis. This postencephalitic form of parkinsonism, however, differs substantially from idiopathic PD. Postencephalitic parkinsonism (also termed "encephalitis lethargica," or "von Economo's disease") lacks Lewy bodies but has extensive neurofibrillary tangle formation as its pathologic characteristic; it also differs clinically, with features not seen in idiopathic PD, such as oculogyric crises (forced dystonic up-gaze) and behavioral abnormalities. Antigens of influenza A have been found in the brains of patients with postencephalitic parkinsonism, and the behavior of this virus in animal experiments has made a strong case that it could have been responsible for postencephalitic parkinsonism. The last cases of von Economo's disease developed more than 60 years ago. Only extremely rare sporadic cases with similar clinical features have occurred since, and it is not known whether these cases bear any relationship to the original disorder. Rarely, encephalitic infections with other viruses (such as western equine encephalitis, coxsackievirus B_2, measles, and Japanese B encephalitis) have been associated with the subsequent development of parkinsonism.

Attempts to culture organisms from PD brains have not met with success. Also, experiments in which extracts from PD brains were injected into nonhuman primates did not provide evidence for transmissibility.

The possibility of an infective agent causing PD is still tenable, however. The elucidation of other neurodegenerative diseases initiated by prions has expanded the possibilities of how a condition like PD could develop without evidence of infection by conventional organisms. In Jakob-Creuzfeldt disease, for example, dropout of neurons occurs within the striatum as a site of predilection. Sporadic as well as hereditary predisposition can be associated with the prion disorders. There has been no evidence for a prion mechanism in PD, however.

Recently, an animal model of parkinsonism was developed by systematically infecting mice with certain strains of *Nocardia asteroides*. This organism has an apparent tropism for the substantia nigra, manifesting an extensive localized infection within days of inoculation. This nonlethal brain infection resolves spontaneously, but persistent motor impairment develops in most mice. This disorder is levodopa-responsive, and on postmortem examination, striatal dopamine is lost and substantia nigra intraneuronal inclusions resembling Lewy bodies are present. Further studies are under way by the author, who is using this model to learn whether it might conceivably be analogous to the human disease.

SELECTED READING

Beal MF. Aging, energy, and oxidative stress in neurodegenerative diseases. *Ann Neurol* 1995; 38: 357-366.

Ben-Shlomo Y. How far are we in understanding the cause of Parkinson's disease? *J Neurol Neurosurg Psychiatry* 1996; 61: 4-16.

Calne DB. Is idiopathic parkinsonism the consequence of an event or a process? *Neurology* 1994; 44: 5-10.

Golbe LI. The epidemiology of Parkinson's disease, in *Parkinson's Disease: The Treatment Options* (LeWitt PA, Oertel WH eds), Martin Dunitz Publishers, London, 1999, pp. 63-77.

Hubble JP, Cao T, Hassanein RE, Neuberger JS, Koller WC. Risk factors for Parkinson's disease. *Neurology* 1993; 43:1693-1697.

Jenner P. Oxidative mechanisms in nigral cell death in Parkinson's disease. *Mov Disord* 1998; 13 Suppl 1: 24-34.

Kopin IJ, Markey SP. MPTP toxicity: implications for research in Parkinson's disease. *Ann Rev Neurosci* 1988; 11: 81-96.

Le WD, Rowe DB, Jankovic J, Xie W, Appel SH. Effects of cerebrospinal fluid from patients with Parkinson disease on dopaminergic cells. *Arch Neurol* 1999; 56: 194-200.

McGeer EG, McGeer PL. Neurodegeneration and the immune system, in *Neurodegenerative Diseases* (Calne DB ed), W. B. Saunders Company, Philadelphia, 1994, pp. 277-300.

Pollanen MS, Dickson DW, Bergeron C. Pathology and biology of the Lewy body. *J Neuropathol Exp Neurol* 1993; 52: 183-191.

Polymeropoulos MH, Lavedan C, Leroy E, Ide SE, Dehejia A, Dutra A, Pike B, Root H, Rubenstein J, Boyer R, Stenroos ES, Chandrasekharappa S, Athanassiadou A, Papapetropoulos T, Johnson WG, Lazzarini AM, Duvoisin RC, Di Iorio G, Golbe LI, Nussbaum RL. Mutation in the alpha-synuclein gene identified in families with Parkinson's disease. *Science* 1997; 276: 2045-2047.

Quinn N, Critchley P, Marsden CD. Young onset Parkinson's disease. *Mov Disord* 1987; 2: 73-91.

Tanner CM, Ottman R, Goldman SM, Ellenberg J, Chan P, Mayeux R, Langston JW. Parkinson disease in twins: an etiologic study. *JAMA* 1999; 281: 341-346.

Youdim MB, Ben-Shachar D, Riederer P. The possible role of iron in the etiopathology of Parkinson's disease. *Mov Disord* 1993; 8: 1-12.

7

Medication Strategies for Slowing the Progression of Parkinson's Disease

J. Eric Ahlskog, PhD, MD

CONTENTS

The recognition that levodopa therapy is capable of reversing the symptoms and signs of Parkinson's disease (PD) revolutionized the treatment of this disorder. It was soon recognized, however, that the disease continues to progress despite levodopa treatment. This progression is not rapid but occurs over many years. Potentially, there are three avenues of progression, and they occur to variable degrees (Table 7-1). In the first are levodopa treatment complications, including motor fluctuations and involuntary movements (dyskinesias). These motor fluctuations reflect a short-duration response to a dose of levodopa, with the improvement in motor symptoms lasting from 1 hour to a few hours and then "wearing off." A transition from one state to the other may occur very rapidly in some patients; this has been termed the "on-off" response. These levodopa treatment complications typically occur later in the course of the disease, at least in most cases. The second avenue of progression is a declining motor response to levodopa therapy; treatment may become much less effective with the passage of time. This is not necessarily a problem for all patients, and many have a relatively preserved levodopa response that persists for many years. The third avenue of progression is the development of problems not related to motor control. Dementia occurs in approximately 10% to 30% of patients with advancing disease. Psychosis may also develop with advancing disease, occurring as delusions, hallucinations, or inappropriate behavior. This may be provoked by medications, but in some cases psychosis is independent of drug therapy. Symptoms of dysautonomia, which may be present to a mild degree early in the course, often become

From: *Parkinson's Disease and Movement Disorders:*
Diagnosis and Treatment Guidelines for the Practicing Physician
Edited by: C. H. Adler and J. E. Ahlskog © Mayo Foundation
for Medical Education and Research, Rochester, MN

Table 7-1
Progression of Parkinson's Disease

Problem	Description
Levodopa treatment complications	• Dyskinesias (chorea, dystonia) • Short-duration motor responses (wearing off, on-off)
Declining response to levodopa	• Less effective despite optimization of dosage
Nonmotor problems	• Dementia • Psychosis • Dysautonomia (constipation, neurogenic bladder, orthostatic hypotension) • Sleep disorders (rapid eye movement behavior disorder, disrupted sleep cycles)

substantial problems with advancing disease states. Included are constipation, neurogenic bladder, and orthostatic hypotension; these disorders are often further exacerbated by the medications for parkinsonism. Certain sleep disorders also may become pronounced with disease progression, such as rapid eye movement sleep behavior disorder, in which patients act out their dreams while asleep.

The problems of advancing PD have been a focus of research over the past two decades. This interest has generated a variety of treatment recommendations to slow disease progression and limit treatment complications (Table 7-2). There has been an intense focus on levodopa therapy, with expressed concerns that this medication might be contributing to the progression of PD. Thus, physicians have been admonished to limit the use of levodopa treatment and specifically to delay levodopa therapy, restrict the dose, and substitute other drugs (levodopa-sparing strategy) whenever possible. Other medication strategies have also been proposed, including administration of antioxidant vitamins, selegiline (the monoamine oxidase [MAO]-B inhibitor), or amantadine (an antagonist of the neurotransmitter glutamate).

These proposed treatment strategies have major implications for the treatment of PD. Levodopa therapy (carbidopa-levodopa; Sinemet) remains the foundation of symptomatic treatment for this condition. If levodopa is truly potentially toxic, the clinician is in a precarious position, in that the most potent medication could be inducing long-term harm. These strategies also add another level of complexity to treatment.

Do these proposed treatment recommendations have clinical validity? This chapter focuses on the clinical evidence bearing on whether these medication strategies slow disease progression and reduce treatment complications. Obviously, these issues are fundamental to the treatment of patients with PD.

ETIOLOGIC HYPOTHESES WITH TREATMENT IMPLICATIONS FOR PARKINSON'S DISEASE

The previous chapter discusses the various etiologic hypotheses proposed to explain the development of PD. Three of these have generated much interest in the research community and have been the basis for most of the therapeutic recommendations for slowing disease progression (Table 7-3). These three hypothetical models

Table 7-2
Recommendations for Slowing the Progression
of Parkinson's Disease and Limiting Treatment Complications

Rationale	Proposal
Concerns about levodopa toxicity (based on oxidant stress hypothesis)	• Delay initiation of levodopa therapy • Limit the dosage of levodopa • Substitute other drugs for levodopa, such as direct-acting dopamine agonists (bromocriptine, pergolide, pramipexole, ropinirole)
Oxidant stress may be a causative factor	• Antioxidant vitamins
Monoamine oxidase may generate toxic by-products	• Selegiline
Excitotoxicity related to the neurotransmitter glutamate	• Amantadine • Early administration of levodopa or dopamine agonist therapy

are (1) the oxidant stress hypothesis, (2) the 1-methyl-4-phenyl-1,2,3,6-tetra hydropyridine (MPTP) model, and (3) the glutamate excitotoxicity hypothesis. These are briefly reviewed below.

Oxidant Stress Hypothesis

Why the substantia nigra should be selected for degeneration among patients with PD has been a matter of debate. This midbrain nucleus contains dopamine, which is metabolized by MAO, yielding hydrogen peroxide as a by-product. Hydrogen peroxide in sufficiently large concentrations and in the absence of effective intracellular control mechanisms can generate potentially cytotoxic oxyradicals as by-products. Furthermore, dopamine is prone to auto-oxidation, again yielding oxidative products.

The substantia nigra also appears to have the right microenvironment to facilitate the generation of potentially toxic oxyradicals. This midbrain nucleus contains high concentrations of iron, which can serve as an electron donor, thought important for the generation of radical oxygen species. The substantia nigra also contains neuromelanin, which can facilitate oxyradical reactions. In addition, several postmortem studies of PD substantia nigra have found reduced activities of enzymes critical for antioxidant defense, including catalase and glutathione peroxidase. However, it has been debated whether such reduced enzymatic activity in an area of cellular degeneration might simply represent an epiphenomenon.

The underlying tenet that dopamine is central to the degeneration of the substantia nigra has obvious treatment implications. Therapeutic replenishment of cerebral dopamine with levodopa therapy may be adding fuel to the neurodegenerative fire. Such concerns have led to repeated admonitions regarding levodopa administration, including recommendations to delay therapy whenever possible and to limit the dose. Use of direct-acting dopamine agonist drugs that can substitute for levodopa has also been advocated. Such direct agonists as bromocriptine (Parlodel), pergolide

Table 7-3
Hypotheses Generating Treatment Strategies

Hypothesis	Details
Oxidant stress	• Substantia nigra dopamine prone to auto-oxidation and enzymatic oxidation • Substantia nigra microenvironment optimal for facilitating generation of oxyradicals: 　　High iron content 　　Neuromelanin 　　Deficiency of enzyme activity 　　　for oxidative detoxification (?)
1-Methyl-4-phenyl-1,2,3,6-tetrahydropyridine (MPTP) model	• Injected MPTP induces permanent parkinsonism in humans • Similar effect in animals • Animals protected by pretreatment with selegiline, a monoamine oxidase type B inhibitor • No current evidence linking MPTP exposure to idiopathic Parkinson's disease
Glutamate excitotoxicity	• Glutamate is the neurotransmitter in widespread projections to striatum • Excessive glutamate activity potentially toxic • The mitochondrial dysfunction in Parkinson's disease may predispose to glutamate toxicity

(Permax), pramipexole (Mirapex), and ropinirole (Requip) are not converted into dopamine and are not metabolized by reactions yielding oxidative by-products; they also reduce dopamine turnover.

The oxidant stress theory of PD has also generated other treatment recommendations for slowing disease progression. These have included early initiation of the medication selegiline (L-deprenyl; Eldepryl); this drug inhibits one of the two major MAO isozymes, MAO-B, potentially reducing oxidative by-products generated from dopamine deamination. The use of antioxidant vitamins, alpha-tocopherol (vitamin E) and ascorbate (vitamin C), has also been recommended. Finally, other agents known to have antioxidant properties, such as bromocriptine and pramipexole, have also been proposed for use as a neuroprotective treatment strategy.

MPTP, Animal Models, and Implications for Parkinson's Disease

A small epidemic of irreversible parkinsonism occurred among illicit drug users in California a number of years ago. Investigations revealed that a synthetic street drug was the culprit, and the specific chemical responsible was MPTP. Clinically, there was substantial resemblance to idiopathic PD, including a response to levodopa therapy. Animal models using injected MPTP revealed that the substantia nigra was selectively damaged by a product of MPTP, the pyridinium ion, MPP^+. The generation of MPP^+ from MPTP occurs via the enzyme MAO-B. Animals could be protected from MPTP toxicity and the development of parkinsonism by pretreatment with a MAO inhibitor, including the selective MAO-B inhibitor selegiline.

These experiences led to speculation that MPTP or a similar substance might contribute to the development of idiopathic PD in humans. If so, disease progression might be prevented by administering selegiline early in the course of PD. There is, however, no current evidence that MPTP or some similar agent contributes to idiopathic PD. Thus, the relevance of the MPTP model to idiopathic PD remains speculative.

Glutamate Excitotoxicity

Glutamate is an excitatory neurotransmitter found in many cerebral circuits, including the massive projection from the cortex to the neostriatum (caudate and putamen). Obviously, the caudate and putamen are also sites of the dopaminergic projections from the substantia nigra, the site of degeneration among patients with PD.

Glutamate may have potential for toxicity if the activity of this neurotransmitter is excessive. Glutamate receptors are linked to calcium ion channels, and glutamate receptor activation induces an influx of calcium into the cell. Under normal physiologic conditions, this triggers appropriate biologic events. Excessive glutamate activity, however, could lead to toxic levels of intracellular calcium. Such glutamate excitotoxicity has been proposed to have a role in several neurodegenerative disorders, including Huntington's disease, amyotrophic lateral sclerosis, and PD.

Patients with PD may be predisposed to glutamate excitotoxicity because of mitochondrial dysfunction. As discussed in the previous chapter, mitochondrial function among patients with PD is deficient, primarily complex I of the respiratory chain. Such deficient oxidative phosphorylation could result in inadequate generation of adenosine triphosphate, which is necessary for maintenance of cell membrane polarization. Membrane depolarization, which can occur with mitochondrial insufficiency, is known to facilitate glutamate-triggered calcium entry, potentially to toxic levels. Whether this sequence of events actually occurs among patients with PD is unknown.

This PD glutamate excitotoxicity model implies that chronic administration of glutamate antagonist drugs might slow disease progression. Unfortunately, the most potent drugs of this type are not tolerated as long-term therapy. It has recently been recognized, however, that the antiparkinsonian drug amantadine is a weak antagonist of one of the major glutamate receptors (the N-methyl-D-aspartate, or NMDA, receptor). Might administration of this drug slow progression of PD?

Of additional interest are studies demonstrating that glutamate activity in the striatum is normally inhibited by the nigrostriatal dopaminergic system. In animal models, glutamate activity increases after experimental destruction of the dopaminergic nigrostriatal pathway. Could the complications of advancing PD be accelerated by initial dopaminergic neuron loss leading to increased striatal glutamate activity? If this were the case, early administration of dopamine-active drugs, such as levodopa or a dopamine agonist, might actually slow disease progression.

SLOWING DISEASE PROGRESSION
AND REDUCING LONG-TERM TREATMENT COMPLICATIONS
Evidence for Strategies Based
on Concerns About Levodopa Toxicity

BACKGROUND

The oxidant stress hypothesis predicts that levodopa therapy should be toxic. Levodopa administration increases cerebral dopamine that is subject to auto-oxidation and enzymatic (MAO) oxidation, potentially yielding oxyradicals. Indeed, levodopa is toxic when added to certain cell cultures. Furthermore, biochemical evidence of oxidant stress is present in the substantia nigra of patients with PD examined post mortem.

GENERAL CAVEATS, OXIDANT STRESS, AND LEVODOPA TOXICITY

Although the oxidant stress theory is the most popular hypothesis for the etiopathology of PD, the validity of this theory is far from proven. The cited postmortem evidence of oxidative damage in PD brains may simply represent a final common and nonspecific pathway of the degenerative disease process, that is, an epiphenomenon. Furthermore, in vitro studies demonstrating levodopa toxicity may not be directly relevant to living organisms. Cells in culture obviously lack the normal homeostatic and protective mechanisms found in vivo. In such an artificial situation, substances not normally toxic could be cytodestructive. For example, the antioxidant substance vitamin C (ascorbate) is also toxic when added to cell cultures.

Treatment strategies implied from the oxidant stress hypothesis are obviously appropriate if there is support from clinical evidence in human patients. In the paragraphs that follow, clinical evidence for the proposed neuroprotective strategies derived from the oxidant stress hypothesis is evaluated.

DELAYING LEVODOPA THERAPY

Numerous retrospective studies have addressed whether early, as opposed to later, initiation of levodopa therapy predisposes to subsequent treatment complications, such as motor fluctuations (e.g., wearing-off and on-off phenomena), dyskinesias, dementia, and psychiatric symptoms. The seminal paper focusing attention on this issue identified early treatment as a predisposing factor in the development of clinical fluctuations (Lesser et al., 1979). This paper, however, has been criticized because of patient selection bias. Obviously, patients with an initially more benign clinical course are more likely to begin levodopa therapy later. Conversely, patients destined for rapid progression are more likely to initially present with a more ominous clinical state; for these patients, levodopa therapy is likely to be started earlier. An additional six retrospective studies have also assessed this issue, and most did not find a significant relationship between early levodopa treatment and motor fluctuations.

Other treatment complications have also not been convincingly linked to early initiation of levodopa treatment. Several studies failed to find a significant relationship between levodopa-induced dyskinesias and the duration of levodopa therapy. Nonmotor symptoms, including dementia and psychiatric problems, appear to be unrelated to the early initiation of levodopa treatment or the duration of therapy.

Another investigational approach to this issue has focused on patients from the era just before levodopa became available, circa 1970. These patients had various durations of parkinsonism before levodopa was introduced. This factor circumvents the problem of patient selection bias inherent in the retrospective postlevodopa era studies (sicker patients more likely to be treated earlier). Patients from the immediate prelevodopa era had different durations of parkinsonism before beginning levodopa therapy, simply because the drug was not available. Analysis of these patients indicated that decline in motor scores over years of follow-up was directly related to the duration of symptoms and not to the duration of levodopa treatment (Markham and Diamond, 1986).

LIMITING LEVODOPA DOSAGE

Do patients fare better in the long term when a ceiling is set on the administered dose of levodopa? This issue cannot be evaluated retrospectively without substantial bias, simply because patients who are doing worse receive larger doses of medication. Only one study has evaluated this question prospectively (Poewe et al., 1986). In this trial, patients were randomized to receive either low-dose levodopa therapy or "maximum tolerated doses." After 6 years of evaluation, parkinsonian motor symptoms were less well controlled among patients in the low-dose group. Moreover, the frequency of dyskinesias and motor complications was only modestly less in the low-dose group. The investigators concluded that patients treated with levodopa monotherapy do better when the dose is adjusted according to their needs and gain little by arbitrary restriction of the dosage.

INITIATING TREATMENT WITH A DIRECT-ACTING DOPAMINE AGONIST MEDICATION

Is there any merit in delaying levodopa treatment and initiating, instead, treatment with a direct-acting dopamine agonist drug? These medications, which include bromocriptine, pergolide, pramipexole, and ropinirole, are clinically less potent than levodopa. They are unlikely, however, to induce motor fluctuations or dyskinesias when administered as monotherapy. Several prospective trials have compared bromocriptine or pergolide monotherapy with levodopa. Unfortunately, about two-thirds of the patients randomized to agonist monotherapy (bromocriptine or pergolide) in these studies dropped out within 3 years, precluding any meaningful analysis. A major reason for this high dropout rate was poor or declining efficacy.

Animal models and in vitro studies have suggested that several of the dopamine agonist medications (bromocriptine, pergolide, and pramipexole) might have neuroprotective effects. This possibility has never been clinically assessed, however, in patients with PD. On the basis of findings in the clinical trials cited in the previous paragraph, any major neuroprotective effect of bromocriptine or pergolide seems unlikely. In these studies, symptom progression continued with bromocriptine or pergolide monotherapy, and these medications were inadequate after the first 3 years in many or most patients.

A recent investigation compared the rate of parkinsonian motor deterioration after 1 year of bromocriptine therapy with that after 1 year of carbidopa-levodopa therapy. At the end of this blinded, prospective trial, patients were examined 2 weeks after the medications had been discontinued, and the motor scores were compared with baseline values. There was no difference between the two groups, indicating that the rate of deterioration was similar whether bromocriptine or levodopa therapy was administered.

On the basis of current evidence, substituting a dopamine agonist for levodopa as the initial therapy confers no proven clinical advantage. On the other hand, long-term use of these dopamine agonist drugs without levodopa therapy is inadequate in many patients.

ADJUNCTIVE TREATMENT WITH A DIRECT-ACTING DOPAMINE AGONIST DRUG

Lower doses of levodopa are possible if a dopamine agonist medication is concurrently administered. Retrospective analyses have suggested a lower frequency of later clinical fluctuations, dyskinesias, and other motor complications when the initial regimen is a dopamine agonist drug plus levodopa (Rinne, 1986). These studies, however, have been criticized because of substantial patient selection bias. Several subsequent prospective trials have yielded mixed results. An important question that is unanswered in any of these studies is whether the outcomes would be the same if administration of the agonist were simply started once motor complications developed; that is, is there some narrow window of opportunity that is lost if adjunctive agonist therapy is deferred?

A European multicenter trial (Przuntek et al., 1992) reported significantly increased mortality in patients who had been randomized to levodopa monotherapy rather than to combined levodopa and bromocriptine treatment. The very high dropout rate, however, made it impossible to interpret the data. The groups were also dissimilar, with disproportionate preexisting cardiovascular disease in the levodopa monotherapy group.

In summary, no convincing evidence exists to support early initiation of adjunctive dopamine agonist drug therapy as a means of delaying motor complications or as a levodopa-sparing strategy. However, despite this paucity of clinical evidence, many clinicians favor initial agonist therapy among patients with early-onset PD. Patients with young-onset PD are particularly prone to clinical fluctuations and dyskinesias. The argument is that there is little to lose while more definitive clinical evidence is awaited.

IS LEVODOPA TRULY TOXIC? OTHER EVIDENCE CASTING DOUBT

Multiple additional lines of evidence argue against substantial levodopa toxicity, as described below.

1. The appearance of postmortem microscopic pathologic findings did not change after levodopa was introduced. The extent of the degeneration and the pathologic markers are no different among patients treated with levodopa for years than among those who died in the prelevodopa era.
2. The advent of levodopa therapy was associated with significantly reduced mortality, documented in multiple studies.
3. Neuronal degeneration among patients with PD is not confined to dopamine-containing cells. Degeneration is also prominent in certain nondopaminergic nuclei (e.g., nucleus basalis of Meynert, dorsal motor nucleus of the vagus). If dopamine plays the pivotal role, such nondopaminergic neurons should be spared.
4. Levodopa treatment of other conditions has not resulted in substantia nigra damage on postmortem examination. Included are patients treated with levodopa for vascular parkinsonism, dystonia, or essential tremor.

5. Levodopa does not damage substantia nigra in mice given large doses for extended periods.
6. Administered levodopa is oxidized in the circulation but without evidence of oxidant stress, as measured by serum malondialdehyde concentrations. Serum levels of this lipid peroxidation product are increased in conditions known to generate oxidant stress, such as diabetes mellitus, but are not increased after administration of levodopa to patients with PD.
7. Superoxide dismutase (SOD) is a major antioxidant enzyme present in substantial concentrations within the nigrostriatal system. If the nigra hangs in tenuous oxidative balance, as proposed by the oxidant stress hypothesis, defective SOD should tip the scales, resulting in PD. However, the SOD genetic mutation sufficient to cause another neurodegenerative disease (familial amyotrophic lateral sclerosis) fails to damage the nigrostriatal system.

CONCLUSIONS, STRATEGIES DESIGNED TO LIMIT HYPOTHETICAL LEVODOPA TOXICITY

The oxidant stress theory of PD is hypothetical, and it is unclear to what extent, if any, oxidant stress has a primary role in causation. Clinical evidence that can be brought to bear does not support strategies designed to delay or limit the amount of levodopa administered. Similarly, clinical evidence, to date, does not demonstrate a convincing long-term advantage from early initiation of a direct-acting dopamine agonist drug.

Direct evidence of levodopa toxicity is lacking. In fact, it can be argued that early levodopa treatment may actually have a neuroprotective effect (see the discussion below in the section on the glutamate excitotoxicity model). The issue of levodopa toxicity obviously is of critical importance to the treating clinician. Currently, the accumulated clinical evidence argues for using doses of levodopa appropriate to the patient's needs.

Evidence for Antioxidant Agents

The oxidant stress hypothesis suggests that the substantia nigra is in a state of precarious oxidative balance. Hence, any strategies that favor detoxification of oxidant products might favor the course of the disease. Consequently, administration of the antioxidant vitamins E (alpha-tocopherol) and C (ascorbate), as well as other antioxidant substances, has been proposed.

VITAMIN E (ALPHA-TOCOPHEROL)

Prospective evaluation of high-dose (2,000 mg/day) alpha-tocopherol therapy in 400 patients revealed no effect on disease progression over that in patients receiving placebo (Parkinson Study Group, 1993). Also of relevance is the clinical experience with vitamin E deficiency states. Vitamin E malabsorption is known to result in a neurologic syndrome. However, this syndrome is a spinocerebellar degeneration rather than parkinsonism. A recent study reporting a mild reduction in the rate of progression of Alzheimer's disease with vitamin E therapy is of unclear relevance to PD. Thus, administration of vitamin E to patients with PD is not supported by hard evidence.

OTHER ANTIOXIDANT AGENTS

No controlled trials have evaluated vitamin C or any other antioxidant substances. Nonetheless, some clinicians prescribe vitamin C to patients with PD, on faith.

The antiparkinsonian drug bromocriptine has recognized antioxidant properties. Prospective trials, however, reveal continued progression of disease despite bromocriptine monotherapy.

Selegiline does have substantial antioxidant properties with the capacity to reduce oxidant stress through several mechanisms. This drug is discussed in the next section.

Evidence for Selegiline

BACKGROUND

The initial interest in selegiline was driven by the MPTP model of PD, with the recognition that this drug could block the MPTP neurotoxic effects in animals. If a toxin similar to MPTP contributes to idiopathic PD, selegiline might be able to inhibit this process, assuming similar conversion to the active toxin via MAO-B. Parenthetically, no MPTP-like toxin has been identified, to date, as a likely cause of idiopathic PD.

There is also reason to predict that selegiline might have a neuroprotective effect in PD through other mechanisms. First, selegiline potentially reduces oxidant stress via inhibition of oxidative deamination of dopamine by MAO-B. Second, selegiline tends to inhibit dopamine auto-oxidation, similarly reducing potentially toxic oxidative by-products. Third, selegiline increases the activity of certain antioxidant enzymes, including SOD. Finally, evidence from in vitro studies suggests that selegiline may have a unique capacity to rescue cultured neurons from apoptotic cell death. Certainly, these mechanisms suggest that selegiline should have substantial potential for slowing the progression of PD.

CLINICAL SELEGILINE TRIALS SUGGESTING A NEUROPROTECTIVE EFFECT, AND LIMITATIONS

The interest in selegiline as a potential neuroprotective agent led to a large multicenter trial (Parkinson Study Group, 1993), designated by the acronym "DATATOP" (deprenyl and tocopherol antioxidant therapy of parkinsonism). New and untreated patients were randomized to selegiline (deprenyl) or placebo and followed for up to 2 years. A symptomatic effect from selegiline monotherapy was not expected on the basis of previous experience. To document that selegiline (or placebo) was not simply treating symptoms, 30-day "wash-in" and "wash-out" phases were designed into the protocol. Any symptomatic effect should be obvious within the first month of treatment and documented by an improved motor score at the end of the 30-day wash-in phase, compared with baseline. Similarly, any symptomatic effect should resolve after a 30-day wash-out period. The primary end point in this double-blinded study was progression to the point at which both clinician and patient agreed that levodopa therapy was necessary.

The results revealed that selegiline therapy was associated with a significantly reduced likelihood of progression to the end point. Unfortunately, this study was confounded by a symptomatic effect apparent 30 days after initiation of selegiline therapy (wash-in). This investigation was further confounded by a too brief wash-out interval. At the time the study was designed, it was not recognized that the half-life of selegiline's effect on brain MAO-B is 40 days, obviously longer than the 30-day wash-out period. Also of concern, the significant benefit of selegiline was detected only in the first 12 months of the study. On the other hand, confining the analysis to patients without improved scores at the time of the 30-day wash-in examination revealed that selegiline therapy was associated with a significantly lower risk of reaching the study end point.

A subsequent Scandinavian multicentered trial, modeled after the DATATOP investigation, found a modest benefit from selegiline, which could have been due to a symptomatic, rather than a true, neuroprotective effect (Palhagen et al., 1998). There was a small yet statistically significant beneficial effect on the rate of clinical motor progression, but whether this represents a true neuroprotective effect is open to speculation.

An additional investigation, which enrolled patients beginning levodopa therapy, found less deterioration of parkinsonism with adjunctive selegiline treatment (Olanow et al., 1995). In this 1-year study, examinations done after 2 months of drug wash-out revealed less decline from the baseline motor scores with selegiline. This study and the DATATOP trial described above initially led to widespread use of selegiline as a presumed neuroprotective agent.

CLINICAL TRIALS CASTING DOUBT ON SELEGILINE AS A NEUROPROTECTIVE AGENT

The DATATOP investigators recognized the limitations of their earlier study and continued with follow-up of their patients in two subsequent trials. It was predicted that the 1 to 2 years of early use of selegiline in the initial DATATOP trial slowed progression and that the patients would reach subsequent progression milestones later than those who had been randomized to placebo. In these subsequent phases of the investigation, all patients received selegiline; hence, any symptomatic effects should be similar in the two groups.

Approximately 300 patients had not reached the study end point in the original DATATOP trial and continued to be followed in this second phase. Contrary to expectations, patients who had received selegiline early had the same or perhaps even a slightly greater risk of progression to the end point (Parkinson Study Group, 1996a).

Patients who had reached the end point in the initial DATATOP trial were also followed after beginning levodopa plus selegiline therapy (Parkinson Study Group, 1996b). It was predicted that the early initiation of selegiline therapy in the previous DATATOP study would be associated with fewer levodopa motor complications in this subsequent study. Selegiline, however, did not confer any long-term benefits, and the frequency of dyskinesias, clinical fluctuations, and gait freezing was similar in patients who had received selegiline rather than placebo in the previous DATATOP trial. This confirmed results of an earlier retrospective study.

A recent British study found a significantly higher mortality rate associated with long-term selegiline treatment (Lees, 1995). However, confounding factors cast doubt on this finding, and this report runs counter to the experience of most clinicians who have frequently prescribed selegiline.

CONCLUSIONS

The initial great enthusiasm for selegiline as a possible neuroprotective agent for patients with PD has not been supported by recent studies. The short-term benefit with selegiline monotherapy may simply represent a symptomatic effect rather than a countering of the underlying disease process. Long-term benefits have not been detected in controlled clinical trials.

What, then, is the role for selegiline in treating patients with PD? Although occasional physicians choose to "hedge their bets" and start selegiline therapy in new patients, this is no longer the prevailing practice. Perhaps the major place for selegiline is as adjunctive therapy with levodopa. It potentiates the effects of levodopa by blocking one of the routes by which dopamine is degraded (MAO-B). In fact, it was for this indication that selegiline was approved by the Food and Drug Administration.

Evidence for Strategies Based on the Glutamate Excitotoxicity Model

BACKGROUND

The neurotransmitter glutamate, contained in widespread (glutamatergic) projections to the striatum, may be contributing to disease progression. As described above, defective mitochondrial function occurring in PD may predispose to glutamate excitotoxicity. Furthermore, the activity of glutamatergic input to the striatum is normally inhibited by nigrostriatal dopaminergic projections. In PD, dopaminergic modulation of the corticostriatal glutamatergic system is necessarily compromised. The corticostriatal glutamatergic system is excessively active after experimental depletion of dopamine in animal models. In the MPTP model of PD, pretreatment with a glutamate receptor inhibitor partially protects against MPTP-induced parkinsonism.

Antagonists of glutamate receptors have been developed. Unfortunately, none of the more potent drugs is suitable for long-term treatment because of side effects.

AMANTADINE AS A POSSIBLE NEUROPROTECTIVE AGENT

Amantadine has been available for symptomatic treatment of PD for approximately as long as levodopa. Its mechanism of antiparkinsonian action was debated for many years. Recently, inhibition of glutamate receptors (NMDA receptor subtype) has been documented, although amantadine is not highly potent in this capacity.

Might amantadine have a neuroprotective effect by virtue of its antiglutamate effect? A single retrospective study (Uitti et al., 1996) linked administration of this drug to increased longevity in patients with PD. This retrospective analysis obviously is not sufficient as a basis for any hard clinical decisions but raises speculation that amantadine or some other glutamate antagonist might have a favorable effect on the course of PD. Without any more firm clinical evidence, however, it is difficult to recommend its routine use for this purpose.

EARLY USE OF LEVODOPA OR DOPAMINE AGONIST DRUGS FOR NEUROPROTECTIVE THERAPY

As mentioned above, the activity of the glutamatergic corticostriatal system is normally inhibited by the dopamine projections from the substantia nigra. Does the dopamine depletion that occurs with PD unleash glutamate excitotoxicity? If so, early initiation of levodopa or dopamine agonist drug therapy might slow disease progression.

Several studies have compared early versus delayed levodopa therapy, as discussed above. The retrospective trials did not demonstrate consistent advantages to early therapy; they were, however, confounded by patient selection factors (see "Delaying Levodopa Therapy" in this chapter). Patient selection should not be a major factor in studies from the eras immediately before and after levodopa, in which levodopa treatment was delayed by unavailability. Follow-up of a large group of patients from the early levodopa trials revealed increased mortality associated with longer (more than 3 years) durations of parkinsonism before levodopa availability, when controlling for disease duration (Diamond et al., 1987). As mentioned, multiple studies have found that increased longevity was associated with the availability of levodopa shortly after it was introduced about 1970. These arguments and findings are in contrast to some of the writings cited above, raising concerns that early use of levodopa therapy may be detrimental.

SUMMARY

Current popular theories about the cause of PD have generated various proposals for slowing disease progression and limiting treatment complications. These have included limiting or delaying use of levodopa, substituting other drugs for levodopa, and administering selegiline, antioxidants, or amantadine. There is, however, no compelling clinical evidence from trials in patients to support the widespread use of any of these treatment strategies as a means to slow PD progression.

Of critical importance to both patients and clinicians are the frequently heard admonitions to limit and delay levodopa therapy because of concerns about levodopa toxicity. These guidelines sometimes run contrary to what would appear to be optimal symptomatic treatment of patients. Without any compelling clinical proof of levodopa toxicity, one cannot recommend compromising the control of patients' parkinsonian motor symptoms in deference to this strategy. On the basis of the current data, it seems most reasonable and in the patients' best interests to use whatever doses of levodopa are appropriate for the patients' needs. Obviously, there would be no purpose in cavalierly prescribing excessive quantities of levodopa if they were not indicated; however, the dosage of levodopa should be adjusted so that patients are kept well within the mainstream of life whenever possible.

Other drugs that have been proposed for neuroprotective therapy, including selegiline, amantadine, and antioxidants (e.g., vitamins E and C) may be prescribed for this purpose, but on faith. Evidence supporting use of these drugs as neuroprotective treatment is preliminary or hypothetical at best.

SELECTED READING

Ahlskog JE. Treatment of early Parkinson's disease: are complicated strategies justified? *Mayo Clin Proc* 1996; 71: 659-670.

Beal MF. Does impairment of energy metabolism result in excitotoxic neuronal death in neurodegenerative illnesses? *Ann Neurol* 1992; 31: 119-130.

Blandini F, Porter RH, Greenamyre JT. Glutamate and Parkinson's disease. *Mol Neurobiol* 1996; 12: 73-94.

Diamond SG, Markham CH, Hoehn MM, McDowell FH, Muenter MD. Multi-center study of Parkinson mortality with early versus later dopa treatment. *Ann Neurol* 1987; 22: 8-12.

Fahn S, Cohen G. The oxidant stress hypothesis in Parkinson's disease: evidence supporting it. *Ann Neurol* 1992; 32: 804-812.

Lees AJ. Comparison of therapeutic effects and mortality data of levodopa and levodopa combined with selegiline in patients with early, mild Parkinson's disease. Parkinson's Disease Research Group of the United Kingdom. *BMJ* 1995; 311: 1602-1607.

Lesser RP, Fahn S, Snider SR, Cote LJ, Isgreen WP, Barrett RE. Analysis of the clinical problems in parkinsonism and the complications of long-term levodopa therapy. *Neurology* 1979; 29: 1253-1260.

Markham CH, Diamond SG. Long-term follow-up of early dopa treatment in Parkinson's disease. *Ann Neurol* 1986; 19: 365-372.

Olanow CW, Hauser RA, Gauger L, Malapira T, Koller W, Hubble J, Bushenbark K, Lilienfeld D, Esterlitz J. The effect of deprenyl and levodopa on the progression of Parkinson's disease. *Ann Neurol* 1995; 38: 771-777.

Palhagen S, Heinonen EH, Hagglund J, Kaugesaar T, Kontants H, Maki-Ikola O, Palm R, Turunen J. Selegiline delays the onset of disability in de novo parkinsonian patients. Swedish Parkinson Study Group. *Neurology* 1998; 51: 520-525.

The Parkinson Study Group. Effects of tocopherol and deprenyl on the progression of disability in early Parkinson's disease. *N Engl J Med* 1993; 328: 176-183.

The Parkinson Study Group. Impact of deprenyl and tocopherol treatment on Parkinson's disease in DATATOP subjects not requiring levodopa. *Ann Neurol* 1996a; 39: 29-36.

The Parkinson Study Group. Impact of deprenyl and tocopherol treatment on Parkinson's disease in DATATOP patients requiring levodopa. *Ann Neurol* 1996b; 39: 37-45.

Poewe WH, Lees AJ, Stern GM. Low-dose L-dopa therapy in Parkinson's disease: a 6-year follow-up study. *Neurology* 1986; 36: 1528-1530.

Przuntek H, Welzel D, Blümner E, Danielczyk W, Letzel H, Kaiser HJ, Kraus PH, Riederer P, Schwarzmann D, Wolf H, Überla K. Bromocriptine lessens the incidence of mortality in L-Dopa-treated parkinsonian patients: prado-study discontinued. *Eur J Clin Pharmacol* 1992; 43: 357-363.

Rinne UK. Dopamine agonists as primary treatment in Parkinson's disease. *Adv Neurol* 1986; 45: 519-523.

Snyder SH, D'Amato RJ. MPTP: a neurotoxin relevant to the pathophysiology of Parkinson's disease. The 1985 George C. Cotzias lecture. *Neurology* 1986; 36: 250-258.

Uitti RJ, Rajput AH, Ahlskog JE, Offord KP, Schroeder DR, Ho MM, Prasad M, Rajput A, Basran P. Amantadine treatment is an independent predictor of improved survival in Parkinson's disease. *Neurology* 1996; 46: 1551-1556.

8

Initial Symptomatic Treatment of Parkinson's Disease

J. Eric Ahlskog, PHD, MD

Treatment of the patient with recent-onset Parkinson's disease (PD) is often a very gratifying experience. Although we do not have medications with proven efficacy for slowing disease progression, we do have good drugs for symptomatic treatment. The physician's role is to keep patients within the mainstream of life. Patients whose lives are compromised obviously need to be treated, whereas others with more minor symptoms (e.g., isolated tremor) may not require therapy. The patient should participate in the decision. Some symptoms and signs that the physician may regard as minor may be troublesome to the patient. For example, someone in the public eye might be quite disabled by parkinsonian tremor that might not bother others. Clearly, however, persons curtailing their activities because of parkinsonism should be treated with adequate medications and doses.

As discussed in previous chapters, a deficiency of nigrostriatal dopamine underlies many of the motor symptoms and signs of PD. Thus, symptomatic treatment is primarily directed at reversing this dopamine deficiency state. In addition, drugs that block cholinergic transmission (anticholinergic medications) are mildly beneficial against some parkinsonian symptoms and signs (notably tremor, rigidity, and dystonia). More recently, it has been recognized that medications blocking glutamate receptors also possess some antiparkinsonian benefits. The medication amantadine, which has been available for many years, perhaps mediates its antiparkinsonian effects via inhibition of glutamate neurotransmission.

From: *Parkinson's Disease and Movement Disorders:*
Diagnosis and Treatment Guidelines for the Practicing Physician
Edited by: C. H. Adler and J. E. Ahlskog © Mayo Foundation
for Medical Education and Research, Rochester, MN

MEDICATION OPTIONS

A variety of treatment options are available to the patient with newly diagnosed PD. These include less potent drugs, such as amantadine (Symmetrel), selegiline (deprenyl; Eldepryl), and the anticholinergic agents trihexyphenidyl (Artane) and benztropine (Cogentin). Medications with more substantial potency include the direct-acting dopamine agonist drugs bromocriptine (Parlodel), pergolide (Permax), pramipexole (Mirapex), and ropinirole (Requip). However, the most efficacious medication is the drug that has been around for approximately 30 years, levodopa, which is formulated with carbidopa. Thus, carbidopa-levodopa (Sinemet) remains the foundation of treatment for PD.

Arguments for Initiating Treatment
With a Less Potent Drug

If carbidopa-levodopa is the most potent medication, why not administer it to everyone as the initial drug? The primary arguments for deferring levodopa treatment are related to concerns about levodopa toxicity and later motor complications plus the concept of saving the best for later. The arguments for levodopa toxicity are addressed in detail in the previous chapter, and, as described, there is no compelling clinical evidence that levodopa is toxic. In fact, multiple studies have demonstrated that increased longevity was linked to the introduction of levodopa therapy three decades ago. Since levodopa treatment is most effective during the first several years of use, some authors have proposed that these best responses can be saved for later by deferring treatment. There is no evidence supporting this strategy, and carrying it out may result in lost opportunities. Most likely, the decline in response to levodopa therapy over time predominantly reflects disease progression rather than long-term administration of levodopa. Patients occasionally hear or read that levodopa stops working after the first 2 to 5 years of treatment. This is not the case in patients with typical PD. After years of parkinsonism and levodopa treatment, the responses are often less dramatic, consistent, or long lasting; however, the drug typically continues to provide substantial benefit.

Concerns about later-developing levodopa motor complications (dyskinesias and fluctuating motor responses) also persuade some clinicians to initiate treatment with a dopamine agonist medication (bromocriptine, pergolide, pramipexole, or ropinirole). These motor complications typically do not develop during the first several years of levodopa treatment. In some patients, they are never a clinically significant problem.

Dopamine agonist drugs, when administered without levodopa, typically do not provoke dyskinesias or result in clinical fluctuations. Unfortunately, they are also much less effective than levodopa therapy, and usually patients cannot be maintained on these alone after the first 3 to 4 years of treatment. Since dyskinesias and fluctuations are usually not a problem during the first few years of levodopa therapy, the typical patient does not gain any obvious advantage by starting treatment with dopamine agonist monotherapy.

As discussed in the previous chapter, controversial studies have suggested that levodopa motor complications are less frequent if a dopamine agonist drug is administered concurrently with levodopa therapy (begin with one, and then add the other). No one questions that this combination levodopa-dopamine agonist therapy is an effective treatment strategy once these motor complications develop. However, should clinicians start combination therapy before the motor complications develop when symptomatic treatment is first begun? Is there an early window of therapeutic opportunity that is subsequently lost? We have no clinical data that adequately address these questions.

Special Case: Young-Onset Parkinson's Disease

Patients with young-onset PD are particularly prone to the development of levodopa-related dyskinesias and fluctuating motor responses. In one series of patients with PD with onset before age 40, both of these motor complications developed in nearly 100% within 6 years after the start of levodopa therapy. Thus, patients with young-onset PD continue to respond to levodopa, but these responses are more erratic and complicated by involuntary movements. Many clinicians regard the patient with young-onset PD as a special case and take a different initial medication strategy from that with the more typical older patient. They argue that it is probably appropriate to "hedge your bets" when it comes to these younger patients. Thus, even though clinical proof is lacking, initial treatment of young-onset PD with a dopamine agonist medication is often favored in hopes of deferring or reducing later motor complications.

If one subscribes to the "hedging your bets" approach, what are the age ranges to restrict this strategy? It seems appropriate for patients with symptoms starting before age 40, on the basis of published studies such as that described above. However, should we extend this approach to those with symptom onset between ages 40 and 50 or between 50 and 60? Since no direct clinical data bear on this issue, the decision is left to individual clinicians and their own instincts. Among patients with PD symptoms starting after age 60, initiating therapy with carbidopa-levodopa is appropriate.

Another Special Case: Minimal Symptoms

Occasional patients with newly diagnosed PD have only very mild motor symptoms. If these do not interfere with their occupation, other daily activities, or social life, less potent drugs or no treatment may be considered. Among the less potent medications are amantadine, selegiline, and the anticholinergic agents. The advantage of these three drugs is their ease of use. Each can be started at or close to the usual maintenance dose. In contrast, levodopa and dopamine agonist drugs require several weeks of dosage titration. Parenthetically, most clinicians do not choose an anticholinergic drug as first-line treatment because of the side effects, especially in the elderly.

Arguments for Levodopa as Initial Therapy

Levodopa is the most potent antiparkinsonian medication and the drug most likely to return patients with PD to the mainstreams of their lives. In addition, levodopa (formulated with carbidopa) is often better tolerated than therapeutic doses of dopamine agonist medications. Agonist drugs tend to be more complex to use; treatment must be initiated with tiny and nontherapeutic doses, with several weeks of escalation to reach even a modestly therapeutic dose. Although levodopa therapy also requires escalation, the therapeutic responses come sooner and the incremental levels are fewer. Finally, patients experiencing substantial benefits from medication are more likely to remain with the treating physician than to slip out of the medical care community; conversely, patients experiencing poor efficacy or side effects may develop a nihilistic attitude and not seek further medical attention until much later in the course.

Guidelines for Choosing the Initial Drug

As the previous chapter indicates, no medication is known to slow the progression of PD; hence, medical treatment focuses on reversing the symptoms. The primary initial treatment decision is whether to use carbidopa-levodopa monotherapy or to use either dopamine agonist monotherapy or combination therapy with levodopa. For most patients with PD, carbidopa-levodopa monotherapy is an appropriate initial choice. Clearly, this is true for patients older than 60.

For patients whose symptoms began before age 40, either of two options is appropriate. One could treat with a dopamine agonist alone for as long as symptoms can be managed before adding levodopa therapy. Alternatively, one could use combination therapy, starting with one drug first (levodopa or the dopamine agonist) and then adding the other. However, patients whose PD symptoms are poorly controlled should not have arbitrary dosage restrictions that compromise their lives.

For patients whose symptoms began between the ages of 40 and 60, it is the clinician's choice whether to use carbidopa-levodopa monotherapy as the initial strategy or the alternatives described above. In my practice, most patients between the ages of 45 and 60 are managed initially with carbidopa-levodopa but with a low threshold for adding a dopamine agonist drug. If control cannot be obtained with 600 to 750 mg of immediate-release carbidopa-levodopa daily (or a comparable amount of controlled-release carbidopa-levodopa), a dopamine agonist can be added before the levodopa dose is increased. Patients younger than 45 are managed the same as those with young-onset PD.

In the sections that follow, specific dosages and dosing strategies for each drug are provided. Much of the focus is on carbidopa-levodopa, since this drug is the foundation of symptomatic treatment.

LEVODOPA FOR INITIAL TREATMENT

Replenishment of cerebral dopamine with levodopa should not only reverse parkinsonian motor symptoms but also provide confirmatory diagnostic evidence in favor of idiopathic PD. A good response to levodopa therapy implies a nigrostriatal dopamine deficiency state that is characteristic of this disorder. Although favorable levodopa responses may occasionally occur with some of the other parkinsonian syndromes, such as progressive supranuclear palsy, the benefit is often partial or short-lived.

Carbidopa

When levodopa therapy was first introduced in the late 1960s, nausea and vomiting were frequent problems. This was related to the premature conversion of levodopa to dopamine (via the enzyme dopa decarboxylase) outside the brain. Circulating dopamine does not cross the blood-brain barrier but does penetrate into the brain stem chemoreceptive trigger zone (the brain's nausea center, where the blood-brain barrier is patent). Thus, nausea from levodopa therapy is due to dopamine stimulation of brain stem receptors rather than a direct effect on the gastrointestinal mucosa. To counter this, carbidopa was introduced in the early 1970s and formulated with levodopa. Carbidopa inhibits dopa decarboxylase (also called "L-aromatic-amino-acid decarboxylase"). However, carbidopa does not cross the blood-brain barrier and, hence, blocks only the conversion of levodopa to dopamine outside the brain. This effect counters the tendency for levodopa to induce nausea in most patients and also permits lower doses of levodopa.

Patients vary in the amount of carbidopa required to prevent nausea. In most, lower doses are sufficient. In some patients, however, doses higher than available with the conventional formulations of carbidopa-levodopa are required to prevent nausea. Thus,

Table 8-1
Carbidopa-Levodopa Formulations

Formulation	Carbidopa-levodopa, mg	Tablet color
Immediate-release	25/100	Yellow
	10/100	Blue
	25/250	Blue
Controlled-release	25/100	Pink
	50/200	Tan (peach)

patients experiencing troublesome nausea with their initial levodopa treatment may benefit from supplementary carbidopa therapy (see below). On the other hand, once patients are found to tolerate carbidopa-levodopa, the carbidopa dose can essentially be ignored. Greater or lesser amounts of carbidopa do not substantially affect the clinical response to levodopa, provided that nausea is not present. Since carbidopa does not cross the blood-brain barrier, an overdose of carbidopa is not a concern, even with higher supplementary doses.

Carbidopa-Levodopa Formulations

Carbidopa-levodopa is available as both an immediate-release formulation and a controlled (sustained)-release preparation (Sinemet CR). The different formulations are shown in Table 8-1. The 10/100 and 25/250 formulations are appropriate in patients not experiencing nausea, whereas the other formulations, with a greater ratio of carbidopa to levodopa, are advisable in those with nausea and in those receiving initial therapy (unknown whether nausea will be a problem). The tablets differ in color according to formulation.

Patients often ask whether a generic preparation is as good as the brand name tablet. For patients beginning therapy, the generic drug should be sufficient. However, patients with more advanced disease who are very sensitive to the effects of levodopa may experience a different response if generic carbidopa-levodopa is substituted for the brand name preparation (generic formulations may vary in drug content). No generic formulations of the controlled-release preparations are currently available.

Outside the United States, another dopa decarboxylase inhibitor, benserazide, is available, formulated with levodopa. Benserazide is very similar in effect and potency to carbidopa. Hence, carbidopa-levodopa and benserazide-levodopa are almost interchangeable in their effects. Benserazide-levodopa is marketed under the brand name Madopar.

Carbidopa-Levodopa: Take Dose With Meals or on an Empty Stomach?

Levodopa crosses the blood-brain barrier via a saturable transport mechanism. This transport is shared with other large neutral amino acids (e.g., phenylalanine, tyrosine). Presumably, a similar transport mechanism exists within the brain and is necessary for passage of levodopa into brain cells. This transport can be inhibited by large neutral amino acids from dietary protein. This phenomenon is of major consequence to patients with more advanced PD who are very sensitive to levodopa. Such patients often experience loss of the levodopa effect coinciding with a meal (containing protein). Although this loss may be of lesser consequence to patients beginning treatment, it nonetheless seems wise to maximize the cerebral bioavailability of levodopa. Hence, when beginning immediate-release carbidopa-levodopa therapy, most patients do best taking the medication on an empty stomach in three doses, each about an hour before a meal. This schedule allows levodopa passage into the brain before dietary amino acids can competitively inhibit transport. Pharmacists occasionally attach a label to patients' carbidopa-levodopa vials advising them to take the drug with food. Patients should be instructed to ignore these directions unless they have severe problems with nausea (see below).

How best to administer the controlled-release formulation of carbidopa-levodopa in relation to meals is less clear. Levodopa is released very slowly from the controlled-release tablet, which has a greater tendency to remain in the stomach unless gastric emptying is facilitated by ingested food. Levodopa absorption from the stomach is nil, and passage into the small intestine is necessary for entry into the circulation. Hence, it is sometimes recommended that the controlled-release formulation be taken with food. Nonetheless, passage of levodopa from the controlled-release formulation across the blood-brain barrier still is inhibited by dietary protein. Thus, taking this preparation with food may facilitate entry into the blood stream but at the expense of impediment at the blood-brain barrier by dietary amino acids. Given these potentially offsetting effects, it is unclear whether controlled-release carbidopa-levodopa should be administered with meals or on an empty stomach.

Start Therapy With Immediate-Release or Controlled-Release Formulation?

Controlled-release carbidopa-levodopa was initially developed for patients with more advanced PD who had short-duration responses to levodopa and clinical fluctuations. Subsequently, some authors advocated this formulation for initial treatment in patients with newly diagnosed PD. It was argued that the slower release of levodopa from this formulation might more closely replicate the actual physiologic release of dopamine within the brain. However, both formulations release levodopa in somewhat of a pulsatile fashion, albeit slightly more sustained with the controlled-release formulation.

Some have also proposed that initial treatment with the controlled-release preparation might lead to a lower subsequent frequency of levodopa-related clinical fluctuations (short-duration responses) and levodopa-induced dyskinesias. This possibility was assessed in two large multicenter trials that failed to identify any distinct advantage for the sustained- or controlled-release formulation as initial therapy. The subsequent frequency of dyskinesias and clinical fluctuations among patients treated with the immediate-release formulation was similar to that among those treated with the sustained-release drug.

Arguments in favor of the immediate-release formulation include lower cost (available as generic drug) and less erratic absorption. Absorption is of concern in patients with poor responses despite higher doses of the controlled-release formulation. For patients in whom one must be certain that adequate drug is getting to the brain, the immediate-release formulation is preferable.

Administration of controlled-release carbidopa-levodopa is conventionally initiated on a twice daily schedule in contrast to the three times a day regimen for the immediate-release formulation. This frequency is an argument in favor of the controlled-release formulation. However, most newly treated patients with PD do not have short-duration responses or clinical fluctuations with levodopa therapy. Hence, twice a day treatment with immediate-release carbidopa-levodopa might be just as efficacious in many patients as three times a day dosage, assuming a similar total daily dose.

Table 8-2

Carbidopa-Levodopa Initiation and Escalation

• Increase the dose weekly, as tolerated, until substantial improvement occurs or the ceiling dose is reached.

• If several doses are equally effective, choose the lowest.

Formulation	Starting dose	Weekly increments (as necessary)	Usual ceiling dose[a]	Maximum dose tried if no response[b]
Immediate-release	½ or one 25/100 tablet thrice daily (1 hour before each meal)	Add ½ 25/100 tablet to all doses	2½ or three 25/100 tablets thrice daily	Four 25/100 tablets three or four times daily for at least 1 week
Controlled-release	One 25/100 tablet twice daily	Add one 25/100 tablet to both doses (half-tablet increments if nausea)	Four to six tablets twice daily	…

[a]Point of diminishing returns in new patients.

[b]For patients who have *no* response with escalating dosage, use the immediate-release formulation and administer it on an empty stomach, gradually increasing the dose. Patients who do not have a response probably do not have Parkinson's disease but instead have another parkinsonian syndrome, such as striatonigral degeneration.

121

In summary, arguments for initiating therapy with the controlled-release formulation rather than the immediate-release drug are not highly compelling. When cost is a factor, perhaps the immediate-release formulation is preferable. The better bioavailability of the immediate-release formulation also allows more certain titration in patients with suboptimal responses. If an inadequate response were to occur with the controlled-release formulation, one could not be certain that suboptimal bioavailability was not contributing.

Immediate-Release Carbidopa-Levodopa for Initial Treatment

The immediate-release formulation of carbidopa-levodopa is a reasonable choice for initial treatment in most patients. The starting dosage of this medication is usually one-half to one 25/100 tablet taken three times a day, 1 hour before each meal. Since a week is required before the full cumulative effect of a change in levodopa dosage is known, no further increments should be made until at least a week has elapsed. Thereafter, the dosage can be adjusted upward on a weekly schedule, guided by the response (Table 8-2). Patients beginning treatment usually find that their symptoms can be well controlled with doses between one tablet and two and one-half tablets three times a day. Ultimately, if several doses are equipotent, the lowest dose can be maintained. Somewhat higher individual doses, perhaps up to three to three and one-half tablets three times a day, can be tried in patients who experience no response to lower doses. Newly treated patients who do not have a response to such doses most likely have some condition other than idiopathic PD, such as striatonigral degeneration (a form of multiple system atrophy; see Chapter 17).

In newly treated patients, the timing of the doses is not very critical, except in relation to the effects of meals. Later in the course, however, short-duration clinical responses develop in many patients, and the timing of the doses then becomes critical.

Some patients with recently developing PD have difficulty sleeping because of parkinsonism-related akathisia (restlessness) or other motor problems. The conventional daytime levodopa dosage usually extends the therapeutic benefits through the night, thereby countering the insomnia. However, occasional such patients may require a fourth dose of carbidopa-levodopa at bedtime.

Controlled-Release Carbidopa-Levodopa as the Initial Formulation

Physicians preferring controlled-release carbidopa-levodopa as initial therapy may apply the same principles as in the previous section. Twice daily dosage is conventional. It is unclear how to regulate dosage in relation to meals. That being the case, patients may choose to take their two doses with meals but switch to taking them on an empty stomach if it seems that the response is suboptimal. Since up to 2 hours are required before the controlled-release preparation takes effect, an interval of about 2 hours between the dose and meals may be necessary. The beginning controlled-release formulation is usually 25/100, taken twice daily. This dose can be increased weekly by full-tablet increments, as shown in Table 8-2. If nausea is a problem, the weekly increments can be half tablets. The point of diminishing returns is somewhere around 400 mg of levodopa as the controlled-release formulation (i.e., four 25/100 tablets) taken twice a day. Because the controlled-release preparation is about 15% to 20% less bioavailable than the immediate-release formulation, some patients may require a slightly higher total daily dose. Patients with no response to controlled-release therapy despite doses of 800 to 1,000 mg/day should probably switch to taking the immediate-release formulation to make certain that bioavailability is not a factor.

POTENTIAL SIDE EFFECTS OF LEVODOPA
Nausea

Most patients newly treated with carbidopa-levodopa do not experience nausea. When nausea occurs but is mild, it typically resolves with continued administration. Simple measures to counteract nausea include adequate fluid intake and soda crackers or dry bread (low protein) with each dose of carbidopa-levodopa. Patients who have severe difficulties with nausea may try taking their doses with meals.

In occasional patients, the amount of carbidopa contained in each tablet (25 mg) is inadequate and supplementary carbidopa (Lodosyn) is necessary to counter nausea. Supplementary carbidopa is now available for prescription use. It is supplied in 25-mg tablets; one or two of these administered before each dose of carbidopa-levodopa often markedly attenuates the nausea.

Patients refractory to these strategies may be given trimethobenzamide (Tigan) in a dose of 250 mg up to three times daily. In our experience, this is the only antiemetic available in the United States that does not worsen parkinsonism. Other commercially available antiemetics, including prochlorperazine (Compazine) and metoclopramide (Reglan), block dopamine receptors and can exacerbate parkinsonism. Outside the United States, domperidone (Motilium) is available. This drug has substantial antiemetic potency but without the potential to induce parkinsonism, since it does not readily cross the blood-brain barrier.

Dyskinesias

Involuntary movements in PD are predominantly of two major types, dystonia and chorea. Dystonia is characterized by a tonically contracted muscle-tension state, such as foot inversion, curled or dorsiflexed toes, or neck deviation. Often, this is painful and may be experienced as cramps. Such dystonia is frequently a component of parkinsonism in untreated or undertreated patients and usually responds to appropriate levodopa therapy.

Chorea, on the other hand, usually reflects an effect of excessive levodopa. Chorea is characterized by rapidly flowing and randomly patterned movements having somewhat of a dancing quality. Chorea is not seen in untreated PD. Occasionally, chorea and dystonia occur together, also typically reflecting a levodopa effect. Such choreiform movements are much more common later in the course of PD but occasionally develop early in the treatment. In rare patients, the movements develop with low doses of carbidopa-levodopa. Chorea is not dangerous but may interfere with a patient's function or be a source of embarrassment. If that is the case, one can lower the dose of levodopa, although to do so may worsen parkinsonism.

If the minimum therapeutic dose of levodopa induces troublesome chorea in a newly treated patient, certain dosage strategies may be tried. Since choreiform dyskinesias tend to last only a few hours after each dose, one strategy is to give lower doses more frequently. In new patients, the total daily dose of levodopa is more directly linked to the antiparkinsonian effect. Thus, lower doses taken more frequently may be efficacious, capitalizing on the cumulative, long-duration effect. Doses can also be taken at times when the dyskinesias may not be a problem for the patient, such as late in the day, when out of the public eye, or perhaps even at bedtime, when the patient can sleep through the dyskinesias. Dyskinesias do not occur in sleeping patients (although if severe, they can prevent sleep). Parenthetically, in more advanced disease, these strategies are not as effective, since the long-duration effect of carbidopa-levodopa often becomes less important than the more immediate (short-duration) symptomatic effects.

Orthostatic Hypotension

Symptomatic orthostatic hypotension is an occasional problem in newly treated patients with PD. If it is present, the combined effects of multiple factors are usually responsible. Patients with PD typically have at least a mild degree of dysautonomia; consequently, some have a tendency toward orthostatic hypotension, and this is exacerbated by levodopa treatment. Meals may contribute through mesenteric pooling of the circulation. Finally, other medications, including cardiac and antihypertensive drugs (e.g., nitrates, calcium channel blockers) and diuretics, may add to the potential for orthostatic hypotension.

Patients beginning carbidopa-levodopa treatment should have blood pressure (BP) documented at baseline while sitting and standing. If the initial standing BP is low normal, caution should be exercised and an occasional BP reading assessed as carbidopa-levodopa therapy is initiated and escalated. The best time to check BP is after breakfast, which is often the time of maximum orthostatic hypotension (due to the combined effects of the first morning levodopa dose and breakfast). A review of the patient's entire medication regimen should also be done at baseline if BPs are low. Typically, patients do well if the standing BP is at least 90/60.

Patients thought to have symptomatic orthostatic hypotension should be closely queried about the exact character of their symptoms. Some patients use the term "dizziness" for imbalance, which obviously has nothing to do with BP. Also, elderly patients frequently experience vestibular symptoms (benign positional vertigo) and use the term "dizziness" when referring to these symptoms.

If symptomatic orthostatic hypotension occurs early in the course of the disease, multiple system atrophy is a possibility rather than idiopathic PD. Additional tests, such as a thermoregulatory sweat test and autonomic reflex screening, may be appropriate in some patients (see Chapters 11 and 17).

Patients with symptomatic orthostatic hypotension while receiving carbidopa-levodopa can often be managed with supplementary treatments, as described in the chapter on autonomic complications of PD (Chapter 11). Usually, the initial step involves increasing dietary salt (and correspondingly fluids), with an aim toward maintaining the BP above 90/60.

Cognitive and Behavioral Side Effects

Psychosis or impaired cognition is a rare initial treatment complication of carbidopa-levodopa except in patients who have preexisting dementia, who are very old, or who are taking other psychoactive drugs. For example, an elderly patient with a maintenance regimen of selegiline may have hallucinations when carbidopa-levodopa is added. On the other hand, the same patient may be able to tolerate carbidopa-levodopa without selegiline. Patients with dementia or in whom cognitive or behavioral problems seem likely to develop are usually best managed by carbidopa-levodopa monotherapy rather than a combination of drugs.

Rashes

Rashes related to carbidopa-levodopa therapy are very rare. When they occur, the yellow coloring in the 25/100 immediate-release tablet is most likely the culprit. Switching to any of the other formulations with a differently colored tablet should be sufficient.

Table 8-3
Dopamine Agonist Medications

• Start with low (subtherapeutic) dose and escalate the dose over weeks.
• Administer maintenance doses three times daily, one dose with each meal.

Drug	Formulations, mg	Starting dose, mg	Range of typical target dose in new patients, mg/day[a]
Bromocriptine (Parlodel)	2.5, 5.0	1.25 daily	15-30
Pergolide (Permax)	0.05, 0.25, 1.0	0.05 daily	1.5-3.0
Pramipexole (Mirapex)	0.125, 0.25, 0.5, 1.0, 1.5	0.125 thrice daily	3.0-4.5
Ropinirole (Requip)	0.25, 0.5, 1.0, 2.0, 5.0	0.25 thrice daily	6-24

[a]These drugs have dose-related effects, and higher doses are sometimes administered, if tolerated.

DOPAMINE AGONIST DRUGS AS INITIAL TREATMENT

As previously discussed, early administration of a dopamine agonist medication may be appropriate in the following situations:

- When used as a "hedge" against later-developing motor complications (dyskinesias, fluctuations) in *young-onset PD*, with parkinsonism starting before age 40 years. As stated, this is done on faith, since there is no clinical proof that this early dopamine agonist strategy is effective.
- When the treating clinician perceives that is it wise to extend this hedging strategy to slightly older patients, such as those in the 40 to 50 range or even a little older.

Several facts about the four available direct-acting dopamine agonist drugs should be kept in mind. None is as potent as carbidopa-levodopa, but they share dopaminergic side effects, such as the potential for nausea, orthostatic hypotension, and psychosis. They are unlikely to induce choreiform dyskinesias when administered as *monotherapy*, but they contribute to the dyskinesia potential when given with levodopa therapy.

These drugs have several drawbacks as initial treatment. First, complicated dose escalation regimens involving multiple tablet sizes are required during the advance to the anticipated maintenance doses. Also, raising the dosage to therapeutic levels takes longer than it does with levodopa therapy. In addition, certain side effects, such as psychosis and orthostatic hypotension, are more likely than with clinically equivalent levodopa doses. Rare patients may experience first-dose symptomatic orthostatic hypotension. These drugs are also more expensive than immediate-release carbidopa-levodopa in clinically effective doses. Thus, even among patients with young-onset PD, one has to balance the relative advantages and disadvantages of early agonist therapy, recognizing that proof of this strategy is lacking.

As shown in Table 8-3, four dopamine agonist medications are currently available in the United States. Bromocriptine and pergolide have been available the longest. Both are ergot drugs with the potential for side effects related to the ergoline structure. These ergot side effects are not common but must be kept in mind, including the potential for vasoconstriction (e.g., Raynaud's phenomenon or angina, in susceptible persons). In addition, severe pleuropulmonary inflammatory-fibrotic disorders can develop rarely, similar to those with other ergot drugs, such as methysergide (Sansert). Pramipexole and ropinirole are the newer dopamine agonist drugs that do not share the ergoline chemical structure and hence are not subject to ergot side effects.

Pergolide, pramipexole, and ropinirole appear to have fairly similar efficacy, although only limited head-to-head clinical comparisons of these drugs have been done. Comparative clinical studies have suggested that bromocriptine may be slightly less efficacious than the other three agonist medications. Bromocriptine is the most expensive of the four drugs and has no advantages over the others.

Specific Treatment Strategies for Early Dopamine Agonist Administration

For the patient with younger-onset PD, early use of one of the dopamine agonist drugs might entail one of three strategies:

1. Using agonist monotherapy for as long as motor symptoms can be controlled, before starting levodopa therapy.
2. Initiating the agonist first, followed shortly thereafter by carbidopa-levodopa.
3. Beginning carbidopa-levodopa therapy first, escalating a little short of optimum, and then adding the agonist.

The doses of these dopamine agonist medications must be very low at first and increased slowly to avoid side effects, especially nausea. The dosage guidelines are shown in Table 8-3. The very low doses used for the first 2 to 3 weeks are unlikely to result in perceptible benefit, and patients should be apprised of this. Occasional patients become discouraged during this escalation phase and discontinue the medication if not informed. As shown in Table 8-3, the escalation strategies are somewhat complicated, and patients typically need written or printed instructions. For each of the drugs, such printed instructions are available from representatives of the pharmaceutical company.

If dopamine agonist *monotherapy* is chosen, the doses often need to be near the higher ends of the daily dosage ranges shown in Table 8-3. For combination therapy with carbidopa-levodopa, dopamine agonist doses anywhere in the range may be appropriate (the clinical response should guide therapy). Generally, approximately 6 to 8 weeks are required to advance the dopamine agonist dosage to the higher ends of the ranges shown in Table 8-3. For the low ends of these ranges, about 4 to 6 weeks should be sufficient. Typical carbidopa-levodopa doses administered as combination therapy are one to two 25/100 tablets three times daily of the immediate-release preparation or one to three 25/100 tablets twice daily of the controlled-release formulation.

INITIAL TREATMENT WITH OTHER MEDICATIONS

Selegiline, amantadine, or an anticholinergic drug is occasionally used for initial treatment of patients with very mild symptoms. None is very potent, but they are easy to use because complex escalation strategies are not required.

Selegiline

Selegiline blocks one of the dopamine degradative pathways, inhibiting the B form of monoamine oxidase (MAO-B). When administered with carbidopa-levodopa, it enhances the levodopa effects.

Several years ago, selegiline was favored as the initial drug for treatment of PD on the basis of preliminary evidence suggesting neuroprotective properties. As described in the previous chapter, however, subsequent evidence argued against a substantial neuroprotective effect, and most clinicians have stopped prescribing selegiline as an initial medication. It does have a slightly beneficial effect when used as monotherapy, but most patients do not find this very gratifying.

Treatment with selegiline may be started in the conventional dose of 5 mg twice daily, taken in the morning and at noon. However, the effects of selegiline are similar whether 5 mg is given daily or twice daily; therefore, many clinicians prescribe this drug as a single daily dose, taken in the morning. The drug can cause insomnia, hence the reason for administration early in the day.

Patients receiving selegiline should not be given meperidine (Demerol), because of the potential for severe, albeit rare, interactions with symptoms of fever, agitation, and fluctuating blood pressure. Interactions of selegiline with the selective serotonin reuptake inhibitors (e.g., fluoxetine) and with tricyclic antidepressant drugs have also been reported rarely, although these drugs are often used concurrently without problems.

Unlike the nonselective MAO inhibitors used to treat psychiatric disorders, selegiline does not require a special diet. Conventional doses of selegiline do not substantially block MAO type A; hence, it has no potential for inducing a hypertensive crisis in this dosage range.

Amantadine

Amantadine has a mild symptomatic effect, most appreciated during the first few months of treatment. Most patients cannot be maintained on amantadine monotherapy beyond perhaps a year or so. The mechanism of action was debated for many years, but current evidence suggests that the symptomatic effect may be mediated through blockade of the N-methyl-D-aspartate subtype of glutamate receptor within the brain. Glutamate neurotransmission occurs in a number of brain regions, including the basal ganglia.

Administration is started with one 100-mg tablet daily and then advanced to two or three tablets daily. The drug has few side effects, the most frequent being swelling and redness of the legs, perhaps with livedo reticularis. This benign problem resolves with dosage reduction or discontinuation. Amantadine, along with the anticholinergic drugs, is less expensive than the other antiparkinsonian agents.

A single study, described in the preceding chapter, suggested that amantadine may have a neuroprotective effect. There is no additional supporting evidence for this; hence, it is not in wide use for this purpose.

Anticholinergic Drugs

Anticholinergic medications reduce parkinsonian tremor, rigidity, and dystonia. They are no more effective than levodopa therapy for treatment of these symptoms, however, and do not substantially relieve the other motor manifestations of PD, such as bradykinesia, gait disorders, and loss of automatic movements. They have prominent side effects (dry mouth, memory impairment, visual blurring, constipation, and urinary hesitancy) that often offset the rather modest benefits. They are rarely tolerated in clinically sufficient doses in elderly patients. In the present era, with multiple medications available for the treatment of PD, the role for anticholinergic agents has markedly diminished.

The two primary anticholinergic agents still in use for PD are trihexyphenidyl and benztropine. With each, there is a dose-related symptomatic effect. Trihexyphenidyl therapy is started with a low dose, such as one-half of a 2-mg tablet once or twice a day, and then the frequency is increased to three times a day. If tolerated, the dosage can be increased to a whole tablet three times a day. Higher doses can be used if side effects allow. Similarly, benztropine administration can begin with a dose of one-half of a 0.5-mg tablet once or twice a day, increasing in frequency to three times a day and progressing to a whole tablet three times daily. If tremor control is insufficient on these doses and if no other treatment options are available, higher doses can be used, as tolerated.

NONMEDICATION THERAPIES

The goal of medical therapy for PD is to keep patients active and within the mainstream of life. For those whose activity level is compromised by parkinsonism or who have a lifelong history of a sedentary lifestyle, a physical therapy program is often advisable. Aerobic exercise is important in countering many of the complications of aging, and this is especially true in patients with PD. Stretching programs for rigid limbs improve function and counter pain. These issues are discussed in more detail in Chapter 14.

Patients with newly diagnosed PD often are well-informed about the latest surgical treatments (pallidotomy, deep brain stimulation, and cerebral transplantation) and may inquire if they are candidates. Given the inherent risks and the incomplete and inconsistent benefits with the currently available brain operations in contrast to the typical benefits with conventional medications, medical treatment in new patients is the better choice.

SUMMARY

In most patients with PD, carbidopa-levodopa therapy is the best option for initial symptomatic treatment. Either the immediate- or the controlled-release formulation is an appropriate choice; however, lower expense and more consistent absorption may favor the immediate-release formulation. Initial treatment with a dopamine agonist drug (bromocriptine, pergolide, pramipexole, or ropinirole) alone or concurrently with levodopa therapy is reasonable in patients with young-onset PD. However, proof that early administration of these agonist drugs has any long-term advantages is lacking.

SELECTED READING

Adler CH, Sethi KD, Hauser RA, Davis TL, Hammerstad JP, Bertoni J, Taylor RL, Sanchez-Ramos J, O'Brien CF. Ropinirole for the treatment of early Parkinson's disease. The Ropinirole Study Group. *Neurology* 1997; 49: 393-399.

Ahlskog JE. Treatment of early Parkinson's disease: are complicated strategies justified? *Mayo Clin Proc* 1996; 71:659-670.

Block G, Liss C, Reines S, Irr J, Nibbelink D, The CR First Study Group. Comparison of immediate-release and controlled release carbidopa/levodopa in Parkinson's disease. A multicenter 5-year study. *Eur Neurol* 1997; 37:23-27.

Olanow CW, Koller WC. An algorithm (decision tree) for the management of Parkinson's disease: treatment guidelines. American Academy of Neurology. *Neurology* 1998; 50 (Suppl 3):S1-57.

Quinn N, Critchley P, Marsden CD. Young onset Parkinson's disease. *Mov Disord* 1987; 2:73-91

Shannon KM, Bennett JP Jr, Friedman JH for the Pramipexole Study Group. Efficacy of pramipexole, a novel dopamine agonist, as monotherapy in mild to moderate Parkinson's disease. *Neurology* 1997; 49: 724-728.

9

Advancing Parkinson's Disease and Treatment of Motor Complications

Ryan J. Uitti, MD

CONTENTS

HISTORY-TAKING IN ADVANCING PARKINSON'S DISEASE
EXAMINATION OF PATIENTS WITH ADVANCING DISEASE
SPECIFIC THERAPY FOR MOTOR PROBLEMS IN ADVANCING DISEASE
CONCLUSION
NOTES ON SPECIFIC PHARMACOLOGIC THERAPEUTICS
SELECTED READING

This chapter addresses issues related to the diagnosis and treatment of motor dysfunction in advancing Parkinson's disease (PD). The format, designed for use by the clinician, identifies keys to a pertinent history and physical examination and suggests treatment strategies for specific clinical situations. A discussion of other practical issues follows this section.

A variety of treatment algorithms have been published in recent years. All have their own biases, and the suggestions outlined here are no different in that regard. Some algorithms have been created with a particular pharmacologic treatment in mind and therefore may emphasize one class of medications at the expense of others. It is important to recognize that more than one treatment algorithm may prove beneficial for a given clinical condition; therefore, clinicians using flexible algorithms may be of greatest value to their patients.

HISTORY-TAKING IN ADVANCING PARKINSON'S DISEASE

"My Parkinson's Disease Is Much Worse"

Patients with PD (and their family members) frequently return to their physicians with concerns that the disease has advanced. Many have read extensively, and the lay literature on PD often catalogues the myriad symptoms and signs that may arise in the course of the disease. Thus, patients may be convinced that evidence of advancing disease has developed despite a relatively short passage of time.

From: *Parkinson's Disease and Movement Disorders:*
Diagnosis and Treatment Guidelines for the Practicing Physician
Edited by: C. H. Adler and J. E. Ahlskog © Mayo Foundation
for Medical Education and Research, Rochester, MN

Table 9-1
Questions to Ask the Patient About "Advancing"
Parkinson's Disease

- When do you take your medications? With food, milk?
- When do you get up and go to sleep? Are doses spaced equitably throughout the waking hours?
- What individual doses and preparations of medication do you take (regular [immediate]-release or controlled [sustained]-release; generic)?
- Do you miss doses of medication (compliance problems)?
- Have you started to take any new medications (e.g., dopamine receptor blockers)?
- Has anything else happened recently (e.g., bladder infections, pneumonia)?
- Are you sleeping well?
- How is your mood?

Patients are usually unaware of the relative time course for advancement of disease. It is unusual for perceptible changes in signs of disease to occur over a few weeks. Rather, the expectation for significant advancement in severity of disease is typically many months or years.

Table 9-1 contains a short list of questions that need to be asked when patients complain that PD is advancing. Although parkinsonism inevitably advances in severity, there may be other explanations for *apparent* (but not true) progression of symptoms and signs. It is important to recognize these "other explanations"; the problems may be solved simply, to the relief of both patient and physician (Table 9-2).

Other Explanations for Apparent Worsening

Adverse Effects Induced by Antiparkinsonian Medications

Patients may believe that nausea and weight loss represent worsening of their disease. Such side effects often resolve within several weeks of the beginning of levodopa or dopamine agonist therapy and are usually minimized by starting with low doses that are gradually increased. Additionally, if the drug given is inadvertently switched to a levodopa tablet preparation with a smaller relative amount of carbidopa (e.g., from carbidopa-levodopa 25/100 to 10/100 or 25/250), patients may experience nausea. Although sometimes encountered early in the course of treatment with levodopa, nausea is an unusual new side effect in patients who have taken the medication for several years.

Some medications may be less well tolerated as patients grow older. Anticholinergic medications (such as trihexyphenidyl and benztropine) are prime examples of drugs that are typically better tolerated by younger persons. With advancing age, patients become more susceptible to confusion, hallucinations, urinary retention, constipation, dry mouth, and other effects than in their younger years.

Hypotension induced by antiparkinsonian medications may occur in aging patients with PD-related dysautonomia. It is not unusual for previously hypertensive persons to no longer require antihypertensive pharmacotherapy after several years of parkinsonism

Table 9-2
Other Explanations for "Advancing" Parkinson's Disease

- Adverse effects induced by antiparkinsonian medication
- Timing of medication administration; compliance
- Changes in formulation of antiparkinsonian medication
- Other medications
- Other changes: infection, depression, exercise, sleep

(and treatment with levodopa or dopamine agonists). Previous practices for treating hypertension (e.g., avoiding salt) also exacerbate orthostatic hypotension in these persons. Discontinuing these practices often easily corrects orthostatic hypotension. More refractory symptomatic hypotension can be treated as described in Chapter 11.

CHANGES IN TIMING OF MEDICATION ADMINISTRATION AND COMPLIANCE

Patients who report abrupt symptomatic deterioration may have changed the way in which they take their medications. If they are confused, they may not be taking their medications at all. Compliance is often a problem for geriatric patients taking multiple medications of any kind. Additionally, patients who have benefitted from levodopa taken on an empty stomach may inadvertently switch to taking doses with food. Well-meaning pharmacists often affix labels to bottles of carbidopa-levodopa instructing that the medication be taken with food. This practice may reduce the bioavailability of levodopa, as discussed in the previous chapter. The routine of taking levodopa with food arose when the active drug, levodopa, was available only in isolation (without carbidopa) in the late 1960s and early 1970s. Most patients taking levodopa alone experienced nausea and vomiting. With the addition of carbidopa (combination tablets have been available since 1973), most patients could tolerate the medication without vomiting. This tremendous improvement in tolerability of levodopa led to the selection of the brand name, Sinemet, from *sin* (Latin) meaning *without* and *emet* (Greek) meaning *emesis*. Sinemet ("without emesis") was a medication that no longer needed to be taken concurrently with food. Hence, if patients take carbidopa-levodopa with food or milk (contains protein) less levodopa may be delivered to the brain. Consequently, it is important to ask whether levodopa is taken "on an empty stomach."

It is also helpful to ask patients to describe their medication regimen, reporting not only the amounts of medicine taken but also the *time* of day when taken. The act of reciting the regimen allows the physician to ascertain who (patient, spouse, or caregiver) is directing the medication program. If the regimen is described with uncertainty, great potential exists for poor compliance or erratic dosage of antiparkinsonian medications. A new caregiver may not understand the principles of sequential dosing of levodopa throughout the day, often believing that the total quota for the day is all that needs to be accomplished. It is surprisingly frequent for patients to understand "three times per day" to mean 8 a.m., 4 p.m., and midnight (or bedtime). Patients may be correctly taking controlled-release carbidopa-levodopa twice a day but mistakenly take the doses on awakening and retiring. Consequently, these patients receive essentially one dose during their waking hours. Although most patients with mild and early disease can be relatively cavalier about the timing of levodopa administration without experiencing a decline in response, this is usually not the case for those with longer-standing disease.

A clear "short-duration response" to levodopa may also be unmasked by changes in timing of levodopa intake. "Short-duration response" refers to symptomatic benefit that is pronounced for several hours immediately after administration of a dose of medication. Advancing disease frequently is accompanied by the emergence of an obvious short-duration response to levodopa (discussed under "Truly Advancing Parkinson's Disease," below). However, alterations in the timing of medication administration in such patients may give them the false impression that the response to levodopa has drastically changed. For example, patients may delay their initial dose of antiparkinsonian medication because of other drugs (e.g., alendronate [Fosamax] on awakening, nothing by mouth for the next hour, and levodopa later with breakfast). A delay in the intake of levodopa could understandably result in more obvious parkinsonian symptoms during the morning hours without reflecting substantially advancing disease.

CHANGES IN THE FORMULATION OF ANTIPARKINSONIAN MEDICATION

Levodopa is available in immediate-release and controlled-release (Sinemet CR) formulations. Switching between the two may lead to worsening of parkinsonism despite no actual change in the severity of underlying disease. At face value, patients switching from two tablets of immediate-release carbidopa-levodopa 25/100 to a single tablet of controlled-release carbidopa-levodopa 50/200 would seem to be taking identical amounts of carbidopa and levodopa. However, the controlled-release preparation is substantially less bioavailable than the immediate-release formulation. Switching to the controlled-release preparation without taking into account the lower bioavailability frequently results in a decline in functioning. The approximate equivalent of 150 mg of immediate-release levodopa is 200 mg of levodopa in the controlled-release tablet. Controlled-release levodopa tablets that are crushed have the same bioavailability as regular formulation levodopa. Chewing or crushing these tablets therefore occasionally leads to side effects related to larger effective doses (patients may crush them to obtain a quicker response time for action).

In addition to formulation differences, most antiparkinsonian medications (levodopa and dopamine agonists) come in several dosage sizes. Consequently, inadvertent mixups of tablets may result in altered responses. For example, controlled-release 25/100 tablets may be inadvertently substituted for the 25/100 immediate-release formulation. In the United States, all immediate-release carbidopa-levodopa 25/100 tablets are yellow, whereas the 25/100 and 50/200 controlled-release tablets are pink and peach, respectively (often a difficult color distinction for patients). Unfortunately, pergolide 1.0-mg tablets are also pink. Finally, both the 10/100 and the 25/250 carbidopa-levodopa tablets are blue. This situation is fertile ground for confusion.

Immediate-release carbidopa-levodopa is available as a generic formulation. Rarely, patients may experience a mild decline in function if a switch is made from brand-name to generic preparation.

OTHER MEDICATIONS

A patient experiencing nausea may be inadvertently given antinauseant medications that can cause or exacerbate parkinsonism. Metoclopramide (Reglan) and prochlorperazine (Compazine) are classic examples of dopamine receptor blocking agents that exacerbate parkinsonism. Patients receiving neuroleptic agents can also

Table 9-3
Truly Advancing Parkinson's Disease

- Increases in severity of symptoms
- Postural instability
- Difficulty with speech or ignition of gait
- Motor fluctuations
- Medication side effects of advancing disease
  (e.g., dyskinesias)
- Dementia

---

experience a substantial worsening of parkinsonian symptoms despite no change in the underlying disease.

OTHER MEDICAL PROBLEMS, SUCH AS INFECTION, DEPRESSION, AND SLEEP DISTURBANCE

Patients do not deteriorate "over the weekend" because of PD per se. Acute changes in motor function may occur with infections (e.g., urinary tract infections, pneumonia) or the onset of other illnesses. Patients who are sleep-deprived or depressed may also experience temporary worsening of their parkinsonian symptoms. Consequently, acute changes in parkinsonism necessitate a vigilant search for an explanation other than merely PD.

In conclusion, unless one asks questions that address the aforementioned possibilities, treatable problems may be ignored or mistaken for advancing disease.

Truly Advancing Parkinson's Disease ("I Am Slower, Stiffer, and Shaking More")

Advancing symptoms and signs of parkinsonism may be all too obvious to patients and their families and friends (Table 9-3). Key questions that quickly corroborate symptoms of advancing disease are listed in Table 9-4.

INCREASING SEVERITY OF CARDINAL SYMPTOMS, GAIT INSTABILITY, SPEECH DYSFUNCTION

Resting tremor is often more apparent, but bradykinesia is typically more disabling. Patients complain that tasks now take two or three times as long to perform as in the past. Rigidity may be severe, to the point of pain. Slowing of gait, with hesitation in initiating stride and making turns (gait "ignition" difficulties and "freezing"), commonly occurs with advancing disease. Deteriorating balance and falls are the most dangerous potential problems for most patients with PD. Falls are generally preceded by an unmistakable sense of unsteadiness. Spouses and caregivers often note deterioration in volume and clarity of a patient's speech (hypokinetic dysarthria).

MOTOR FLUCTUATIONS

Variability in motor function (sometimes from near-normal to near-immobilization) is a consequence of the levodopa short-duration response that substitutes for the long-duration response in advancing disease. Periods when patients have good motor function-

Table 9-4
Questions to Ask the Patient
About Advancing Parkinson's Disease

• Do you need help with any normal daily tasks, such as dressing (buttoning, putting on stockings)?

• How much longer than "normal" does it take to get ready for the day?

• Do you feel stable on your feet? Any falls (make certain these were not syncopal)?

• Does your ability to move about change greatly at different times of the day (motor fluctuations, wearing off, "on-off" phenomenon)?

• Do you have involuntary movements other than tremor (dyskinesias)?

• Do you experience hallucinations?

• How is your memory (corroborate answer with caregiver)?

---

ing are termed "on" times. "Off" periods are times with poor motor functioning when the medication is having no or minimal efficacy. Patients may also experience other disorders of movement, particularly in "off" states. For example, dystonia may occur, particularly in younger patients, during the early morning hours, often involving the toes, foot, and ankle (characteristically with toe and foot flexion or cupping and inversion at the ankle). Such "off" dystonia often responds to antiparkinsonian therapy.

"WEARING OFF," "ON-OFF," AND DYSKINESIAS

With advancing disease, the response to levodopa (pharmacodynamics) frequently changes. Whereas in mild disease motor improvement is not time-locked to an individual dose of levodopa, this occurs with advancing disease. Patients may take a dose of levodopa and recognize, minutes later, that levodopa has begun "to work," producing an "on" response. After several more hours (or minutes), they may recognize when levodopa stops working, returning them to an "off" state. Typically, with further advancing disease, an individual levodopa dose effect produces shorter "on" periods. Thus, this response profile, usually several hours in length, is termed the "short-duration response" to levodopa. The "long-duration response" to levodopa refers to motor function benefit obtained from the drug that persists for days. When individual dose responses are punctuated by "off" states before the next dose of levodopa, the patient is experiencing a "wearing-off" phenomenon. Wearing off most likely primarily represents advancing disease (and less endogenous levodopa or dopamine storage capacity) but also changes in the dopamine receptor in response to long-term levodopa therapy. Rarely, patients report having abrupt changes in their ability to move about, analogous to turning off a light switch. This phenomenon is termed "on-off." Equally abruptly, motor functioning may be restored after several minutes.

Antiparkinsonian medications, particularly levodopa, may also cause dyskinesias. The typical levodopa-induced dyskinesias are predominantly those of chorea, but dystonia may also be a component. Dyskinesias can be restricted to small regions of the body, for example, the orofacial musculature, or be generalized. Dyskinesias vary from mild, giving an appearance of mild fidgeting, to pronounced, with wild ballistic movements.

Although gait freezing (feet stuck to floor) is usually reflective of the underlying parkinsonism, it can also be induced by medications. Higher doses of a dopamine agonist and, rarely, even levodopa therapy may do this.

Consequently, it is imperative to ask patients "How does your day go?" in an attempt to outline the typical course of their day. If they can recognize patterns in the degree of functioning at a given time of day, these may coincide with doses of medication. It may be helpful to provide patients with a diary form (with 30-minute time blocks) so that they can outline their daily experience ("on" with good functioning, "off" with poor motor function, and "on with disabling dyskinesias") over the course of several days, including doses of medication and times administered. It is also important to know if and when hallucinations or confusion occurs. With this information, patterns of response can be deciphered and a logical treatment plan formulated.

When the physician tries to determine what movement troubles patients are experiencing, it is helpful to demonstrate what is meant by "dyskinesias," "tremor," and similar terms. Correct use of terminology is critical, since wrong use by patients (e.g., calling dyskinesias "tremor") results in incorrect treatment.

ADVERSE EFFECTS OF ANTIPARKINSONIAN MEDICATION

The response to medications may change with advancing disease. For example, some patients who could tolerate large amounts of levodopa at one time may become intolerant of such doses, with disabling dyskinesias. In aged patients, anticholinergic medications once tolerated may ultimately produce delirium, hallucinations, or apparent dementia.

DEMENTIA

Dementia in PD occurs in about 25% of patients. Patients most commonly have "subcortical" dementia, presumably related to relative dysfunction in basal ganglia–cortical connections on the basis of the same neurodegenerative process (see Chapters 12 and 20). Dementia can develop from other causes (see Chapters 12 and 20) and therefore requires evaluation to detect potentially reversible contributing factors. Unfortunately, dementia is often far more disabling than motor dysfunction related to parkinsonism and is predictive of reduced life expectancy in PD. The antiparkinsonian medications may induce hallucinations or delusions in demented patients, and medication adjustments may be necessary in this situation.

EXAMINATION OF PATIENTS WITH ADVANCING DISEASE

The examination of a patient with advancing disease ideally confirms what is gleaned from the historical inquiry. Examination points are outlined in Chapter 4. Several points related to the patient with advancing disease are reviewed here (Table 9-5).

It is important to verify with patients that their definitions of tremor and dyskinesia, for example, are correct. The examiner may have observed these signs while obtaining a history from the patient. If resting tremor and dyskinesias are not initially apparent, it may be useful to ask the patient to perform rapid alternating movements, such as tapping sequential digits to the thumb or twisting the hand, "as if one were screwing in a light bulb." During these sequences, dyskinesia is often readily apparent in the contralateral limb or orofacial musculature. Similarly, contralateral resting tremor may become apparent for the first time while performing these maneuvers.

Table 9-5
Examination of Advancing Parkinson's Disease:
Noteworthy Components

• Observe tremor and levodopa-induced dyskinesias

• Discuss with patient and agree on definitions of movement disorder terminology

• Assess postural instability; use "pull test"

• Measure standing blood pressure (early detection of potential for orthostatic
    hypotension)

• Identify other abnormalities, such as extraocular movement dysfunction

The "pull test" (see Chapter 4) is an essential part of the examination of every patient with PD, particularly those with advancing disease, who may not be aware of relative instability. Unfortunately, there are too many examples of first falls producing serious injury. Obviously, patients and caregivers (witnessing the pull test) who are aware of instability may take measures to reduce the risk of falling.

The examination of patients with advancing disease should include not only motor assessment but also measurement of sitting and standing blood pressures. Occasional patients with complaints of exhaustion and weakness have symptoms related to orthostatic hypotension. Systolic blood pressures of less than 90 mm Hg should alert one to this possibility. Extraocular movements should be recorded to identify early evidence of another parkinsonian syndrome rather than PD (e.g., progressive supranuclear palsy, as discussed in Chapter 16).

At times it is not possible to ascertain what a patient is trying to report, and the initial examination may take place at a time when such symptoms are not present. In these situations, prolonged office monitoring (over the course of several hours) may be extremely helpful. This is done by having the patient come to the office in an "off" state (often in the morning before the first dose of medications) and then continuing the monitoring after the typical doses of antiparkinsonian medications. Subsequent brief serial examinations may characterize the response to carbidopa-levodopa and adjunctive medications. In this manner, the physician may see the severity of an "off" state, timing and magnitude of the best "on" effect, peak-dose dyskinesia, and duration of single-dose effect, with these captured by several serial 30-second motor examinations. The physician can also reinforce descriptive terminology during this exercise, helping to facilitate useful discussions of motor fluctuations in the future.

SPECIFIC THERAPY
FOR MOTOR PROBLEMS IN ADVANCING DISEASE

Once the nature of a patient's motor problem has been ascertained, appropriate treatment strategies can be formulated (Table 9-6). The following discussion describes treatment strategies for specific motor problems.

Short Duration Responses to Antiparkinsonian Therapy: Treatment Issues

INADEQUATE RESPONSE

Patients may experience a suboptimal response to levodopa therapy. Such patients may report that they never experience adequate control of their parkinsonian symptoms

Table 9-6
Short-Duration Responses
to Antiparkinsonian Therapy: Treatment Issues

- Inadequate response

- "Wearing off"

- Dyskinesias
    Chorea
    Dystonia
    Peak-dose dyskinesia
    Biphasic dyskinesia

- "Freezing"

- "On-off" phenomenon

- Insomnia

throughout the course of the day, although with some fluctuation. It is paramount first to establish compliance with and proper administration of antiparkinsonian therapy. If patients are experiencing lack of response to a typical dose of levodopa, the problem may be related to meals or snacks high in protein (e.g., dairy products). This difficulty can be corrected by insisting that the doses be taken without food.

For most patients with an inadequate short-duration response to levodopa, the first therapeutic maneuver is simply to increase the dosage of the medication. The individual doses of levodopa can be increased (by 25 to 50 mg for all doses) each week until a beneficial response occurs or side effects develop. Note that the peak response to immediate-release carbidopa-levodopa is fully developed 60 to 90 minutes after a dose; for the controlled-release preparation, it is apparent by 120 to 150 minutes following a dose. The motor response at these times of peak effect provides the index of whether the dose is adequate. The individual dose of levodopa (immediate-release) can be increased up to 350 mg. It is exceedingly unusual for individual doses higher than this to produce more improvement (more commonly producing no greater benefit or intolerable side effects).

WEARING OFF

The wearing-off phenomenon implies that the levodopa benefit is short-lived and the effect is lost before the next dose. There are several treatment approaches to this problem. Intuitively, one might expect this problem to respond to simply increasing the individual doses of levodopa. This approach, however, usually fails to substantially increase the duration of the levodopa response in patients already taking moderately high or high doses (i.e., individual doses of 200 mg or greater). A higher dose may then only result in side effects. In general, if the patient is experiencing a very good response during the levodopa "on" phase, the most effective strategy is to maintain that optimal dose and simply shorten the duration between doses to match the duration of the levodopa response. This plan usually requires adding extra daily doses to provide continuous daytime coverage.

A second strategy is to substitute controlled-release carbidopa-levodopa for the immediate-release preparation. The controlled-release effect lasts 60 to 90 minutes longer, although it requires up to 2 hours to become active. As already discussed, there is no milligram to milligram correspondence between these two preparations, and the individual doses of the controlled-release formulation must be 30% to 50% greater to produce a response comparable to that of the immediate-release formulation. Some patients find the responses to the controlled-release formulation more erratic, and switching to this formulation typically is not effective in those with very short levodopa responses (e.g., less than 2 hours) or those with wide variation between "on" and "off" states.

A third treatment approach is to add a direct-acting dopamine agonist after the carbidopa-levodopa dosage has been adjusted. These drugs, which include bromocriptine (Parlodel), pergolide (Permax), pramipexole (Mirapex), and ropinirole (Requip), are less potent than levodopa but have a several-hour response duration that may help smooth out the motor fluctuations in patients with short-duration levodopa responses. Initiation and dosage escalation strategies are outlined at the end of this chapter. Typically, doses of these drugs begin very low (subtherapeutic) and then are increased to the maintenance dose over several weeks. A concomitant reduction in the levodopa dosage may be necessary once the maintenance dose of the agonist drug is approached.

The addition of a catechol-*O*-methyltransferase (COMT) inhibitor, such as tolcapone (Tasmar), may also improve control of motor symptoms in patients with the wearing-off phenomenon. COMT is one of the dopamine degradative enzymes, and inhibition of this enzyme increases levodopa potency. Tolcapone is given in an initial dose of 50 to 100 mg two or three times a day, and the dose can be increased up to 200 mg three times a day. This drug increases "on" time but also may increase levodopa dyskinesias, requiring adjustment of the levodopa dose. It may also induce diarrhea and abdominal cramps. Tolcapone is potentially hepatotoxic, and rare cases of fatal fulminant liver failure have been reported (on the order of 1 in 20,000). Frequent assessment of serum liver function tests (aspartate transaminase and alanine transaminase) is strongly advised. Another COMT inhibitor, entacapone (Comtan), which appears to have no substantial hepatotoxicity, recently became available for prescription use. It is administered in a dose of 200 mg with each dose of carbidopa-levodopa (up to eight doses per day).

Patients with very rapid swings in their levodopa response may occasionally benefit from taking levodopa in liquid form. The advantages are the quick onset of the effect, often about 20 minutes, and the possibility of close titration of dosage. The disadvantages are the inconvenience of mixing the liquid formulation (must be done by the patient or caregiver) and the short duration of the effect, which is 60 to 90 minutes. For patients with a very narrow therapeutic window (i.e., little room between doses of levodopa that induce improvement and those that produce dyskinesias), liquid carbidopa-levodopa may allow better hour-to-hour control. Patients must be willing to accept, however, the administration of doses at 60- to 90-minute intervals and the inconvenience of mixing and carrying the liquid formulation. Only a few patients experience sufficient benefit to justify the impracticality of liquid carbidopa-levodopa. The recipe for mixing and administering the liquid preparation is given at the end of this chapter.

Other adjunctive drugs are also occasionally tried for treatment of levodopa-related motor fluctuations. These less potent medications, which include selegiline, amantadine, and anticholinergic agents, are typically of modest benefit, at best, in treating short-duration levodopa responses. A note of caution: occasional patients with fluctuations who have received maintenance therapy with one of these drugs may experience substantial deterioration if administration is discontinued, especially if rapidly terminated. Dosage of these drugs is also addressed at the end of this chapter.

DYSKINESIAS

Treatment of dyskinesias requires determination of the nature of the dyskinesia. Dyskinesias typically take the form of chorea or dystonia and can occur simply on the basis of PD or as the result of antiparkinsonian medications. Choreiform dyskinesias are rapidly flowing, dance-like movements that are unpredictable (i.e., not stereotyped). Dystonic dyskinesias tend to be stereotyped and appear as a sustained contraction of muscles, often producing prolonged posturing. Dystonia affecting the ankle (with inversion at that joint) and toe cramps are particularly common and may occur as presenting features of PD. Dystonia in the absence of chorea is typically a parkinsonian symptom. On the other hand, chorea with or without dystonia is almost always a medication-induced effect. Another clue to whether dyskinesia is a component of the underlying parkinsonism or is medication-induced is when it occurs with respect to the time of levodopa administration. Patients who awaken in the morning with painful dystonia involving the ankle or foot, not having taken any medication, do not have medication-induced dyskinesia. These patients typically experience resolution of dystonia if an early morning dose of levodopa is administered. In contrast, patients with levodopa-induced dyskinesias have a roughly "time-locked" relationship between levodopa administration and development of dyskinesia. In their mildest form, levodopa-induced dyskinesias are nearly imperceptible, perhaps becoming apparent only when the patient concentrates or performs specific voluntary motor activities and quickly disappearing on relaxation. When severe, choreiform dyskinesias can take on ballistic proportions. Prolonged dystonic posturing may be sufficiently pronounced to induce cramp-like pain.

PEAK-DOSE DYSKINESIA

The most common form of levodopa-induced dyskinesia is peak-dose chorea (Fig. 9-1). Occurring at the time of peak levodopa effect, this dyskinesia most likely reflects high concentrations of striatal dopamine. Many patients actually feel their best when their medication dosage is adjusted to the point at which choreiform dyskinesias are present. In fact, choreiform dyskinesias, if mild, may be imperceptible to the patient.

If peak-dose dyskinesia is severe or bothersome, several treatment options are possible. First, the amount of levodopa taken per dose may be reduced. Reducing the amount of levodopa by 25- to 50-mg decrements every 3 days eventually leads to resolution of dyskinesia. However, significant reductions in levodopa doses often result in the development of intolerable parkinsonian signs and symptoms. Patients taking controlled-release carbidopa-levodopa may achieve better control by switching to an immediate-release formulation that affords greater flexibility in timing and size of doses. Patients receiving selegiline may try discontinuing this medication (without making any other changes) to experience reduction in levodopa-induced dyskinesia. Because of the long duration of the selegiline effect (up to several weeks), the drug can be withdrawn abruptly, without tapering.

Motor fluctuations characterized by alternating, disruptive dyskinesias and intolerable "off" states refractory to levodopa adjustment require the addition of another antiparkinsonian medication. A dopamine agonist (bromocriptine, pergolide, pramipexole, or ropinirole) is a good choice at this juncture (Fig. 9-2). Specific comments on initiation and dosage of these medications are found at the conclusion of this chapter.

In general, dopamine agonists have a less robust antiparkinsonian effect than levodopa, but this effect may be longer lasting than short-duration levodopa effects. Dopamine ago-

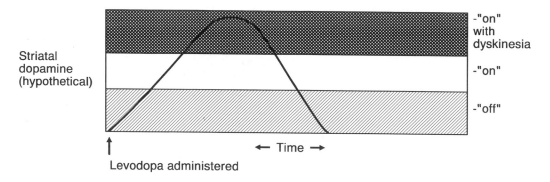

Fig. 9-1. Levodopa-induced peak-dose dyskinesia.

nists are less likely than levodopa to cause dyskinesias but more likely to cause other side effects, including hallucinations, confusion, agitation, and orthostatic hypotension.

Patients may not tolerate adjunctive therapy with a dopamine agonist because of a failure to follow the guidelines illustrated in Figure 9-2. Dopamine agonists should be introduced gradually without initially changing the optimal levodopa regimen (section A in Figure 9-2). Only after dyskinesias become more prominent or other signs of excessive motor effects are appreciated should levodopa doses be reduced by 25- to 50-mg decrements (section B in Figure 9-2). The dosages of levodopa and dopamine agonist can then be stabilized for optimal benefit (good motor function with less severe levodopa-induced dyskinesias).

Common mistakes when adjunctive therapy is started with a dopamine agonist are premature reductions in levodopa intake (causing increasing parkinsonism symptoms), too rapid escalation of the dose of dopamine agonist (causing side effects), and use of an insufficient amount of dopamine agonist (leading to no improvement in symptoms). Guidelines and specific details for individual dopamine agonists are at the end of the chapter.

Occasional patients experience a reduction in dyskinesias with the addition of amantadine (100 mg two or three times daily), an antagonist at glutamate *N*-methyl-D-aspartate (NMDA) receptors. Patients who do not experience substantial improvement of dyskinesias with the approaches suggested above may be good surgical candidates (see Chapter 13).

Peak-dose dyskinesias related to levodopa may be precipitated by initiation of therapy with other antiparkinsonian medications, such as selegiline or COMT inhibitors (including tolcapone and entacapone). In these circumstances, reduction of levodopa doses (by 25- to 50-mg decrements) may lead to resolution of dyskinesias if continued administration of selegiline or a COMT inhibitor is desired.

BIPHASIC DYSKINESIA

Biphasic dyskinesia is an uncommon temporal pattern of dyskinesia that requires a different treatment approach than peak-dose dyskinesia. Patients with this pattern experience a transient dyskinetic state twice during each levodopa cycle: just as the levodopa effect is starting to take effect and again when the levodopa response begins to wear off. The initial dyskinetic phase typically lasts 10 to 30 minutes, and the last dyskinetic period is usually a little longer, up to 90 minutes. In between, a satisfactory motor "on" response is present (Fig. 9-3). Patients displaying this pattern may experience dyskinesias through-

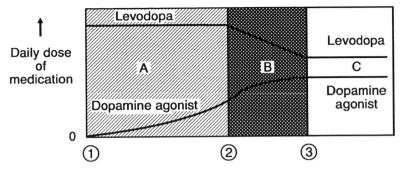

① Initiate dopamine agonist; gradually increase dose

② Begin to lower levodopa by 25- to 50-mg decrements, each dose (e.g., 200 mg t.i.d.➡175 mg t.i.d)

③ Maintain stable doses of levodopa and dopamine agonist

A Motor fluctuations: "off" and levodopa-induced dyskinesia

B Increased levodopa-induced dyskinesia

C Improved "on" time with fewer motor fluctuations

Fig. 9-2. Treatment plan for peak-dose dyskinesia: introduction of a dopamine agonist to a levodopa regimen.

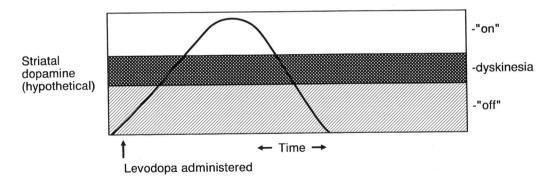

Fig. 9-3. Biphasic dyskinesia. Conceptually, there are separate thresholds for levodopa-induced dyskinesias and the "on" response. When the striatal dopamine level surpasses the "on" threshold, dyskinesias are overridden.

out their levodopa response cycle if their levodopa dose is sufficiently reduced, as may be inferred from Figure 9-3. Biphasic dyskinesia is also termed the "D-I-D response" for "dyskinesia-improvement-dyskinesia."

Some patients have a favorable response to shortening of the dose interval of levodopa therapy with overlapping of dose effects (i.e., dovetailing of dose responses before the effects of levodopa wear off). For the overlapping strategy to be effective, the levodopa doses must be high enough for the effects to pass from the intermediate dyskinetic range into the improvement range, where the dyskinetic tendency is essentially overridden (Fig. 9-3); thus, occasional patients benefit from an increment in dosage. Switching from sustained-release to immediate-release levodopa and carefully readjusting the dosage is also often helpful. Patients with biphasic dyskinesia usually find that dyskinesias are unavoidable at the onset and end of pharmacotherapy for the day. If the aforementioned measures do not improve motor function for most of the day, addition of a dopamine agonist, a COMT inhibitor, or liquid levodopa (with frequent doses) can be tried. In the most refractory cases, high-dose dopamine agonist monotherapy is an alternative. If these measures fail, patients often have resolution of symptoms with surgical therapy in the form of pallidotomy (see Chapter 13).

Freezing

"Freezing" refers to the inability to initiate movement, typically involving hesitancy in arising from a seated position, initiating gait, or turning. Freezing can occur as an "off"-phase phenomenon or reflect insufficient medication. In rare cases, it can occur as a peak-dose effect ("peak-dose freezing"). Delineation of the timing of freezing may help to determine the ideal treatment plan. Peak-dose freezing, usually related to higher doses of dopamine agonist drugs, may be minimized by dose reductions. More aggressive dosing may reduce freezing in patients receiving inadequate levodopa. If freezing is restricted to the levodopa "off" phase, the strategies described earlier in this chapter are appropriate (e.g., shortening the carbidopa-levodopa interdose interval, addition of a dopamine agonist or COMT inhibitor). Patients "trapped" in a frozen "off" state who need a quick response can try dissolving their usual carbidopa-levodopa dose in a glass of carbonated beverage and drinking the entire contents. This will result in a more rapid "on" response (e.g., 20 minutes); however, the "on" state will last only about 60 to 90 minutes. Nonetheless, this rescue therapy buys them time.

Apomorphine is a fast- and short-acting dopamine agonist that is available in sublingual, subcutaneous, and intranasal formulations throughout much of the world (but continues to be an investigational compound in the U.S.). Apomorphine may be helpful by reversing abrupt freezing and "off" states within 5 to 10 minutes of administration.

Frequently, gait freezing does not respond to simple pharmacologic changes. A variety of other measures are sometimes useful in helping patients counteract freezing. For example, many patients can initiate stride by stepping over another person's foot (or some other object) or by merely imagining an object in front of the feet. Patients are often aware that they do not freeze at the bottom of a flight of stairs, so that imagining they are marching in place or walking up stairs may allow stride to begin.

"On-Off" Phenomenon

Some patients treated with levodopa experience unpredictable episodes of an abrupt change from a relatively mobile state to a state with seemingly little or no benefit from medication, the "on-off" phenomenon. These episodes may be dramatic, occasionally brief, and difficult to treat effectively. In some patients, more aggressive levodopa therapy, perhaps with adjunctive medications, can more consistently keep the clinical response in the "on" phase. Some patients identify factors related to abrupt "off" periods, especially intake of dietary protein. In these instances, patients may redistribute dietary protein, usually to a time late in the day, to minimize such "off" periods.

"Rescue therapy" with liquid carbidopa-levodopa, as described above, may be helpful to individuals experiencing rapid transitions to their "off" state. Patients can even mix a carbidopa-levodopa solution beforehand and carry that in a flask in anticipation of need. If used the same day it is mixed and if not exposed to extreme heat, no preservative is necessary. Surgical therapy (medial pallidotomy or deep brain stimulation therapy; see Chapter 13) is also an option for patients with disabling "off" states.

INSOMNIA

Patients with increasing parkinsonian symptoms overnight without continuing doses of medication may experience insomnia as the result of uncomfortable rigidity and immobility. This complaint can be addressed by administration of a bedtime dose of controlled-release levodopa. Since this formulation often takes up to 2 hours to become effective, it can be an advantage in patients whose last evening levodopa dose wears off early in the sleep cycle. If patients awaken in an uncomfortable state with prominent parkinsonian symptoms, use of immediate-release levodopa is appropriate to obtain a quick response. Overnight doses of levodopa may also serve to improve morning functioning. Rarely, bedtime levodopa administration leads to insomnia, primarily when it induces dyskinesias. Other reasons for sleep disruption in PD are discussed in Chapter 10.

Other Issues

COST

With the increasing availability of antiparkinsonian medications comes the propensity for polypharmacy. Many published algorithms suggest an early polypharmaceutical approach, frequently with use of newer and more expensive medications. These suggestions are made without any definitive data to argue, for example, that symptomatically successful monotherapy with levodopa is any less beneficial for patients than combination therapies. The cost of polypharmacy can be prohibitive for many patients, particularly those on fixed incomes. The monthly cost of a regimen containing carbidopa-levodopa, a dopamine agonist, and a monoamine oxidase inhibitor can easily exceed the cost of housing for many patients. As a consequence, many patients cannot afford such regimens or become noncompliant in use of medications in an attempt to cut expenses. At present, no compelling data support the contention that the long-term outcome is any better with combination therapy and minimizing levodopa. Consequently, for patients with economic constraints, it seems prudent to optimize the carbidopa-levodopa regimen before beginning more expensive combination therapy.

COMPLIANCE

Many patients have difficulties with compliance. Multiple dose regimens, with one or more medications, necessarily make compliance more difficult. Some patients have trouble taking carbidopa-levodopa systematically without meals to remind them and on this account often skip doses. It may be worthwhile to try medication doses with meals for persons in whom compliance is a problem. Should the patient note deterioration in the effect of medication, individual doses can be increased or a return made to nonmeal administration. For patients who require carbidopa-levodopa doses at short intervals, for example, every 2 to 3 hours, a pillbox timer may help remind them of their schedule.

DRUG INTERACTIONS

Interactions between medications may also complicate the care of patients with advancing PD. The propensity for combination therapy with levodopa, dopamine agonists, and other antiparkinsonian drugs to lower blood pressure and induce nausea, confusion, or hallucinations must be kept in mind. In the older patient with advanced disease, carbidopa-levodopa alone may be preferable, especially if cognitive and behavioral side effects become a problem.

CLINICAL TRIALS AND RESEARCH

Despite the best attempts, some patients do not benefit sufficiently from the treatment strategies outlined above. These persons may be referred to tertiary care centers, where they may be eligible for therapeutic trials of experimental agents or considered for surgical treatment. Many patients also recognize the potential for participating in research as brain donors to a brain bank registry. Such donations may be invaluable to the advancement of research and accumulation of new information to guide rational treatment in the future. Discussion of these issues is often surprisingly well received by patient and family alike, because they realize their potential role in improving the future for other patients with PD.

CONCLUSION

The levodopa era has now extended for more than 30 years, and the pharmacologic armamentarium for treatment of PD has expanded greatly over the past decade. However, the proposed treatment strategies are becoming increasingly complicated, with levodopa used in conjunction with other agents, including dopamine agonists and COMT and monoamine oxidase inhibitors. Recognition of specific clinical problems and the side effect profiles of the available therapeutic agents are helpful in tailoring therapy for the patient with PD.

NOTES ON SPECIFIC PHARMACOLOGIC THERAPEUTICS

Pharmacologic measures are one means for treating patients with advancing PD. Education, physical therapy (Chapter 14), and psychiatric-psychologic therapy (Chapter 12) are potentially helpful and should be considered in conjunction with pharmacologic treatment. When appropriate, surgery may also be an option (Chapter 13).

The brief discussion that follows outlines specific information on current medications for treatment of advancing PD. The antiparkinsonian pharmacologic armamentarium is extensive and includes the medications listed in Table 9-7. Indications for these medications were discussed earlier in the chapter in reference to specific motor problems in advancing PD.

Anticholinergic Medications

Anticholinergic medications, such as trihexyphenidyl (Artane) and benztropine (Cogentin), generally provide some reduction in tremor, rigidity, and dystonia. They can be used in younger persons with less risk of side effects. In general, it is wise to avoid using anticholinergics in elderly patients. Adverse reactions such as confusion, hallucinations,

Table 9-7

Antiparkinsonian Medications for Advancing Disease

Class of drug	Specific examples	Comments
Levodopa	Carbidopa-levodopa (Sinemet, Sinemet CR)	Best symptomatic therapy; can be dissolved to make liquid form
Dopamine agonist	Bromocriptine (Parlodel), pergolide (Permax), pramipexole (Mirapex), ropinirole (Requip)	Less potent than levodopa; longer duration of effects, which allows smoothing of levodopa fluctuations
Catechol-O-methyltransferase (COMT) inhibitor	Tolcapone (Tasmar), entacapone (Comtan)	Maintains plasma levodopa concentration
Monoamine oxidase (MAO) inhibitor	Selegiline (deprenyl, Eldepryl)	Selective MAO type B inhibition
Anticholinergic	Trihexyphenidyl (Artane), benztropine (Cogentin)	Be wary of central nervous system side effects (hallucinations, confusion)
Anti-influenza drug	Amantadine (Symmetrel)	N-methyl-D-aspartate (NMDA) glutamate receptor antagonist

dry mouth, urinary retention, and constipation occur fairly frequently and require diligent surveillance.

Administration of trihexyphenidyl is typically started at a dose of 1 mg per day or twice a day, and the amount can be increased every few days, by 1-mg increments, to 2 to 4 mg three times a day, as required and tolerated. Occasionally, trihexyphenidyl is helpful for reducing intolerable drooling. Scopolamine patches and amitriptyline can be used for the same purpose.

Amantadine

This agent was introduced for symptomatic treatment in the late 1960s, roughly the same time that levodopa became available. Its effect on symptoms was not as great as that of levodopa, and the drug was consequently overshadowed. Recently, it has been recognized as helpful in reducing levodopa-induced dyskinesias. Amantadine is usually well tolerated by patients and produces a mild but long-term benefit (typical dose, 100 mg two or three times a day). Side effects include ankle edema and livedo reticularis (skin mottling); these clear with discontinuation of the drug.

Carbidopa-Levodopa

This is the best agent for symptomatic relief. All patients with advancing parkinsonism should have an aggressive trial with levodopa. With respect to the short-duration levodopa response, there is a dose-related effect up to about 300 mg in single doses (with carbidopa in the immediate-release formulation). Higher single doses are not expected to result in better control of symptoms, although there are rare patients who are best treated with slightly higher doses. This corresponds to about 500 to 600 mg of levodopa in the controlled-release formulation. The "ceiling" total daily dose for levodopa is whatever is necessary for optimal control of symptoms without troublesome side effects.

Controlled-release preparations may be given less frequently and may reduce motor fluctuations. Absorption of levodopa from the controlled-release tablet is approximately 80% of that from immediate-release tablets. Because of this difference and because levodopa from the controlled-release formulation is released more slowly, comparable peak dose effects require an individual dose of the controlled-release preparation that is 30% to 50% higher than that of immediate-release levodopa. For example, if one is switching from one and one-half 25/100 immediate-release tablets each dose, a controlled-release dose of 50/200 or perhaps even a little higher is required for a similar peak effect.

Splitting controlled-release tablets in half partially preserves the sustained-release property, but peak levodopa concentrations are hastened somewhat. If a patient chews or crushes this type of tablet, the sustained-release properties are lost and the patient is subsequently taking the equivalent of regular tablets. This characteristic can be used to advantage if, for example, the patient forgets to take a dose and requires prompt amelioration of symptoms; crushing and taking half of a controlled-release 50/200 tablet is essentially the same as taking a single 25/100 immediate-release tablet.

In Europe, levodopa is formulated with benserazide, another dopa decarboxylase inhibitor, which has a pharmacology that is essentially identical to carbidopa. Benserazide-levodopa (Madopar) is also formulated as a sustained-release preparation (Madopar HBS [hydrodynamically balanced system]), which behaves pharmacologically like controlled-release carbidopa-levodopa.

In summary, when changing from immediate-release to controlled-release levodopa preparations, patients need to take individual levodopa doses that are 30% to 50% greater and spaced at dosage intervals that are 60 to 90 minutes longer than those with the immediate-release regimen.

Liquid Levodopa

Occasional patients benefit from long-term treatment with liquid levodopa. These are patients with marked motor fluctuations and short levodopa responses (e.g., 2 hours or less). The advantages are better potential for dosage titration and quicker responses (takes effect in about 20 minutes). The disadvantages are the brevity of the "on" responses (60 to 90 minutes) and the inconvenience of mixing and carrying the solutions.

For such long-term treatment, dosage is typically done by volume of the carbidopa-levodopa solution. If five immediate-release 25/100 carbidopa-levodopa tablets are dissolved in 500 mL of liquid (juice, water, etc.), each millimeter corresponds to 1 mg of levodopa. Patients switching from, for example, two 25/100 carbidopa-levodopa tablets every 2 hours can take 200 mL of this solution at 60- to 90-minute intervals instead. Subsequently, they can make small adjustments by volume to optimize the motor response. If patients anticipate storing this solution for more than 1 day (up to about 3 days total), they can add ascorbic acid (vitamin C) in a concentration of 2 mg/mL and refrigerate the solution overnight.

As already discussed, liquid levodopa can be used solely as rescue therapy for sudden incapacitating motor "off" states. Although a premixed solution, as described above, can be used, many patients find it easier to simply dissolve their tablets in some liquid and drink the entire contents. They need to understand, however, that even though they are ingesting their usual carbidopa-levodopa dose, the "on" response will last only 60 to 90 minutes. Hence, they will need to take another dose in tablet form perhaps 30 to 60 minutes later.

Dopamine Agonists

Four dopamine agonists are currently available for prescription use. These medications have different profiles of dopamine receptor agonism, but at present, little in the way of clinical data differentiates one from another. All four of these drugs are conventionally administered three times daily (with meals), although bromocriptine and pergolide are initiated as a single daily dose before being built up to three times daily. Typical starting doses for these agents follow.

Bromocriptine
    Initiation: 1.25 mg once daily
    Escalation: Increase to 2.5 mg 3 times daily over 2-3 weeks; then add 2.5 mg to the daily
        dose every 3-7 days
    Typical daily maintenance doses: 15-40 mg
    Available tablet sizes: 2.5 mg (tablet), 5 mg (capsule)
Pergolide
    Initiation: 0.05 mg once daily
    Escalation: Increase to 0.25 mg 3 times daily over 2 weeks; then add 0.25 mg to the total
        daily dose every third day
    Typical daily maintenance doses: 1.5-6 mg
    Available tablet sizes: 0.05, 0.25, 1 mg

Pramipexole
    Initiation: 0.125 mg 3 times daily, maintained for 1 week
    Escalation: Double the dose weekly for the next 2 weeks (to 0.5 mg 3 times daily); then increase weekly by adding 0.75 mg to the daily dose
    Typical daily maintenance doses: 3-6 mg
    Available tablet sizes: 0.125, 0.25, 0.5, 1, 1.5 mg
Ropinirole
    Initiation: 0.25 mg 3 times daily, maintained for 1 week
    Escalation: Add 0.75 mg to the daily dose each week until up to 1 mg 3 times daily; then can increase each week by adding 0.75 to 1.5 mg to the daily dose
    Typical daily maintenance doses: 4.5-24 mg
    Available tablet sizes: 0.25, 0.5, 1, 2, 5 mg

The companies supplying the drugs have starter sample kits available that cover the first 2 to 3 weeks for initiation of pergolide, pramipexole, and ropinirole.

Adverse effects from dopamine agonists include cognitive-behavioral abnormalities, such as hallucinations, confusion, and agitation. These are typically dose-related and resolve with medication reduction or discontinuation. Orthostatic hypotension and nausea occur in some patients. Occasionally, vasoconstrictive side effects occur with use of the ergot dopamine agonists bromocriptine and pergolide. Rarely, inflammatory-fibrotic reactions affecting the pleuropulmonary and pericardial tissues develop in patients treated with these ergot preparations. Daytime sleep attacks affecting automobile driving have been described among patients taking pramipexole or ropinirole.

Selegiline

Selegiline (deprenyl, Eldepryl) is a selective monoamine oxidase type B inhibitor that was introduced as an adjunctive agent to levodopa and was hoped would have neuroprotective properties. That is, this drug, unlike the previously mentioned medications for symptoms, was proposed to arrest or slow the natural progression of PD, as discussed in Chapter 7. Unfortunately, recent studies have failed to provide convincing evidence of any neuroprotective effect. Controversy continues to surround the drug's use. It is administered in doses of 5 mg once or twice daily (taken early in the day to avoid insomnia). There is agreement that the drug has mild but measurable symptomatic benefit in parkinsonism, primarily when used as an adjunct to levodopa therapy. If one discontinues the use of selegiline (a medication with significant expense), the dosages of other antiparkinsonian medications may have to be increased to make up for the loss of symptomatic benefit, since selegiline potentiates the effect of levodopa. The monoamine oxidase type B inhibition in the brain from selegiline therapy lasts for many weeks, and the symptomatic deterioration consequently may be delayed. However, because of this long-lasting effect, selegiline can be withdrawn abruptly without significant ill effect.

Catechol-O-Methyltransferase Inhibitors

Tolcapone and entacapone are selective COMT inhibitors that may be helpful adjuncts to levodopa in the treatment of PD. These drugs can increase "on" time by reducing the wearing-off phenomenon. Because they tend to exacerbate levodopa-induced dyskinesias, levodopa dose reductions may be necessary. The other most common side effects of COMT inhibitors are diarrhea and abdominal cramps and pain. Tolcapone can be administered in doses of 100 or 200 mg three times a day; some patients benefit from as little as 50 mg twice a day as an initial dose. As discussed, because of hepatotoxicity, frequent assessment of serum liver function tests is strongly advised. Entacapone is conventionally administered in a dose of 200 mg with each carbidopa-levodopa dose, up to eight doses daily. To date, entacapone has not been associated with clinically significant hepatotoxicity.

SELECTED READING

Ahlskog JE. Medical treatment of later-stage motor problems of Parkinson disease. Mayo Clin Proc 1999; 74:1239-1254.

Ahlskog JE, Muenter MD, McManis PG, Bell GN, Bailey PA. Controlled-release Sinemet (CR-4): a double-blind crossover study in patients with fluctuating Parkinson's disease. Mayo Clin Proc 1988; 63: 876-886.

Lieberman A, Ranhosky A, Korts D. Clinical evaluation of pramipexole in advanced Parkinson's disease: results of a double-blind, placebo-controlled, parallel-group study. Neurology 1997; 49: 162-168.

Olanow CW, Koller WC. An algorithm (decision tree) for the management of Parkinson's disease: treatment guidelines. American Academy of Neurology. Neurology 1998; 50 (Suppl 3): S1-57.

Parkinson Study Group. Entacapone improves motor fluctuations in levodopa-treated Parkinson's disease patients. Ann Neurol 1997; 42: 747-755.

Rajput AH, Martin W, Saint-Hilaire MH, Dorflinger E, Pedder S. Tolcapone improves motor function in parkinsonian patients with the "wearing-off" phenomenon: a double-blind, placebo-controlled, multicenter trial. Neurology 1997; 49: 1066-1071.

Rascol O, Lees AJ, Senard JM, Pirtosek Z, Montastruc JL, Fuell D. Ropinirole in the treatment of levodopa-induced motor fluctuations in patients with Parkinson's disease. Clin Neuropharmacol 1996; 19: 234-245.

Uitti RJ, Ahlskog JE. Comparative review of dopamine receptor agonists in Parkinson's disease. CNS Drugs 1996; 5: 369-388.

Uitti RJ, Ahlskog JE, Maraganore DM, Muenter MD, Atkinson EJ, Cha RH, O'Brien PC. Levodopa therapy and survival in idiopathic Parkinson's disease: Olmsted County project. Neurology 1993; 43: 1918-1926.

Uitti RJ, Rajput AH, Ahlskog JE, Offord KP, Schroeder DR, Ho MM, Prasad M, Rajput A, Basran P. Amantadine treatment is an independent predictor of improved survival in Parkinson's disease. Neurology 1996; 46: 1551-1556.

10 Sleep and Parkinson's Disease

Cynthia L. Comella, MD, ABSM

CONTENTS

To sleep—perchance to dream: ay, there's the rub,
For in that sleep of death what dreams may come
When we have shuffled off this mortal coil,
Must give us pause.
—Shakespeare, *Hamlet*

The motor problems of Parkinson's disease (PD) typically take center stage during visits to the physician's office. Another factor contributing to the disruption of the lives of patients (and families) is sleep disturbance, which affects 75% to 95% of patients. Patients and caregivers frequently do not realize the importance of reporting sleep disturbances to their physicians unless the disturbance is severe. Without particular attention to the amount and quality of sleep, many patients with PD may unnecessarily endure potentially treatable sleep disorders for prolonged periods. An increased awareness and more aggressive treatment of the nighttime problems endured by patients with PD promote a better quality of life for both patients and caregivers.

From: *Parkinson's Disease and Movement Disorders:*
Diagnosis and Treatment Guidelines for the Practicing Physician
Edited by: C. H. Adler and J. E. Ahlskog © Mayo Foundation
for Medical Education and Research, Rochester, MN

Table 10-1
Sleep Disturbances in Parkinson's Disease

Source	Effects
Parkinsonian symptoms	Insomnia and fragmentation due to akathisia, tremor, stiffness, rigidity, pain, difficulty turning over in bed
Medications	Sleep fragmentation, nightmares, insomnia, hallucinations, daytime sleepiness
Associated conditions	Depression: insomnia Dementia: sleep fragmentation, nightmares, hallucinations, excessive somnolence
Conditions possibly associated	Sleep apnea, rapid eye movement sleep behavior disorder

Sleep disturbance related to PD can arise from different sources. Some of the frequently occurring causes are listed in Table 10-1. As the disease progresses, the frequency and severity of associated sleep disturbances tend to increase. In patients with advanced PD, total sleep time is less than in normal elderly controls, with reductions in stages 3 and 4 sleep and frequent nocturnal arousals. Rapid eye movement (REM) sleep is also affected. Some investigators report that the total duration of REM sleep is decreased, but the percentage of REM time does not differ from that of controls. Untreated patients with PD also have abnormal movements during sleep, with increased blinking, blepharospasm, and tremor. These movements occur during any sleep stage but are more common in the lighter stages.

FRAGMENTATION AND DELAYED ONSET OF SLEEP

A common complaint of patients with PD is fragmented sleep, characterized by frequent nocturnal awakenings. If severe, fragmented sleep leads to a reduction in total night sleep time, daytime fatigue, and excessive daytime sleepiness. Patients "nod off" during daytime sedentary activities and complain that although they may fall asleep quickly, they awaken frequently during the night. These nocturnal awakenings may be prolonged. Patients may initially have difficulty explaining why they awaken, and often they report the need to urinate. A directed interview by the physician can clarify whether it is indeed a urologic problem or whether when awakened, the patient has tremor, stiffness, or difficulty rolling over in bed, which leads to discomfort and awakening. Some patients may experience feelings of restlessness and a need to physically move, consistent with nocturnal akathisia.

A survey study in the United Kingdom of 220 patients with PD found that 215 of those interviewed reported nocturnal sleep disruption; 76% reported that their sleep was disrupted during the night by frequent awakenings. The most frequent disturbance during the night was the need to urinate, with most reporting at least two nightly visits to the bathroom, and one-third of the patients got up at least three times a night. The most frequent motor symptom experienced during the night was the inability to turn over in bed. Painful nocturnal leg cramps associated with PD occurred in 55% of the patients. Nocturnal occurrences of tremor and foot dystonia were also common complaints. These types of sleep complaints are related to the symptoms of PD and tend to occur more frequently as PD progresses.

INSOMNIA SECONDARY TO UROLOGIC SYMPTOMS

In both men and women, age-related urologic abnormalities, including prostatic enlargement in men, may be responsible for urinary frequency and nocturia. Autonomic dysfunction secondary to PD may also contribute. Treatment depends on the type of urinary disorder, and consultation with a urologist is pivotal in determining the approach. Several agents are available for treatment of either hypotonic or hypertonic bladder dysfunction. Peripherally acting cholinergic agents may alleviate symptoms of the hypotonic bladder. Conversely, peripherally acting anticholinergic agents may relieve the urinary symptoms of a hypertonic bladder (see Chapter 11 for medications and doses). Proper diagnosis is paramount in making the correct choice of treatment. One caveat, however: Some patients with PD perceive that it is nocturia that disrupts their sleep when in fact fragmented and light sleep due to parkinsonism leads to frequent awakenings, which then allow urinary urgency to enter conscious awareness. Nocturnal control of parkinsonism may improve sleep in these patients, independent of treatment of urinary symptoms.

INSOMNIA RELATED TO PARKINSONISM

Insomnia due to parkinsonism per se is common. In more advanced, treated PD, the effect of daytime medications wanes in the evening and during the night with a consequent increase in muscle rigidity, slowness, stiffness, and akathisia. Nighttime PD symptoms can be treated by appropriate choice and timing of levodopa preparations (Table 10-2). First, difficulty initiating sleep due to parkinsonism should respond to an evening dose of carbidopa-levodopa, timed so that it will be in effect at bedtime. Second, for those with later awakening due to parkinsonism, controlled-release carbidopa-levodopa (Sinemet CR) administered at the last minute before bedtime may allow sustained sleep. This formulation has a delayed onset of effect, often taking up to 2 hours to become active. This is an advantage at bedtime, with the antiparkinsonian effect beginning around the time that the effect of the evening dose is wearing off. Also, the longer duration of response with the controlled-release formulation may be adequate to cover most of the night. If the effect from the bedtime controlled-release dose is insufficient to last the entire night, a dose of immediate-release carbidopa-levodopa (strategically placed on the nightstand with a glass of water) should be adequate to continue the effect for the rest of the night. Immediate-release carbidopa-levodopa is preferred for middle-of-the-night administration because of its more rapid onset of effect.

Evening and nighttime doses of carbidopa-levodopa usually must be similar to the daytime dose for the optimal sleep-sustaining effect. One must bear in mind that patients adding a dose of controlled-release carbidopa-levodopa at bedtime require 30% to 50% more levodopa to equate with immediate-release carbidopa-levodopa.

Some patients awaken only occasionally during the night because of parkinsonism. Routine bedtime controlled-release carbidopa-levodopa may not be necessary, whereas a nightstand dose of the immediate-release formulation can be made available for use as needed.

Table 10-2
Levodopa Treatment Strategies for Insomnia Secondary to Parkinsonism

Condition	Carbidopa-levodopa treatment[a]	Time of dose	Comment
Initial insomnia (unable to fall asleep)	Add evening dose of immediate-release or controlled-release	At least 1 hour before bedtime	Up to 1 hour to take effect
		At least 2 hours before bedtime	Up to 2 hours to take effect
Frequent awakening 2 to 5 hours after falling asleep	Controlled-release	Bedtime	Delayed effect advantageous
Later awakening, middle of night or several hours after bedtime dose of controlled-release carbidopa-levodopa	Immediate-release	As needed, on awakening	Need quick effect (i.e., not controlled-release)

[a]Doses should be approximately equivalent to effective daytime doses.

154

Recently, adjunctive treatment with tolcapone (Tasmar) at bedtime was proposed to sustain sleep in patients with PD. This drug prolongs the duration of action of levodopa by preventing enzymatic degradation of levodopa by catechol-*O*-methyltransferase. Tolcapone is effective only in patients receiving levodopa. Some patients may require a reduction in their evening and nighttime levodopa therapy because of excessive responses with the addition of tolcapone. Patients maintained on tolcapone require monitoring of serum liver function tests, as described in Chapter 9.

In contrast to the levodopa preparations, the direct-acting dopaminergic agonists have not been adequately evaluated for usefulness in the treatment of nocturnal PD symptoms. Some patients report that these agents are more likely to exacerbate insomnia than to relieve it.

ADVERSE EFFECTS OF DOPAMINERGIC AGENTS ON SLEEP ARCHITECTURE

Although nocturnal use of levodopa improves sleep in many patients with PD, it is known to adversely affect sleep architecture. Parenteral infusions of dopaminergic agents have been shown to delay the onset of REM sleep or acutely inhibit a continuing REM period. With long-term dopaminergic treatment, some investigators observed that REM sleep continues to be suppressed and that discontinuing dopaminergic therapy results in a REM rebound lasting up to 10 days in some patients. Others reported that REM is suppressed in the period immediately after the initiation of levodopa therapy but that REM is restored to baseline levels with long-term treatment. And yet another group reported that REM time actually increases with prolonged therapy. In many reports, the clinical staging of the patients and their motor response to dopaminergic agents are not indicated. The clinical significance of these findings has yet to be fully elucidated.

INSOMNIA SECONDARY TO OTHER FACTORS

As discussed, sleep onset insomnia may be due to parkinsonism. However, other causes, such as medications, concurrent restless legs syndrome, depression, and anxiety, may also be responsible in some patients.

Of the medications used for the treatment of PD, selegiline is the one most likely to result in sleep onset insomnia. Selegiline is metabolized into l-amphetamines and may delay sleep onset if taken in the late afternoon or evening. Anticholinergic agents, dopamine agonists, and amantadine may also cause an alerting effect. Withdrawal from benzodiazepines and other sedatives may induce rebound insomnia unless tapering is slow. Certain types of antidepressant medications, especially the serotonin reuptake inhibitors, are usually best avoided in the evening hours (although occasional patients find these sedating).

Restless legs syndrome (RLS) is marked by restless feelings or dysesthesias, often localized to calf muscles, that occur at rest in the evening and diminish with walking. RLS has a circadian occurrence, with most patients reporting symptoms in the evening hours before bed. RLS and periodic limb movement disorder are associated. The latter occurs primarily during non-REM sleep and is marked by an intermittent involuntary rhythmic movement of the legs that can result in arousals and awakenings if severe (see Chapter 30). In patients with RLS without PD, small doses of a dopamine agonist drug, such as pergolide, pramipexole, or ropinirole, may provide relief. Although usually occurring as an idiopathic disorder, RLS can occur in association with PD. The direct-acting dopamine receptor agonists mentioned above may be tried in patients with co-occurring PD and RLS, with use of the dosage strategy suggested in Chapter 30. This strategy, however, may not be effective in patients with PD who are taking levodopa, because the overriding effect of the more potent levodopa may upregulate the dopamine receptors, making them insufficiently responsive to the less potent dopamine agonist drugs. Hence, RLS symptoms in patients with PD may require administration of frequent doses of levodopa to initiate and sustain sleep, perhaps with a strategy similar to that summarized in Table 10-2. As discussed in Chapter 30, the disadvantage of levodopa therapy for RLS is the potential for rebound (RLS symptoms occurring at the end of the effect of a single levodopa dose) and augmentation (RLS symptoms occurring earlier in the evening).

An additional cause of fragmented sleep with delayed sleep onset is depression. Depression is estimated to affect approximately 40% of patients with PD. In patients who do not have PD, depression often causes an associated sleep disturbance. An early REM period is one of the characteristic polysomnographic features. Treatment of depression can relieve sleep disturbance in patients with and without PD alike. The tricyclic antidepressants (e.g., amitriptyline and nortriptyline), with their sedating side effects, may be very beneficial when administered in small doses in the evening hours. These agents may, however, exacerbate RLS symptoms and increase periodic limb movement disorder. In addition, in predisposed patients, these antidepressants may initiate or worsen nighttime hallucinations.

In patients with PD who have situational anxiety or acute insomnia with no other contributing factor, short-term use of zolpidem (Ambien) immediately at bedtime may provide relief, often without causing rebound insomnia after discontinuation. Melatonin in small doses (less than 3 mg) 1 to 2 hours before bedtime may be beneficial, although specific deficits in this hormone have not been demonstrated in PD.

SLEEP APNEA

Sleep apnea has been inconsistently associated with idiopathic PD. Patients who are excessively sleepy during the day or who have periods of apnea, loud snoring, or gasping and choking during sleep should have polysomnography for evaluation of sleep apnea. Sleep apnea can cause nocturnal hypoxia, cardiac arrhythmias, and sleep fragmentation. Patients with significant sleep apnea may present solely with excessive daytime sleepiness. Sleep apnea may be central (airflow ceases because respiratory muscles are not activated), obstructive (airflow is reduced or stopped despite respiratory muscle effort), or mixed (containing elements of both). Obstructive sleep apnea is most effectively treated by continuous positive airway pressure. Central sleep apnea is more difficult to treat and may require nocturnal administration of oxygen or even tracheostomy at night.

Multiple system atrophy is a neurodegenerative disorder marked by parkinsonism, autonomic instability, and cerebellar abnormalities (see Chapter 17). Patients with multiple system atrophy may have life-threatening sleep apnea with glottic closure resulting in nocturnal stridor. This variant of obstructive sleep apnea may require tracheostomy if continuous positive airway pressure proves ineffective. Recognition is critical, because patients with multiple system atrophy, if untreated, may experience sudden death during sleep.

RAPID EYE MOVEMENT SLEEP BEHAVIOR DISORDER

REM sleep behavior disorder (RBD) has particular relevance to PD and parkinsonian syndromes. Although dreams may occur during other stages of sleep, the dreams during normal REM sleep are vivid and more internally organized. Normal REM sleep is marked by cortical desynchronization similar to that in the waking state, rapid eye movements, and cardiorespiratory irregularities. Normal REM sleep is also characterized by skeletal muscle atonia (except for muscle twitches), which prevents "acting out" one's dreams. The anatomical areas involved in the generation and maintenance of REM sleep lie within the brain stem. In animal models, lesions in these brain stem areas in the pons result in motor behaviors during REM sleep. In humans, the occurrence of RBD is likewise thought to reflect damage to these REM-related areas.

RBD was first described in 1985 by Schenck and colleagues, who observed abnormal nocturnal behaviors occurring mostly in elderly men who were, in effect, acting out their dreams. With greater recognition came diagnostic criteria for RBD, as defined by the International Classification of Sleep Disorders, which include movement of limbs or body associated with dream mentation and at least one of the following: potentially harmful sleep behaviors, dreams that appear to be "acted out," and sleep behaviors that disrupt sleep continuity. The dreams recalled during episodes of RBD are usually aggressive or violent. Polysomnography, although not required for clinical diagnosis, demonstrates excessive chin muscle tone and limb jerking during REM sleep. Occasionally, complex, vigorous, and sometimes violent behaviors also occur.

Episodes of RBD tend to be sporadic. A flurry of episodes every night may be followed by a period of quiescence. In a patient with RBD, dreams of being chased, threatened, or trapped lead to such behaviors as punching, choking, kicking, or leaping out of bed. Not only these patients but also the bed partners are susceptible to injury. Furthermore, the sudden and violent nocturnal outbursts in a person who may be quiet and mild while awake may frighten the family. Usually, only specific questioning by the physician reveals these nighttime episodes. The patient seldom remembers the episodes, and the caregiver may believe that the episodes are not part of a neurologic syndrome.

Recently, RBD was recognized as common among patients with PD, appearing to be a component of the parkinsonian neurodegenerative process. In fact, RBD symptoms may precede those of PD by several years. RBD also occurs frequently in other parkinsonian syndromes, such as multiple system atrophy and diffuse Lewy body disease.

RBD is treatable in most patients. Small doses of clonazepam, such as 0.25 mg, may reduce the RBD symptoms. Higher doses, such as 0.5 to 1 mg at bedtime, can be tried if lower doses fail. Excessive sedation and clumsiness are potential side effects. Patients refractory to clonazepam therapy should have their sleeping arrangements appropriately modified to prevent falls out of bed and injury to the sleep partner.

SLEEP DISTURBANCE AND HALLUCINATIONS SYNDROME

In as many as 33% of medically treated patients with PD, visual hallucinations are induced by their dopaminergic agents. Factors reported to be associated with the occurrence of hallucinations include duration of levodopa therapy, presence of dementia, use of anticholinergic drugs or amantadine, age at onset of PD, duration of PD, and abnormal Minnesota Multiphasic Personality Inventory scores. Nausieda and associates observed that patients with PD who had dopaminergic-induced hallucinations had preexisting sleep complaints, particularly awakenings during the night. These investigators postulated that sleep disturbance and hallucinations are on a continuum, with sleep disruption as the initial event.

Nocturnal hallucinations range from inconsequential to severe, frightening, and disruptive of the nocturnal life of the entire family. Before assuming, however, that all bizarre nocturnal behavior represents hallucinations or psychosis, one must consider whether the behavior is actually that of RBD, which requires a different treatment strategy. A patient acting out dreams may appear to the bed partner to be hallucinating. On closer scrutiny, however, the patient may not be truly awake (i.e., dreaming). Uncertainty may require polysomnography for resolution.

The treatment of nocturnal hallucinations begins with attention to medications. Discontinuing administration of drugs with central effects not essential to the patient's well being is the first approach. Those to consider for elimination are anticholinergic agents, anxiolytics, centrally active pain medications (including codeine compounds), antidepressants, and other drugs that can be withdrawn safely. Providing a night light to increase visual stimulation during the night may be useful. Reducing the dopamine drug dose and avoiding evening doses may be beneficial but may also result in a worsening of motor symptoms. Shifting from the direct dopamine agonists to shorter acting levodopa preparations may also be helpful. If a reduction in the dopaminergic drugs results in an exacerbation of PD symptoms, the atypical dopamine receptor antagonists may be considered. Clozapine has been shown to be effective for hallucinations in PD without causing a deterioration in PD motor symptoms. The major adverse effect from clozapine is the development of agranulocytosis. Monitoring of this potentially life-threatening side effect requires weekly blood cell counts for the first 6 months of therapy. If the patient's condition is stable after 6 months, blood tests are continued at biweekly intervals. Recent experience with quetiapine (Seroquel) suggests it may similarly have substantial antihallucinatory efficacy without exacerbating parkinsonism. Olanzapine (Zyprexa) may also be used for this purpose, although it may exacerbate parkinsonism in occasional patients. Neither quetiapine nor olanzapine requires monitoring of blood cell counts. Further details on antipsychotic agents can be found in Chapter 12.

DAY-NIGHT REVERSAL

Excess daytime napping followed by nocturnal wakefulness is a frequent complaint of patients or their caregivers. In addition to the specific sleep disorders described above, other factors may promote this "reversal" of the normal day-night rhythms. Sedentary patients with PD who largely remain indoors during the day may doze frequently. This tendency may be exacerbated by antiparkinsonian and other medications that cause sleepiness after a dose. These intermittent periods of sleep, especially if they occur during the afternoon and evening, may impede the ability to fall asleep at night.

Treatment of day-night reversal relies on the restoration of normal sleep patterns. Good sleep hygiene includes an established bedtime and wake-up time in addition to exposure to adequate light during the day and darkness at night. Indoor lighting does not have sufficient lux to promote normal circadian rhythm. Exposure to sunlight or its equivalent during the day is needed. This can be accomplished by frequent trips outside, keeping window shades open during the day, or exposure to a light source specifically designed to provide the needed light strength. Physical exercise appropriate to the patient's level of functioning is essential to maintaining flexibility and promotes daytime wakefulness. Strenuous exercise, however, should be avoided for 3 to 4 hours before sleep. Structured daytime activities, preferably out of the home, can promote daytime wakefulness. A hot bath about an hour before bedtime may be relaxing and reduce the delay in sleep onset by increasing body temperature and allowing the patient to fall asleep during the cooling phase. A light bedtime snack and avoiding fluid intake in the evening may alleviate nighttime hunger and reduce nocturnal urination. Relaxation techniques may also be useful in reducing nighttime stress and muscle tension.

Sleepiness after antiparkinsonian medications is sometimes difficult to control. Reducing the dose may lessen postdrug sleepiness but may also lead to an unacceptable increase in parkinsonian symptoms. If the patient is receiving a direct dopamine agonist, changing to another agonist may occasionally provide relief. A few patients have reported the sudden onset of an irresistible urge to sleep associated with pramipexole. This is alleviated by withdrawing the drug or substituting another available agonist.

The administration of daytime stimulant medications is reserved for patients who are unresponsive to other medication adjustments. The amphetamine metabolites of selegiline may increase alertness. Small doses of methylphenidate (Ritalin) administered during the day may be helpful but can also disrupt nighttime sleep. A recently approved wakefulness-promoting agent, modafinil (Provigil; 200 mg administered once daily in the morning), is an alternative to methylphenidate.

SLEEP BENEFIT IN PARKINSON'S DISEASE

The preceding sections have dealt with sleep disruption. Of interest, however, is the effect of a good night's sleep on parkinsonian motor symptoms. Factor and colleagues found that 43.6% of 78 patients with PD reported that morning was the "best time of day" for motor performance. Such patients could delay their first dose of medications because of this sleep benefit. Other surveys, with perhaps more advanced patients, found a lower percentage of patients with sleep benefit. In general, however, this effect lasts 2 to 3 hours after nocturnal sleep. The factors involved in sleep benefit remain to be elucidated.

SELECTED READING

Apps MC, Sheaff PC, Ingram DA, Kennard C, Empey DW. Respiration and sleep in Parkinson's disease. *J Neurol Neurosurg Psychiatry* 1985; 48: 1240-1245.

Askenasy JJ, Yahr MD. Reversal of sleep disturbance in Parkinson's disease by antiparkinsonian therapy: a preliminary study. *Neurology* 1985; 35: 527-532.

Carter J, Carroll S, Lannon MC, Vetere-Overfield B, Barter R. Sleep disruption in untreated Parkinson's disease. *Neurology* 1990; 40 Suppl 1: 220.

Chokroverty S. An approach to a patient with sleep complaints, in *Sleep Disorders Medicine: Basic Science, Technical Considerations, and Clinical Aspects* (Chokroverty S ed), Butterworth-Heinemann, Boston, 1995, pp. 181-186.

Comella CL, Nardine TM, Diederich NJ, Stebbins GT. Sleep-related violence, injury, and REM sleep behavior disorder in Parkinson's disease. *Neurology* 1998; 51: 526-529.

Comella CL, Ristanovic R, Goetz CG. Parkinson's disease patients with and without REM behavior disorder (RBD): a polysomnographic and clinical comparison (abstract). *Neurology* 1993; 43 Suppl: 561p.

Comella CL, Tanner CM, Ristanovic RK. Polysomnographic sleep measures in Parkinson's disease patients with treatment-induced hallucinations. *Ann Neurol* 1993; 34: 710-714.

Factor SA, McAlarney T, Sanchez-Ramos JR, Weiner WJ. Sleep disorders and sleep effect in Parkinson's disease. *Mov Disord* 1990; 5: 280-285.

Ferini-Strambi L, Franceschi M, Pinto P, Zucconi M, Smirne S. Respiration and heart rate variability during sleep in untreated Parkinson patients. *Gerontology* 1992; 38: 92-98.

Hogl BE, Gomez-Arevalo G, Garcia S, Scipioni O, Rubio M, Blanco M, Gershanik OS. A clinical, pharmacologic, and polysomnographic study of sleep benefit in Parkinson's disease. *Neurology* 1998; 50: 1332-1339.

Munschauer FE, Loh L, Bannister R, Newsom-Davis J. Abnormal respiration and sudden death during sleep in multiple system atrophy with autonomic failure. *Neurology* 1990; 40: 677-679.

Nausieda PA, Weiner WJ, Kaplan LR, Weber S, Klawans HL. Sleep disruption in the course of chronic levodopa therapy: an early feature of the levodopa psychosis. *Clin Neuropharmacol* 1982; 5: 183-194.

Schenck CH, Bundlie SR, Mahowald MW. Delayed emergence of a parkinsonian disorder in 38% of 29 older men initially diagnosed with idiopathic rapid eye movement sleep behaviour disorder. *Neurology* 1996; 46: 388-393.

Schenck CH, Mahowald MW. Polysomnographic, neurologic, psychiatric, and clinical outcome report on 70 consecutive cases with REM sleep behavior disorder (RBD): sustained clonazepam efficacy in 89.5% of 57 treated patients. *Cleve Clin J Med* 1990; 57 Suppl: S9-S23.

Silber MH, Ahlskog JE. REM sleep behavior disorder in parkinsonian syndromes. *Sleep Res* 1992; 21: 313.

Silber MH, Ahlskog JE. REM sleep behavior disorder and Parkinson's disease (abstract). *Neurology* 1993; 43 Suppl: A338.

van Hilten B, Hoff JI, Middelkoop HA, van der Velde EA, Kerkhof GA, Wauquier A, Kamphuisen HA, Roos RA. Sleep disruption in Parkinson's disease. Assessment by continuous activity monitoring. *Arch Neurol* 1994; 51: 922-928.

11

Autonomic Complications of Parkinson's Disease

Bradley C. Hiner, MD

CONTENTS

OVERVIEW
ORTHOSTATIC HYPOTENSION
GASTROINTESTINAL DYSAUTONOMIA
UROLOGIC PROBLEMS
THERMOREGULATORY DYSFUNCTION
ROLE OF AUTONOMIC TESTING
SUMMARY
SELECTED READING

OVERVIEW

Although not as severe as the autonomic failure of the Shy-Drager and multiple system atrophy disorders, autonomic dysfunction nevertheless frequently causes problems in patients with Parkinson's disease (PD). James Parkinson recognized this in 1817, and his observations were borne out more than 150 years later by studies showing that more than 90% of patients with PD experience autonomic symptoms during the course of the disease. Orthostatic hypotension, constipation, and bladder difficulties must commonly be dealt with in the outpatient clinic, and occasionally patients are seen because of dysphagia, impotence, or bouts of vasomotor instability.

Some of these disorders are common in the elderly, reflecting changes in autonomic function that occur with normal aging. Others are unique to the patient with PD, such as drooling and paroxysmal sweating that may occur with "on-off" fluctuations. That the patient with PD has dysautonomia is not surprising, since degenerative changes in the autonomic nervous system have been well described. Lewy bodies have been detected in the hypothalamus, sympathetic chain ganglia, and myenteric plexus. Making matters worse, many of the medications used in PD have autonomic side effects—notably ortho-

From: *Parkinson's Disease and Movement Disorders:*
Diagnosis and Treatment Guidelines for the Practicing Physician
Edited by: C. H. Adler and J. E. Ahlskog © Mayo Foundation
for Medical Education and Research, Rochester, MN

Table 11-1
Effects of Medications for Parkinson's Disease on Autonomic Function[a]

	Increased tendency to orthostatic hypotension	Constipation	Bladder retention
Levodopa	+	+	0
Bromocriptine	+	+	0
Pergolide	+	+	0
Pramipexole	+	+	0
Ropinirole	+	+	0
Trihexyphenidyl, benztropine	+	++	++
Amantadine	+	+ (or diarrhea)	+
Tolcapone	+	+ (or diarrhea)	0

[a]In addition, sexual libido is occasionally increased by the dopamine-active medications; anticholinergic drugs may contribute to impotence. Sweating is decreased with the anticholinergic agents, whereas bouts of diaphoresis may occur as the levodopa effect is wearing off or during periods of medication-induced dyskinesias.

static hypotension and constipation (Table 11-1). Sorting out the relative contribution by these three factors—aging, PD, and medications—can be a challenging task.

Ideally, all clinicians would have ready access to a diagnostic autonomic laboratory, but this is not always available or practical. A commonsense approach for the clinic is necessary. Knowing what questions to ask, which tests to order and when to order them, and, most importantly, having strategies to deal with disorders of autonomic function all go a long way toward improving the quality of the patient's life and easing the course of the clinic day.

ORTHOSTATIC HYPOTENSION

Patient History

Lightheadedness or faintness after standing quickly is readily identified as a symptom of orthostatic hypotension (OH), but many patients do not offer a clear description of their symptoms. Instead, they may describe vague dizziness or the "weak and woozies" with fatigue, lethargy, and slowing of mentation. In contrast, the patient with vestibulopathy almost always describes dizziness as a sensation of motion, whether classic vertiginous spinning or the less commonly appreciated sensations of tilt, swaying, and impulsion. (A clinical aside: It is not usually helpful or important to determine whether the vertiginous movement is sensed as subjective [internal] or environmental.) Features that help differentiate OH from vestibulopathy are dimming or graying-out of vision, tunnel vision, and loss of color vividness occurring as blood pressure decreases. Some experience orthostatic headaches (usually occipital) accompanied by aching pain of the neck and shoulders in a "coat hanger" distribution.

It is important to remember, however, that occasional patients with vestibular disorders become vasovagal during episodes of vertigo. As a result, hypotensive presyncopal symptoms may become mixed with labyrinthine symptoms. Similarly, occasional patients with

Table 11-2
Primary Causes of Dizziness Among Patients
With Parkinson's Disease

- Orthostatic hypotension
- Cardiogenic or vasovagal presyncope
- Vestibulopathy (benign positional vertigo common in elderly persons)
- Disequilibrium

OH experience a transient sense of "things going around" or "head swimming," possibly from a decrease in perfusion of the labyrinthine arteries.

After syncopal (hypotensive) and vestibular causes, the third most common type of dizziness in the elderly population is disequilibrium (Table 11-2). Parkinsonism per se is often associated with gait instability, although this is a late-stage problem in idiopathic PD. Aging itself may be associated with disequilibrium and in some cases reflects "multiple sensory deficits." These persons feel unsteady, off balance, or "floaty" while walking, commonly reaching out to steady themselves on walls or furniture. Often, they are anxious and have a fear of falling. The patient with multisensory deficit typically is elderly and has mild-to-moderate neuropathy (usually loss of vibratory and joint position sense in the feet), some loss of vision and hearing, and perhaps some cognitive decline along with other health problems of aging, such as arthritis and general muscular deconditioning. Patients who complain of being dizzy all day very often are in this category. One must bear in mind, however, that advancing PD is a "falling disease"; postural reflexes that are already impaired in PD can be exacerbated by OH, leading the patient to describe symptoms in terms of disequilibrium with worsening gait and balance.

Commonly overlooked are the circumstances in which OH symptoms are likely to occur, such as in the postprandial period or during early morning hours when the patient is making the first visit to the bathroom. Prolonged standing, exercise, exposure to heat (particularly a hot bath), and drinking alcohol all predispose a person to orthostasis and form the basis for some simple preventive treatment strategies.

Reviewing the patient's medication list may identify offenders that contribute to low blood pressure (BP). Antihypertensives, neuroleptics, and tricyclic antidepressants can all interfere with normal cardiovascular responses to standing. Unfortunately, this list also includes several medications used to treat PD, especially the dopamine agonists bromocriptine (Parlodel), pergolide (Permax), pramipexole (Mirapex), and ropinirole (Requip). Because patients may be particularly prone to OH during the first few weeks of therapy, starting with low doses and carefully titrating up can minimize this adverse effect. For the most part, this approach is well accomplished by following the manufacturer's guidelines as outlined in the package insert or *Physicians' Desk Reference*.

Determining Blood Pressure

Ideally, the patient should rest supine for at least 5 minutes before standing to establish a stable baseline BP. OH is defined as a *symptomatic* drop in systolic pressure of more than 30 mm Hg *after 1 minute of standing*. Our practice is to record heart rate and BP supine, immediately after standing, and 1 minute after standing. If the trend is downward

Table 11-3
When Is Orthostatic Hypotension
Clinically Significant?[a]

- Systolic blood pressure in standing position, < 100 mm Hg

    plus

- Orthostatic symptoms: lightheadedness, syncope, blurred vision, lethargy, headaches

---

[a]In patients with Parkinson's disease, the tendency to orthostatic hypotension may vary greatly throughout the day. Orthostatic hypotension is most prominent after meals and during the levodopa "on" response.

but is not yet significant at the 1-minute mark, we continue taking readings every minute up to 5 minutes (or until symptoms occur). In some patients, this quick bedside check is not adequate, and a tilt table study is required to observe delayed OH that may occur gradually over the tilt period (we use head-up tilt at 65° for 30 minutes).

Many elderly persons have a postural drop in BP without symptoms. Furthermore, a drop in systolic pressure of 30 mm Hg may not be significant in a patient whose baseline supine reading is 160. In general, to be significant, the decline in BP must be below 100 systolic and cause relevant symptoms that are relieved by lying back down (Table 11-3). Some older patients may tolerate a drop in BP that would result in syncopal symptoms in younger persons. This circumstance most likely reflects a shift to the left in the cerebrovascular autoregulation curve, resulting in maintained cerebral blood flow despite changes in the perfusion pressure. Even when asymptomatic, however, a systolic decrease of greater than 30 points and below 100 mm Hg suggests at the very least that the patient has "orthostatic vulnerability," which under different circumstances (i.e., outside the clinic) could cause syncopal symptoms. Not all patients with a decrease in BP on standing need to be treated for OH. The drop must cause symptoms, and to warrant treatment, the symptoms should be severe enough to compromise quality of life.

If the bedside orthostatic examination reveals a decrease in BP but little or no change in heart rate (i.e., no compensatory reflex tachycardia), the patient may have an autonomic disorder. On the other hand, an excessive increase in heart rate (>110 beats/minute or an increase exceeding one-third of the resting heart rate) suggests either "postural orthostatic tachycardia syndrome" or relative dehydration with intravascular volume depletion (or muscular deconditioning with peripheral venous pooling or any other condition causing decreased cardiac filling pressures). Finally, abrupt bradycardia or even sinus arrest during hypotension suggests the neurocardiogenic syncope syndrome (also known as "malignant vasovagal syndrome").

Identifying these very different causes of low BP is crucial, as treatment for each is quite different. For example, a beta-blocker such as metoprolol is very effective for neurocardiogenic syncope but would be disastrous for the patient who is volume-depleted, is anemic, or has neurogenic OH. Although a full discussion of the evaluation of OH is beyond the scope of this chapter, a complete blood cell count (to rule out anemia) and general chemistry tests should be done. Abnormal electrolyte findings may point to Addison's disease and lead to determination of morning and afternoon serum cortisol values.

Important note: Patients undergoing a tilt table study for evaluation of syncope may be given an infusion of isoproterenol to provoke neurocardiogenic syncope. Isoproterenol (Isuprel) may cause severe hypertension in patients taking a catechol-O-methyl transferase inhibitor such as tolcapone (Tasmar) or entacapone (Comtan). This medication must be withdrawn at least 24 hours before isoproterenol is administered to avoid this potentially serious complication.

Finally, patients with severe OH and other clinical features of dysautonomia, such as early-onset impotence or incontinence (*especially* if they do not have a tremor, do not have a response to levodopa up to 1,500 mg/day, or have respiratory stridor—all hallmarks of multiple system atrophy), should be referred to a laboratory that performs autonomic testing. Should there be evidence of widespread autonomic failure, the patient may well have multiple system atrophy and not routine idiopathic PD. Making this distinction has important implications for the patient, because the underlying pathologic features and clinical courses are quite different.

Treatment of Orthostatic Hypotension

Nonpharmacologic Strategies (Table 11-4)

1. Administration of antihypertensive agents and, if possible, other medications known to interfere with normal BP responses to standing should be discontinued. Many patients are receiving drug regimens that are continually "carried over," and simply reviewing the medication list periodically may permit some of these to be discontinued. It may be preferable to allow the BP to remain somewhat higher than normal (e.g., systolic, 140 to 160 mm Hg; diastolic, 90 to 100 mm Hg) to provide a "cushion" that protects against orthostasis. This protocol requires close cooperation with the patient's internist or primary care physician.

2. Patients should be told about risk factors that predispose to OH, such as prolonged standing, getting up too quickly after prolonged recumbency, excess heat from the environment, and the immediate postexercise state. The patient is instructed to sit on the edge of the bed before standing and flex leg muscles by wiggling the feet and to march in place or contract thigh and calf muscles (tighten kneecaps or buttocks) to diminish peripheral venous pooling. If warning symptoms of presyncope occur, the patient should sit or lie down and prop the feet up or sit and lean forward with head between legs. Squatting is an effective countermaneuver, and in some cases simply crossing the legs or putting one leg up on a low stool or other support is all that is needed.

3. Because BP is normally higher later in the day, patients with OH should schedule activities for the afternoon rather than the morning, when BP is usually lowest.

4. Patients with PD have an exaggerated decrease in BP after eating and may find it helpful to rest 30 minutes after a meal. Those with more severe autonomic failure may benefit from smaller, more frequent meals. Alcohol intake should be minimized, and one or two cups of caffeinated coffee with breakfast may help alleviate morning symptoms of OH.

5. Patients may choose to avoid activities that involve straining, such as heavy lifting. Prolonged coughing, straining at stool, and even vigorous singing or playing a wind instrument may result in hypotension from diminished venous return and Valsalva effects.

6. The head of the bed should be elevated 10° to 30° (usually 8 to 10 inches). Just using one or two extra pillows is unsatisfactory. Studies have shown that the reverse Trendelenburg position affects renal hemodynamics, resulting in less salt and fluid loss during sleep.

Table 11-4
Nonpharmacologic Treatment of Orthostatic
Hypotension

•Discontinue administration of unnecessary hypotensive drugs (e.g., antihypertensives, diuretics)
•Educate about risk factors: prolonged standing, standing quickly, hot environments, meals, straining-type exercise
•Elevate head of bed 10° to 30°
•Increase dietary salt (salt tablets, if necessary)
•Suggest thigh- or waist-high, fitted compression stockings

This effect not only mitigates early morning orthostasis but also diminishes the supine hypertension that often occurs in autonomic failure (or results from its treatment). A footboard may be needed to prevent sliding out of bed. In some cases, patients find it easier to sleep in a recliner, and on occasion a hospital bed may be necessary.

7. If cardiac condition allows, dietary sodium should be liberalized. Salt intake may be increased up to 2 g/day. Some patients find that this can be accomplished by drinking five servings of salty soup (e.g., bouillon) per day or by simply being allowed to use the salt shaker more often. Salt tablets (Thermotabs) are used by some specialists in autonomic disorders before considering pharmacotherapy. Each tablet contains 179 mg of sodium, and a typical dose regimen begins with one half tablet three or four times daily, which can be increased up to one to two tablets four times a day. Edema or congestive heart failure may occur, so that even something as simple as salt supplementation must be done with care.

8. Compression stockings, such as Jobst, may be considered. To be effective, they must be either panty-hose length or thigh high. "Medium compression" or "custom-fit" should be specified. Thromboembolic disease support hose are not effective. The Jobst stockings are uncomfortable and unwieldy to put on and take off, but they are effective and do not cause systemic side effects.

PHARMACOLOGIC STRATEGIES (TABLE 11-5)

1. Fludrocortisone (Florinef), a salt-retaining mineralocorticoid, is given in a starting dose of 0.1 mg/day or, in more severe cases, twice a day. Escalation may be accomplished at weekly or biweekly intervals. On rare occasions, the effective dose may be as high as 1.0 mg/day in divided doses. The maximal clinical response may take up to 1 to 2 weeks. Because hypokalemia may occur, serum potassium concentration should be determined regularly: after 2 weeks initially, at 1 month, and every 3 months thereafter. Fludrocortisone is relatively contraindicated in some patients with congestive heart failure or severe cardiomyopathy. If there are uncertainties, a cardiologist should be consulted.

2. Pressors: If fludrocortisone alone is insufficient, a sympathomimetic agent is added. The best of those currently available is midodrine (ProAmatine), a selective, peripherally acting α-adrenergic agonist. At the start, 2.5 mg is given three times a day (before breakfast, before lunch, and in the midafternoon). The dose is increased by titration as necessary, guided by the response. The effective dose may be as high as 10 mg three times a day and, on occasion, up to 40 mg/day. Unfortunately, the drug is available in 2.5- and 5-mg tablets only, and the expense may be prohibitive for some patients (our pharmacy quotes a 1999 monthly cost of $265.86 for 10 mg three times a day). Alternatives include ephedrine, 25 to 50 mg three times a day, and phenylpropanolamine, 25 to 50 mg three times a day. Occasional patients do well with methylphenidate, 10 to 20 mg three times

Table 11-5
Major Pharmacologic Strategies for Treatment
of Orthostatic Hypotension

•Salt-retaining mineralocorticoid: fludrocortisone (Florinef)
•Pressor: midodrine (ProAmatine)
•Other pressors: ephedrine, phenylpropanolamine

a day, or dextroamphetamine, 5 mg three times a day, although these drugs are more problematic because of their Schedule II status.

3. If a third drug is needed, indomethacin, 25 to 50 mg three times a day, is added. The presumed mechanism is inhibition of vasodilatory prostaglandins. The effects are usually modest (increase of 5 to 10 mm Hg at best) but may be enough to control symptoms of cerebral hypoperfusion.

4. Others: Erythropoietin (4,000 units subcutaneously twice weekly for 6 weeks), desmopressin, and somatostatin have been described in the literature, but all pose problems that make routine use impractical.

GASTROINTESTINAL DYSAUTONOMIA

Constipation

A frustrating problem for both patient and physician, constipation is a common complaint in PD (Table 11-6). Gut hypomotility, physical inactivity due to generalized hypokinesia, and medication side effects all contribute to the problem. In addition, there is some evidence that dystonia of the skeletal muscles of the pelvic floor can interfere with defecation. Pathologic studies have shown degenerative changes of the myenteric plexus, including Lewy bodies, in patients with PD. Nearly all medications used in PD can slow bowel motility, especially the anticholinergics (e.g., trihexyphenidyl [Artane], benztropine [Cogentin]), so that a preemptive approach is often best.

1. The diet should be improved. Bulk is increased with bran or by using psyllium preparations (e.g., Metamucil) or methylcellulose products (e.g., Citrucel), and the importance of adequate fluid intake is emphasized. The patient is encouraged to drink up to eight glasses (tumblers) of water a day. The amount of fresh fruits and vegetables in the diet is increased, and low-fiber foods, such as fast foods and certain breads and baked goods, are avoided.

2. Regular exercise is encouraged: at least 15 to 20 minutes every day. Walking, using a stationary bicycle, and swimming are reasonable options for some; the patient must find an activity that he or she enjoys or the effort will be short-lived.

3. Administration of anticholinergics should be discontinued (by tapering). Rarely does a patient's quality of life depend on trihexyphenidyl or benztropine, and if it turns out that it does, administration can always be reinstated. Other medications, such as oxybutynin and tricyclic antidepressants, also have anticholinergic activity. Calcium channel blockers, especially verapamil, can exacerbate constipation.

4. A stool softener can be added: docusate sodium, once or twice daily.

5. Osmotic laxatives, such as milk of magnesia (30 mL/day) or lactulose (10 to 20 mL up to four times daily), are often effective and well tolerated. Alternatively, capsules combining docusate with a mild laxative (Peri-Colace) may be tried. If these are ineffective, a stimulant laxative like bisacodyl (Dulcolax), 1 or 2 tablets or suppositories a day, is needed.

Table 11-6
Options for the Treatment of Constipation

First-line

- Dietary bulk
- Increased fluid intake (e.g., eight glasses daily)
- Regular exercise
- Discontinue taking of anticholinergic drugs, if possible
- Stool softeners (e.g., docusate sodium)
- Osmotic laxatives: milk of magnesia, lactulose

For more refractory constipation

- Stimulant laxative (e.g., bisacodyl)
- Cisapride (prokinetic agent)
- Mineral or tap water enemas

6. Cisapride, a prokinetic agent, stimulates gastric motility at all levels of the gastrointestinal tract. The usual dose of 20 mg three times a day and at bedtime may be useful in recalcitrant cases. Because the drug has cholinomimetic properties, PD motor symptoms could theoretically worsen, but this has not been a general observation.
7. Enemas, with mineral or tap water, should be reserved for patients who do not have a response to the measures listed above. If necessary, enemas may be repeated up to four times during a single day. Enlisting the aid of a visiting nurse may be helpful.

Which patients should be considered for further diagnostic evaluation? Those with severe, recalcitrant constipation to the point of obstipation (toxic megacolon has been reported) and those who may have gut dysmotility as part of a more widespread autonomic disorder (e.g., Shy-Drager syndrome or multiple system atrophy) are candidates for referral to a motility specialist. Gastrointestinal motility studies, including manometry and scintigraphic transit time measurements, may be helpful in identifying these cases. In addition, autonomic testing of cardiovascular reflexes and preferably also of thermoregulatory sweating can help to identify other visceral denervation and thus place the gut motility problem in its proper perspective of generalized autonomic insufficiency.

Dysphagia

Dysphagia is uncommon in mild-to-moderate PD, but the incidence increases with advancing severity of the disease. Dr. Parkinson observed that "so much are the actions of the muscles of the tongue, pharynx, etc. impeded by impaired action and perpetual agitation, that the food is with difficulty retained in the mouth until masticated; and then as difficultly swallowed." As usual, his was an accurate description of oropharyngeal dysfunction appearing as an inability to propel the food bolus from the mouth to the esophagus. He also noted the drooling (sialorrhea) of saliva that results not from hypersecretion but from impaired swallowing. Not only can dysphagia compromise the nutritional status of the patient but also it may occasionally be responsible for lack of medication effect. Rare patients without a response to levodopa may have, in reality, never effectively swallowed their medication, with bits of pills ending up down the front of their hospital gowns.

Because the muscles that initiate the swallow are skeletal, they are subject to the same on-off clinical fluctuations that occur with generalized motor function. Thus, many patients may be dysphagic only during "end-of-dose" deterioration, and they may be helped by simple adjustments in the timing of doses, such as taking medications 20 to 30 minutes before meals. Unfortunately, "off" periods are not always so predictable, and bulbar impairment often is resistant to pharmacotherapy. There may also be silent aspiration with pooling of material in the valleculae and pyriform sinuses. Esophageal motility may become impaired on rare occasions to the point of achalasia.

If dysphagia is limited to the early, or oropharyngeal, phase of the swallow, measures such as double swallowing, using a chin tuck, and increasing the consistency of the food (commercial thickening products are available) may be helpful. Referral may be made to a speech pathologist, otolaryngologist, or physiatrist skilled in swallowing disorders. A videofluoroscopic swallowing study is the diagnostic procedure of choice.

In advanced parkinsonism, swallowing may be so impaired that placement of a feeding gastrostomy is a consideration. For those with advanced disease, this poses end-of-life and quality-of-life issues. If severe immobility or dementia has rendered the patient incapable of self-care, many families elect not to place a feeding tube. In other cases, placement of a percutaneous esophagogastrostomy tube can augment strength not only by improving nutrition but also by improving delivery of parkinsonian medication to the gastrointestinal tract.

Excessive Drooling

As noted above, drooling is thought to be the result not of autonomic salivary gland hyperactivity but rather of impaired swallowing. Nevertheless, the problem can sometimes be helped by anticholinergic medications like trihexyphenidyl, 2 to 5 mg three times a day, or benztropine, 0.5 to 1.0 mg three times a day. If these prove ineffective, one may try glycopyrrolate (Robinul), 1 to 2 mg three or four times a day. Although all anticholinergic agents may cause dizziness, confusion, and short-term memory impairment, glycopyrrolate may be less likely to do so because its quaternary ammonium structure impedes its ability to cross the blood-brain barrier. Surgical options include tympanic neurectomy, which interrupts salivary nerves to the parotid glands via a middle ear approach, and rerouting of the parotid ducts so that saliva runs down the throat rather than into the mouth. Rarely, submandibular glands are removed.

UROLOGIC PROBLEMS

Urinary urgency, urinary frequency, and nocturia are frequent problems in the geriatric population. Disorders like stress incontinence in women with weakened pelvic floor musculature and obstructive uropathy in men from prostatic hypertrophy increase problems for parkinsonian patients already having trouble getting to the bathroom in time because of rigidity and bradykinesia. In addition, PD can alter supraspinal reflexes governing micturition, resulting in a neurogenic bladder that contracts either too much or too little.

Empirical treatment of neurogenic bladder symptoms may be appropriate in some patients, but formal urologic studies may be necessary to exclude intrinsic bladder lesions (cystoscopy) or to guide pharmacologic management (urodynamics). Distinction between hyperactive and hypoactive detrusor states is often difficult without urodynamic studies.

Hyperactive Detrusor

Detrusor instability of this type is the predominant abnormality of bladder function in PD. Normal micturition is a coordinated series of events, involving relaxation of the striated urethral muscles, contraction of the detrusor (smooth muscle), and opening of the bladder neck allowing urine to pass. An increase in detrusor muscle tone results when inhibitory centers in the brain (frontal lobe, basal ganglia, and upper brain stem form one important loop pathway) governing detrusor activity are affected by parkinsonian pathology. Excessive detrusor contractility results in voiding before the bladder is full, initially as nocturia that may ultimately lead to daytime urge incontinence and nocturnal enuresis.

Urinalysis and urine culture should be done first with any change in bladder habits, because older patients may not experience dysuria despite infection. An empirical trial of medication based on historical features may then be considered. The goal of treatment is to suppress the detrusor reflex and increase bladder capacity, and the following peripherally acting anticholinergics are the mainstays of therapy.

> Tolterodine (Detrol), 2 mg twice a day
> Oxybutynin, 5 mg 3 or 4 times daily
> Propantheline, 15 to 30 mg every 4 to 6 hours
> Hyoscyamine, 0.15 to 0.3 mg at bedtime or up to four times a day
> Imipramine, 10 to 25 mg at bedtime

Tolterodine is a new anticholinergic drug with the unique properties of high potency at bladder muscarinic receptors but low affinity for salivary gland binding sites. Thus, although equipotent to oxybutynin in decreasing detrusor hyperactivity, it lacks the systemic side effects (especially dry mouth) that frequently limit the use of the other drugs. Patient tolerability is paramount, because 5 to 8 weeks of treatment may be required before maximum clinical efficacy occurs with any of the medications. All these medications have the potential to cause memory and cognitive impairment, to which patients and their families need to be alerted.

Flavoxate hydrochloride (Urispas) is reported to directly inhibit bladder smooth muscle at a dose of 100 to 200 mg three or four times a day. Its efficacy has been questioned however; a recent study found that it had no advantage over placebo in the treatment of detrusor hyperreflexia.

A final note on the common co-occurrence of detrusor hyperreflexia and prostatism in the male parkinsonian patient: Studies have shown generally poor surgical results in this group, with transurethral prostatic resection providing relief of symptoms in only about 50% of patients. Careful selection of patients with urodynamically proven prostatic obstruction is recommended.

Hypoactive Detrusor

The hypoactive form is less common than the hyperactive form. The patient with decreased detrusor tone experiences difficulty initiating urination, a weak stream, and incomplete bladder emptying. In women, leakage of urine may occur during coughing or laughing. In men, symptoms mimicking an outlet obstruction from prostatic hypertrophy may occur, including difficulty initiating urination, hesitancy, and a weak stream.

Anticholinergic parkinsonian medications should be discontinued, if possible, because they decrease detrusor contraction and may cause urinary retention. Giving an α-adrenergic blocking agent may be considered to decrease tone in the bladder neck and thus outlet resistance. Because these agents may induce OH, initial doses should be small, followed by careful titration and monitoring of standing BP.

> Terazosin (Hytrin): 1 mg at bedtime; may be increased in 1- or 2-mg increments every 3 to 5 days up to the usual dose of one 5-mg tablet once or twice daily.
>
> Doxazosin (Cardura): 1 mg at bedtime; may be increased every 3 to 5 days in 2-mg increments up to the usual dose of 4 to 8 mg daily.
>
> Prazosin (Minipress): 1 mg at bedtime, which is increased gradually to the usual dose of 1 to 2 mg two or three times a day.
>
> Tamsulosin (Flomax): 0.4 mg at bedtime; if response failure, increase to 0.8 mg once daily.

Tamsulosin is the newest of these medications and may be the best choice for patients with PD because of less potential for exacerbating orthostatic hypotension. If the patient is not able to tolerate an α-adrenergic blocker, stimulating bladder contractility with a parasympathomimetic agent, such as bethanechol chloride (Urecholine), 25 mg three times daily, may be useful.

Again, formal urologic studies (urodynamics and cystoscopy) may be necessary in many cases to guide therapy. This takes the guesswork out of "hyper or hypo" detrusor function, and the consultant can assess for other causes of bladder dysfunction. An additional role for urodynamic studies has recently been described, that of differentiating PD from multiple system atrophy. In PD, detrusor hyperreflexia with urgency to void but without chronic retention is more common. In contrast, multiple system atrophy is more likely to cause a hypoactive detrusor with low urethral pressure and chronic retention.

Male Impotence

The prevalence of erectile dysfunction in PD is unknown. Patients do not often bring up the issue, and physicians frequently do not ask. Attaining and maintaining an erection may be the result of iatrogenic, endocrine, vascular, urologic, or psychologic causes. A review of the patient's medications may provide the answer, because antihypertensives (especially beta-blockers), antidepressants (particularly the selective serotonin reuptake inhibitors), and any medication with sedative properties can cause impotence. A medical evaluation for diabetes and vascular and endocrinologic disorders should be followed by urologic evaluation. If organic disorders are ruled out, psychogenic factors, such as depression and anxiety, need to be considered.

Finally, a trial of medication may be initiated. Sildenafil (Viagra) is a new agent that has been found to be highly effective for impotence. One study documented that 78% of patients receiving this medication reported better erections. It acts by potentiating the smooth muscle relaxation mediated by nitric oxide in the corpora cavernosa that produces erections. Visual disturbances, such as light sensitivity and a bluish tinge, have been reported. The medication is contraindicated in patients with coronary artery disease. Other options are yohimbine, 5 mg three times a day, and intracavernous injections of papaverine.

THERMOREGULATORY DYSFUNCTION

Anhidrosis, hyperhidrosis, heat or cold intolerance, and accidental hypothermia are examples of disordered thermoregulation that may occur in PD. Parkinsonian pathologic features have been found in areas of the central nervous system responsible for temperature regulation, primarily the hypothalamus.

Hyperhidrosis

The most commonly encountered problem is hyperhidrosis, occurring as paroxysms of drenching sweats when the patient's motor status deteriorates from "on" to "off." As tremor, rigidity, and bradykinesia set in, profuse sweating breaks out. The best strategy is to keep the patient above the "off" phase threshold, which may be accomplished in a variety of ways. For some, it may be as simple as decreasing the time between doses of levodopa. Using adjunctive medications such as a dopamine agonist (bromocriptine, pergolide, pramipexole, or ropinirole), catechol-*O*-methyltransferase inhibitor (tolcapone, entacapone), or monoamine oxidase type B inhibitor (deprenyl) are other options. Anticholinergic agents, such as propantheline (Pro-Banthine) and glycopyrrolate, are occasionally helpful, although many patients are unable to tolerate the side effects (dry mouth, blurred vision, urinary retention). The use of topical aluminum hydrochloride (the active ingredient in most antiperspirants) is advocated by some, but most patients find it impractical because of the diffuse distribution of their sweating.

Other patients experience excessive sweating as a consequence of levodopa "on" period dyskinesias. Reducing medications may help but may be difficult in patients with narrow therapeutic windows. These patients may benefit from a trial of a beta-blocker (e.g., propranolol, 10 to 20 mg four times a day), with monitoring for significant OH.

Hypothermia

Changes in body temperature are detected in the anterior hypothalamus by neurons with temperature-dependent firing rates. This information is conveyed to nuclei in the posterior hypothalamus that are responsible for initiating appropriate effector responses (like shivering when exposed to cold or sweating when hot). This homeostatic system is disrupted by PD in such a way that patients may become susceptible to accidental hypothermia, defined as a drop in body temperature of 2°C or greater. Some studies have suggested that core temperature in PD is lower than normal, and others have demonstrated absence of shivering (impaired heat production) and defective peripheral vasoconstriction (impaired heat conservation) during exposure to cold. Striking rigidity and bradykinesia unresponsive to parkinsonian medications may occur during a hypothermic episode, along with mental dulling and even encephalopathy. There may be characteristic diphasic and triphasic waves on electroencephalograms during severe and prolonged bouts of hypothermia. As hypothermia is easily overlooked as a cause of clinical deterioration, physicians practicing in northern climates need to bear this in mind during the winter months.

ROLE OF AUTONOMIC TESTING

Autonomic testing typically consists of a battery of cardiovascular responses supplemented, when available, by other techniques, such as sweat testing. Cardiovascular testing generally consists of measuring heart rate responses to various stimuli (e.g., deep breathing, Valsalva's maneuver) and blood pressure responses to a stress such as sustained handgrip or forearm immersion in ice water (cold pressor test). Although an oversimplification, tests of heart rate variability assess parasympathetic (cardiovagal) function, whereas BP responsiveness reflects sympathetic vasomotor tone. In diabetic patients with autonomic neuropathy, the parasympathetic nerves are thought to be affected first, and the course of the neuropathy can be tracked over time, which is important for instituting measures that can lower morbidity and mortality (e.g., operative risks are much higher in diabetic patients with autonomic neuropathy than in those without). This same schema unfortunately does not apply to the PD population, because parasympathetic involvement is an early finding in some studies but sympathetic function is affected sooner in others.

Autonomic testing generally shows mild or, at most, moderate impairment in PD and is usually not necessary for management of the problems discussed in this chapter. It can be helpful, however, in two specific circumstances: (1) differentiating PD from the more malignant multiple system atrophy (Shy-Drager syndrome) and (2) discerning whether OH is neurogenic (i.e., from autonomic neuropathy) or non-neurogenic (from dehydration, medication side effect, or other cause). Often an educated guess can be made on the basis of clinical features and observations, such as what the heart rate does in response to an orthostatic BP determination (discussed earlier). Finally, patients with PD can, like anyone else, have autonomic neuropathy from such causes as diabetes, amyloid, alcohol, nutritional deficiencies, chemotherapeutic agents (especially cisplatin and vincristine), renal failure, and carcinoma (remote effect). Fortunately, this association is infrequent.

Differential Diagnosis

It is important to differentiate multiple system atrophy (MSA) from idiopathic PD accompanied by autonomic symptoms. The MSA syndromes include Shy-Drager syndrome (parkinsonism with cerebellar and corticospinal tract signs and early, marked autonomic failure), striatonigral degeneration (levodopa-refractory parkinsonism and dysautonomia), and olivopontocerebellar atrophy (syndrome of prominent cerebellar ataxia and autonomic insufficiency). These disorders generally begin at a younger age, have more complications, and have a less favorable prognosis for survival than typical PD, which usually does not substantially shorten life span (see Chapter 17).

If testing documents abnormalities on more than 50% of the measures of cardiovascular reflex, especially if supplemented with other modalities like thermoregulatory sweat testing, the patient probably has an MSA. Severe autonomic dysfunction or autonomic dysfunction that precedes parkinsonism in addition to certain clinical features, such as inspiratory stridor and poor response to levodopa, suggests MSA (see Chapter 17).

Testing for Plasma Catecholamines

Determination of plasma catecholamines in the supine and standing positions is of little value in the routine evaluation of PD. Its main role is in differentiating MSA from "idiopathic orthostatic hypotension," also known as pure autonomic failure or Bradbury-Eggleston syndrome. Similarly, whether the dysautonomia is predominantly preganglionic or postganglionic, afferent or efferent is not of much practical significance in the evaluation and treatment of PD or MSA.

SUMMARY

Patients with PD are more likely to have autonomic dysfunction than age-matched controls. However, if severe autonomic impairment is present, the diagnosis may be one of the MSA syndromes rather than PD. Autonomic testing is usually not needed to manage the most common autonomic problems in PD. Practical approaches in assessment and treatment of these common problems can be incorporated into a primary care outpatient practice.

SELECTED READING

Aminoff MJ. Other extrapyramidal disorders, in *Clinical Autonomic Disorders: Evaluation and Management* (Low PA ed), Little, Brown and Company, Boston, 1993, pp. 527-536.

Bannister R, Mathias CJ. Clinical features and investigation of the primary autonomic failure syndromes, in *Autonomic Failure: A Textbook of Clinical Disorders of the Autonomic Nervous System* (Bannister R, Mathias CJ eds), 3rd ed, Oxford University Press, Oxford, 1992, pp. 531-547.

Chancellor MB, Blaivas JG. Parkinson's disease, in *Practical Neuro-Urology: Genitourinary Complications in Neurologic Disease* (Chancellor MB, Blaivas JG eds), Butterworth-Heinemann, Boston, 1995, pp. 139-148.

Kaufmann H. Neurally mediated syncope and syncope due to autonomic failure: differences and similarities. *J Clin Neurophysiol* 1997; 14: 183-196.

Koller WC, Vetere-Overfield B, Williamson A, Busenbark K, Nash J, Parrish D. Sexual dysfunction in Parkinson's disease. *Clin Neuropharmacol* 1990; 13: 461-463.

Low PA. Autonomic nervous system function. *J Clin Neurophysiol* 1993; 10: 14-27.

Low PA. Clinical evaluation of autonomic function, in *Clinical Autonomic Disorders: Evaluation and Management* (Low PA ed), Little, Brown and Company, Boston, 1993, pp. 157-168.

McLeod JG, Tuck RR. Disorders of the autonomic nervous system: Part 1. Pathophysiology and clinical features. *Ann Neurol* 1987; 21: 419-430.

Olanow CW, Koller WC. An algorithm (decision tree) for the management of Parkinson's disease: treatment guidelines. *Neurology* 1998; 50 Suppl 3: S1-S5.

Quinn N. Multiple system atrophy: the nature of the beast. *J Neurol Neurosurg Psychiatry* 1989; Suppl: 78-89.

Tanner CM, Goetz CG, Klawans HL. Autonomic nervous system disorders in Parkinson's disease, in *Handbook of Parkinson's Disease* (Koller WC ed), 2nd ed, Marcel Dekker, New York, 1992, pp. 185-215.

12 Treatment of Cognitive Disorders and Depression Associated With Parkinson's Disease

Erwin B. Montgomery, Jr., MD

COGNITIVE COMPLICATIONS

James Parkinson's description in 1817 of the disease that bears his name is still one of the best descriptions of this disorder. He was wrong, however, on one important feature: Parkinson said that the senses and intellect are uninjured. We now recognize that as many as 20% of patients with Parkinson's disease (PD) have substantially impaired cognitive function.

Parkinson's oversight, however, is defensible. Most likely, many patients with PD in the 1800s did not live long enough to manifest the symptoms of dementia. Also, they were probably so overwhelmed by the tremor, rigidity, bradykinesia, and akinesia that dementia went unnoticed. James Parkinson did not have the medical drugs and neuro-surgical options now available to us that prolong the lives of patients with PD but induce their own problems. Clearly, the more aggressively that physicians can treat PD, the more complications that have to be faced, particularly those that affect cognitive function. These complications are the result of a now-altered natural history of disease as well as a consequence of our treatments.

From: *Parkinson's Disease and Movement Disorders: Diagnosis and Treatment Guidelines for the Practicing Physician* Edited by: C. H. Adler and J. E. Ahlskog © Mayo Foundation for Medical Education and Research, Rochester, MN

Minor Cognitive Impairment
Versus Dementia and Psychosis

It is becoming increasingly clear that the basal ganglia are involved in a number of cognitive functions. These include use of skill, attention, estimation of time, and executive functions, such as goal selection and ability to modify behavior in response to changing conditions. Many of these functions are impaired in patients with PD. Similar to the slowness of movement (bradykinesia) experienced by patients with PD, slowed thinking (bradyphrenia) may also affect them. Cognitive impairments of these types are typically not disabling. They stand in contrast to the dementia and psychotic symptoms experienced by a minority, perhaps 20%, of patients with PD. This chapter confines discussion of cognitive impairment to dementia and psychosis.

Recognition of Significant Cognitive Dysfunction

Physicians with an active PD practice frequently encounter patients with significant dementia or psychosis that interferes with the quality of life, shapes therapeutic management, and has the potential to be dangerous. The first responsibility (often overlooked) is to recognize that there is a significant problem, which may be obscured by the slow and insidious onset. Sometimes family members or caregivers attribute cognitive changes to willful behavior, and this misunderstanding often breeds resentment and may lead to inadequate care of the patient. Physicians should routinely inquire about cognitive problems, such as changes in personality, increased dependence, fearfulness, becoming withdrawn, confusion, disorientation, hallucinations, and delusions. It is very important to recognize cognitive impairment, because it is the leading reason patients with PD are placed in long-term care institutions.

When hallucinations occur, they are predominantly visual. Hallucinations can occur in young patients and patients with relatively early disease. In these patients, the hallucinations may be an isolated problem. Delusions may be very disruptive of family life; sometimes they take on sexual overtones. Even with explanation that these delusions are related to the disease or medications, many family members find them difficult to accept. Prominent hallucinations or delusions early in the course of the disease raise suspicion of diffuse Lewy body disease (see Chapter 20).

Some patients are adept at masking symptoms. They may deny problems or continually defer questions and issues to family members. More formal testing may be needed to identify dementia. The Mini-Mental State Examination is a relatively good screening instrument that can be administered in a few minutes during the course of an office visit (Table 12-1). A score of less than 24 is considered abnormal. However, physicians may wish to use a slightly higher cutoff score to signal the possibility of early problems. Formal neuropsychologic testing may be necessary in certain cases, such as determination of competence or detection of possible pseudodementia due to depression. Neuropsychologic testing results may also serve as a baseline for subsequent testing to document deterioration.

Differential Diagnosis

The physician should avoid jumping to the conclusion that any cognitive problems are simply due to PD. Dementia or psychosis can develop in patients with PD for the same reason as in anyone else. These causes include:

1. Antiparkinsonian medications
2. Other prescription or over-the-counter medications
3. Alcohol or drug withdrawal (including anti-PD medications)
4. Infections
5. Major organ failure (e.g., kidney, liver, heart, or lung)
6. Concurrent Alzheimer's disease (see Chapter 20)
7. Metabolic or hormonal abnormalities (e.g., hypothyroidism)
8. Vitamin deficiencies (e.g., B_{12})
9. Subdural hematoma
10. Stroke
11. Depression (pseudodementia)

Table 12-1
Mini-Mental State Examination

Maximum score is 30; < 24 is conventional cutoff for dementia. The maximum score for each subtest is shown in parentheses.

Orientation
>   Ask the patient to provide information about the time and place of testing: year, season, date, day, month, state, county, town, building or hospital, floor. If from out of state, the patient may be asked for county of residence. (10)

Registration
>   Ask the patient if you may test memory, and then say the names of three unrelated objects, about 1 second for each. Ask the patient to repeat them, scoring one point for each. (3) The first repetition determines the score, but if the patient is unable to say them back, continue to repeat them up to six trials as a basis for the recall subtest, below.

Attention and Calculation
>   Serial sevens: start at 100 and count backwards by sevens; stop after five answers. Alternative: spell "world" backwards, scoring the number of letters in correct order. (Maximum for either = 5)

Recall
>   Ask for the names of the three objects from the Registration subtest. (3) Note that if the patient was unable to name all three after the maximum six trials, recall cannot be meaningfully tested.

Language
>   Ask the patient to
>   • Name a pencil and watch after being shown each. (2)
>   • Repeat the phrase "no ifs, ands, or buts." (1)
>   • Follow the three-part spoken command "Take this paper in your right hand, fold it in half, and put it on the floor." (3)
>   • Read and obey the printed command "Close your eyes." (1)
>   • Write a sentence; it must contain a subject and a verb and be sensible, but grammar and punctuation need not be correct. (1)
>   • Copy a design (intersecting pentagons). (1)
>   The maximum score on this language subset is 9.

Adapted from Folstein MF, Folstein SE, McHugh PR, *J Psychiatr Res* 1975; 12: 189-198. By permission of Pergamon Press Ltd.

Patients with cognitive problems sometimes decompensate at night (sundowning). However, abnormal nocturnal behavior and hallucinations or delusions must be differentiated from vivid dreams and rapid eye movement sleep behavior disorder, in which patients act out their dreams (see Chapter 10).

Laboratory Evaluation

The laboratory evaluation depends on the clinical picture. Obviously, if the history or examination suggests an underlying systemic disorder (e.g., infection, connective tissue disease) or structural lesion (e.g., tumor), this possibility needs to be pursued. Rapidly developing cognitive decline suggests a cause other than neurodegenerative disease. However, with slowly developing cognitive decline and with no suspicions raised by history or examination, a workup is still appropriate. This might include the following studies, which should exclude most treatable causes of cognitive impairment.

1. Computed tomography head scan may be sufficient in many cases and is often adequate to exclude tumor, intracranial hemorrhage, and hematoma. Magnetic resonance imaging may be additionally helpful in assessing normal pressure hydrocephalus, multi-infarct dementia, and inflammatory, infectious, or diffuse neoplastic disorders affecting the brain.
2. Blood determinations include complete blood cell count, erthyrocyte sedimentation rate, liver and renal functions, vitamin B_{12} and folate levels, and thyroid battery. Lyme titers, syphilis serology, human immunodeficiency virus (HIV) testing (there have been recent reports of parkinsonism associated with HIV infection), and urine toxicology screens may be justified, depending on the clinical context.

Cerebrospinal fluid examination may be considered in subacutely developing or rapidly progressive dementia or suspected central nervous system infection. It is a good screening instrument to assess the possibility of any infectious, inflammatory (e.g., connective tissue disease), or infiltrating neoplastic (e.g., lymphomatosis) disorder. An electroencephalogram is rarely necessary, except for possible Creutzfeldt-Jakob disease or continuing seizures (e.g., complex partial status epilepticus).

Basic Principles

Physicians should avoid undue pessimism. Psychosis often responds to medical management. Even if the cognitive impairment is not due to some easily reversible abnormality, modifications in the home and counseling of the family can help to improve the situation and prevent complications.

Not all patients have hallucinations or delusions that are of sufficient severity to warrant medications. Many patients retain insight and recognize that the hallucinations and delusions are not real. Other patients can be comforted by calm and gentle reassurance. Keeping a low light on at bedtime and avoiding strange new environments and stressful conditions may help. If the hallucinations and delusions are emotionally painful or would cause endangerment to self or others, the hallucinations and delusions should be treated directly.

Choice of Antiparkinsonian Therapeutic Agents

The first approach to treating the psychotic complications is to avoid them. When choosing the medications for treatment of parkinsonian motor symptoms, one should consider the greater risk for such complications in some patients with PD (e.g., the very elderly or those with mild cognitive impairment at baseline). Obviously, all drugs directed at brain mechanisms carry risk for psychosis and confusion, including the antiparkinsonian drugs. Not all antiparkinsonian medications, however, have the same degree of risk. Those most likely to induce hallucinations, paranoia, or confusion are the anticholinergic medications, including trihexyphenidyl (Artane) and benztropine (Cogentin). Anticholinergics are best avoided in patients older than 65. Amantadine (Symmetrel) is included in this category because its side effects are similar to those of the other anticholinergics, although in this group amantadine is probably better tolerated.

The next group, in terms of highest risk for psychotic symptoms, consists of the direct dopaminergic agonists (e.g., pergolide [Permax], bromocriptine [Parlodel]). Indeed, studies have suggested that the incidence of hallucinations and decline on the Mini-Mental Status Examination are correlated with the use of dopaminergic agonists. It is unclear if the newer agonist drugs, pramipexole (Mirapex) and ropinirole (Requip), have different potential for cognitive and behavioral side effects, since comparative studies are lacking. There is some evidence that these newer dopamine agonists have a lower risk of cognitive impairment, estimated to be 15% for pramipexole and 8% for ropinirole.

The next group in descending order of risk is levodopa compounds. Psychotic side effects may be more frequent with the controlled-release formulation of carbidopa-levodopa (Sinemet CR) than with the immediate-release formulation.

Selegiline (Eldepryl), a selective monoamine oxidase type B inhibitor, is relatively well tolerated for monotherapy in terms of psychotic side effects. However, adjunctive use of selegiline with levodopa increases the bioavailability of levodopa, which in turn can increase the risks and side effects associated with levodopa therapy. For some patients, a small amount of selegiline is converted to amphetamine, and this may account for some of these effects. Rare complications of agitation and mania associated with selegiline use have been reported. This drug should be discontinued in hallucinating or delusional patients.

Demented patients or those with greater potential for psychosis who require treatment of parkinsonian motor symptoms are usually best managed with carbidopa-levodopa monotherapy. The ratio of benefit to cognitive side effects is greatest with this strategy.

Antipsychotic Treatment

The development of psychotic symptoms in patients with PD requires special treatment strategies. The typical antipsychotic medications (neuroleptics) are problematic because of the risk of exacerbated parkinsonian disability.

Certainly, the first attempt to treat psychotic symptoms is to use antiparkinsonian medications with the least risk of hallucinations and paranoia and in the lowest effective doses. However, if this fails and if the patient's parkinsonian symptoms demand continuing antiparkinsonian medical therapy, the addition of a neuroleptic drug to control the psychotic symptoms needs to be considered.

Two new drugs, olanzapine (Zyprexa) and quetiapine (Seroquel) produce a much lower incidence of extrapyramidal symptoms than conventional neuroleptics. Olanzapine has some potential to exacerbate parkinsonism, however; quetiapine appears to be less likely to do so and is probably the better choice. Quetiapine administration is usually started at one-half of a 25-mg tablet at bedtime, and the dose can be increased as needed. The range of quetiapine dosage for treatment of PD-related psychosis has not yet been fully established but may be between 12.5 and 200 mg twice daily. Olanzapine dosage is usually started at one-half to one 5-mg tablet at bedtime and can be increased to 10 to 15 mg once daily, if necessary.

The novel antipsychotic medication clozapine (Clozaril) is an appropriate choice for refractory psychosis. It is very effective at controlling hallucinations, often with very low doses. The risk of exacerbating PD symptoms is nil. However, clozapine is associated with a 1.8% risk of reversible agranulocytosis, which can be fatal. This necessitates weekly blood tests, which add to the expense and difficulty of its use. Other side effects are confusion, sedation, and increased salivation. Dosage typically starts with one-fourth to one-half of a 25-mg tablet each night. The dose can be increased as rapidly as every 3 days until the patient's condition is under control, with caution taken to avoid excessive sedation. Psychosis in many patients with PD can be controlled with as little as 25 to 50 mg of clozapine at bedtime. Doses up to 100 mg/day are occasionally necessary.

Nonpharmacologic Management and Care of the Caregiver

Nonpharmacologic management options are also available to help with the behavioral problems and to reduce stress on the caregiver. Establishing simplified daily routines and reducing novelty and stressors benefit confused and easily agitated patients. Modifications of the home may be helpful, such as putting labels on drawers, placing commonly used items in full sight, removing obstacles, and minimizing home clutter.

It is critical to consider the caregiver in any treatment strategy. Caregivers often develop psychologic reactions to the patient's illness that can be counterproductive, and this is particularly true if cognitive problems develop. Caregivers are then deprived of one of the few rewards of still caring for the patient, that is, social communication and companionship. Some caregivers who were in a more dependent role when the patient was healthy find it difficult to assume the more dominant role when the patient becomes more disabled. Some caregivers continue to defer decisions and responsibility to the patient when faced with the patient's significant cognitive decline.

Sustained nighttime sleep for the patient with PD is also critical for the caregiver. A night undisrupted by the patient's needs can give the caregiver the rest needed to provide for the patient during the day. Strategies for treatment of sleep disorders experienced by patients with PD are outlined in Chapter 10.

Sometimes, significant behavioral problems appear only after the motor symptoms have been relieved. Impulsiveness may not be a problem if the patient is bradykinetic enough to be unable to act out those impulses. Once the patient is more active, impulsiveness and lack of insight can contribute to significant problems for the caregiver and endanger the patient as well. Caregivers and perhaps the physician may be tempted to leave the patient more motorically disabled for the sake of behavioral control, creating an ethical dilemma. This situation can be avoided by providing or enlisting support for the caregiver.

It is also important for the clinician to address safety issues when appropriate. Cognitive impairment superimposed on bradykinesia may require that driving be prohibited. Use of certain power tools may be similarly dangerous. Finally, the clinician may need to counsel the family about taking business and financial matters out of the hands of the patient.

DEPRESSION

Depression is a common problem for patients with PD. Not only can the physical disabilities and socioeconomic burdens of the disease cause depression but also the neurochemical substrates of PD predispose to depression. Several studies have shown that depression often is not correlated with the severity of disability. Indeed, significant depression can antedate the clinical diagnosis. Therefore, depression should be considered a medical condition and part of the symptom complex of PD. For some patients, relating the depression to a medical condition may improve insurance coverage.

Depression should be aggressively treated. Some depression may be relieved by treatment of the motor symptoms, and specific treatment of mild depression may be deferred until the motor symptoms are controlled. Sometimes PD symptoms lessen with relief of depression. In addition, major depression in patients with PD is associated with greater impairment of cognitive functions.

Diagnosis

Although common among patients with PD, depression is often difficult to diagnose. The psychomotor retardation of depression can be mistaken for the bradykinesia of PD. The pseudodementia of depression could be mistaken for a dementia related to PD. Additionally, the stereotypical notion of the symptoms and signs of depression may interfere with the diagnosis. Elderly depressed persons may not have the dysphoria, such as crying spells, seen in younger persons, nor are the elderly as likely to have suicidal ideation. Conversely, it is not unusual for elderly persons, particularly those with chronic disease, to think about death or even welcome it. Inquiry about depression should be routine in evaluating patients with PD.

Just as with cognitive problems, the physician should consider the differential diagnosis of depression. The depression may be related to some other illness such as hypothyroidism or underlying malignant disease. The depression may be situational and related to the burdens the disease imposes on the patient or the patient's family. Medications prescribed for other conditions, such as propranolol (Inderal) or benzodiazepines for anxiety, may contribute to depression. Each possibility in the differential diagnosis may require slightly different treatment.

Treatment

Approaches to the treatment of depression include both psychologic measures and medications. Studies have repeatedly shown the effectiveness of either approach but have also pointed to a much greater efficacy when psychologic and medical treatments are coupled. The potential benefits of psychologic therapy, including those that follow, should not be overlooked.

1. Psychologic therapy can help the patient reestablish a sense of self-worth in the face of declining functional capacities.
2. It can help the patient maintain positive relationships with caregivers and family members despite increasing dependency.
3. It can help divert the patient from nihilistic thinking to more positive approaches to problem-solving.

ANTIDEPRESSANT MEDICATIONS

There are a wide variety of antidepressant medications, each with its distinctive advantages and disadvantages. The choice of class of antidepressants depends greatly on the patient's overall condition and specific needs. Amoxapine (Asendin) should be avoided because its antidopamine effects can exacerbate parkinsonism.

Bupropion (Wellbutrin) is theoretically advantageous in patients with PD because of its dopaminergic effects; it blocks dopamine reuptake. However, some find that bupropion has not been a consistently effective antidepressant. Many physicians treating depression, particularly in older patients, favor the selective serotonin reuptake inhibitors (SSRIs), such as fluoxetine (Prozac), sertraline (Zoloft), and paroxetine (Paxil). These compounds have proven to be very effective antidepressants, including for treatment of PD-related depression, and have a side effect profile that is much more benign than those of many others. Occasional side effects that could be troublesome include excessive agitation and sleep disturbance.

A potential disadvantage of the SSRIs is the possibility of increasing the parkinsonian symptoms. Although the exact risk for worsening parkinsonism is unknown, this side effect is probably infrequent and should not otherwise dissuade the physician from using SSRIs with careful observation.

The major concern about SSRIs in the treatment of PD is related to patients taking selegiline. There have been isolated reports of "serotonergic crises" linked to concurrent administration of these two drugs, manifested as combinations of confusion, disorientation, fever, tremor, myoclonus, diarrhea, and flushing. This reaction, potentially very serious, was fatal in one instance. The incidence of this complication is not known with certainty but most likely is extremely low. A survey of physicians caring for patients treated concurrently with selegiline and SSRIs showed that only 0.24% of patients had any evidence of serotonergic crisis and only 0.04% had any serious problems.

The decision whether to use selegiline concurrently with SSRIs is difficult. One has to weigh the potential advantages and problems. The advantages of the SSRIs are efficacy and high level of tolerance. The advantages of selegiline are the possibility of reducing the rate of progression of the disease, although this objective is controversial (see Chapter 7), and the ability to increase the bioavailability of levodopa and dopamine, which can reduce motor symptoms for some patients.

The traditional tricyclic antidepressants include amitriptyline (Elavil) and nortriptyline (Pamelor). These medications may pose a risk, particularly in older parkinsonian patients, of hallucinations, confusion, memory disturbance, and orthostatic hypotension. However, the anticholinergic properties of the traditional tricyclic antidepressants may be beneficial in selected cases; they have a mild antitremor effect and tend to reduce urinary urgency in patients with a hyperactive bladder. They are also sleep aids.

ELECTROCONVULSIVE THERAPY

Electroconvulsive therapy (ECT) can be valuable for treatment of depression associated with PD. It has the advantage of rapid onset of action, unlike medications, which may take several weeks to produce an effect. Moreover, parkinsonian symptoms, in addition to the depressive symptoms, often decrease with ECT. Thus, PD itself is not a contraindication to ECT. Patients with cognitive impairment, however, may have difficulty tolerating ECT, since transient (i.e., several days) confusion is common in such persons. Nevertheless, our practice is to aggressively treat depression, and we do not hesitate to use ECT if necessary.

SELECTED READING

Bayles KA. *Communication and Cognition in Normal Aging and Dementia* (Bayles KA, Kaszniak AW eds), Little, Brown and Company, Boston, 1987, pp. 325-359.

Bayles KA, Tomoeda CK, Wood JA, Montgomery EB Jr, Cruz RF, Azuma T, McGeagh A. Change in cognitive function in idiopathic Parkinson disease. *Arch Neurol* 1996; 53: 1140-1146.

Folstein MF, Folstein SE, McHugh PR. "Mini-mental state." A practical method for grading the cognitive state of patients for the clinician. *J Psychiatr Res* 1975; 12: 189-198.

Graham JM, Grunewald RA, Sagar HJ. Hallucinosis in idiopathic Parkinson's disease. *J Neurol Neurosurg Psychiatry* 1997; 63: 434-440.

Graybiel AM. The basal ganglia and cognitive pattern generators. *Schizophr Bull* 1997; 23: 459-469.

Kuzis G, Sabe L, Tiberti C, Leiguarda R, Starkstein SE. Cognitive functions in major depression and Parkinson disease. *Arch Neurol* 1997; 54: 982-986.

Lane RM. SSRI-induced extrapyramidal side-effects and akathisia: implications for treatment. *J Psychopharmacol* 1998; 12: 192-214.

Moellentine C, Rummans T, Ahlskog JE, Harmsen WS, Suman VJ, O'Connor MK, Black JL, Pileggi T. Effectiveness of ECT in patients with parkinsonism. *J Neuropsychiatr Clin Neurosci* 1998; 10: 187-193.

Richard IH, Kurlan R, Tanner C, Factor S, Hubble J, Suchowersky O, Waters C. Serotonin syndrome and the combined use of deprenyl and an antidepressant in Parkinson's disease. Parkinson Study Group. *Neurology* 1997; 48: 1070-1077.

Ripich DN. Functional communication with AD patients: a caregiver training program. *Alzheimer Dis Assoc Disord* 1994; 8 Suppl 3: 95-109.

Tamaru F. Disturbances in higher function in Parkinson's disease. *Eur Neurol* 1997; 38 Suppl 2: 33-36.

Tandberg E, Larsen JP, Aarsland D, Laake K, Cummings JL. Risk factors for depression in Parkinson disease. *Arch Neurol* 1997; 54: 625-630.

13 Surgical Treatment of Parkinson's Disease

Kathleen M. Shannon, MD

CONTENTS

PALLIDOTOMY
THALAMOTOMY
SUBTHALAMOTOMY
DEEP BRAIN STIMULATION
CEREBRAL TRANSPLANTATION
GENE THERAPY
CONCLUSIONS
SELECTED READING

The gentleman . . . is seventy-two years of age. About eleven . . . years ago, he first perceived weakness in the left hand and arm, and soon after found the trembling commence. In about three years afterwards, the right arm became affected in a similar manner: and soon afterwards the convulsive motions affected the whole body. . . . About a year since, on waking in the night, he found that he had nearly lost the use of the right side. . . . During the time of their having remained in this state, neither the arm nor the leg of the paralytic side was in the least affected with the tremulous agitation; but as their paralyzed state was removed, the shaking returned.

—James Parkinson, 1817

As early as James Parkinson's pivotal clinical description of paralysis agitans, it was recognized that acquired structural lesions in the motor pathways have the potential to relieve symptoms of the disease. Subsequently, surgical ablation of deep brain structures became the predominant treatment for Parkinson's disease (PD). Ablative operations were largely abandoned, however, when specific dopaminergic pharmacotherapy became available. In recent years, there has been a renewed focus on surgery, including the targeting of lesions at certain well-defined central nervous system loci and the emergence of techniques that favor protection or replacement of at-risk cell populations. Surgical treatment of PD has a checkered history. Nonetheless, patients continue to be attracted to surgical strategies that are glamorized by clinician proponents and the media.

From: *Parkinson's Disease and Movement Disorders:*
Diagnosis and Treatment Guidelines for the Practicing Physician
Edited by: C. H. Adler and J. E. Ahlskog © Mayo Foundation
for Medical Education and Research, Rochester, MN

This discussion of the surgical treatment of PD focuses on procedures that palliate an abnormal physiologic state by destroying or inducing dysfunction in brain structures (ablative operations) and on experimental techniques that attempt to replace degenerated neurons or deliver pharmacologic agents with nurturing or healing properties (restorative operations). Ablative procedures include pallidotomy, thalamotomy, and deep brain stimulation. Brain cell transplantation and gene therapy are examples of restorative procedures.

PALLIDOTOMY

Background

In 1952, Cooper inadvertently created the first pallidal lesion in a patient with PD by ligating an injured anterior choroidal artery during a neurosurgical procedure for an unrelated condition. The improvement in contralateral signs led him to investigate choroidal artery ligation and, subsequently, chemopallidotomy as therapy for PD. Other investigators applied stereotactic techniques to the procedure. Leksell and coworkers found that benefits were more consistent with a stereotactic approach to a site in the posterior portion of the globus pallidus interna (GPi). The recent resurgence of interest in pallidotomy was spurred by anecdotal reports and open-label trials of the Leksell technique and by the recognized limitations of long-term pharmacotherapy for PD. Better neurosurgical techniques and the ability to more accurately place lesions by use of electrophysiologic mapping of the globus pallidus have increased confidence in the procedure.

Rationale

Animal models suggest that the GPi is overly active in PD and that excessive inhibitory GPi outflow produces pathologic degrees of inhibition of thalamocortical pathways. This thalamocortical inhibition is believed to underlie bradykinesia in PD. In animals rendered parkinsonian by the specific neurotoxin 1-methyl-4-phenyl-1,2,3,6-tetrahydropyridine, surgical lesions that reduce pallidal outflow improve behavioral measures of parkinsonism. Surgical ablation of the GPi would thus be expected to reduce bradykinesia in human PD.

Description

The GPi is localized by stereotactic technique through use of magnetic resonance imaging (MRI) or computed tomography (CT) or both. At some centers, the lesion site is confirmed by microelectrode recordings on the basis of the characteristic firing patterns of pallidal cells and their response to passive and active limb movements. Several passes of the microelectrode can create a map of the sensorimotor GPi. Macroelectrode or microelectrode stimulation may then be used to determine that the planned lesion does not encroach on the visual pathways or internal capsule. Most surgeons produce a test (transient) lesion before actually destroying the targeted tissue, typically by heating the electrode tip. Some have proposed using the gamma knife for the surgical ablation, a technique that does not require any direct surgical incision through the skull. However, without the invasive neurophysiologic techniques to accurately localize the site, lesion placement may be less consistent; hence, gamma knife surgery has not gained widespread acceptance for this purpose.

No pallidotomy studies have compared the procedure with a sham operation. Among open-label trials of unilateral pallidotomy in patients with advanced disease, most authors have reported 15% to 70% improvements in measures of tremor, rigidity, and bradykinesia, particularly in the unmedicated, or "off," state. The improved motor function occurs principally contralateral to the lesion; however, more modest and sometimes transient benefits appear ipsilateral to the lesion and in such midline functions as gait and postural control. The effects of pallidotomy on symptoms in the levodopa "on" state are more variable. A striking benefit of pallidotomy is the alleviation of levodopa-induced dyskinesias on the sides contralateral to and, less prominently, ipsilateral to the lesion. This freedom from dyskinesia may substantially improve the quality of "on" time. On-period levodopa-resistant symptoms, such as gait freezing, do not respond to pallidotomy. The improvement in parkinsonian function does not generally allow a major reduction in the dose of antiparkinsonian agents. In a single study in which patients were randomly assigned to unilateral microelectrode-guided pallidotomy or best medical management, results were significantly better in the surgical group. The beneficial effects of pallidotomy appear to persist for at least 4 years.

The use of bilateral pallidotomy is controversial. Careful study has suggested that improvements following the second of two staged pallidotomies are less striking that those achieved after the first. Bilateral procedures may be associated with cognitive decline and severe speech disability.

Appropriate Candidates

The prototypical patient most likely to benefit from unilateral pallidotomy is relatively young and healthy, has quite asymmetrical idiopathic PD with a robust levodopa response complicated by motor fluctuations, and has troublesome dyskinesias on the side of maximal involvement. The magnitude of response to levodopa may predict the response to surgery. There are no absolute restrictions based on age or stage of disease. Patients with cognitive disability or significant psychiatric dysfunction are considered poor candidates for this approach. There has been some discussion that pallidotomy performed early in the course of the illness might avoid disability due to progressive bradykinesia and circumvent the development of treatment-related motor fluctuations and dyskinesia. Without data supporting this contention, however, this practice cannot be recommended. Patients with bilaterally symmetrical and severe disease are likely to be inadequately helped by unilateral pallidotomy and should seek other surgical approaches. Pallidotomy does not generally significantly reduce the need for antiparkinsonian agents, although control of dyskinesia may allow more effective dopaminergic pharmacotherapy.

Side Effects

A number of adverse effects of pallidotomy have been reported, including cortical or lobar intraparenchymal hemorrhage, ischemic stroke, hemianopsia, frontal lobe syndrome, hemiparesis, cognitive decline, and speech disorders. In most centers, the risk of significant permanent injury is less than 5%. At some centers, however, significant adverse effects have been reported in as many as one-third of patients. Although the surgical risks vary widely, most complications occur during the surgeon's initial experience with the procedure. The surgical risk may be greater in older patients. Systematic neuropsychologic assessments suggest that verbal fluency suffers after dominant hemisphere pallidotomy and that 25% to 35% of patients have modest, but clinically significant, personality changes of a frontal lobe character. Patients generally report, however, that the benefits of surgery outweigh the mild cognitive and behavioral changes. A number of centers have reported weight gain, sometimes substantial, after surgery. Bilateral placement of lesions carries a greater risk of morbidity.

Limitations

The benefits of pallidotomy may be significant, but normal function is not restored in patients with PD. In some series, improvements in ipsilateral and midline functions, such as gait and balance, are transient or modest. In the long run, symptoms ipsilateral to the surgical site worsen. Patients continue to require doses of antiparkinsonian agents similar to those required preoperatively. Most surgeons avoid doing bilateral pallidotomies because of the much less favorable risk-to-benefit ratio.

THALAMOTOMY

Background

In the 1950s and 1960s, ventrolateral thalamotomy grew in favor over pallidotomy because of better efficacy in treating parkinsonian tremor. The advent of levodopa therapy, however, largely spelled the end of stereotactic thalamotomy except in rare patients with horrific tremor poorly controlled with medications. Although interest in thalamotomy has increased a bit in parallel with interest in pallidotomy, its incomplete antiparkinsonian benefit has kept its use at lower levels.

Rationale

The neuroanatomical substrates responsible for tremor are less well understood than those for bradykinesia. Tremor appears to result from disinhibition of a facilitatory neuronal loop extending from the cerebral cortex to the striatum, to the globus pallidus, to the lateral ventral thalamic nuclear mass, and then to the cortex. Creation of a lesion in this loop should ameliorate tremor.

Description

Thalamotomy is performed under light sedation and local analgesia with MRI- or CT-guided stereotactic technique. The target is better defined if electrode recording is used to identify characteristic cellular activity. The lesion is produced in the nucleus ventralis lateralis, just anterior to the hand representation in the sensory (ventralis posterior) thalamus. The target is moved slightly laterally for leg tremor. A test lesion precedes the creation of a permanent lesion.

In series published before the widespread use of levodopa, the reported success rate of unilateral thalamotomy for contralateral tremor was between 70% and 90%. Although more recent series have been heavily weighted with patients in whom levodopa failed, similar success has been reported. With current technology, up to 90% of patients remain free of tremor at 2 years after surgery, and the mortality and permanent morbidity have ranged from 0.5% to 8%. No prospectively gathered long-term data are available on the efficacy of thalamotomy. However, data from retrospective series suggest that many patients have good tremor control 10 or more years after surgery. Some authors have suggested that the procedure has a salutary effect on disease progression; however, these contentions should be viewed with some skepticism in the absence of carefully performed prospective studies. Thalamotomy also tends to reduce contralateral levodopa-induced dyskinesias, although probably not as consistently or effectively as pallidotomy. Similar to the experience with pallidotomy, bilateral thalamotomy is associated with an increased risk of cognitive or speech dysfunction and is generally discouraged.

Appropriate Candidates

The ideal candidate for thalamotomy has PD that is characterized by severe, unilateral, medication-resistant tremor and does not have significant speech or gait disturbance. Patients older than 65 or 70 are considered poorer candidates for the procedure. Significant dementia or psychiatric disease should be considered relative contraindications. Patients with severe bilateral tremor may have unilateral thalamotomy followed by implantation of a thalamic stimulator on the opposite side or by bilateral thalamic stimulation (see below).

Side Effects

Major side effects of thalamotomy performed with current stereotactic techniques are relatively infrequent and include hand weakness or hemiparesis, speech disorders, and abulia. Transient confusion has also been reported. Rare instances of hemiballismus have developed after thalamotomy. There is a small risk of hemorrhage with every stereotactic procedure. Bilateral thalamotomy may result in severe dysarthria or muteness as well as cognitive dysfunction.

Limitations

Although thalamotomy may control tremor and rigidity, bradykinesia, dysequilibrium, and gait dysfunction are not relieved. Although some surgeons perform staged bilateral thalamotomies, most believe that the increased risk of significant speech disorder does not justify the procedure.

SUBTHALAMOTOMY

Background

A number of subthalamotomies were performed in the 1960s. Although results similar to those with thalamotomy were reported, the response rate was lower and the magnitude of response often less for this procedure. Significant postoperative hemiballismus was also reported.

Rationale

The pathologically increased activity of the GPi in patients with PD has been described above. The subthalamic nucleus (STN) exerts an excitatory influence on the GPi, potentially contributing to the pathologic overactivity of this nucleus. Thus, excessive activity of the subthalamic-GPi circuit probably contributes to the development of parkinsonian symptoms. Unilateral STN lesions in parkinsonian monkeys dramatically reduce contralateral parkinsonism and modestly benefit midline functions, such as facial expression. Various degrees of hemiballismus or hemichorea occur in the improved limb but often are not severe enough to impair voluntary movement. Additional support for the STN as a potential surgical target in PD comes from anecdotal reports of decreases in parkinsonism after STN hemorrhage.

Description

In general, the procedures followed to produce a stereotactic lesion in the STN are similar to those for placing a lesion in the thalamus or GPi. The small size of the STN and its close proximity to GPi efferents and the substantia nigra make microelectrode recording essential for accurate lesion localization. Among 10 patients treated with unilateral thermolytic lesions of the STN, axial and contralateral parkinsonism was reduced, whereas postoperative dyskinesias developed in only 1 patient. Data on this procedure are limited, however.

Appropriate Candidates

Appropriate candidates for subthalamotomy are patients with levodopa-responsive PD in whom medical management is insufficient and motor fluctuations have become disabling. Unless the risk of exacerbation of levodopa-induced dyskinesias is better defined, patients with severe levodopa-induced dyskinesias probably should not undergo the procedure. Indeed, because of the encouraging results of bilateral STN stimulation, as described below, and because of limited experience with subthalamotomy compared with other ablative lesions, this procedure probably should be abandoned in favor of bilateral STN stimulation.

Side Effects

As with any stereotactic deep brain procedure, there is a risk of mechanical, ischemic, or hemorrhagic damage to surrounding structures. Ballistic involuntary movements may develop contralateral to the side of surgery.

Limitations

The procedure has had insufficient study. Series are small, and there are no controlled data. Additional study is required to fully outline the risks and potential benefits of subthalamotomy.

DEEP BRAIN STIMULATION

Background

During electrical stimulation to assist with placement of lesion-producing electrodes for stereotactic thalamotomy, Benabid and colleagues noted that high-frequency thalamic stimulation effectively suppressed parkinsonian tremor. They proposed that long-term stimulation be used as a substitute for a permanent lesion. They first successfully applied the technique contralateral to unilateral thalamotomy in patients with bilateral tremor and subsequently as a primary intervention for tremor.

Stimulation of the pallidal or subthalamic nucleus is a logical extension of pallidotomy or subthalamotomy. Analogous to thalamic stimulation as a substitute for thalamotomy, stimulation of the pallidal or subthalamic nucleus reproduces the effect of a permanent ablative lesion in these nuclei. The presumed advantage, however, is that this allows for a graded, reversible effect and bilateral surgical procedures.

Rationale

The goal of deep brain stimulation is the same as that for deep brain lesions: to reduce the activity of overactive brain structures. Although the mechanism remains unknown, it is theorized that high-frequency electrostimulation inhibits neuronal firing either by inducing depolarization blockade or by facilitating the action of inhibitory interneurons. Deep brain stimulation has been used as an approach to every deep brain site for which lesion placement results in therapeutic benefit. Although data are incomplete, the prevailing view is that deep brain stimulation can be safely performed bilaterally.

Description

Deep brain stimulators are implanted under local anesthesia in unmedicated patients. The site of placement is localized by stereotactic techniques. Microelectrode recording further confirms the area for implantation via observation of well-described specific neuronal firing patterns and their response to limb movements. Macroelectrode stimulation determines that the stimulating electrode is not placed too close to sensitive brain regions.

The pulse generator is implanted in the subclavicular chest wall, similar to placement of a cardiac pacemaker. Stimulation settings can be optimized by an external computerized device that programs stimulation frequency, pulse width, and intensity and also allows the choice of the four stimulation sites near the electrode tip. Patients can turn the stimulation on or off with a handheld magnet. A number of implantation sites have been suggested for PD, but only thalamic stimulation is currently approved by the Food and Drug Administration. Trials of bilateral thalamic, unilateral and bilateral GPi, and bilateral STN stimulation are under way. As with deep brain lesions, the effects of stimulation occur contralateral to the implantation and maximal therapeutic effect requires bilateral implantation.

With any of the stimulator surgical interventions, mechanical failure is possible, and there have been a number of reports of broken leads and other mechanical difficulties. The batteries must be replaced periodically, as often as every 2 years.

Thalamic Stimulation

The first brain target studied for long-term stimulation was the ventralis intermedius nucleus (VIM) of the thalamus. In the series by Benabid and associates, 88% of 91 patients with PD continued to have a good response to stimulation at 6 months. Rigidity, pain, and levodopa-induced dyskinesias were reduced in some patients, but there was no effect on akinesia. The treatment was effective for up to 11 years. In a multicenter study in which raters were unaware of whether the stimulator was on or off, 58% of patients with PD had complete relief of tremor and an additional 38% had partial tremor relief with the stimulator turned on. Favorable results in clinical trials led the Food and Drug Administration to approve VIM stimulation as a relatively safe and effective therapy for parkinsonian rest tremor. The complications are similar to those of thalamotomy, although the relative frequencies have yet to be adequately compared.

The effects of VIM thalamic stimulation are limited to tremor control, and patients with progressing PD may eventually become disabled by akinesia and impairment of postural reflexes, requiring more aggressive pharmacotherapy or additional surgery. Levodopa-induced dyskinesias may also decrease with this procedure. A postmortem study of a patient with PD who had been treated with a VIM stimulator for 43 months before death showed small areas of gliosis and spongiosis surrounding the track and an accumulation of T lymphocytes above the active tip of the electrode. Whether these low-grade pathologic findings have implications for the long-term outcome of deep brain stimulation is open to speculation.

Pallidal and Subthalamic Nuclei
Deep Brain Stimulation

Although not currently approved for treatment of PD symptoms, stimulating the GPi or the STN offers a broader spectrum of antiparkinsonian efficacy than thalamic stimulation. Preliminary data suggest that STN stimulation may be the preferred procedure for the bradykinetic symptoms of PD. Bilateral STN stimulation improves motor scores in both the "on" and the "off" states of levodopa. On the surface, this result compares favorably with pallidotomy or GPi stimulation; with these pallidal procedures, the improvement is primarily restricted to "off"-state symptoms. Controlled comparisons of these procedures are very limited, however. Somewhat surprisingly, levodopa-induced dyskinesias also may be attenuated with STN stimulation. This occurs through reduction in the levodopa dosage, permitted by the antiparkinsonian effect from STN stimulation. Some initial reports of bilateral STN stimulation, however, suggest a potential for mild cognitive or personality changes.

CEREBRAL TRANSPLANTATION

Rationale

In the simplest sense, cerebral transplantation aims to replace dead neurons. The ultimate goal is to introduce a new cell population that will make afferent and efferent connections with host brain cells and function like the neurons they replace. PD is ideally suited for such an approach because the degenerative process underlying most of the motor symptoms is well localized; a single cell type is lost, and the motor symptoms respond to replacement of dopamine, the neurotransmitter synthesized by those cells. Experimental evidence in animal models of PD suggests that transplanted fetal cells survive in large numbers, integrate with the host brain, forming afferent and efferent connections, and produce behavioral effects. Transplantation of dopaminergic neurons into the brains of patients with PD might relieve symptoms by providing a passive source of dopamine or might reestablish a more normal neurologic circuitry and regulated dopamine release.

Experience

From several perspectives, valuable lessons can be learned from the history of transplantation in PD. The first transplantations performed for PD were autologous grafts of adrenal medullary tissue. After initial reports of marked benefit by Madrazo and associates, more than 300 adrenal-brain transplantations were performed worldwide before it became apparent, largely by analysis of data submitted to a voluntary registry, that the procedure had high morbidity and only modest, transient benefits. This experience indicated the need for controlled trials with standardized study designs and outcome measures.

Transplantation of fetal midbrain (containing dopaminergic substantia nigra) to the striatum of patients with PD followed adrenal-brain transplantation, but these early trials, too, were not standardized or controlled. Some "improvements" in PD symptoms and signs involved transplantation techniques that we now know from animal studies should not have produced viable or anatomically integrated grafts. Thus, the early experience with transplantation has raised the specter of investigator bias, placebo effects, and nonspecific responses to the surgery, underscoring the need for controlled clinical trials.

Considerable animal data have become available from which to extrapolate optimal transplantation technique with respect to selection of donor tissue, donor age, amount of tissue implanted, preparation of tissue, and sites chosen. Inadequacies in the early transplantation experience were often related to poor cell survival, implantation of inadequate numbers of cells, or implantation into inappropriate regions of the host brain.

Current Status

There have been a number of pilot studies of human fetal substantia nigra transplantation into the striatum of patients with PD. In one such open-label clinical trial, fetal cells from the ventral mesencephalon of six to eight donors between 5 and 9 weeks of postconceptional age were stereotactically implanted into the postcommissural putamen bilaterally in staged operations. Patients were immunosuppressed with cyclosporine perioperatively and for 6 months postoperatively. Significant and "meaningful" benefit was seen in graft recipients. These changes included reduced time spent "off," reduced disability when "off," improvements in "on" time, and reduced dependence on antiparkinsonian medications. These improvements followed a delayed onset and gradual development, as expected from a transplant effect. Positron emission tomography studies confirmed a progressive increase in fluorodopa uptake postoperatively, consistent with viable grafts. Most importantly, autopsy studies in two patients demonstrated viable dopaminergic neurons with projections crossing the graft-host interface and innervating the host brain. The graft showed no evidence of the neurodegenerative process of PD. Although these preliminary results are encouraging, parkinsonian symptoms were only partially relieved by the procedure and only 5% to 10% of grafted neurons survived transplantation. Before the technique can become an accepted therapy for PD, more robust cell survival and clinical effect must be demonstrated. For the technique to become widely used, it must also be accessible to physicians outside major research centers. At present, the technique requires a very high level of expertise. The two fatalities reported may have been due to operative technique; one patient had acute obstruction of the ventricular system and the other had development of an intraparenchymal cyst.

Porcine fetal cells have been proposed as an alternative to human fetal brain tissue for transplantation in PD. The potential advantage is that there is a limitless source of cells raised and harvested in a controlled environment. The technique has several potential drawbacks, particularly the need for long-term immunosuppression in xenograft recipients and the unknown risks of zoonosis. A phase I trial of unilateral grafts of porcine fetal substantia nigra cells in 12 patients with PD showed that the procedure could be done safely. Despite relatively poor cell survival documented by a single autopsy case, clinical benefits have been apparent in some patients. Demonstration of efficacy requires a larger controlled trial of bilateral transplantation.

Problems

The problems associated with transplantation can be divided into those associated with the scientific method and those associated with societal concerns. There are seemingly endless scientific questions related to how best to restore dopaminergic circuitry in the human brain. Some, such as optimal donor age, volume of cells, graft location, and methods of enhancing cell survival, can be answered in part in the animal laboratory. However, only controlled clinical trials will provide a believable answer to the question of clinical utility of methods developed in the laboratory. Patients selected for such heroic studies are likely to be young and exceptionally healthy and not really reflective of the PD population at large. Multicenter clinical trials with more lenient eligibility requirements would be required to better define the potential benefits for the average patient with PD undergoing the procedure outside highly specialized academic centers. Scientific rigor demands that such studies be adequately controlled, but the use of sham or placebo surgical arms remains controversial. Prospective controlled studies of human fetal substantia nigra transplants are currently under way, and the results will help better define the usefulness of human fetal transplantation.

Societal questions relate to the ethics of obtaining and using donor fetal tissue as well as the societal cost-to-benefit ratio of using technologically advanced and very expensive methods to treat neurodegenerative disease in a largely elderly population. Spontaneous abortions are not a realistic source of tissue; tissue from induced abortions is required. The acquisition of sufficient quantities of high-quality tissue has been difficult, even for small clinical trials, and would doubtless be a major obstacle to more widespread use of the procedure. Despite diligent adherence to local, state, and federal laws and to ethical guidelines for the use of human fetal tissue as outlined by the National Institutes of Health, antiabortion activists would certainly be galvanized by such high-profile use of human fetal tissue. The cost of large-scale use of this technology would tax the health care system.

On the other hand, xenografts (e.g., porcine tissue) represent a potentially inexhaustible source of homogeneous tissue raised and harvested under rigorously controlled environmental conditions. However, the use of xenografts raises several alarming considerations, notably a more acute risk of a robust immunologic response to the graft and the risk of zoonotic disease, to which the "mad cow" experience has made society more vigilant. Donor pathogens may include bacteria, prion-like molecules, parasites, and viruses. Introducing such pathogens into humans might threaten the health of the patient as well as the population at large. Indeed, porcine cells have recently been demonstrated to harbor C-type retrovirus particles that can infect human cells. Although there have as yet been no cases of human infectious disease following implantation of a porcine graft, such pathogens may have a very long latent period.

Prospects for the Future

Human fetal cell transplantation in its current form is unlikely to become an accepted therapy for PD. Development of a renewable source of pluripotent human cells (such as genetically modified established cell lines or autologous cells, or stem cells) and enhancements of technique that greatly increase cell survival and improve the distribution of cells into the denervated brain would make the procedure a more viable treatment. With these limitations in mind, there is an active effort to develop alternative sources of cells for neurotransplantation.

GENE THERAPY

Preclinical studies of gene therapy suggest the potential usefulness of this family of techniques for PD. Gene therapy techniques for PD could proceed along several lines: (1) replacement of a defective gene, (2) introduction of a gene product that may protect against or reverse neurodegeneration, and (3) introduction of a gene to control the synthesis of dopamine. Since the role for inheritance in typical PD remains unknown, only the second and third theoretical constructs apply to PD at present.

There are two basic approaches to gene therapy. Ex vivo gene therapy involves the in vitro genetic modification of cells, which are then transplanted into the organism. Alternatively, genes could be directly transferred to target tissues in vivo by use of viral vectors. An advantage of ex vivo gene therapy is that the expression of the transgene can be assessed before implantation. The transgene can be introduced into cells in culture by plasmid or retroviral vectors.

One ex vivo strategy has been the transfection into cell lines of genes coding for neurotrophic factors, such as brain-derived neurotrophic factor, glial cell-derived neurotrophic factor, and ciliary neurotrophic factor. Theoretically, these might slow the neurodegenerative process or facilitate neuronal repair mechanisms. In animal studies, transplantation of cells engineered to express neurotrophic factors has partially protected against the effects of specific nigral neurotoxins. Another potential use of neurotrophin-expressing cells is as a co-graft to increase the survival of human fetal tissue.

The second ex vivo strategy involves transplantation of cell lines transfected with dopamine synthetic enzymes. Once transplanted, these can become a continuing source of dopamine. However, it would be preferable that they also become integrated into the host brain and form synaptic connections to allow physiologic control of dopamine stimulation at the receptor. For such integration to occur, the choice of cell line would be critical.

The theoretically ideal cell for ex vivo gene therapy should be easy to obtain and culture, express transgenes for a long duration, and survive in the brain without excessive immunologic response or excessive mitotic potential. Skin fibroblasts, Schwann cells, astrocytes, central nervous system progenitor cells, and central nervous system cells transformed by temperature-sensitive oncogenes have been studied. Using neural cells is theoretically advantageous, because they have the potential to make afferent and efferent connections within the host brain. However, even non-neural cells have shown promise in animal models of PD.

In vivo gene transfer could have advantages over ex vivo procedures because host central nervous system cells are inherently integrated into the neuroanatomical framework and may have a greater capacity than implanted cells to synthesize, package, and secrete the desired products. This technique also avoids the intrusion posed by transplanted cells on the normal brain architecture and allows for delivery of intracellular proteins. A number of viral vectors have been investigated for this procedure, including herpes simplex type 1, adenovirus, adeno-associated virus vector, retrovirus, and lentivirus. They differ along several lines: (1) potential to infect postmitotic cells, (2) efficiency of infection, (3) long-term expression of the transgene, (4) tendency to revert to wild type, and (5) subsequent potential for cytotoxicity and immune response. As an alternative to viral vectors, double-stranded DNA plasmids containing an expression construct can be used, but these appear less promising at present. In vivo gene therapy has been used in animal models to deliver neurotrophic factors and dopamine synthetic enzymes. Potential technical drawbacks to the use of in vivo gene therapy include (1) difficulty spreading the therapeutic agent throughout the affected brain regions while minimizing injection-related trauma and (2) overcoming the effects of concentrated pockets of therapeutic agent at each injection site, with the potential for local toxicity. At present, gene therapy for PD remains on the distant horizon.

CONCLUSIONS

Failings in our ability to adequately control advanced PD have become obvious over more than two decades of pharmacotherapy. In the rosiest outcomes, surgical techniques offer the promise of permanent therapeutic benefit when medications have proven inadequate. Surgical procedures also hold the potential to injure patients. Unfortunately, there is a reluctance to hold surgical procedures to the highest standard of proof, the randomized, placebo-controlled, double-blind study. It appears that ablative surgical procedures are effective in the treatment of certain well-selected patients, with a focus on certain clinical problems (e.g., thalamotomy for tremor, pallidotomy for levodopa-induced dyskinesias). Deep brain stimulation may ultimately supplant ablative surgery, but such a conclusion is premature at present. Preliminary evidence of the efficacy of transplantation in the treatment of advanced PD awaits confirmation in controlled trials now under way. Gene therapy offers the promise of a restorative technique free from the social and ethical constraints associated with the use of human fetal tissue. Thus, surgery for PD is a work in progress.

SELECTED READING

Baron MS, Vitek JL, Bakay RA, Green J, Kaneoke Y, Hashimoto T, Turner RS, Woodard JL, Cole SA, McDonald WM, DeLong MR. Treatment of advanced Parkinson's disease by posterior GPi pallidotomy: 1-year results of a pilot study. *Ann Neurol* 1996; 40: 355-366.

Freeman TB, Olanow CW, Hauser RA, Nauert GM, Smith DA, Borlongan CV, Sanberg PR, Holt DA, Kordower JH, Vingerhoets FJG, Snow BJ, Calne D, Gauger LL. Bilateral fetal nigral transplantation into the postcommissural putamen in Parkinson's disease. *Ann Neurol* 1995; 38: 379-388.

Goetz CG, Stebbins GT III, Klawans HL, Koller WC, Grossman RG, Bakay RA, Penn RD. United Parkinson Foundation Neurotransplantation Registry on adrenal medullary transplants: presurgical, and 1- and 2-year follow-up. *Neurology* 1991; 41: 1719-1722.

Kang UJ. Potential of gene therapy for Parkinson's disease: neurobiologic issues and new developments in gene transfer methodologies. *Mov Disord* 1998; 13 Suppl 1: 59-72.

Koller W, Pahwa R, Busenbark K, Hubble J, Wilkinson S, Lang A, Tuite P, Sime E, Lazano A, Hauser R, Malapira T, Smith D, Tarsy D, Miyawaki E, Norregaard T, Kormos T, Olanow CW. High-frequency unilateral thalamic stimulation in the treatment of essential and parkinsonian tremor. *Ann Neurol* 1997; 42: 292-299.

Limousin P, Krack P, Pollak P, Benazzouz A, Ardouin C, Hoffmann D, Benabid AL. Electrical stimulation of the subthalamic nucleus in advanced Parkinson's disease. *N Engl J Med* 1998; 339: 1105-1111.

Pahwa R, Wilkinson S, Smith D, Lyons K, Miyawaki E, Koller WC. High-frequency stimulation of the globus pallidus for the treatment of Parkinson's disease. *Neurology* 1997; 49: 249-253.

14 Adjunctive Therapies in Parkinson's Disease: Diet, Physical Therapy, and Networking

Padraig E. O'Suilleabhain, MB
and Susan M. Murphy, MD

CONTENTS

DIET
REHABILITATION
PSYCHOLOGIC AND SOCIAL SUPPORT
SELECTED READING

Pharmacologic and surgical advances in recent decades have resulted in increasingly successful symptom relief for persons with Parkinson's disease (PD). Diet management and physical therapies as well as psychologic and social support continue to have important complementary roles in the optimal management of the patient with PD. In this chapter, the principles and practice of these aspects of care are addressed.

DIET

Caloric Needs

Dietary recommendations often are the same for a patient with PD as for an unaffected adult of the same age. In persons older than 51 years, caloric requirements are about 20 kcal/kg per day for basal energy expenditure and approximately half again as much for typical daily physical activities. For a 77-kg man and a 63-kg woman performing usual activities, average requirements are 2,300 and 1,900 kcal/day, respectively, with individual variability of about 20%. Despite bradykinesia and resulting immobility, which might be expected to decrease caloric needs in PD, weight loss is prevalent, and a greater percentage of patients are below ideal body weight than are controls. The loss may reflect either lower caloric intake, for reasons discussed below, or a higher energy expenditure than expected. Increased metabolism by rigid muscles is a partial explanation for higher expenditure, as is the metabolic cost of severe dyskinesias and tremor.

From: *Parkinson's Disease and Movement Disorders:*
Diagnosis and Treatment Guidelines for the Practicing Physician
Edited by: C. H. Adler and J. E. Ahlskog © Mayo Foundation
for Medical Education and Research, Rochester, MN

Reasons for an insufficient caloric intake are as follows:

- Depression. Major depression occurs in perhaps 20% of patients with PD. Lesser degrees of dysphoria are at least as common and may occur, for example, during periods of adjustment to advancing physical limitations. Depression may remain undetected, because the physical manifestations of altered affect may be mistaken for the signs of PD. Whether pharmacologic (see Chapter 12) or psychologic and social interventions are used for treatment, the physician should recognize all forms of depression as risk factors for an inadequate diet.
- Bradykinesia, dyskinesia, tremors, and dementia. These abnormalities may impair the ability to feed oneself as well as to shop for and prepare food. The assistance of family members, meals-on-wheels, and home-help services can circumvent these limitations.
- Nausea and aversion to food as side effects of medications. These effects may be countered by slow titration of doses, supplemental carbidopa (see Chapter 8), and, on occasion, selected antinausea medications.
- Early satiety due to delayed gastric emptying.
- Dysphagia. Although prevalent in PD, dysphagia rarely causes malnutrition on its own. If dysphagia is an early symptom, a parkinsonism-plus syndrome should be considered as an alternative diagnosis. A swallowing evaluation and videofluoroscopic swallowing study may help identify the specific swallowing deficiency. An occupational therapist can advise on treatment strategies, such as neck positioning, modifications in food consistency, and airway protection. The consistency of food may need to be softened because of either dysphagia or easy fatigability of the masticatory muscles. Sialorrhea due to infrequent swallowing that does not improve sufficiently with levodopa and related medications can be relieved by propantheline or other peripherally acting anticholinergic agents, although the consequent drying of the oral mucosa may not be favored by many patients.

Timing of Meals and Protein Restrictions

Several aspects of levodopa pharmacokinetics are affected by food. First, gastric emptying tends to be slowed by PD-related dysautonomia, and emptying is further delayed by large meals, particularly meals with high fat and protein contents. Thus, medication absorption (and thus the latency to pharmacologic response) is delayed. Meals also tend to prevent administered levodopa from reaching its site of action. There are several reasons for this, including premature decarboxylation or degradation of levodopa with prolonged gastric time. Second, dietary iron appears to chelate levodopa, inhibiting its absorption; iron supplements, when indicated, are best taken separated in time from medications. Third, and most important, passage of levodopa from the intestine into the bloodstream and then from the bloodstream into the brain occurs via a saturable transport system shared with other large neutral amino acids. Thus, levodopa transport tends to be competitively inhibited by dietary amino acids. Consequently, a subtherapeutic response may occur when levodopa administration overlaps with intake of dietary protein. In the most sensitive patients with fluctuating responses to levodopa, a protein bolus can effectively turn off the levodopa motor response, even when consumed some hours after the medication.

The effect of dietary protein is usually not an important issue in patients with early-stage PD taking immediate-release carbidopa-levodopa three times a day. To minimize any potential problems, we simply ask all patients to take levodopa on an empty stomach, 1 hour before each meal. This advice may contradict the pharmacist's instructions to take the medication with food. Although it is acknowledged that side effects such as nausea are less likely when levodopa is taken with food, patients generally tolerate the medication on its own if administration begins with a small dose (one 25/100 carbidopa-levodopa tablet or, if necessary, ½ or ¼ tablet) that is gradually increased. A cracker or dry bread may be eaten, if necessary, because it has minimal interaction with levodopa.

Whereas administration of immediate-release carbidopa-levodopa an hour before meals should be sufficient to assure adequate passage to the brain, the situation is more complex with the controlled-release formulation (Sinemet CR). Some recommend that controlled-release carbidopa-levodopa be taken with meals because levodopa entry into the bloodstream from this formulation is increased with mealtime dosage. However, the inhibitory effect of dietary amino acids on levodopa passage across the blood-brain barrier probably outweighs any advantages from mealtime doses. Others advocate administration of the controlled-release formulation at least 2 hours before each meal, since the time required for this formulation to be released into the circulation, cross the blood-brain barrier, and be taken up by brain cells can be 2 hours. Although perfectly logical, this timing of administration has yet to be assessed in any comparative controlled trial.

Further dietary adjustment may be warranted for levodopa-related motor complications, such as an inadequate "on" effect or "on-off" fluctuations. Although there are a number of causes of suboptimal levodopa responses, pharmacokinetic interactions of food with levodopa are among the most common remediable causes of response fluctuations in our experience. This is most clearly the case when the patient notices a good response while fasting, for example, the first thing in the morning, but inadequate responses when the doses are taken during or after mealtimes. In such instances, we focus on educating the patients (and spouses or other caregivers) on the potential interactions between meals and levodopa. With this knowledge, they can shift mealtimes when necessary to guarantee a predictable empty-stomach levodopa response for social or other important occasions. They also learn to avoid between-meal protein snacks. For most levodopa-treated patients with fluctuations, this is the most practical strategy.

Two major dietary adjustments have also been recommended for patients with troubling meal-levodopa interactions: protein redistribution and protein restriction. For redistribution, the protein intake over 24 hours is not changed, but a larger fraction of the whole is taken in the evening. Low-protein foods are eaten at breakfast and lunch, with high-protein foods avoided (examples in Table 14-1). Daytime levodopa responsiveness is thus improved at the expense of a diminished response in the evening. A dietitian with experience in managing PD is necessary to help design this diet for the individual patient.

The protein restriction diet, also sometimes tried in certain patients, consists of limiting protein intake. For healthy adults, the recommended daily intake of protein is 0.8 g/kg per day. The most commonly proposed protein restriction diet reduces protein intake to 0.6 g/kg per day. Without any intervention, Americans tend to eat far more than 0.8 g/kg per day, so that a less aggressive restriction diet involves simply limiting intake to 0.8 g/kg per day. In patients with a diminished or fluctuating response to levodopa, a 2-week trial of the protein-restricted diet is reasonable. The dietitian can again help in instructing the patient on how to maintain a healthy diet under these protein limitations. To provide the same number of calories as before, the dietary composition requires more carbohydrate and fat. The patient, family, and physician should be alert to the possi-

Table 14-1
Foods High or Low in Protein

Foods high in protein

    Meat
    Fish
    Eggs, egg substitutes
    Cheese
    Milk
    Yogurt
    Custard
    Ice cream
    Peanut butter
    Nuts
    Soybeans
    Sunflower seeds
    Legumes

Foods low in protein

    Fruits and juices
    Vegetables
    Potatoes
    Rice
    Pasta
    Cereals
    Sweets
    Fats, bakery goods

bility of malnutrition, both at baseline evaluation and at subsequent visits when weight changes are reviewed. Three-month follow-up studies indicate that malnutrition and weight loss are usually not a problem. However, many patients discontinue these diets despite an improved pharmacologic response because the meals are less palatable and less satisfying.

Vitamins

The use of vitamin supplements in PD and in other neurodegenerative conditions has been promoted, but evidence for a neuroprotective effect remains unconvincing. If oxygen free radicals are confirmed to be pathogenic in these diseases, as believed by many, antioxidant vitamins E and C would theoretically be of benefit. Vitamin E (α-tocopherol) in doses of 2,000 units a day was assessed to determine whether progression of PD could be delayed. In the Deprenyl and Tocopherol Antioxidant Therapy of Parkinsonism (DATATOP) study, no benefit over placebo was apparent for this vitamin either alone or in combination with deprenyl, an inhibitor of monoamine oxidase type B. In a similar study involving patients with Alzheimer's dementia, the primary end point of institutionalization, death, or severe dementia was mildly delayed by vitamin E, although the statistical methods in this study have been questioned. No controlled clinical trials have assessed whether vitamin C or β-carotene (the common dietary form of vitamin A) has any effect on the progression of PD. Choline supplements, taken by some in an attempt to increase levels of the neurotransmitter acetylcholine, appear to exacerbate PD symptoms, if anything.

Patients often ask if they should take a multivitamin. In the general population, a healthy, diverse diet most likely provides all necessary vitamins and nutrients. However, plasma levels of vitamin A, vitamin E, iron, and zinc were shown to be low in some patients with PD. Therefore, vitamin supplements are appropriate if there is any doubt about the quality of the diet. An inexpensive multivitamin with 100% of the recommended dietary allowance of the common vitamins makes most sense.

Occasionally, one sees the recommendation that patients with PD avoid pyridoxine (vitamin B_6). This is a throwback to the original experience with levodopa without carbidopa. Pyridoxine facilitates the premature decarboxylation of levodopa to dopamine outside the brain, compromising its effectiveness. This was of concern when levodopa was first introduced. However, once the drug began to be formulated with carbidopa (which blocks dopa decarboxylase), pyridoxine intake became of no concern.

Dietary Supplementation to Prevent Osteoporosis

Osteoporosis is a relevant concern among patients with PD. Osteoporosis is a later-life disorder, often developing concurrently with PD. Although osteoporosis is frequently thought of as a disease of women, in fact, men are also at risk. Patients with PD may be at greater risk for this condition and also for untoward consequences due to falls. Recent studies have suggested that vitamin D may be insufficient in many patients with PD. This appears to be a result of both decreased intake of milk (and other sources of vitamin D and calcium) and changes in vitamin D metabolism. Also, some patients with PD restrict milk intake because of inhibitory effects of the protein content on the levodopa response. Osteoporosis may also be exacerbated in some patients with PD because of limited physical activity.

In general, we favor vitamin D and calcium supplementation in senior patients with PD as well as in certain younger patients. The necessary amount of daily vitamin D is contained in a single over-the-counter multivitamin tablet. One tablet daily should be sufficient, but in some cases two tablets, doubling the daily vitamin D intake, may be administered. The recommended daily calcium intake is 1,200 mg for men and premenopausal women and 1,500 mg for postmenopausal women. This should be in the form of elemental calcium; calcium supplements, such as Tums tablets, display the elemental calcium content on the label. These supplements should be accompanied by appropriate recommendations for exercise.

A nuclear medicine bone density study is an appropriate consideration, especially in postmenopausal women or others who seem to be especially at risk for complications from osteoporosis. Additional drug therapy, such as estrogens in women, may then be considered on the basis of the outcomes.

Dietary Treatment of Constipation

Constipation occurs in patients with PD as a consequence of pathologic involvement of enteric and perhaps central autonomic system neurons. Medication side effects and decreased physical activity may contribute to constipation. To manage this complaint, a series of measures may be tried with a minimum goal of a bowel movement every 3 days without excessive straining. First, fluid intake can be increased to 2.5 L/day, physical activity encouraged, and foods with a higher proportion of fiber (fruits, vegetables, legumes, bran, and whole-grain foods) suggested. Some patients find that a morning cup of prune juice warmed in the microwave is a helpful adjunct. A bulk-forming agent containing psyllium and methylcellulose may be added at this stage, followed by a stool softener such as docusate. More refractory constipation may require lactulose solution (10 g/15 mL; 1 to 4 tablespoons) or, alternatively, the stimulant laxative bisacedyl (oral tablet or suppository). Those having no stools in the preceding 3 days may require an enema to avoid obstipation and impaction.

Dietary Treatment of Orthostatic Hypotension

Some patients with PD experience symptoms of orthostatic hypotension, a problem that is exacerbated by antiparkinsonian medications. Standing blood pressures that are frequently less than 90/60 mm Hg require treatment. Increased fluid (eight glasses a day); liberalized dietary salt and, if necessary, one salt tablet twice a day; and caffeine may be beneficial. Unnecessary antihypertensive medications are, of course, eliminated, the head of the bed is raised, and patients are encouraged to arise gradually from a lying to a seated to a standing position. Fludrocortisone and adrenergic agonists are used if these measures fail; this topic is discussed in more detail in Chapter 11.

Dietary Factors as Cause or Cure

The search for environmental agents as a cause of PD was given impetus with the recognition of toxin-induced syndromes that have similarities to PD. These included levodopa-responsive parkinsonism with loss of substantia nigra cells after illicit injection of the meperidine analogue 1-methyl-4-phenyl-1,2,3,6-tetrahydropyridine (MPTP). Guam disease, which has clinical overlaps with parkinsonism, dementia, and amyotrophic lateral sclerosis, has been linked with a neurotoxin found in cycad plants, although the association is controversial.

Despite many investigative efforts, however, no dietary neurotoxin has been identified as a risk factor in epidemiologic studies of PD, although a number of leads have been raised by such studies. Analyses of food diaries in PD (recorded prospectively in one study) have not turned up any important associations. Neither dietary omissions nor dietary commissions in the years preceding onset of the disease were found to impressively increase the risk of PD. Among persons in whom PD later developed, intake of carbohydrates and animal fats may have been somewhat higher and of legumes and niacin somewhat lower, but the importance of these findings is uncertain. Intakes of antioxidants, such as vitamins E, C, and A, were generally no different among groups in most but not all studies.

Thus, there is no evidence that any food prevents the development of parkinsonism, cures the disease, or even slows its progression. Certain plants containing levodopa or other dopamine precursors can relieve symptoms. For example, $\frac{1}{4}$ lb of fava beans per day is estimated to provide a pharmacologic dose of levodopa. This "natural" medication cannot be recommended in preference to pills, which have standardized constituents and are coadministered with carbidopa, which reduces side effects. Bananas contain a significant amount of dopamine, but ingested dopamine (in contrast to levodopa) does not gain access to the brain. Foods containing belladonna alkaloids or other nonsynthetic anticholinergic agents might be expected to reduce some PD symptoms. Indeed, the modest palliative effects of belladonna were recognized in the years before discovery of levodopa.

REHABILITATION

Impairments in motor control associated with PD result in diminished functional abilities. Physical interventions aimed at limiting this functional decline are an important part of the therapeutic program for the patient with PD. The overall rehabilitative goal is to maximize the patient's functional independence and thus enhance quality of life at all stages of the disease. The design of a rehabilitation program for an individual patient is best done by a physician skilled in neurorehabilitation in consultation with other professionals, such as a physical therapist, occupational therapist, speech therapist, and recreational therapist. The program is tailored to each patient on the basis of the results of an initial multidisciplinary assessment. The types of disability identified by this evaluation can then be assigned priority according to the input of the patient and caregiver, with the high-priority goals being addressed first.

Assessment

A unique problem in PD is hour-to-hour and day-to-day variability, which must be considered when a functional assessment is done. The use of standard assessment tools, when available, facilitates communication among professionals and allows for objective measures of change over time. The initial assessment should include the following characteristics:

- Bradykinesia and rigidity. Established scales, such as the Unified Parkinson's Disease Rating Scale (Fahn S. *Recent Developments in Parkinson's Disease*. Raven Press, New York, 1986), can be used to assess the degree of bradykinesia and rigidity and how these symptoms interfere with activities of daily living.
- Balance. Simple measures, such as single leg stance time, may be used or more complex tasks, such as the Berg Balance Test (Berg K, Wood-Dauphinee S, Williams JI, Gayton D. Measuring balance in the elderly: preliminary development of an instrument. *Physiother Can* 1989; 41: 304-311).
- Gait. Evaluation should include the ability to negotiate obstacles, to turn and to step backwards. Useful assessment tools are the Get-Up and Go Test (Mathias S, Nayak US, Isaacs B. Balance in elderly patients: the "get-up and go" test. *Arch Phys Med Rehabil* 1986; 67: 387-389) and the 3-minute walking distance.
- Fine motor tasks, for example, writing. Samples of the patient's writing can provide a very useful objective measure of manual dexterity over time.
- Joint range of motion.
- Muscle strength.
- Swallowing and cognition. These may be evaluated in addition if clinically indicated.

Specific Therapeutic Approaches

EXERCISE

Rigidity results in a decrease in the available range of motion of joints. The characteristically flexed posture of the parkinsonian patient lends itself to the development of potentially disabling flexion contractures. Contractures, in turn, can have a detrimental effect on balance and increase the risk of falls. Thus, the patient with PD should be instructed to move all major peripheral joints and the spine through a full range of motion daily. Range of motion exercises in combination with gentle, prolonged stretches of major muscle groups, including back extensors and hip, knee, and ankle flexors, maintain joint range of motion and prevent contractures.

General aerobic conditioning should be encouraged for persons of all ages. A stationary bike or rowing machine that trains reciprocal movements is useful for patients with PD. Engaging in aerobic exercise also provides the opportunity for socialization. One goal of exercise instruction by a physical therapist is to provide the patient with a home program of daily exercises. This exercise program should begin early in the course of the disease so that it has become a daily habit by the time the disease has progressed to the point that exercise is more difficult. Periodic reevaluations by the physical therapist as the disease progresses allow for modification of the exercise program to meet the changing needs of the patient and caregivers.

BALANCE AND FALLS

Multiple factors, from postural alterations to postural hypotension, can result in impaired balance and increased risk of falls in the patient with PD. Balance exercises may be introduced from an early disease stage to become part of the daily habit of exercise. For those in the earlier stages of disease, higher level balance training, such as that provided by tai chi, may be recommended. Simpler techniques, such as weight shifting while standing at the kitchen and bathroom sinks, are also useful.

In general, canes and standard walkers have not been found to be very useful in improving balance or reducing the risk of falls in most persons with PD. If a gait aid is considered, a weighted rolling walker with brakes is the most useful. The walker should be high enough to promote an upright posture.

Patients are more likely to fall when stooping to retrieve objects. A wheeled trolley or reacher is therefore useful. For all persons at risk of falling, it is important that they and their caregivers receive instruction on techniques for getting up off the floor after a fall.

TRANSFERS AND GAIT TRAINING

Ease of transfers can be facilitated by raising the level of the bed, chair, and toilet seat and by installing grab rails. Training in a rocking motion may help the patient rise from a sitting position. Rotating the trunk back and forth in bed before attempting bed transfers can be useful. Satin sheets and satin pajamas also facilitate movement in bed. Ultimately, equipment such as an electric chair lift and a hospital bed with head-raising controls may be necessary.

Freezing episodes during walking may be overcome by maneuvers such as voluntarily goose-stepping or making a high steppage motion over a caregiver's shoe or a real or imaginary line on the floor. Another trick for the patient with gait freezing is to carry a long L-handled cane for inverted use. The handle placed on the floor in front of the patient provides a target over which a step can be made to initiate walking (see Figure 4-1). A laser pointer held by the patient can also be used to generate a visible target on the floor for foot placement. Sometimes a caregiver can use a gentle side-to-side rocking motion to elicit the gait pattern again. Relaxation techniques may be helpful during freezing episodes.

Slow initiation of movement is one of the most disabling features of PD. Multiple techniques have been tried to facilitate movement. One of the most useful is training in a rhythmic pattern to music or another auditory cue, such as handclapping or the ticking of a metronome. This technique encourages a more automatic pattern to alternating movements. Training in exaggerated limb movements, for example, arm swing, is also helpful. When shuffling is a problem, training in a high-step gait, such as walking over small boxes, is worthwhile.

When walking in public becomes difficult, a manual wheelchair may be liberating for both patient and caregiver. Input from a physical therapist to ensure correct seating is important. A slight recline may discourage kyphotic posturing.

ACTIVITIES OF DAILY LIVING

Many problems with activities of daily living are related to reduced manual dexterity and deficits in rapidly repeated movements, such as teeth brushing. An experienced occupational therapist can guide the patient in the choice of most appropriate adaptive equipment (e.g., an electric toothbrush). Dressing can be made easier by replacing buttons with hook-and-loop (Velcro) fastenings or using zipper hooks. Women may find it easier to leave a brassiere fastened and then put it on over the head. The risk of spills when eating may be reduced by the use of weighted and large-handled cups and utensils, plate guards, and nonslip mats. Exercises in fine motor coordination skills and in practical skills, such as handwriting, are useful.

COMMUNICATION

With a soft voice and undemonstrative face and body, the person with PD may have difficulty gaining entry into conversation. In addition, the delay in conversational pace associated with bradykinetic speech produces "openings" in conversation that the able-bodied person may feel compelled to fill, adding to the frustration of the patient. Speech and language therapists recognize the multifaceted character of communication and emphasize communicative behavioral skills as well as voice production in their therapeutic interventions. A useful therapeutic approach is to focus on prosodic intonation—the variations in pitch and stress that add to the communicative content and interest of speech. The fluency of speech can be impaired in either of two ways: repeated hesitancies or a festinating pattern. As with other aspects of motor control (e.g., gait), the use of pacing strategies, such as teaching the patient to tap out the words, may be useful.

VOCATIONAL NEEDS

For the younger patient with PD, maintaining gainful employment can be increasingly challenging. Subtle cognitive impairment may become obvious in the workplace while the patient continues to function well socially. Formal psychologic evaluation, including psychometrics, may be indicated. Workplace adaptations to accommodate disease-related impairments may include equipment to support writing (e.g., built-up pens) and typing (e.g., electronic keyboards) and devices for power mobility (e.g., scooter).

AVOCATIONAL NEEDS

Social isolation is an unfortunate and all too common consequence of PD. The masked facies, drooling, and tremor can be embarrassing and socially disabling. Patients need encouragement to maintain social contacts and leisure activities. The recreational therapist may help devise imaginative solutions to barriers to pleasurable social activities, for example, clamps to hold small components in benchwork and weighted hand tools to facilitate painting.

DRIVING

Retaining driving skills is an important issue for persons with PD. Slowed reaction times and difficulties performing dual tasks may affect competency. When safety becomes a concern, the therapist can provide a screening driving evaluation and arrange for on-the-road testing with a driving instructor in a dual-controlled vehicle. There are significant state-to-state variations in the reporting of medical conditions likely to impair driving skills. Most states encourage, but do not require, physicians to report their concerns. Increasingly, states are also providing indemnity for the reporting physician against potential charges of breach of confidentiality. It is the responsibility of the physician to become familiar with the regulations in the patient's state.

PSYCHOLOGIC AND SOCIAL SUPPORT

Support Groups

PD support groups serve a number of purposes. It is generally encouraging for the patient with newly diagnosed disease to meet with others facing the same issues and adapting to the same disease. A group of peers at a similar stage of disease progression are most likely to be beneficial to one another. Although support groups are not for everyone, many patients benefit from the solidarity and from learning of others' problems and their solutions. Through shared experiences, the group can learn optimal uses of social services and other local resources. Networking introduces one to persons naturally empathetic who can provide encouragement and help keep one informed about advances and research studies in which one might want to participate. In addition, these organizations work to educate the public at large and promote changes in public policy that affect patients with PD. The local branches of patient advocacy and informational associations can be obtained from the central offices listed in Table 14-2 or in many cases from their Web sites. If there is no support group in the local region, these central offices can advise on how to establish one.

Family

Caring for a chronically ill person is physically and emotionally draining, and associated expenses can place an extra burden on family relationships. Spouses and other family members of patients with advanced PD appreciate the physician's sensitivity to these facts. Demands on family members can be decreased through social service or volunteer agencies that can provide home visits, hardware, or transportation services. Caregivers may benefit from vacations from the responsibilities, confident in the patient's care in a temporary respite setting. Through support groups, family members can also meet people with similar experiences.

Internet

Information on PD (and everything else one can imagine) has been made available to increasing numbers of people by way of the Internet. A computer-literate patient can research topics of interest, exchange information in what is effectively a geographically unrestricted support group, or collectively organize for political aims, all from the comfort of home. Motivated patients and family members can also learn of clinical trials that are available in their geographical area. At its best, the Internet facilitates patients' input into the management of their condition.

As the medium is undergoing rapid evolution, no specific Web site addresses are given here. Hundreds of sites are found by navigating with hyperlinks, using "Parkinson's disease" as a starting point on any of the common search services. The ease with which information can be put in the public forum can, however, make it difficult and time-consuming to find the gold in the dross. Much of the medical information provided in some medically oriented discussion groups is unconventional and based on limited evi-

Table 14-2
National Parkinson's Disease Patient Organizations

Parkinson's Disease Foundation
William Black Medical Building
Columbia Presbyterian Medical Center
710 West 168th Street, 3rd Floor
New York, NY 10032-9982
(800) 457-6676

Parkinson's Disease Foundation Midwest Regional Office
833 West Washington Boulevard
Chicago, IL 60607
(312) 733-1893

National Parkinson Foundation, Inc.
Bob Hope Parkinson Research Center
1501 NW 9th Avenue/Bob Hope Road
Miami, FL 33136-1494
(800) 327-4545

The American Parkinson Disease Association, Inc.
1250 Hylan Boulevard, Suite 4B
Staten Island, NY 10305-1946
(800) 223-2732

dence. Therefore, physicians may need to remind patients that the truth of Web-based information is highly variable without quality control. Instruments intended to rate sites providing health information are themselves unverifiable and thus may produce more harm than good. Among the Internet sites found in one of our recent searches were some that featured MD-endorsed "natural" agents, with claims for efficacy based on anecdotal rather than scientific data. Such agents claim to help a wide range of the most prevalent conditions; for example, one agent was touted for multiple sclerosis, dementia, PD, attention-deficit disorder, and fibromyalgia. Most Americans use complementary or alternative agents and overwhelmingly do not confide in their physicians about doing so. The Internet introduces a public willing to try these agents to a marketplace willing to sell them; as always in these situations, "buyer beware."

SELECTED READING

Weight loss in PD

Beyer PL, Palarino MY, Michalek D, Busenbark K, Koller WC. Weight change and body composition in patients with Parkinson's disease. *J Am Diet Assoc* 1995; 95: 979-983.

See JA, Ahlskog JE, Palumbo PJ. Weight loss in Parkinson's disease (abstract). *J Am Diet Assoc* 1995; 95 Suppl: A60.

Toth MJ, Fishman PS, Poehlman ET. Free-living daily energy expenditure in patients with Parkinson's disease. *Neurology* 1997; 48: 88-91.

Effect of protein on levodopa effect

Karstaedt PJ, Pincus JH. Protein redistribution diet remains effective in patients with fluctuating parkinsonism. *Arch Neurol* 1992; 49: 149-151.

Nutt JG, Woodward WR, Carter JH, Trotman TL. Influence of fluctuations of plasma large neutral amino acids with normal diets on the clinical response to levodopa. *J Neurol Neurosurg Psychiatry* 1989; 52: 481-487.

Riley D, Lang AE. Practical application of a low-protein diet for Parkinson's disease. *Neurology* 1988; 38: 1026-1031.

Tsui JK, Ross S, Poulin K, Douglas J, Postnikoff D, Calne S, Woodward W, Calne DB. The effect of dietary protein on the efficacy of L-dopa: a double-blind study. *Neurology* 1989; 39: 549-552.

Research into possible dietary causes of PD

Hellenbrand W, Boeing H, Robra BP, Seidler A, Vieregge P, Nischan P, Joreg J, Oertel WH, Schneider E, Ulm G. Diet and Parkinson's disease. II: A possible role for the past intake of specific nutrients. Results from a self-administered food-frequency questionnaire in a case-control study. *Neurology* 1996; 47: 644-650.

Vitamins in PD

Logroscino G, Marder K, Cote L, Tang MX, Shea S, Mayeux R. Dietary lipids and antioxidants in Parkinson's disease: a population-based, case-control study. *Ann Neurol* 1996; 39: 89-94.

Parkinson Study Group. Impact of deprenyl and tocopherol treatment on Parkinson's disease in DATATOP subjects not requiring levodopa. *Ann Neurol* 1996; 39: 29-36.

Parkinson Study Group. Impact of deprenyl and tocopherol treatment on Parkinson's disease in DATATOP patients requiring levodopa. *Ann Neurol* 1996; 39: 37-45.

Osteoporosis

Ishizaki F, Harada T, Katayama S, Abe H, Nakamura S. Relationship between osteopenia and clinical characteristics of Parkinson's disease. *Mov Disord* 1993; 8: 507-511.

Sato Y, Hotta N, Sakamoto N, Matsuoka S, Ohishi N, Yagi K. Lipid peroxide level in plasma of diabetic patients. *Biochem Med* 1979; 21: 104-107.

Rehabilitation and physical therapy

Cedarbaum JM, Toy L, Silvestri M, Green-Parsons BA, Harts A, McDowell FH. Rehabilitation programs in the management of patients with Parkinson's disease. *J Neurol Rehabil* 1992; 6: 7-19.

Cohen AM, Weiner WJ. *The Comprehensive Management of Parkinson's Disease.* Demos Publications, New York, 1994.

Support services and the Internet

Culver JD, Gerr F, Frumkin H. Medical information on the Internet: a study of an electronic bulletin board. *J Gen Intern Med* 1997; 12: 466-470.

Jadad AR, Gagliardi A. Rating health information on the Internet: navigating to knowledge or to Babel? *JAMA* 1998; 279: 611-614.

Whetten-Goldstein K, Sloan F, Kulas E, Cutson T, Schenkman M. The burden of Parkinson's disease on society, family, and the individual. *J Am Geriatr Soc* 1997; 45: 844-849.

C

PARKINSONISM BUT NOT PARKINSON'S DISEASE (OTHER AKINETIC-RIGID SYNDROMES)

The signs characteristic of Parkinson's disease (PD) are termed *parkinsonism*. Not all patients with parkinsonism have PD, however. Neither do all patients with parkinsonism necessarily have substantia nigra pathology, Lewy bodies, or dopamine deficiency states. Parkinsonism may be secondary to destructive lesions of the basal ganglia and its connections or be induced by certain medications (primarily dopamine blocking or depleting drugs). Parkinsonian clinical signs may also be present in neurodegenerative disorders with neuropathologic characteristics quite different from those of PD.

When a new patient with parkinsonism is seen, the possibility of secondary parkinsonism should also not be forgotten. The medications should be reviewed, with attention to drugs that potentially interfere with dopamine neurotransmission (especially neuroleptics and antiemetics). If there are atypical features, a head scan may be appropriate.

Neurodegenerative disorders resembling PD often provide a clinical challenge to even seasoned physicians. Parkinsonism is typically a prominent component of progressive supranuclear palsy (PSP), multiple system atrophy (MSA), and corticobasal degeneration. It also frequently occurs in familial spinocerebellar atrophy (SCA), in which the parkinsonian features may overshadow the cerebellar signs in occasional patients. Furthermore, the primary dementing neurodegenerative disorders, including Alzheimer's disease, are often associated with mild parkinsonian signs.

Distinctions among the different neurodegenerative conditions with parkinsonian features are primarily made from the history and examination. Usually, certain red flags tip off the clinician that the disorder at hand may be something other than PD. Clinical clues pointing toward specific disorders include down-gaze palsy (PSP), early falls (PSP or MSA), prominent dysautonomia (MSA), corticospinal tract signs (PSP, MSA, or SCA), cerebellar ataxia (MSA or SCA), and early dementia (Alzheimer's disease or diffuse Lewy body disease).

When prominent parkinsonian features are present in neurodegenerative conditions other than PD, symptomatic treatment with drugs used in PD may still be effective. Levodopa therapy is the most appropriate initial choice in most cases. If a trial of escalating doses of carbidopa-levodopa, as described in Chapter 8, fails, other antiparkinsonian drugs (e.g., dopamine agonists) are unlikely to prove beneficial, although there are exceptions. If there is uncertainty whether the disorder is truly PD or another neurodegenerative disorder, a trial of levodopa may help answer the question; marked improvement suggests PD, whereas levodopa failure favors another disorder.

With this background information in mind, we now address other conditions resembling or masquerading as PD.

15 Secondary Causes of Parkinsonism

Eric S. Molho, MD
and Stewart A. Factor, DO

CONTENTS

Secondary parkinsonism implies that one or more of the cardinal features of tremor, bradykinesia, rigidity, and postural instability are present and are due to a specific and potentially identifiable cause (e.g., drugs, toxins). In contrast, the primary parkinsonian syndromes, including Parkinson's disease (PD), are idiopathic, with no currently accepted cause (Table 15-1). Although most cases of parkinsonism encountered in the clinic are primary and due to an idiopathic neurodegenerative process, it is obviously important to recognize when there is a specific cause, especially when that cause can be treated.

The list of recognized causes of secondary parkinsonism is long and varied. Hence, we have divided the material into broad categories of drug-induced, toxic, structural, infectious, metabolic, and genetic causes. The important examples of each are discussed in detail, and the remaining entities are presented in tabular form for easy reference. Since parkinsonism is the common denominator among all these entities, we emphasize the additional clinical features that are most distinctive and useful in contrasting each entity with idiopathic PD.

From: *Parkinson's Disease and Movement Disorders:*
Diagnosis and Treatment Guidelines for the Practicing Physician
Edited by: C. H. Adler and J. E. Ahlskog © Mayo Foundation
for Medical Education and Research, Rochester, MN

Table 15-1
Definition of Parkinsonian Terms

Term	Definition
Parkinsonism	The clinical picture that includes combinations of rest tremor, bradykinesia, rigidity, and postural instability. The manifestations imply dysfunction in extrapyramidal systems, which could have a variety of causes.
Parkinson's disease	Levodopa-responsive parkinsonism, pathologically associated with substantia nigra degeneration and microscopic Lewy bodies. In life, substantia nigra Lewy body degeneration is implied if parkinsonian signs are present without signs of other motor systems (e.g., cerebellar, corticospinal tract) and there is a robust response to levodopa therapy.
Primary parkinsonism	Parkinsonism occurring as a neurodegenerative disorder with a definite cause. This category includes Parkinson's disease and other neurodegenerative disorders with prominent parkinsonian features, such as progressive supranuclear palsy and multiple system atrophy. These disorders are covered in other chapters.
Secondary parkinsonism	Parkinsonism that has been induced by a specific and potentially identifiable cause, such as drugs, toxins, or trauma.

Clinical clues raising the possibility of secondary parkinsonism are listed in Table 15-2. First, young age at onset suggests that close diagnostic scrutiny is warranted. Idiopathic PD only rarely appears before age 30. Second, an atypical mode of onset and progression warrants consideration of other causes. Idiopathic PD generally has an insidious onset and progresses slowly but relentlessly over time. Acute or subacute onset raises a question of a secondary cause, as does rapid progression. If on neurologic examination one finds nonparkinsonian features, such as true paresis, sensory loss, hyperreflexia, ataxia, or early cognitive-psychiatric disturbance, a secondary cause should be considered, although the primary parkinsonism syndromes other than PD (e.g., multiple system atrophy, diffuse Lewy body disease) can be associated with these features. Finally, a dramatic beneficial response to dopamine replacement therapy is almost synonymous with idiopathic PD. Failure to respond raises diagnostic questions.

DRUG-INDUCED PARKINSONISM

This form of secondary PD was initially recognized in the 1950s when the first neuroleptic medications were developed for the treatment of psychosis. Since then, the number of antipsychotic drugs has proliferated, and drug-induced parkinsonism has correspondingly increased in frequency to become the most common form of secondary parkinsonism (Table 15-3). The incidence is currently thought to be as high as 10% to 20% in patients receiving neuroleptic agents. These drugs block brain dopamine receptors, resulting not only in the therapeutic antipsychotic effect but also in extrapyramidal side effects, such as parkinsonism and acute dystonia. This complication of therapy has

Table 15-2
Clues Suggesting Secondary Parkinsonism

Onset at young age (< 40 years)
Sudden onset
Rapid progression
Symptoms beginning after initiation of certain medications (see Table 15-3)
Static course
Atypical features on examination (nonparkinsonian)
Poor response to antiparkinsonian medications

Table 15-3
Drugs Causing Parkinsonism

Dopamine antagonist drugs (neuroleptics, antiemetics)
    Phenothiazines (e.g., chlorpromazine, prochlorperazine)
    Butyrophenones (e.g., haloperidol)
    Thioxanthenes (e.g., thiothixene)
    Dibenzazepines (e.g., loxapine)
    Substituted benzamides (e.g., metoclopramide)

Dopamine depletors
    Reserpine
    Tetrabenazine
    Alpha-methyldopa

Calcium channel blockers with dopamine antagonist activity
    Flunarizine, cinnarizine

Other medications reported to cause or worsen parkinsonism

Diltiazem	Fluoxetine and other selective serotonin reuptake inhibitors
Captopril	Procaine
Lithium	Meperidine
Phenytoin	Amphotericin B
Amiodarone	Lovastatin
Bethanechol	Phenelzine
Pyridostigmine	Cytosine arabinoside (ara-C)
Tacrine	

been described with all classes of neuroleptic medications (Table 15-3). Dopamine-blocking antiemetic agents, including prochlorperazine (Compazine) and metoclopramide (Reglan), similarly can induce reversible parkinsonism.

With the introduction of the "atypical" neuroleptic agent clozapine, it has become apparent that the antipsychotic effect need not be invariably associated with the neuroleptic effect. Clozapine is an extremely effective antipsychotic medication that has no potential for inducing parkinsonism or other extrapyramidal side effects. Recently, three other "atypical" neuroleptic drugs were introduced: risperidone (Risperdal), olanzapine (Zyprexa), and quetiapine (Seroquel). Claims of a low incidence of extrapyramidal side effects were made with the release of each of these. However, it is now well known that risperidone and olanzapine can cause

Table 15-4
Risk Factors for Drug-induced Parkinsonism

Female
Older age
Greater drug potency
Higher dose
Genetic predisposition (?)
Previous brain injury (?)

parkinsonism in some persons, and our experience has been that each can worsen the motor features of patients with PD. It is still unclear whether quetiapine has similar potential.

Not all patients are equally susceptible to the parkinsonian side effects of neuroleptic medications (Table 15-4). Indeed, not all neuroleptic agents are equally capable of causing these side effects. The problem is twice as likely to develop in women as in men, and the incidence increases with age. Generally, the risk increases with the dopamine blocking potency of a neuroleptic medication and the size of the dose being used. Genetic susceptibility and previous brain injury may also contribute.

Although neuroleptic-induced parkinsonism can appear identical to the idiopathic form, certain clinical features may be useful in separating the two. Usually, neuroleptic-induced parkinsonism is characterized by symmetrical symptoms and signs, in contrast to idiopathic PD, which is often asymmetrical. Neuroleptic-induced parkinsonism may occasionally be associated with the "rabbit syndrome," or tremor of the mouth and jaw giving rise to a peculiar chewing motion. In contrast, jaw tremor in idiopathic PD is typically without the more prominent chewing movements. Concurrent tardive dyskinesia, such as classic oral-buccal-lingual dyskinesia, may also point to a neuroleptic (or antiemetic) cause. The clinical course of drug-induced parkinsonism, unlike that of idiopathic PD, may remain static or even recede over time. In many cases, it is difficult to separate drug-induced from primary parkinsonism without withdrawing the drug. If the signs are due to a dopamine antagonist drug, they often resolve within several weeks, although in exceptional cases resolution has taken several months.

Drugs that interfere with the ability of the brain to store dopamine can also induce parkinsonism. These include reserpine, once a commonly prescribed antihypertensive agent, and tetrabenazine. Tetrabenazine is available for prescription use in most countries outside the United States. Both reserpine and tetrabenazine are used in the treatment of tardive and Tourette's syndromes.

Flunarizine and cinnarizine are calcium channel blockers that have been widely prescribed in Europe and South America as a treatment for vertigo and other ailments. Although these drugs are not currently available in the United States, clinicians should be aware that they are well-established causes of parkinsonism and other extrapyramidal syndromes. Several other medications have been implicated in the literature as causes of parkinsonism (Table 15-3). However, these reports are rare, and in some cases the link is rather tenuous.

TOXIC PARKINSONISM

Several environmental and manmade toxins are now recognized as causes of parkinsonism (Table 15-5). Unique among these is 1-methyl-4-phenyl-1,2,3,6-tetrahydropyridine

Table 15-5
Toxic Causes of Irreversible Parkinsonism

1-methyl-4-phenyl-1,2,3,6-tetrahydropyridine (MPTP)
Manganese
Cyanide
Carbon monoxide
Carbon disulfide
Methanol
Petroleum products

(MPTP). In the early 1980s, *irreversible* parkinsonism rapidly developed in several persons after self-injection of a homemade synthetic narcotic containing this impurity. Unlike other toxins listed, MPTP is a very specific poison that kills nigral dopaminergic neurons, resulting in pathologic and clinical features that are virtually identical to those of idiopathic PD. Similar to patients with idiopathic PD, those with MPTP-induced parkinsonism respond to levodopa therapy. With the recognition that MPTP was the cause of parkinsonism in these unfortunate illicit-drug users, the scientific community sought evidence that this substance or an MPTP analogue might contribute to idiopathic PD. Despite extensive investigative efforts, however, no evidence has surfaced implicating MPTP or its analogues in other cases of parkinsonism. The herbicide paraquat is structurally similar to MPTP, but toxic exposure typically does not result in parkinsonism, perhaps because it does not reach the brain. No additional cases of MPTP-induced parkinsonism have been reported since the miniepidemic in the early 1980s.

The most common cause of toxic parkinsonism is manganese poisoning. Continual or massive exposure is required, such as that in manganese miners or industrial workers. Documented cases have predominantly been outside the United States, in countries where manganese mining and smelting are major industries, such as Brazil, Australia, and South Africa. In miners, the clinical syndrome often begins with psychiatric symptoms, such as irritability, emotional instability, compulsive behaviors, and psychosis (locura manganica, or manganic madness). A slowly progressive extrapyramidal syndrome usually follows, with varying proportions of parkinsonism and dystonia. A peculiar gait, with extension of the spine, flexed elbows, and toe walking, is characteristic and has been termed "cock walk." The diagnosis of manganism obviously rests on obtaining a history of significant environmental exposure to this toxin. Magnetic resonance imaging (MRI) can also be helpful, typically showing paramagnetic signal changes associated with metal deposition in the pallidum, striatum, and substantia nigra. Presumably, because postsynaptic pallidal and striatal neurons are involved, response to treatment of symptoms with levodopa and other agents is poor. The results of chelation therapy have also been unsatisfactory.

Other rare, but important, causes of toxic parkinsonism may be suspected on the basis of the circumstances in which they occur along with associated neurologic deficits and radiographic findings. Cyanide intoxication, whether accidental or suicidal, is often fatal. Among survivors, however, a delayed-onset akinetic-rigid syndrome may develop. Tremor is generally minor, if present at all. Cyanide is a mitochondrial poison that causes widespread tissue anoxia, especially affecting the basal ganglia. MRI scans typically show bilateral signal changes in the globus pallidus (decreased on T1 and increased on T2 sequences) consistent with necrosis. Since the injury is downstream from the dopaminergic nigrostriatal system, results of levodopa therapy are poor.

Acute carbon monoxide poisoning may also damage the basal ganglia, with parkinsonism developing in some of those who survive the initial insult. The parkinsonism may occur acutely or be delayed up to several weeks. Other neurologic consequences are confusion, emotional lability, psychosis, pyramidal tract signs, and akinetic mutism. Parkinsonian symptoms can remain static or progress over time and are occasionally associated with choreoathetosis and other hyperkinesias. As with cyanide, the primary basal ganglia injury is bilateral pallidal necrosis, which may be apparent on brain imaging. Surprisingly, there have been reports of a beneficial symptomatic response to levodopa and bromocriptine in these patients.

Carbon disulfide is used as an organic industrial solvent and as a fumigant in the grain industry. Continual occupational exposure may result in parkinsonism, often in conjunction with other neurologic features, including cerebellar signs, peripheral neuropathy, hearing loss, cranial neuropathies, brain stem dysfunction, and, rarely, choreoathetosis. The MRI findings may be normal or show central demyelination. Levodopa and other antiparkinsonian agents have been generally unsuccessful.

Acute methanol intoxication occurs when the liquid is ingested as a substitute for, or an adulterant of, ethanol. The usual result is severe metabolic acidosis and coma. In survivors, parkinsonism may develop within a few days of the initial exposure. Blindness also often occurs, as a consequence of retinal toxicity. Unlike damage caused by the previously mentioned toxins, the brunt of cerebral injury is borne by the putamen rather than the globus pallidus, and computed tomography (CT) scanning or MRI may reveal bilateral putaminal abnormalities. Transient parkinsonism has also been reported in alcohol withdrawal.

PARKINSONISM
DUE TO STRUCTURAL LESIONS OF THE BRAIN

Obviously, structural brain damage, both diffuse and focal, can cause parkinsonism (Table 15-6). Generally, the diagnosis should be fairly straightforward. Some patients have clinical features that clearly differentiate the disorder from idiopathic PD, such as acute onset or pure unilateral distribution (e.g., stroke or hemorrhage). Others have associated clinical features, such as fever, meningismus, and signs of raised intracranial pressure, that occur with brain abscesses. Additionally, almost all have diagnostic radiologic findings.

Atherosclerotic Parkinsonism

The concept of atherosclerotic or vascular parkinsonism encompasses three clinical entities. The most straightforward type is a single cerebral infarction specifically located in the basal ganglia or its connections, resulting in contralateral parkinsonian symptoms, such as resting tremor, rigidity, and bradykinesia. Usually, an acute strokelike onset with strict unilaterality of signs and symptoms, perhaps with initial improvement, is followed by a static course over time. Also, the response to anti-PD drugs is often poor. These cases, however, are uncommon. More frequently, strokes in these brain areas are silent clinically or associated with other neurologic symptoms (since the vascular insult may not obey the anatomical boundaries of the basal ganglia).

A second type of atherosclerotic parkinsonism has been attributed to multiple subcortical lacunar infarctions (état lacunaris, or lacunar state). This has been differentiated from the third and more controversial diagnosis of Binswanger's disease (subacute arte-

Table 15-6
Structural Lesions That Cause Parkinsonism

Atherosclerotic
    Cerebral infarction
    Multiple lacunar infarctions
    Binswanger's disease
Hydrocephalus
    Intraventricular obstructive hydrocephalus (IVOH)
    Extraventricular obstructive hydrocephalus (EVOH)
Hemorrhagic
    Intracerebral hemorrhage
    Subdural hematoma
Neoplasm
Granuloma
Abscess
Syringomesencephalia
Traumatic brain injury
    Single head injury
    Dementia pugilistica

riosclerotic encephalopathy), in which widespread and confluent areas of periventricular white matter injury are readily seen on CT scanning and MRI. With both of these conditions, patients are generally older and usually have multiple risk factors for small vessel arterial disease, especially hypertension but also diabetes mellitus. Gait disturbance is often an early feature, and a poor response to anti-PD medications is typical. In Binswanger's disease, dementia is especially common. The clinical features of these two conditions may be contrasted with those of idiopathic PD. In these forms of atherosclerotic parkinsonism, the course is usually stepwise rather than gradual. Parkinsonian signs and symptoms are often restricted to the legs, hence the term "lower body parkinsonism." The gait of atherosclerotic parkinsonism is often associated with a more upright posture, normal or minimally reduced arm swing, and a wide-based stance, in contrast to the narrow-based stance, stooped posture, and reduced arm swing typical of the patient with idiopathic PD. Additional features more typical of vascular parkinsonism are paratonic rigidity (gegenhalten) rather than true cogwheeling rigidity and pseudobulbar symptoms. A magnetic or freezing gait can occur in both conditions, but a true festinating gait is thought to be specific for idiopathic PD. Prominent rest tremor occurs in most patients with idiopathic PD but is uncommon in vascular parkinsonism.

Sometimes, a conclusive diagnosis is difficult in a patient with parkinsonism and cerebrovascular disease. Since both conditions are common, particularly in the elderly, it is certainly possible for a patient to have both idiopathic PD and unrelated strokes. In the present era, parkinsonism due to neurodegenerative disease is much more common than vascular parkinsonism. If there is uncertainty, the most practical approach is to treat the parkinsonism medically and perform the relevant diagnostic tests for treatable causes of cerebrovascular disease (e.g., carotid and other vascular imaging studies, echocardiography, and attention to risk factors such as hypertension). A very favorable response to levodopa therapy suggests idiopathic PD.

Parkinsonism Secondary to Hydrocephalus

Hydrocephalus is defined as an increased volume of the ventricular system. In obstructive hydrocephalus (noncommunicating), the normal flow of cerebrospinal fluid (CSF) from the lateral ventricles to the third or fourth ventricle is internally blocked. As a result, the ventricular system is disproportionally enlarged proximal to the obstruction. On the other hand, if there is no apparent obstruction to flow within the ventricular system and all the ventricles are proportionally enlarged, the term "nonobstructive (communicating) hydrocephalus" is used. Examples of this type include hydrocephalus from subarachnoid hemorrhage, tuberculous meningitis in which intracranial pressure is increased, and normal pressure hydrocephalus (associated with normal intracranial pressure, by definition). It has been proposed that this traditional terminology be discarded, since even in "nonobstructive" hydrocephalus normal flow of CSF is obstructed between the subarachnoid space and the venous system. Thus, the terms "intraventricular obstructive hydrocephalus" (IVOH) and "extraventricular obstructive hydrocephalus" (EVOH) have been proposed and are used here to avoid confusion.

The ventricles also enlarge when there is generalized loss of cerebral tissue, such as that occurring with severe, widespread neurodegenerative disease. This ventricular enlargement is associated with prominent subarachnoid spaces over the cerebral convexities and has been termed "hydrocephalus ex vacuo." With this condition, clinical symptoms and signs are due to the generalized loss of brain tissue rather than to intraventricular pressure gradients.

Both IVOH and EVOH have been reported to cause parkinsonism. The most prominent feature is often a wide-based gait disorder composed of parkinsonian and ataxic features. Some patients, however, have typical parkinsonian features, including bradykinesia, rigidity, resting tremor, and even dopamine responsiveness.

When the typical features of PD occur in IVOH, the diagnosis should be fairly straightforward. These patients also have signs of raised intracranial pressure, such as headache, nausea, and depressed consciousness. Symptoms usually progress rapidly, and the patients almost always have nonparkinsonian neurologic features, such as corticospinal tract signs and, in later stages, obtundation. These features should lead to emergency CT scanning or MRI of the brain. The treatment is surgical shunting of CSF (ventriculoperitoneal), but some patients benefit from levodopa even after successful shunting.

The situation with EVOH is often more complicated. If there has been a recent hemorrhagic, inflammatory, or infectious process (e.g., subarachnoid hemorrhage or meningitis), EVOH may be a sequela, and the clinical picture usually declares itself. However, the more common form of EVOH, normal pressure hydrocephalus (NPH), may masquerade as parkinsonism or dementia, and there has been much controversy about its accurate diagnosis and management. The classic triad of NPH consists of dementia, incontinence, and gait disturbance, although the complete triad may not be present in all patients. The typical gait disorder has the appearance of a parkinsonian shuffling gait, but usually with a widened base and preserved arm swing. In most cases of NPH, motor symptoms and signs are not present in the upper half of the body. The signs tend to be symmetrical, and rest tremor is not expected. Distinguishing between NPH and hydrocephalus ex vacuo is essential, since the latter is associated with degenerative conditions such as Alzheimer's disease and does not respond to CSF shunting. The distinction is often apparent on CT scans or MRI when the degree of cortical atrophy is compared with the amount of ventricular enlargement. In NPH, the cortical gyri appear full on imaging studies, whereas in hydrocephalus ex vacuo, the patient has prominent cortical atrophy in addition to ventricular enlargement. In cases between these extremes, further diagnostic testing can be helpful, but the findings should not be considered diagnostic. Nuclear medicine ventriculography can demonstrate abnormal CSF dynamics but is rarely convincing enough to rely on alone. Either lumbar puncture with removal of 30 to 50 mL of CSF or a trial of lumbar drainage performed by a neurosurgeon is perhaps the most useful diagnostic test. Clear improvement over hours to days suggests true NPH and has been proposed as a predictor of responsiveness to more permanent shunting. Here, too, levodopa responsivity does not rule out true NPH. In some cases, parkinsonism lingers after surgery, and levodopa can be tried. Some patients who seem to have true NPH may have sufficient response to levodopa to avoid shunt surgery indefinitely.

POST-TRAUMATIC PARKINSONISM

Post-traumatic parkinsonism refers to several distinct relationships between traumatic injury and parkinsonian syndromes. The current state of knowledge generally supports three basic types of post-traumatic parkinsonism. The first is the fairly common situation in which a patient without clinically significant head injury retrospectively reports that either physical or emotional trauma preceded the onset of PD. It is now generally accepted that this type represents the coincidental occurrence of trauma in a person with preclinical or asymptomatic PD and not a cause and effect relationship. Whether the trauma has hastened the onset of symptoms and how this might occur are still debated.

More pertinent to a discussion of secondary parkinsonism is the rarer but also well established concept of a single significant head injury resulting in a progressive form of parkinsonism distinct from idiopathic PD. Here, the head injury should be severe enough to cause loss of consciousness and, by definition, result in structural damage to basal ganglia or brain stem structures. The initial symptoms should occur shortly after the injury and may be unilateral. Eventually, generalized bradykinesia and rigidity with a variable degree of tremor, gait disturbance, postural instability, and other nonparkinsonian features develop, depending on the extent of traumatic brain injury.

The other generally accepted form of post-traumatic parkinsonism is the syndrome resulting from chronic and repeated head injury with multiple concussions, as occurs in boxers. Known as dementia pugilistica (punch-drunk syndrome), this condition generally begins insidiously at the end of a long career or later in retirement. Symptoms include dysarthria, dementia, ataxia, and, in about 40%, parkinsonism. Tremor can be quite prominent. Pyramidal tract signs and seizures may also be seen. Imaging of the brain shows cerebral atrophy, ventricular enlargement, and, in some, a cavum septum pellucidum. An electroencephalogram usually demonstrates nonspecific slowing of brain waves.

Although either of these accepted forms of traumatic PD may respond to levodopa, the results are usually unsatisfactory. The best treatment is still prevention.

INFECTIOUS AND POSTINFECTIOUS PARKINSONISM

Table 15-7 shows that numerous infections of the brain have been reported to cause parkinsonism either acutely or as a chronic postinfectious disorder. Historically, the

Table 15-7
Reported Infectious Causes of Parkinsonism[a]

Viral encephalitis
Encephalitis lethargica (postencephalitic)
Other viruses
Acquired immunodeficiency syndrome (AIDS)
Primary human immunodeficiency virus (HIV) infection
Toxoplasmosis
Cryptococcal meningoencephalitis
Progressive multifocal leukoencephalitis
Syphilis
Tuberculosis
Whipple's disease
Mycoplasma pneumoniae
Creutzfeldt-Jakob disease

[a]Abscesses may also cause parkinsonism, as discussed in the section on structural lesions of the brain.

most important was postencephalitic parkinsonism due to the epidemic occurrence of encephalitis lethargica (von Economo's disease) from 1919 to 1926. During the 1920s and 1930s, postencephalitic parkinsonism was actually thought to be more common than idiopathic PD. Encephalitis lethargica was frequently fatal, and survivors often had severe chronic neurologic sequelae. Parkinsonism generally developed as a delayed manifestation and progressed at a variable pace over decades. A number of unusual clinical features made differentiation from idiopathic PD relatively straightforward. These included oculogyric crises, blepharospasm and other dystonic spasms, tics, myoclonus, respiratory irregularities, bulbar symptoms, psychiatric disorders, and bizarre behavioral disturbances. Although some patients had improvement with levodopa, side effects often occurred and at lower doses than in idiopathic PD.

The infectious agent causing postencephalitic parkinsonism has never been established, but it is generally accepted that encephalitis lethargica was a viral illness. Since the disappearance of the epidemic form of this disease more than 50 years ago, other viral encephalitides have been reported to cause parkinsonism. However, these reports are rare, and no particular current virus seems to be capable of causing an illness similar to encephalitis lethargica. On the other hand, it is not unreasonable to expect that an illness similar to encephalitis lethargica could occur again, and some awareness of its clinical features and vigilance against its recurrence should be maintained.

Currently, acquired immunodeficiency syndrome (AIDS) is probably the most common infectious condition associated with parkinsonism. An akinetic-rigid syndrome with or without gait instability can be seen with the primary infection as part of the AIDS-dementia complex. Unilateral parkinsonism or single limb involvement is more typical of focal opportunistic infections associated with AIDS, such as toxoplasmosis. A diagnosis of secondary parkinsonism usually can be considered on the basis of the young age of the patient; in many cases, the diagnosis of AIDS has already been established. MRI and lumbar puncture are the most useful diagnostic tests.

In patients with AIDS, one must carefully rule out drug-induced parkinsonism. Patients with AIDS are particularly prone to this complication of dopamine blocking medications, such as neuroleptic antipsychotic agents or metoclopramide or prochlorperazine for gastrointestinal complaints. Treatment of this complication is difficult, and symptoms do not abate with anti-PD medications. The best approach is prevention by assiduous avoidance of these medications in patients with AIDS. In patients with intractable psychosis, one should consider using atypical neuroleptic agents, such as clozapine.

A variety of rare infectious causes of parkinsonism are listed in Table 15-7. In patients with a typical clinical picture of PD at presentation, these do not need to be considered. They enter the differential diagnosis only when there are symptoms or signs of systemic illness, evidence of organ involvement apart from the brain, or unusual neurologic manifestations. When an infectious cause is an appropriate consideration, blood studies, including relevant serologic tests, and MRI of the head followed by CSF examination (if there are no mass lesions on MRI) usually settle the issue. Creutzfeldt-Jakob disease is an exception; however, the very rapid course and prominent and progressive dementia of this disease typically suggest the diagnosis.

HEMIPARKINSONISM-HEMIATROPHY

As the name implies, this disorder is characterized by atrophy of one side of the body (face and limbs in combination or in isolation) together with ipsilateral hemiparkinsonism. The parkinsonism is only slowly progressive and in most cases remains unilateral. Ipsilateral dystonia is also frequent. Many such patients do not have a response to levodopa therapy, although some do. Brain imaging may not demonstrate any corresponding asymmetry. The typical age at symptom onset is in the 40s, earlier than that of most patients with idiopathic PD. Some patients have a retrospectively recognized perinatal injury, but in many, the cause remains unexplained.

PARKINSONISM DUE TO METABOLIC DISEASE

Several metabolic disorders can result in parkinsonism or produce symptoms that may resemble parkinsonism (Table 15-8). Metabolic by-products from liver failure may damage the basal ganglia, resulting in an extrapyramidal syndrome that may include parkinsonism as well as chorea or dystonia. Postural tremor, asterixis, myoclonus, and cognitive impairment are also common in this hepatocerebral degeneration syndrome.

Chronic acquired hypoparathyroidism is associated with basal ganglia calcification. If the disorder is pronounced, extrapyramidal symptoms may develop. A CT head scan suggests the diagnosis.

Severe hyponatremia and too-rapid correction of serum sodium concentration may result in the syndrome of central pontine myelinolysis. In this disorder, the primary insult is to the pontine white matter, resulting in quadriparesis. Extrapontine lesions may occur, however, and rarely, an extrapyramidal syndrome develops, perhaps with akinetic-rigid features.

Finally, patients who move slowly and appear lethargic may have parkinsonism. However, the clinician should not forget that hypothyroidism can also induce such symptoms.

PARKINSONISM DUE TO GENETIC DISORDERS

Parkinsonism may occur as a major or minor component in many inherited disorders. For the practicing clinician, perhaps the most important of these to keep in mind

Table 15-8
Parkinsonism in Acquired Metabolic Disorders

Liver failure (nonwilsonian hepatocerebral degeneration)
Hypoparathyroidism (basal ganglia calcification)
Central pontine myelinolysis
Hypothyroidism (not truly parkinsonian, but slowed behavior)

is Wilson's disease. This diagnosis is worth considering in any patient younger than 55 or 60 years who has an extrapyramidal disorder. Wilson's disease is an autosomal recessive disorder of copper metabolism resulting in damage to the liver and brain that, if untreated, ultimately proves fatal. It usually appears in the second or third decade with symptoms of hepatic, psychiatric, or neurologic dysfunction. Cognitive decline, tremor, and dystonia are perhaps the most common neurologic presentations, but isolated parkinsonism can also occur. The diagnosis is suggested by Kayser-Fleischer rings, a low serum ceruloplasmin value, and increased 24-hour urinary excretion of copper. The MRI shows typical signal changes in the basal ganglia, cerebellum, and brain stem. See Chapter 32 for further discussion of Wilson's disease.

Other inherited disorders in which atypical parkinsonism may occur and lead to diagnostic confusion are shown in Table 15-9. When Huntington's disease develops during adolescence or early adulthood, an akinetic-rigid syndrome, rather than chorea, is often seen. This may be confused with juvenile PD.

Familial basal ganglia calcification occurs with both recessive and dominant inheritance and independent of systemic calcium homeostasis. The calcification is obvious on CT scan (Fig. 15-1) and may extend into other cerebral regions. The clinical picture tends to mirror the findings on the scan. This disorder is to be differentiated from the mild basal ganglia calcification that occurs in a distinct minority of normal persons, perhaps as part of aging; in such persons, the finding is incidental and does not result in neurologic symptoms.

Parkinsonism may also be a component of the clinical features in the dominantly inherited spinocerebellar ataxias (see Chapter 18). Usually, this is overshadowed by cerebellar and corticospinal tract signs; however, parkinsonism occasionally is predominant in patients with spinocerebellar ataxia. This has been well documented, for example, in spinocerebellar ataxia type 3 (Machado-Joseph disease).

In addition, an increasing number of familial parkinsonian disorders have been linked to specific genetic mutations. In these disorders, parkinsonism or parkinsonism-dystonia is a primary and consistent feature. Some of the best-studied of these familial conditions are listed in Table 15-10. For the most part, these are exceptional disorders and are diagnostic considerations only if there is a prominent family history.

Finally, in exceptional cases, parkinsonism is a prominent component of some rare inborn errors of metabolism (e.g., Hallervorden-Spatz disease, certain mitochondrial cytopathies). Such cases are beyond the focus of this text.

INVESTIGATION OF SECONDARY PARKINSONISM

In a patient with typical PD, including bilateral features (often asymmetrical), gradual onset, slow progression over time, and, most importantly, a robust response to levodopa therapy, no additional investigation is needed. However, if clinical clues raise a question of secondary parkinsonism, further studies are necessary. A standard

Table 15-9
Inherited Disorders in Which Parkinsonism May Occur as a Minor or Infrequent Component[a]

Condition	Inheritance	Gene and chromosome (reference)	Comments
Wilson's disease	Recessive	ATP7B gene (copper-transporting ATPase). chromosome 13[b]	Associated with reduced serum ceruloplasmin, increased copper in urine, and ocular Kayser-Fleischer rings. Often, prominent postural tremor.
Huntington's disease	Dominant	Huntingtin gene, chromosome 4[c]	If young at onset, may begin as an akinetic-rigid syndrome.
Familial basal ganglia calcification (multiple forms)	Dominant and recessive forms		Combinations of chorea, parkinsonism, and dementia. Variety of forms include familialhypoparathyroidism, pseudohypoparathyroidism, and disorders not linked to calcium homeostasis.
Spinocerebellar ataxias (SCA)	Dominant		Prominent ataxia, but parkinsonism may be a component, especially with SCA-3

[a]Parkinsonism may also be a minor component, or an atypical major component, of various other rare inherited disorders (e.g.. mitochondrial cytopathies). These uncommon conditions are beyond the scope of this book.
[b]Thomas GR, Forbes JR, Roberts EA, Walshe JM, Cox DW. *Nat Genet* 1995; 9: 210-217.
[c]Andrew SE, Goldberg YP, Kremer B, Telenius H, Theilmann J, Adam S, Starr E, Squitieri F, Lin B, Kalchman MA, Graham RK, Hayden MR. *Nat Genet* 1993; 4: 398-403.

Table 15-10
Inherited Disorders With Parkinsonism as a Major Component

Condition	Inheritance	Gene and chromosome (reference)	Comments
Dopa-responsive dystonia-parkinsonism (Segawa's disease)	Dominant	GTP cyclohydrolase I gene, chromosome 14[a]	Presents with dystonia early in life, with parkinsonism later in adulthood, or adult-onset parkinsonism. Very responsive to even small doses of levodopa.
Dopa-responsive dystonia-parkinsonism (Segawa's Disease)	Recessive	Tyrosine hydroxylase gene, chromosome 11[b]	Similar phenotype to dominant form (above).
Juvenile-onset parkinsonism	Recessive	"Parkin" gene, chromosome 6[c]	Onset, ages 20-30 yr. Levodopa responsive; may have diurnal fluctuations.
Dominantly inherited apathy, central hypoventilation, and Parkinson's syndrome	Dominant[d]		Medication-refractory depression, apathy, weight loss are early signs, with respiratory problems and parkinsonism later.
Italian-Greek familial parkinsonism	Dominant	Alpha-synuclein gene, chromosome 4[e]	Levodopa responsive. The mutation was not found among sporadic cases of Parkinson's disease. A second mutation of the same gene has been reported in German familial parkinsonism.

Disease	Inheritance	Gene/locus	Clinical features
X-linked dystonia-parkinsonism of Philippines ("Lubag")	X-linked	X-chromosome[f]	Typical onset in 30s. Combinations of parkinsonism and dystonia.
Disinhibition-dementia-parkinsonism-amyotrophy complex	Dominant	Tau gene, chromosome 17[g]	Prominent personality and behavior change (frontal lobe type). A disorder known as pallidopontonigral degeneration appears to be due to mutation in same gene.
Rapid-onset dystonia-parkinsonism	Dominant[h]		Acute (hours) to subacute (days, weeks) development of symptoms. Levodopa not very effective.

[a] Ichinose H, Ohye T, Takahashi E, Seki N, Hori T, Segawa M, Nomura Y, Endo K, Tanaka H, Tsuji S, Fujita K, Nagatsu T. Nat Genet 1994; 8: 236-242.

[b] Ludecke B, Dworniczak B, Bartholome K. Hum Genet 1995; 95: 123-125.

[c] Tassin J, Durr A, de Broucker T, Abbas N, Bonifati V, De Michele G, Bonnet AM, Broussolle E, Pollak P, Vidailhet M, De Mari M, Marconi R, Medjbeur S, Filla A, Meco G, Agid Y, Brice A. Am J Hum Genet 1998; 63: 88-94.

[d] Perry TL, Wright JM, Berry K, Hansen S, Perry TL Jr. Neurology 1990; 40: 1882-1887.

[e] Polymeropoulos MH, Higgins JJ, Golbe LI, Johnson WG, Ide SE, Di Iorio G, Sanges G, Stenroos ES, Pho LT, Schaffer AA, Lazzarini AM, Nussbaum RL, Duvoisin RC. Science 1996; 274: 1197-1199.

[f] Wilhelmsen KC, Weeks DE, Nygaard TG, Moskowitz CB, Rosales RL, dela Paz DC, Sobrevega EE, Fahn S, Gilliam TC. Ann Neurol 1991; 29: 124-131.

[g] Hutton M, Lendon CL, Rizzu P, Baker M, Froelich S, Houlden H, Pickering-Brown S, Chakraverty S, Isaacs A, Grover A, Hackett J, Adamson J, Lincoln S, Dickson D, Davies P, Petersen RC, Stevens M, de Graaff E, Wauters E, van Baren J, Hillebrand M, Joosse M, Kwon JM, Nowotny P, Che LK, Norton J, Morris JC, Reed LA, Trojanowski J, Basun H, Lannfelt L, Neystat M, Fahn S, Dark F, Tannenberg T, Dodd PR, Hayward N, Kwok JBJ, Schofield PR, Andreadis A, Snowden J, Craufurd D, Neary D, Owen F, Oostra BA, Hardy J, Goate A, van Swieten J, Mann D, Lynch T, Heutink P. Nature 1998; 393: 702-705.

[h] Dobyns WB, Ozelius LJ, Kramer PL, Brashear A, Farlow MR, Perry TR, Walsh LE, Kasarskis EJ, Butler IJ, Breakefield XO. Neurology 1993; 43: 2596-2602.

Table 15-11
Investigation of Secondary Parkinsonism

History of drug or toxin exposure (especially
   neuroleptics and antiemetics)
Family history
Neuroimaging
        Computed tomography scan
        Magnetic resonance imaging
Laboratory tests
        Serum ceruloplasmin
        Thyroid studies
        Liver function
        Serum ammonia
        Serum calcium, phosphorus, parathyroid
        hormone if basal ganglia calcification
        24-hour urinary copper excretion
Slit-lamp examination (for Kayser-Fleischer rings)
Lumbar puncture (in selected cases)

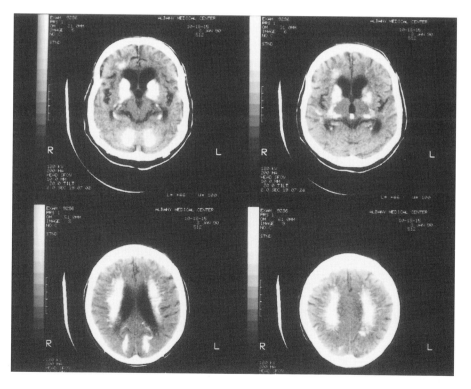

Fig. 15-1. Noncontrast computed tomography scans demonstrate extensive bilateral calcifications.

strategy for such an investigation is outlined in Table 15-11. Other tests may be added as determined by the history and examination. Brain imaging is essential in any patient with a strict unilateral presentation or a sudden onset of symptoms. MRI is generally the imaging procedure of choice because of its ability to demonstrate signal changes in deep brain matter structures, such as the basal ganglia and brain stem. However, a CT scan is a better choice to rule out basal ganglia calcification or acute hemorrhage.

In any patient younger than 55 years at presentation, Wilson's disease should be specifically ruled out. A normal serum ceruloplasmin level and absence of obvious Kayser-Fleischer rings are usually adequate. However, if clinical suspicion is increased by psychiatric disease, liver dysfunction, or prominent dystonic features, further testing is indicated, including a slit-lamp examination, MRI, 24-hour urinary copper excretion, and, sometimes, liver biopsy. A more detailed discussion of the laboratory diagnosis of Wilson's disease can be found in Chapter 32.

Lumbar puncture is necessary only occasionally and is most useful in patients thought to have a central nervous system infection or in the evaluation of EVOH (NPH). Before a lumbar puncture is attempted, a CT scan or MRI should always be done to rule out a mass lesion and to ensure the safety of this procedure.

TREATMENT OF SECONDARY PARKINSONISM

Obviously, if the workup reveals a treatable cause for secondary parkinsonism, attention to this should take precedence. Symptomatic treatment with levodopa therapy, however, is appropriate in any patient with persistent symptoms. The carbidopa-levodopa dosage strategy is the same as that outlined for idiopathic PD in Chapter 8.

Drug-induced parkinsonism deserves special mention. Parkinsonian signs and symptoms can linger up to 6 months after dopamine blocking medications are withdrawn. Patients requiring neuroleptic treatment for psychosis may benefit from a switch to clozapine, olanzapine, or quetiapine, as previously discussed. Clozapine is the best choice in terms of antipsychotic potency and nil potential for inducing parkinsonism. However, it has induced reversible but life-threatening agranulocytosis in rare instances and is sedating. Anticholinergics, such as benztropine and trihexyphenidyl, beginning with a small dose that is gradually titrated upward, or amantadine (up to a dose of 100 mg three times a day), are sometimes added to the conventional neuroleptic agents to treat parkinsonism, but success is limited.

ACKNOWLEDGMENT

This paper was supported by the Albany Medical College Parkinson Research Fund and the Riley Family Chair in Parkinson's Disease (SAF).

SELECTED READING

Factor SA. Posttraumatic parkinsonism, in *Parkinsonian Syndromes* (Stern MB, Koller WC eds), Marcel Dekker, New York, 1993, pp. 95-110.

Friedman JH. Drug-induced parkinsonism, in *Drug-Induced Movement Disorders* (Lang AE, Weiner WJ eds), Futura Publishing Company, Mount Kisco, New York, 1992, pp. 41-83.

Friedman JH. Postencephalitic parkinsonism, in *Parkinsonian Syndromes* (Stern MB, Koller WC eds), Marcel Dekker, New York, 1993, pp. 203-226.

Hurtig HI. Vascular parkinsonism, in *Parkinsonian Syndromes* (Stern MB, Koller WC eds), Marcel Dekker, New York, 1993, pp. 81-94.

Sanchez-Ramos JR. Toxin-induced parkinsonism, in *Parkinsonian Syndromes* (Stern MB, Koller WC eds), Marcel Dekker, New York, 1993, pp. 155-172.

Shannon KM. Hydrocephalus and parkinsonism, in *Parkinsonian Syndromes* (Stern MB, Koller WC eds), Marcel Dekker, New York, 1993, pp. 123-136.

Sweeney PJ. Metabolic causes of parkinsonism, in *Parkinsonian Syndromes* (Stern MB, Koller WC eds), Marcel Dekker, New York, 1993, pp. 195-202.

Tetrud JW, Langston JW. MPTP and Parkinson's disease: one decade later, in *Parkinsonian Syndromes* (Stern MB, Koller WC eds), Marcel Dekker, New York, 1993, pp. 173-194.

Waters CH. Structural lesions and parkinsonism, in *Parkinsonian Syndromes* (Stern MB, Koller WC eds), Marcel Dekker, New York, 1993, pp. 137-144.

16 Progressive Supranuclear Palsy

Mark Stacy, MD

CONTENTS

Progressive supranuclear palsy (PSP) is the most common of a group of idiopathic neurodegenerative disorders classified as the parkinsonism-plus syndromes. As the term "parkinsonism-plus syndrome" implies, patients usually have neurologic symptoms in addition to those of Parkinson's disease (PD). The pathologic involvement in these conditions, including PSP, is more widespread than that in typical PD. The four classic symptoms of PD—bradykinesia, rigidity, resting tremor, and postural instability—primarily result from pathology within the substantia nigra, a component of the basal ganglia pathways. In PSP, additional basal ganglia nuclei are involved, as are other subcortical areas and sometimes the cortex (especially frontal). The clinical hallmarks of PSP—down-gaze paresis, falling early in the course, and other features noted below—reflect this extranigral pathologic involvement.

CLINICAL FEATURES

Although PSP is the most frequently occurring parkinsonism-plus syndrome, it is 10 times less common than idiopathic PD. The condition is more common in men and most often appears during the sixth and seventh decades. Extensive epidemiologic review has not identified any environmental or genetic risk factors associated with PSP.

In the patient with early parkinsonism, the differences between PSP and PD may be subtle, but the patient with PSP may already have visual difficulties (gaze paresis), a greater tendency toward falling, and less tremor than a typical patient with PD. With disease progression, the cognitive, visual, motor, and brain stem symptoms, as well as a poor response to anti-PD therapy, clarify this diagnosis (Table 16-1).

From: *Parkinson's Disease and Movement Disorders:*
Diagnosis and Treatment Guidelines for the Practicing Physician
Edited by: C. H. Adler and J. E. Ahlskog © Mayo Foundation
for Medical Education and Research, Rochester, MN

Table 16-1
Common Symptoms of Progressive Supranuclear Palsy
(PSP) Compared With Those of Parkinson's Disease
(PD)

Symptom	PSP	PD
Bradykinesia	+	+
Rigidity	Axial > limb	Limb > axial
Gait disturbance	+	+
Early falling	+	-
Tremor	-[a]	+
Down-gaze paresis	+	-
Eyelid apraxia	+	-
Emotional incontinence	+	-
Levodopa response	±	+
Corticospinal tract signs	±	-

+, present; -, absent; ±, may or may not be present.
[a]Occurs occasionally but is uncommon.

Supranuclear Gaze Paresis

The most important distinguishing feature of PSP is ocular gaze paresis, and one-third of patients report blurred vision or diplopia early in the disease. Primarily, eye movements are limited, and the most diagnostically specific component is impaired downward gaze. This lack of ability to redirect gaze or follow an object may be overcome by passive head movement while the eyes are fixed on a target in front of the patient (the "doll's eye" maneuver). This movement activates the brain stem oculocephalic reflexes, which are preserved. The designation "supranuclear" implies that the pathologic area is above these brain stem ocular nuclei. The impaired downward gaze specifically limits the ability of patients to walk down stairs and further increases the potential for injury. Because patients have difficulty looking at the dinner table, they often become messy eaters, frequently spilling food on themselves (the "dirty tie sign") and the table. Lateral-gaze paresis also occurs, and patients may experience double vision, often when attempting to accommodate vision from a television set (distant) to a newspaper (near).

Wide-eyed, Unblinking Face

The patient with PSP who has a fixed, staring expression may appear to be uninterested in conversation because of poor eye contact and a blank facial appearance. Some patients become unable to voluntarily open their eyes (apraxia of eyelid opening), occasionally so severe that they are rendered functionally blind.

Early Falls

Severe gait instability with frequent falls early in the course is an important diagnostic clue to the diagnosis of PSP. Although some patients with PSP have a shuffling gait similar to that of PD, others have normal stride length and pivoting turns but are subject to lurching falls. In addition, the patient with PD seems to walk in a protective fashion, whereas many patients with PSP walk as if unconcerned about the risk of injury. At the conclusion of gait assessment during the physical examination, the patient with PSP tends to "fall" into the chair with feet rising in the air (sitting "en bloc"), and the patient's head may hit the wall, in contrast to the patient with PD.

Axial Rigidity

Rigidity is a prominent extrapyramidal sign in PSP, tending, however, to be axial, affecting the neck and trunk, unlike the prominent limb rigidity in PD. In patients with PD, there may be a ratcheting or cogwheeling quality during passive limb movement, reflective of superimposed tremor. In contrast, tremor is often minimal with PSP. In addition, patients with PSP characteristically have back and neck *extension* rather than the neck *flexion* and kyphosis associated with PD.

Supranuclear Speech and Swallowing Problems

The term "supranuclear" implies that symptoms are secondary to degenerating cortical and subcortical neuronal pathways and the "release" of brain stem symptoms. The speech and swallowing problems (and characteristic facial expression) of PSP reflect upper motor neuron lesions, much like the upper motor neuron symptoms in a patient who has recovered from stroke. Although the disorder is more complicated than just corticospinal pathway destruction, increased tone (spasticity) of the palate and larynx produces strained and sometimes hypernasal speech; also, a monotonous quality is typical, reflecting an extrapyramidal (parkinsonian) contribution. Similar to the apraxia of eyelid opening, difficulty initiating speech along with speech arrest may also occur in conjunction with palilalia (repeating of short phrases or sentences). Patients may also become "emotionally incontinent," with laughing or crying out of proportion to (and sometimes without) accompanying feelings of happiness or sadness. Rarely, patients exhibit involuntary vocalizations, such as continuous moaning or humming. Complete aphonia has also been reported, and one patient described could talk only during times of heightened excitement or while asleep. In most instances, the gag reflex is increased. Functionally, chewing difficulties and choking may require dietary modification and, in advanced disease, feeding gastrostomy.

Cognitive Impairment

In their original description, Steele and associates reported mild memory difficulties in this condition, and it is generally accepted that cognitive and personality changes occur with PSP. Most investigators agree that the "subcortical" deficits of PSP are most apparent in tasks requiring sequential movements, conceptual shifts, and information retrieval and that short-term memory is relatively spared. However, because of these difficulties, coupled with visual and motor impairment, data from neuropsychologic testing are often difficult to interpret. Besides intellectual decline, personality changes are apparent, including indifference, decreased anxiety, and agitation. These patients, however, have less depression than those with PD.

DIFFERENTIAL DIAGNOSIS

In patients with no secondary causes of parkinsonism (e.g., medication exposure or strokes), one major clinical distinction is between PSP and PD. Important clues to PSP are severe gait instability with falls within the first 3 years, down-gaze paresis, lack of prominent rest tremor, and a poor response to levodopa. Prominent corticospinal and corticobulbar signs (e.g., emotional incontinence, pseudobulbar affect) argue against PD but are consistent with PSP.

Other parkinsonism-plus syndromes also need to be differentiated from PSP. Multiple system atrophy may include parkinsonism and early falls. Clues to multiple system atrophy rather than PSP are significant and early dysautonomia (orthostatic hypotension, incontinence, and sexual dysfunction) and prominent cerebellar signs. Although down-gaze paresis is not part of multiple system atrophy, it occurs in another parkinsonism-plus syndrome, corticobasal ganglionic degeneration. Corticobasal ganglionic degeneration, however, is highly asymmetrical, in contrast to the usual symmetry of motor signs in PSP. Also, the severe limb apraxia of corticobasal ganglionic degeneration is not expected in PSP.

Other neurodegenerative disorders with prominent dementia, such as Alzheimer's disease, Pick's disease, and diffuse Lewy body disease, may also have features in common with PSP. In contrast to PSP, the dementia of these disorders is typically early and prominent. Also, the motor signs in Alzheimer's and Pick's diseases are later and less pronounced. Diffuse Lewy body disease often has both early and prominent dementia in addition to parkinsonian signs. Hallucinations and delusions are early symptoms in diffuse Lewy body disease but not in PSP, and down-gaze paresis is not expected.

Secondary causes of parkinsonism may result in clinical resemblance to PSP, but the history usually allows these to be distinguished. Included is parkinsonism due to dopamine antagonist drugs (e.g., neuroleptic agents, many antiemetic drugs) or secondary to strokes (stepwise course, brain imaging findings). Normal pressure hydrocephalus often occurs with the triad of parkinsonian gait disorder, dementia, and urinary incontinence; lateral ventricular enlargement without cortical atrophy is identifiable on brain imaging. The parkinsonian signs of normal pressure hydrocephalus tend to be predominantly in the lower limbs, with minimal facial masking (in striking contrast to occurrence in PSP). Whipple's disease is an extremely rare disorder associated with ocular motility problems but usually in conjunction with oculomasticatory myorhythmia, characterized by repetitive, pendular movements of the eyes and jaw.

PATHOLOGY AND PATHOGENESIS

The hallmark of PSP is the globose (globe-shaped) neurofibrillary tangle, in contrast to the Lewy body of PD. These globose neurofibrillary tangles are found predominantly in a subcortical distribution and are associated with neuronal loss and gliosis. Limited numbers of neurofibrillary tangles, resembling the globose variety typical of PSP, are also found in other disorders, such as Pick's disease and corticobasal ganglionic degeneration. However, prominent cortical disease is found in these conditions but not in PSP. Nuclei with prominent involvement in PSP include the substantia nigra, subthalamic nucleus, globus pallidus, superior colliculus, and pretectal area. Perhaps contributing to the cognitive and behavioral changes is marked degeneration in the basal nucleus of Meynert, an area of the brain with a high concentration of cholinergic neurons.

Deficits in striatal dopamine are present in PSP, as in PD. Other neurotransmitters (e.g., γ-aminobutyric acid [GABA], acetylcholine, serotonin, and epinephrine) are also reduced in PSP, a reflection of widespread neurodegenerative changes. Also, striatal dopamine (D_2) receptors are reduced in patients with PSP but not in patients with PD. The incomplete and often poor response to levodopa therapy and dopamine agonist drugs in patients with PSP reflects these more widespread neurodegenerative changes, especially impairment of GABAergic striatal *efferent* neurons expressing dopamine receptors, as opposed to the degeneration of the *afferent* dopaminergic striatonigral neurons in PD.

PROGNOSIS

The prognosis is much worse for the patient with PSP than for the patient with PD. Response to antiparkinsonian medications is inconsistent and unsustained in patients with PSP, and both cognitive and motor abilities decline more rapidly. The median interval from onset of the initial symptom to walking difficulties, usually falling, is 3 months. Ambulatory assistance is typically required by 3 years, and wheelchair confinement is needed by 9 years, if not sooner. Death usually occurs within 10 years, most often because of complications from falls (e.g., hip fracture or subdural hematoma) or dysphagia (e.g., aspiration pneumonia).

EVALUATION AND MANAGEMENT

Although PSP is associated with loss of multiple neurotransmitters, pharmacologic therapies remain disappointing. Levodopa therapy reduces the bradykinesia and rigidity in about one-third of patients, but the benefit diminishes after 2 years. Some investigators have argued that the severe cholinergic striatal interneuron loss in brains of patients with PSP suggests that acetyl cholinomimetic therapy would be of benefit. Although scopolamine, a cholinergic blocking agent, produced gait decline at low doses, treatment with physostigmine, a cholinergic stimulating drug, produced no benefit in patients with PSP. These investigators suggested that drugs with anticholinergic properties be avoided, but it is generally accepted that treatment with amitriptyline or imipramine is helpful with the bothersome symptom of uncontrollable laughing or crying. Idazoxan, a noradrenergic investigational agent, produced moderate improvement in a small number of patients, but sympathomimetic side effects have limited further development of this drug. Blepharospasm (involuntary eye closing) responds to botulinum toxin injection, and dry eyes can be treated with topical lubricants. Physical therapy and occupational therapy are beneficial in many cases. Evaluation should address issues of gait safety, activities of daily living, and home evaluation for assistive devices. Electroconvulsive therapy, adrenal implantation, pallidotomy, and deep brain stimulation are of no benefit.

Patient information about PSP may be obtained from the national support group:

The Society for Progressive Supranuclear Palsy, Inc.
Johns Hopkins Hospital
5065 Outpatient Center
601 North Caroline Street
Baltimore, MD 21287

SELECTED READING

Brucke T, Wenger S, Asenbaum S, Fertl E, Pfafflmeyer N, Muller C, Podreka I, Angelberger P. Dopamine D2 receptor imaging and measurement with SPECT. *Adv Neurol* 1993; 60: 494-500.

Burn DJ, Sawle GV, Brooks DJ. Differential diagnosis of Parkinson's disease, multiple system atrophy, and Steele-Richardson-Olszewski syndrome: discriminant analysis of striatal ^{18}F-dopa PET data. *J Neurol Neurosurg Psychiatry* 1994; 57: 278-284.

Collins SJ, Ahlskog JE, Parisi JE, Maraganore DM. Progressive supranuclear palsy: neuropathologically based diagnostic clinical criteria. *J Neurol Neurosurg Psychiatry* 1995; 58: 167-173.

Ghika J, Tennis M, Hoffman E, Schoenfeld D, Growdon J. Idazoxan treatment in progressive supranuclear palsy. *Neurology* 1991; 41: 986-991.

Golbe LI, Davis PH, Lepore FE. Eyelid movement abnormalities in progressive supranuclear palsy. *Mov Disord* 1989; 4: 297-302.

Golbe LI, Davis PH, Schoenberg BS, Duvoisin RC. Prevalence and natural history of progressive supranuclear palsy. *Neurology* 1988; 38: 1031-1034.

Hauw JJ, Daniel SE, Dickson D, Horoupian DS, Jellinger K, Lantos PL, McKee A, Tabaton M, Litvan I. Preliminary NINDS neuropathologic criteria for Steele-Richardson-Olszewski syndrome (progressive supranuclear palsy). *Neurology* 1994; 44: 2015-2019.

Levy R, Ruberg M, Herrero MT, Villares J, Javoy-Agid F, Agid Y, Hirsch EC. Alterations of GABAergic neurons in the basal ganglia of patients with progressive supranuclear palsy: an in situ hybridization study of GAD67 messenger RNA. *Neurology* 1995; 45: 127-134.

Litvan I, Blesa R, Clark K, Nichelli P, Atack JR, Mouradian MM, Grafman J, Chase TN. Pharmacological evaluation of the cholinergic system in progressive supranuclear palsy. *Ann Neurol* 1994; 36: 55-61.

Litvan I, Campbell G, Mangone CA, Verny M, McKee A, Chaudhuri KR, Jellinger K, Pearce RK, D'Olhaberriague L. Which clinical features differentiate progressive supranuclear palsy (Steele-Richardson-Olszewski syndrome) from related disorders? A clinicopathological study. *Brain* 1997; 120: 65-74.

Litvan I, Mega MS, Cummings JL, Fairbanks L. Neuropsychiatric aspects of progressive supranuclear palsy. *Neurology* 1996; 47: 1184-1189.

Nieforth KA, Golbe LI. Retrospective study of drug response in 87 patients with progressive supranuclear palsy. *Clin Neuropharmacol* 1993; 16: 338-346.

Riley DE, Fogt N, Leigh RJ. The syndrome of "pure akinesia" and its relationship to progressive supranuclear palsy. *Neurology* 1994; 44: 1025-1029.

Stacy M, Jankovic J. Differential diagnosis of Parkinson's disease and the parkinsonism plus syndromes. *Neurol Clin* 1992; 10: 341-359.

Steele JC, Richardson JC, Olszewski J. Progressive supranuclear palsy: a heterogeneous degeneration involving the brain stem, basal ganglia and cerebellum with vertical gaze and pseudobulbar palsy, nuchal dystonia and dementia. *Arch Neurol* 1964; 10: 333-359.

17

Multiple System Atrophy

James H. Bower, MD

CONTENTS

Through the history and neurologic examination, it is usually easy to recognize parkinsonism. However, diagnosing the specific type of parkinsonism is often much more difficult. Also, patients frequently do not understand that there is a difference between parkinsonism and Parkinson's disease (PD). Multiple system atrophy (MSA) is a condition that often imitates PD but has a very different prognosis, so that the early diagnosis of this condition is important for both counseling and treating the patient.

HISTORY AND EPIDEMIOLOGY

In 1900, Dejerine and Thomas described two patients with an akinetic-rigid state as well as ataxia, dysarthria, brisk reflexes, and incontinence. Because the autopsy in one case revealed atrophy of the olives, pons, and cerebellum, they labeled this disorder "olivopontocerebellar atrophy" (OPCA).

In 1960, Shy and Drager described two patients with "a neurological syndrome associated with orthostatic hypotension." Both patients had a rest tremor, reduced facial expression, and rigidity. In addition, they had marked autonomic failure, slurred speech, impaired coordination, distal muscle wasting, and external ocular muscle weakness. Autopsy revealed degenerative changes in the caudate, substantia nigra, olives, locus ceruleus, cerebellum, and intermediolateral cell columns of the spinal cord. Since that report, neurologists have commonly used the term "Shy-Drager syndrome" (SDS) to describe parkinsonism with marked autonomic failure.

Adams and colleagues, in 1961, coined the term "striatonigral degeneration" (SND) to describe an akinetic-rigid state, tremor, brisk reflexes, and dysarthria in three patients.

From: *Parkinson's Disease and Movement Disorders:*
Diagnosis and Treatment Guidelines for the Practicing Physician
Edited by: C. H. Adler and J. E. Ahlskog © Mayo Foundation
for Medical Education and Research, Rochester, MN

Table 17-1
Multiple System Atrophy Subtypes

Syndromes of
- Shy-Drager
- Olivopontocerebellar atrophy (OPCA)
- Striatonigral degeneration

Some also had ataxia and incontinence. In the post-levodopa era, many physicians have used the term "SND" to describe parkinsonism that fails to respond to levodopa therapy.

In 1969, Graham and Oppenheimer recognized the similarity in the clinical and pathologic features of SDS, OPCA, and SND. All three entail various combinations of parkinsonism, cerebellar ataxia, autonomic failure, and spasticity. To avoid "the multiplication of names for 'disease entities' which, in fact, are merely the expression of neuronal atrophy in a variety of overlapping combinations," they proposed the term "multiple system atrophy."

The definition of MSA has been evolving since the term was initially conceived. Many physicians erroneously use "MSA" to describe an unknown degenerative condition involving multiple areas of the central nervous system. The term "nonspecific multiple system degeneration" would be preferable to describe that condition. Because MSA is defined as a specific entity with three subtypes sharing common pathologic characteristics, the term should be reserved for sporadic occurrences of various combinations of parkinsonism, cerebellar ataxia, spasticity, and autonomic dysfunction. It is considered to be a distinct disease, encompassing the syndromes of sporadic OPCA, SDS, and SND (Table 17-1). In 1996, a consensus committee published the currently accepted definition of MSA (Table 17-2).

Related conditions often confused with MSA include idiopathic late-onset cerebellar ataxia, pure autonomic failure, and the spinocerebellar ataxias (SCAs). Idiopathic late-onset cerebellar ataxia is a sporadic degenerative condition consisting of pure cerebellar ataxia without other neurologic symptoms. Pure autonomic failure is a sporadic degenerative disease affecting only the autonomic nervous system. Neither of these conditions involves the combinations of parkinsonism, dysautonomia, and ataxia seen in MSA. However, it is conceivable that some of these cases represent formes frustes of MSA.

The SCAs are a group of conditions that often feature ataxia, parkinsonism, and other heterogeneous neurologic symptoms. However, they are familial diseases and rarely appear with early prominent dysautonomia.

The incidence of MSA in persons older than 50 is 3 cases/100,000 population, so that it is approximately one-sixteenth as common as PD. Because the condition is rare, few epidemiologic data have been collected. The median age at onset has been reported to be between 52 and 66 years, and the median length of survival is 8.5 to 9.5 years. There has been no pathologically proven case of MSA with onset before age 30 years.

CLINICAL FEATURES

Although parkinsonism develops in the vast majority of patients with MSA, occasional patients with dysautonomia and cerebellar ataxia never become parkinsonian. In

Table 17-2
Definition of Multiple System Atrophy

A sporadic (i.e., nonfamilial), progressive adult-onset disorder characterized by autonomic dysfunction, parkinsonism, and ataxia in any combination. The features include:
1. Parkinsonism (bradykinesia with rigidity or tremor or both)
2. Cerebellar or corticospinal signs
3. Orthostatic hypotension, impotence, and urinary incontinence or retention, usually preceding or occurring within the first 2 years of the motor symptoms

Data from The Consensus Committee of the American Autonomic Society and the American Academy of Neurology. *Neurology* 1996; 46: 1470.

a 1997 meta-analysis of autopsy-proven MSA, Wenning and associates noted that 87% of the patients had exhibited parkinsonism. This figure is similar to that in a 1994 clinical series of patients with MSA in which Wenning and associates found that 91% were parkinsonian. In general, the parkinsonism is very similar to that of PD, with some combination of tremor, rigidity, bradykinesia, and postural instability. Some features, however, should alert one to the possibility that the parkinsonism is "atypical," as discussed below.

Most patients with MSA do not have an adequate response to levodopa. A nil response to 1,000 mg/day on an empty stomach is often the first indication that the patient does not have PD. When a patient with MSA does benefit from levodopa, the response is typically modest, short-lived, or severely limited by levodopa-induced decompensation of orthostatic hypotension. Furthermore, the classic peak dose choreic or dystonic extremity dyskinesias of PD are uncommon in MSA. When levodopa-induced dyskinesias develop in MSA, they tend more to be dystonic spasms of the face, head, or neck.

Other "atypical" features often develop in patients with MSA. Early falls, rapid progression, severe hypophonia, pain unrelieved by levodopa, severe levodopa intolerance, prominent antecollis, and contractures can often complicate the course.

Although asymmetrical parkinsonism and a classic rest tremor are more indicative of PD than MSA, they clearly occur in MSA. In fact, 74% of patients with MSA present with asymmetrical parkinsonism and up to 39% have a tremor at rest. The tremor is usually jerky, however, with only 9% having the classic pill-rolling type.

Symptomatic autonomic dysfunction occurs in up to 74% of patients. If isolated impotence is included, the figure increases to 97%. Constipation and male impotence are extremely common in patients with MSA, but these symptoms are nonspecific in this age group and also common in PD. Otherwise, urinary incontinence is the most common feature of dysautonomia. The cause of the urinary incontinence is probably multifactorial. Loss of parasympathetic innervation should theoretically lead only to retention from detrusor denervation. Patients with MSA, however, also have evidence of detrusor hyperreflexia, probably a reflection of loss of inhibitory control from brain stem involvement. In addition, they have sphincter weakness, reflecting disease in Onuf's nucleus in the sacral cord. Their incontinence, therefore, is due to a combination of detrusor hyperreflexia, incomplete bladder emptying, and sphincter weakness.

Symptomatic postural hypotension not due to medications is the next most common feature of dysautonomia. These symptoms preceding or occurring early in the course of parkinsonism are often some of the most reliable clues to MSA. Although clinically troublesome in many patients, the postural hypotension in some remains moderate and does not lead to syncope. Patients with postural hypotension may complain of dizziness, lightheadedness, blurred vision, weakness, fatigue, nausea, cognitive impairment, palpitations, tremulousness, headache, and neck ache.

Urinary retention and fecal incontinence are other symptoms of autonomic dysfunction commonly experienced by these patients. Anhidrosis is also very common but is usually apparent only on formal laboratory testing of sudomotor function.

Cerebellar ataxia is seen in approximately half of all patients with MSA. Dysmetria (incoordination) of the limbs during the finger-nose or heel-shin maneuver is more common than gait ataxia. Eye findings, such as nystagmus, ocular dysmetria, impaired smooth pursuit, and square wave jerks of the eyes, are often seen.

Babinski's signs, hyperreflexia, or spasticity of tone—reflecting pyramidal dysfunction—occurs in almost two-thirds of patients. Other features of MSA are a jerkiness of the outstretched hands (possibly representing postural myoclonus); involuntary deep sighs; slurring, strained, or quivering speech; and Raynaud's phenomenon, with cold, dusky hands. Cognitive impairment is present in only 22% of patients and usually remains mild to moderate.

Respiratory stridor, present in approximately one-third of patients, has been linked to an increased risk of sudden nocturnal death. Paradoxical adduction of the vocal cords producing the stridor, probably secondary to abductor weakness, has been noted in sleeping patients. One autopsy case revealed atrophy of the posterior cricoarytenoid muscles, the abductors of the vocal cords. Interestingly, though, the nucleus ambiguus, which supplies these muscles, showed little cell loss or gliosis. Because patients with MSA have also been shown to have irregular patterns of ventilation, the sudden death is thought to be secondary to a combination of central hypoventilation and obstructive stridor.

PATHOLOGY

The pathologic criteria for MSA include cell loss or gliosis in at least two of several nuclei, including the striatum, substantia nigra, locus ceruleus, pontine nuclei, middle cerebellar peduncles, cerebellar Purkinje's cells, inferior olives, and intermediolateral cell columns. The posterior putamen and substantia nigra are usually the most affected, with the cortex, thalamus, anterior horn cells, nucleus ambiguus, vestibular nuclei, and subthalamic nucleus remaining relatively spared. In general, correlation between the clinical features of the patient and the pathologic findings has been good. Akinetic-rigid cases are associated with severe disease in the substantia nigra and striatum. Basal ganglia involvement beyond the dopaminergic substantia nigra accounts for patients whose parkinsonism is levodopa-refractory. Ataxic cases correlate with cell loss in the inferior olives, pons, and cerebellum. Dysautonomia correlates with disease in the intermediolateral cell columns.

In addition to the cell loss and gliosis, the brains of patients with MSA have glial cytoplasmic inclusions that stain with silver as well as ubiquitin and alpha synuclein histochemistry. These inclusions are present in all patients with MSA. They are, however, also present to a lesser extent in occasional patients with other diseases, such as the inherited SCAs, progressive supranuclear palsy (PSP), and corticobasal degeneration. Their presence is therefore necessary for the diagnosis but is not specific for it. Because of the overlap, the pathologic diagnosis of MSA should be made only if the clinical presentation is consistent with the diagnosis.

WORKUP

The ancillary investigations necessary to diagnose MSA are limited. Usually, a presumptive diagnosis is made in the office. Laboratory, radiologic, and electrophysiologic studies are performed only to aid confirmation or to rule out other treatable conditions. The major clinical distinction is between MSA and certain other neurodegenerative disorders, not only PD but also PSP and the familial SCAs. However, if prominent dysautonomia not due to drugs is present, PSP or familial SCAs are unlikely. Other considerations may also include Wilson's disease, Binswanger's disease (multi-infarct disease), normal pressure hydrocephalus, space-occupying lesions, drugs, and metabolic, toxic, or infectious causes. However, a comprehensive history and examination usually rule out those conditions.

Magnetic resonance imaging of the head can sometimes be helpful. Certainly, it can help rule out conditions such as normal pressure hydrocephalus, Binswanger's disease, and a space-occupying lesion. Furthermore, abnormalities have been reported that may help differentiate MSA from PD. Posterior and lateral putaminal hypointensities with a hyperintense rim at the lateral putaminal edge on T2-weighted images have been reported in MSA. Pontine and cerebellar atrophy can sometimes be seen, but usually when cerebellar ataxia is already obvious clinically.

Autonomic dysfunction may occur in other neurodegenerative disorders, most notably PD; however, pronounced early involvement is almost pathognomonic for MSA, when it occurs in the proper clinical context. Formal autonomic testing (e.g., thermoregulatory sweat test, tilt table test) can be used to help document the extent of dysautonomia. Because the degree of dysautonomia is often used as one of the criteria for differentiating PD from MSA, these tests can sometimes be helpful in making the diagnosis.

Many authors have advocated the use of urethral or rectal sphincter electromyography to confirm the diagnosis of MSA. Onuf's nucleus in the sacral cord is the group of anterior horn cells that supply the rectal and urethral striated sphincter muscles. This nucleus has been shown to be selectively involved in MSA. Results of rectal sphincter electromyography have been reported to be abnormal in most patients with MSA but normal in most patients with PD. This method is controversial, however, because others have not been able to use it successfully to differentiate the two disorders. The discomfort of the procedure and false-positive abnormal electromyographic findings from pelvic operations, obstetrical trauma, and other causes probably limit its utility.

Because of the risk of sudden death, any patient with MSA who has a history of excessive snoring, apnea, or stridor should undergo formal sleep evaluation. If stridor is identified, continuous positive airway pressure is often used and may be adequate to reverse the stridor. Tracheostomy is the definitive procedure but is rarely necessary unless refractory vocal cord paradoxical adduction is found.

TREATMENT

As with most of the neurologic degenerative conditions, no treatment reverses the progression of MSA. Symptomatic therapy, too, is unfortunately limited. Table 17-3 lists the common therapeutic options for symptoms.

Table 17-3
Options for Treatment of Symptoms in Multiple System Atrophy

Symptom or sign	Therapeutic options
Parkinsonism	Carbidopa-levodopa trial (± concurrent treatment of orthostatic hypotension)
Orthostatic hypotension	Elimination of offending medications Increase in fluids Increase in salt (including salt tablets) Elevation of head of bed Compressive stockings Fludrocortisone Midodrine
Bladder incontinence	Bedside commode or urinal Adult diapers or condom catheter Intermittent self-catheterization Anticholinergic agent (if detrusor hyperreflexia) Indwelling or suprapubic catheter Surgery
Impotence	Sildenafil Vacuum pumps Papaverine injections Penile implants

For patients presenting with parkinsonism, a trial of levodopa therapy is warranted. Up to one-third of patients have at least a moderate response. As with PD, treatment can be started at a dose of one-half or one whole 25/100 immediate-release carbidopa-levodopa tablet three times daily on an empty stomach, with slow titration upward. Patients with orthostatic hypotension may have difficulty tolerating even low doses of levodopa because of decompensation of their blood pressure. In those who can tolerate it, the dose can be increased to 250 to 300 mg three or four times daily if necessary. If no reduction in motor symptoms occurs at this dose, the medication can be tapered off or to the lowest beneficial dose. In occasional patients, levodopa therapy reverses parkinsonian motor symptoms but results in symptomatic orthostatic hypotension. In such patients, concurrent treatment directed at blood pressure control may allow levodopa therapy to continue. Rarely is a dopamine agonist, an anticholinergic drug, or amantadine helpful in a patient who has no benefit from levodopa.

Autonomic failure can be treated symptomatically. Orthostatic hypotension is often the most pressing of the various manifestations of the dysautonomia. Therapy is described in Chapter 11. Treatment of MSA-related orthostatic hypotension typically requires aggressive strategies. Thus, besides elimination of all unnecessary hypotensive drugs, increase in dietary salt and fluids, and elevation of the head of the bed to 30°, fludrocortisone is often required in doses up to approximately 0.2 mg twice a day. In many patients, it is also necessary to add the α_1-adrenergic agent midodrine in doses of

2.5 to 10 mg two to four times daily. This regimen may be especially helpful in patients with profound postprandial orthostatic hypotension. Waist-high, fitted compressive hose can also be efficacious but are usually poorly tolerated by the patient (difficult to get on, hot, uncomfortable).

A neurogenic bladder is among the more troublesome problems in patients with MSA. Often, a urodynamic study under the direction of a urologist helps define the characteristics (flaccid or spastic bladder, sphincter dyssynergia). Simple urge incontinence is often treated with the use of a bedside commode or urinal. Males can use a condom catheter. For urinary retention or incomplete emptying, intermittent catheterization is often required. Detrusor hyperreflexia is best treated with an anticholinergic agent, such as oxybutynin (2.5 to 5.0 mg twice to three times a day), propantheline bromide (15 to 60 mg/day), or tolterodine tartrate (1 to 2 mg twice a day). For more advanced incontinence, an indwelling or suprapubic catheter is required. Finally, surgical procedures, such as augmentation cystoplasty, artificial sphincter, and urinary diversion with stoma collection bag, are a last resort in some patients.

Male impotence may lend itself to treatment with sildenafil (Viagra), 50 mg 1 hour before intercourse. Other measures in more refractory patients are vacuum pumps, papaverine injections, and penile implants.

Active patients with anhidrosis may need to be cautioned about overheating during exercise. No pharmacologic treatment is available for anhidrosis. Treatment of constipation is outlined in Chapter 11.

A consultation with a physical medicine specialist may be helpful in the management of patients with MSA. These specialists have expertise in treating dysautonomia and motor disorders.

CONCLUSION

MSA is a degenerative neurologic disease that produces various combinations of parkinsonism, ataxia, autonomic dysfunction, and spasticity. It is the accepted term for the disorder that now includes sporadic OPCA, SDS, and SND. Although MSA often resembles PD clinically, the prognosis is worse, so that differentiation of one from the other is important in counseling and treating a patient.

SELECTED READING

Adams R, van Bogaert L, van der Eecken H. Nigro-striate and cerebello-nigro-striate degeneration. (Clinical uniqueness and pathological variability of presenile degeneration of the extrapyramidal ridigity type). *Psychiat Neurol (Basel)* 1961; 142: 219-259.

Bower JH, Maraganore DM, McDonnell SK, Rocca WA. Incidence of progressive supranuclear palsy and multiple system atrophy in Olmsted County, Minnesota, 1976 to 1990. *Neurology* 1997; 49: 1284-1288.

The Consensus Committee of the American Autonomic Society and the American Academy of Neurology. Consensus statement on the definition of orthostatic hypotension, pure autonomic failure, and multiple system atrophy. *Neurology* 1996; 46: 1470.

Dejerine J, Thomas A. L'atrophie olivo-ponto-cérébelleuse. *Nouvelle Iconographie de la Salpêtrière* 1900; 13: 330-370.

Graham JG, Oppenheimer DR. Orthostatic hypotension and nicotine sensitivity in a case of multiple system atrophy. *J Neurol Neurosurg Psychiatry* 1969; 32: 28-34.

Quinn N. Multiple system atrophy, in *Movement Disorders 3* (Marsden CD, Fahn S eds), Butterworth-Heinemann, London, 1994, pp. 262-281.

Quinn N, Wenning G. Multiple system atrophy. *Curr Opin Neurol* 1995; 8: 323-326.

Sandroni P, Ahlskog JE, Fealey RD, Low PA. Autonomic involvement in extrapyramidal and cerebellar disorders. *Clin Auton Res* 1991; 1: 147-155.

Shy GM, Drager GA. A neurological syndrome associated with orthostatic hypotension. *Arch Neurol* 1960; 2: 511-527.

Wenning GK, Ben Shlomo Y, Magalhaes M, Daniel SE, Quinn NP. Clinical features and natural history of multiple system atrophy. An analysis of 100 cases. *Brain* 1994; 117: 835-845.

Wenning GK, Tison F, Ben Shlomo Y, Daniel SE, Quinn NP. Multiple system atrophy: a review of 203 pathologically proven cases. *Mov Disord* 1997; 12: 133-147.

18

Familial Adult-Onset Spinocerebellar Degenerations

James H. Bower, MD

CONTENTS

The familial spinocerebellar degenerations are a heterogeneous group of disorders with onset in both childhood and adulthood. Their prevalence is estimated to be between 1.5 and 22.1 per 100,000. Although pathologic and clinical classification systems have been proposed in the past, all have had shortcomings. In the past several years, great strides in genetics have radically changed our thinking about the classification, and thus clinical presentation, of these disorders. In this chapter, we concentrate on the adult-onset familial ataxias.

CLINICAL CLASSIFICATION

Before the recent discoveries in genetics, patients with a familial progressive ataxic syndrome were classified according to clinical features. The onset of ataxia was one of the most important differentiating characteristics. The most common diagnoses in patients

From: *Parkinson's Disease and Movement Disorders:*
Diagnosis and Treatment Guidelines for the Practicing Physician
Edited by: C. H. Adler and J. E. Ahlskog © Mayo Foundation
for Medical Education and Research, Rochester, MN

Table 18-1
Harding Classification
of Autosomal Dominant Cerebellar Ataxia (ADCA)

ADCA I	Ataxia with various combinations of ophthalmoplegia, bulbar dysfunction, dementia, dystonia, parkinsonism, choreoathetosis, amyotrophy, optic atrophy, peripheral neuropathy
ADCA II	As above, with retinal degeneration
ADCA III	Pure cerebellar ataxia
ADCA IV	Episodic ataxia

with disorders of childhood onset were Friedreich's ataxia, ataxia-telangiectasia, abetalipoproteinemia, ataxia with selective vitamin E deficiency, and other metabolic conditions. The accompanying signs and symptoms helped lead the clinician to the diagnosis.

Although most genetic ataxias of childhood onset are autosomal recessive, familial disorders with onset after age 25 years are, in general, autosomal dominant conditions. The classification of these disorders was difficult because the clinical presentation of these families was so heterogeneous. The disease often varied greatly across and within families.

The adult-onset familial ataxias commonly appear with multiple system involvement. Pyramidal, extrapyramidal, and bulbar dysfunction resulting in spasticity, dystonia, parkinsonism, dysphagia, dysarthria, and ophthalmoparesis is common. Dementia, optic atrophy, and retinal degeneration can occur. On the basis of clinical characteristics, Harding classified these into four autosomal dominant cerebellar ataxic syndromes, as shown in Table 18-1. This clinical classification scheme, however, is now being replaced by genetic classification.

GENETIC CLASSIFICATION

With the many advances in genetics over the past 10 years, the classification of these conditions is evolving. At least eight genetic mutations involving trinucleotide repeats have been discovered that can cause a progressive ataxic syndrome. They all share many properties.

Currently, 12 neurologic conditions are known to be caused by expanded trinucleotide repeats: spinocerebellar ataxia (SCA) types 1, 2, 3, 6, 7, and 8, Friedreich's ataxia, Huntington's disease, dentatorubropallidoluysian atrophy (DRPLA), spinal and bulbar muscular atrophy, myotonic dystrophy, and fragile X syndrome. SCA types 1, 2, 3, 6, 7, and 8 and Friedreich's ataxia commonly present as progressive ataxias. DRPLA can sometimes appear in this manner.

The trinucleotide repeat diseases are caused by these expanded repeats either in the intron (outside the protein-coding region) or in the exon (within the protein-coding region). Friedreich's ataxia, myotonic dystrophy, and fragile X syndrome have expanded repeats outside the protein-coding region of the gene. Except for SCA 8, the remainder of these disorders are referred to as the "CAG/polyglutamine repeat diseases" because they all have expanded repeats of the trinucleotide cytosine-adenine-guanine (CAG) in the exon. The result is an expanded polyglutamine region within the diseased protein. SCA 8 is characterized by a CTG (cytosine-thymine-guanine) expansion in the exon.

The mechanism through which these abnormal proteins cause neuronal death is unknown. Because most of the CAG/polyglutamine diseases are autosomal dominant (spinal and bulbar muscular atrophy is X-linked), it is believed that the diseased protein from the affected allele has some toxic gain of function over the healthy protein from the wild-type allele. It is hypothesized that some altered protein-protein interaction produces abnormal aggregation. Interestingly, the identified proteins are present in different regions of the cell (e.g., ataxins 1 and 7 are nuclear; ataxins 2, 3, and 6 are cytoplasmic), but on pathologic study, ubiquitin-containing nuclear inclusions are present in all.

Although all these diseases are neurodegenerative and the proteins are expressed diffusely throughout the brain, the neuronal populations affected in each vary greatly. Therefore, some other modifying factors must influence neuronal susceptibility. The expanded polyglutamine proteins may interact with proteins specific for target cells. The length of the repeat, which varies among individuals, may play a role. Somatic mosaicism, in which different cell populations in the same person express proteins with different repeat lengths, may also explain part of this diversity. Lastly, the number of repeats in the normal allele may somehow influence disease expression.

A striking feature of all the trinucleotide repeat ataxias is their clinical heterogeneity. The phenotypic expression within each disease varies tremendously. Longer repeat lengths have been correlated with greater severity and earlier age at onset. The repeat length also tends to increase from one generation to the next, especially with paternal transmission. This change helps to explain the clinical phenomenon of anticipation, in which successive generations have more severe and earlier disease.

At the same time, there is phenotypic similarity and overlap among these genetically different disorders. This clinical overlap underlies the difficulty in clinical classification systems. The genetic classification of these disorders into the different SCAs is evolving rapidly. Table 18-2 summarizes features of the currently recognized familial cerebellar degenerations that arise in adulthood (or, in Friedreich's ataxia, occasionally in adulthood). Clinical diversity within each SCA and similarity across each SCA cannot be overemphasized.

FRIEDREICH'S ATAXIA

Friedreich's ataxia has been traditionally defined as a childhood disease (onset before age 25 years) of progressive gait and limb ataxia and lower extremity areflexia. Because of recent genetic discoveries, however, it now has to be included in a list of familial adult-onset ataxias.

Classically, these children have progressive ataxia, dysarthria, lower extremity areflexia, and proprioceptive sensory loss. Eye findings include dysmetric eye movements, square wave jerks, optic flutter, and optic atrophy. Sensorineural hearing loss, foot deformities (pes cavus), and kyphoscoliosis are common. A hypertrophic cardiomyopathy can lead to congestive heart failure, electrocardiographic changes, and arrhythmias. Up to 15% of patients have diabetes mellitus. Most patients are wheelchair-bound by age 30.

The genetic mutation underlying Friedreich's ataxia, long known to be an autosomal recessive condition, has recently been discovered. This disorder is caused by a

Table 18-2
Genetic Causes of Ataxia in Adults

Genetic classification	Clinical classification	Mutation	Distinguishing characteristics
Friedreich's ataxia	Friedreich's ataxia	GAA repeat, chromosome 9	Areflexia, pes cavus cardiomyopathy
SCA1	ADCA I	CAG repeat, chromosome 6	Bulbar, corticospinal, extrapyramidal signs
SCA2	ADCA I	CAG repeat, chromosome 12	Slow saccades, neuropathy, dementia
SCA3-MJD	MJD/ADCA I	CAG repeat, chromosome 14	Bulging eyes, facial fasciculations, olivary and cerebellar sparing
SCA4	ADCA I	Unknown gene, chromosome 16	Sensory neuropathy
SCA5	ADCA I or III	Unknown gene, chromosome 11	Mild disease
SCA6	ADCA III	CAG repeat, chromosome 19	Isolated ataxia
SCA7	ADCA II	CAG repeat, chromosome 3	Retinal degeneration
SCA8	ADCA I	CTG repeat, chromosome 13	Corticospinal signs, vibratory sense loss
SCA10	ADCA III	Unknown gene, chromosome 22	Isolated ataxia (?seizures)
DRPLA	DRPLA	CAG repeat, chromosome 12	Chorea, myoclonus, seizures, dementia
EA-1	EA-1	K^+-channel gene, chromosome 12	Episodic ataxia, myokymia
EA-2	EA-2	Ca^{++}-channel gene, chromosome 19	Episodic ataxia

ADCA, autosomal dominant cerebellar ataxia (the older clinical classification system); CAG, cytosine-adenine-guanine; CTG, cytosine-thymine-guanine; DRPLA, dentatorubropallidoluysian atrophy; EA, episodic ataxia; GAA, guanine-adenine-adenine; MJD, Machado-Joseph disease; SCA, spinocerebellar ataxia (nomenclature for the genetic classification system).

guanine-adenine-adenine (GAA) trinucleotide repeat on a gene intron on chromosome 9. The normal length is 7 to 34 GAA repeats, but patients with Friedreich's ataxia have lengths of 120 to 1,700 repeats. This gene codes for a mitochondria-targeted protein named "frataxin." Among affected persons, the repeat expansion is present in both alleles or, alternatively, present in only one allele with a point mutation in the other. This leads to the loss of expression of frataxin.

Since identification of the mutation, Friedreich's ataxia with onset after 25 years of age has been recognized, and these occurrences may represent up to 25% of all cases. In general, later-onset disease is less severe and slower in progression. Pes cavus, cardiomyopathy, and areflexia may never develop. This variation is probably related to smaller repeat lengths in these patients. Because Friedreich's ataxia is autosomal recessive, many patients do not have clinically informative family histories.

SPINOCEREBELLAR ATAXIA TYPE 1

SCA1 was the first autosomal dominant, adult-onset ataxia in which a genetic defect was identified. This ataxia is caused by an expanded CAG repeat on a gene exon on chromosome 6. The normal length is 6 to 39 CAG repeats, and the diseased length is 40 to 83 repeats. The gene product is a nuclear protein named "ataxin-1."

Ataxia (limbs, gait, speech, eye movements) is typically prominent. It is often associated with corticospinal tract signs (e.g., hyperreflexia, Babinski's signs) and extrapyramidal features (e.g., parkinsonism, dystonia, and, rarely, chorea). Peripheral nervous system involvement may result in areflexia or amyotrophy. Bulbar signs include ophthalmoparesis and dysphagia. Mild cognitive dysfunction may also occur. Pontine and cerebellar atrophy is often seen on magnetic resonance (MR) or computed tomography imaging. Pathologically, there is cell loss in cerebellar Purkinje's cells and brain stem nuclei and degeneration in the spinocerebellar tracts, ventral pons, middle cerebellar peduncles, and inferior olives. Most patients become wheelchair-bound within 15 years.

SPINOCEREBELLAR ATAXIA TYPE 2

SCA2 was first described in a Cuban family with a progressive ataxic syndrome associated with slow ocular saccades and neuropathy. Typical of all the SCAs, however, the clinical presentation is now known to be diverse. It is caused by an expanded CAG repeat on chromosome 12. The normal length is 15 to 32 repeats, and the diseased length is 35 to 77 repeats. Its gene product has been named "ataxin-2."

The clinical features described above for SCA1 also fit the SCA2 phenotype (as well as SCA3, described below). In some cases, clinical clues point to SCA2, including slow ocular saccades, hyporeflexia, a postural tremor, and dementia.

SPINOCEREBELLAR ATAXIA TYPE 3

The SCA3 phenotype was linked to an expanded CAG repeat on chromosome 14. The normal gene has a length of 12 to 39 repeats, whereas the diseased gene has 61 to 84 repeats. The encoded protein, ataxin-3, is widespread in the brain.

Around the same time that the SCA3 mutation was identified, the mutation for Machado-Joseph disease (MJD) was linked to the same genetic abnormality. They are now thought to be the same disorder. Although the disease is sometimes characterized as SCA3, it is often called "SCA3-MJD" to acknowledge the historical classification.

MJD was originally described in families living in the Azores, a group of islands off Portugal. Subsequently, North American and, later, worldwide families of Azorean descent were identified with this disease. Although it commonly consists of ataxia and marked dysarthria and dysphagia, four types have been described. In type 1, prominent corticospinal tract dysfunction, rigidity, and dystonia occur before age 25 years. Patients with type 2 have ataxia and ophthalmoparesis in early to middle adulthood. In type 3, ataxia and amyotrophy appear in middle to late adulthood. A fourth type, occurring in late adulthood, is characterized by prominent parkinsonism and peripheral neuropathy. It has long been recognized that all four types can be present in the same family.

In general, the onset of SCA3-MJD ranges from childhood through the seventh decade, although onset in the 30s and 40s is most common. Most patients have ataxia with a supranuclear ophthalmoparesis. Corticospinal tract signs and parkinsonism are present to variable degrees. "Bulging eyes" from dystonic lid retraction and facial fasciculations are particularly common and somewhat specific for this genotype. Corticobulbar signs, including dysarthria, dysphagia, dysphonia, facial palsy, ptosis, and lingual atrophy, are common. Amyotrophy and areflexia are also present in some cases.

Pathologically, there is degeneration of the cerebellar afferent and efferent pathways, substantia nigra, subthalamic nucleus, and globus pallidus and of pontine, dentate, and cranial nerve nuclei. In general, there is relative sparing of the Purkinje cells, olives, and posterior columns, accounting for the lack of olivary and cerebellar atrophy on head scans. This radiologic feature is one that may help differentiate SCA3 from SCA1 and SCA2.

SCA3 is currently considered to be the most common autosomal dominant inherited cerebellar ataxia. Its prevalence is particularly high in Portugal and Japan.

SPINOCEREBELLAR ATAXIA TYPES 4 AND 5

SCA4 and SCA5 are less common than the other SCAs, and less has been written about the patients. Patients with SCA4 commonly have a sensory neuropathy and corticospinal tract signs. The disease has been mapped to chromosome 16, but no gene has been identified. Patients with SCA5 tend to have a milder ataxia and dysarthria than those with the other SCAs. The disorder has been mapped to chromosome 11.

SPINOCEREBELLAR ATAXIA TYPE 6

SCA6 has been linked to an expanded CAG repeat on chromosome 19 in a gene encoding for a voltage-dependent P/Q type of calcium channel. The normal allele has 3 to 17 repeats, with the diseased allele carrying 21 to 30 repeats. A different mutation in the same gene produces a different disease, known as episodic ataxia type 2 (EA-2). The two diseases can be distinguished clinically because SCA6 is a progressive ataxic syndrome and the ataxia of EA-2 occurs in discrete episodes lasting days and is often provoked by stress or fatigue. Persistent, relatively mild ataxia develops after many years in typical EA-2. Interestingly, a different point mutation in the same gene causes a third disease: familial hemiplegic migraine.

Onset of SCA6 is usually in middle adulthood, with a pure cerebellar syndrome and signs of ataxia, dysarthria, and nystagmus. Progression is slow, and often little to no bulbar dysfunction, spasticity, or extrapyramidal symptoms develop. Life span usually is normal.

SPINOCEREBELLAR ATAXIA TYPE 7

SCA7 is caused by an expanded CAG repeat on chromosome 3 in a gene encoding a protein named "ataxin-7." Normal alleles carry 7 to 17 repeats, whereas SCA7 alleles carry 38 to 130 repeats. Although the patients are clinically similar to those with SCA types 1, 2, and 3, the distinguishing characteristic of SCA7 is retinal degeneration. Up to 83% of patients have decreased visual acuity, and 28% are blind.

OTHER SPINOCEREBELLAR ATAXIA TYPES

Characterization of additional autosomal dominant SCAs continues. Recently, expansion of a CTG repeat on chromosome 13 was linked to a kindred with generalized ataxia, spasticity, and loss of vibratory sensation, defining SCA8. SCA10 is characterized by a relatively pure cerebellar ataxia (similar to SCA5) and has been linked to chromosome 22. At present, approximately one-third of recognized autosomal dominant SCA kindreds have not been assigned to one of the known SCA types; hence, the nomenclature will continue to expand.

DENTATORUBROPALLIDOLUYSIAN ATROPHY

DRPLA is another CAG/polyglutamine disease that sometimes appears as a progressive ataxia. It is rare in the United States but more common in Japan. Two phenotypes, albeit with overlap, occur as a function of the number of CAG repeats. Patients with longer repeat lengths have ataxia, myoclonus, seizures, and cognitive impairment and at a younger age. In these patients, DRPLA is considered to be one of the progressive myoclonic epilepsies. Patients with shorter repeat lengths have phenotypes similar to those of Huntington's disease, with the adult onset of psychiatric symptoms, chorea, and dystonia. Patients thought to have Huntington's disease in whom the result of Huntington genetic testing is negative should also undergo testing for the DRPLA mutation.

MISCELLANEOUS CHILDHOOD-ONSET ATAXIAS

There are many other causes of progressive ataxia with onset in childhood. In some cases, medical evaluation might not be done before the patients are adults. Ataxia-telangiectasia is one of the more common childhood ataxias. It is a disorder of progressive ataxia associated with oculocutaneous telangiectasias, immune dysfunction, premature aging, and increased risk of cancer. Extrapyramidal signs, especially dystonia and myoclonus, are also seen. The gene for ataxia-telangiectasia has been found on chromosome 11, with more than 100 possible mutations identified. For this reason, genetic testing is impractical. Useful laboratory markers include increased serum alpha-fetoprotein and carcinoembryonic antigen levels and decreased IgA levels.

Mitochondrial diseases, abetalipoproteinemia, and a wide variety of inborn errors of metabolism can also cause childhood ataxia. They rarely cause progressive ataxias in adults and are beyond the scope of this chapter.

EPISODIC ATAXIAS

Two familial disorders take the form of episodic ataxias rather than the progressive ataxias already discussed. Episodic ataxia type 1 (EA-1) is caused by a nonrepeat mutation in a gene on chromosome 12 encoding for a potassium channel. The patients have brief spells of ataxia (seconds to minutes) that are often startle-

or exercise-induced. Facial myokymia is common. The ataxia typically responds to treatment with phenytoin or acetazolamide. Patients with EA-2 have longer spells, lasting days, with ataxia, vertigo, nausea, vomiting, headache, and weakness. These spells may also respond to acetazolamide. As previously mentioned, the mutation has been found to be a nonrepeat mutation in a gene on chromosome 19 encoding a calcium channel. Different mutations in this same gene result in SCA6 and familial hemiplegic migraine.

EVALUATION OF A PATIENT WITH PROGRESSIVE ATAXIA

History and Examination

The approach to a patient with progressive ataxia begins as all investigations do in neurology—with a comprehensive history. Secondary causes need to be explored. Toxic causes to be considered include alcohol abuse, medications (e.g., anticonvulsants, such as phenytoin; benzodiazepines, such as diazepam; and, in general, any psychoactive drug), and environmental toxins (if there has been intensive occupational exposure). Deficiency states should also be considered in appropriate patients (especially vitamins B_{12}, E, and thiamine). Vitamin E deficiency has been identified as the major cause of neurologic symptoms (especially ataxia) in patients with fat malabsorption syndromes.

Clues to immune-mediated causes of ataxia also come from the history. A subacute course (developing over days to weeks) in a patient with a viral illness suggests a possible parainfectious cause. A subacute course is also typical of paraneoplastic cerebellar syndromes. A relapsing-remitting course of neurologic deficits suggests multiple sclerosis, although in rare patients, an isolated progressive ataxic syndrome may occur.

Tumors and other structural lesions of the posterior fossa may also be announced by progressive ataxia. Headache may occur, but often historical clues pointing specifically to tumor are absent.

Prion disease may also be signaled by ataxic disorders. Sporadic cases of Creutzfeldt-Jakob disease should be easy to distinguish because of the rapidly progressive course with dementia. Gerstmann-Sträussler-Scheinker disease may begin as a familial ataxic syndrome. The history of progression to death over just a few years in other family members in addition to prominent dementia may help make the diagnosis.

Ataxia developing acutely suggests a cerebrovascular cause. The history usually makes this diagnosis obvious.

Multiple system atrophy often is manifested as a progressive ataxic syndrome. As discussed in Chapter 17, dysautonomia producing bowel and bladder dysfunction, impotence, and postural hypotension is typically prominent in this disorder. Parkinsonism, too, is common.

Laboratory Studies

IMAGING

Further workup of progressive ataxia depends on the results of the history and examination. An MR head scan is appropriate in essentially all these patients. Clinically relevant structural lesions are apparent, and clues suggesting stroke or demyelinating disease may be apparent. Cerebellar atrophy with or without pontine atrophy suggests a neurodegenerative process.

BLOOD TESTS

Studies to investigate metabolic and toxic causes include complete blood cell count, erythrocyte sedimentation rate, chemistry battery, thyroid studies, and determination of vitamins B_{12} and E levels. In appropriate patients, other blood tests include ceruloplasmin screen for Wilson's disease (in patients younger than 55 years), human immunodeficiency virus (HIV) and syphilis serologic studies, connective tissue disease battery, Lyme disease serology, and anti-Purkinje cell antibodies.

Further testing may be considered in patients with a possible rare inborn error of metabolism: alpha-fetoprotein and immunoglobulin levels (ataxia-telangiectasia); blood smear for acanthocytes (neuroacanthocytosis, abetalipoproteinemia); cholesterol value (abetalipoproteinemia and malabsorption syndromes); phytanic acid level (increased in Refsum's disease); and lactate and creatine kinase concentrations (high in mitochondrial cytopathies). Additional tests for rare inborn errors of metabolism are available for those with other childhood-onset ataxias.

URINE TESTS

Routine urinalysis may suggest clues to a relevant systemic abnormality (e.g., connective tissue disease, vasculitis). If Wilson's disease is considered, a 24-hour urine collection for copper may be ordered (typically increased in this disorder).

CEREBROSPINAL FLUID

Lumbar puncture for cerebrospinal fluid may be appropriate if the differential diagnosis includes certain disorders: multiple sclerosis, inflammatory processes (e.g., vasculitis), infectious causes (e.g., Lyme disease, HIV, syphilis), or neoplasm, either diffuse infiltrating (e.g., lymphoma) or paraneoplastic. Relevant cerebrospinal fluid tests include cells, protein, glucose, oligoclonal bands, IgG index, and synthesis rate; in addition, cytologic, specific serologic, culture, and polymerase chain reaction studies may be done.

AUTONOMIC TESTING

In patients with progressive ataxia and no family history, multiple system atrophy should be considered. As with some of the inherited SCAs, multiple system atrophy is often associated with parkinsonism and corticospinal tract signs. An important clinical clue, however, is prominent dysautonomia, which is not expected in the inherited SCAs. This can be objectively assessed with formal autonomic testing, available at most large medical centers.

GENETIC TESTING

A family history consistent with autosomal dominant inheritance suggests that genetic testing for one of the SCA genes may be diagnostic. These tests are now commercially available. A family history consistent with a recessive disorder or patients with a relevant clinical syndrome may be tested for the Friedreich ataxia gene. Patients without a family history could nonetheless have an inherited disorder. At present, however, genetic testing in most of these sporadic patients is usually nonproductive.

THERAPY

Therapeutic options for patients with ataxia are limited. Although amantadine and buspirone have been reported to be beneficial, their efficacy is modest at best and has never been proven in large double-blind, placebo-controlled studies. Physical therapy and occupational therapy with gait training and aids are appropriate. Speech therapy and communication aids are helpful in certain patients with prominent dysarthria.

CONCLUSION

The familial adult-onset ataxias are a heterogeneous group of disorders that feature a combination of progressive ataxic, bulbar, pyramidal, extrapyramidal, ophthalmologic, and cognitive disturbances. Recent advances in genetics are rapidly changing our understanding and classification of these disorders.

SELECTED READING

Boetz MI, Boetz-Marquard T, Elie R, Pedraza OL, Goyette K, Lalonde R. Amantadine hydrochloride treatment in heredodegenerative ataxias: a double blind study. *J Neurol Neurosurg Psychiatry* 1996; 61: 259-264.

Durr A, Brice A. Genetics of movement disorders. *Curr Opin Neurol* 1996; 9: 290-297.

Harding AE. Inherited ataxias. *Curr Opin Neurol* 1995; 8: 306-309.

Junck L, Fink JF. Machado-Joseph disease and SCA3: the genotype meets the phenotypes. *Neurology* 1996; 46: 4-8.

Klockgether T, Evert B. Genes involved in hereditary ataxias. *Trends Neurosci* 1998; 21: 413-418.

Lou JS, Goldfarb L, McShane L, Gatev P, Hallett M. Use of buspirone for treatment of cerebellar ataxia. An open-label study. *Arch Neurol* 1995; 52: 982-988.

Rosenberg RN. The genetic basis of ataxia. *Clin Neurosci* 1995; 3: 1-4.

Rosenberg RN. Spinocerebellar ataxias and ataxins. *N Engl J Med* 1995; 333: 1351-1353.

Subramony SH. Clinical aspects of hereditary ataxias. *J Child Neurol* 1995; 10: 353-362.

Zoghbi HY. The expanding world of ataxins. *Nat Genet* 1996; 14: 237-238.

19 Corticobasal Degeneration

Bradley F. Boeve, MD

CONTENTS

BACKGROUND

Since the original description in 1967 by Rebeiz, Kolodny, and Richardson of three patients with a progressive asymmetrical akinetic-rigid syndrome and apraxia, numerous other cases have been described under the labels "corticonigral degeneration," "cortical-basal ganglionic degeneration," "corticobasal ganglionic degeneration," and "corticobasal degeneration" (CBD). The last name has been adopted by most current authors. The core clinical features are progressive asymmetrical rigidity and apraxia, with other findings suggesting additional cortical dysfunction (e.g., alien limb phenomena, cortical sensory loss, myoclonus, mirror movements) and basal ganglionic dysfunction (e.g., bradykinesia, dystonia, tremor, choreoathetosis).

The characteristic findings at autopsy are asymmetrical frontoparietal cortical atrophy with basal ganglia and nigral degeneration. Microscopically, swollen neurons that do not stain with conventional hematoxylin-eosin (so-called ballooned, achromatic neurons) are found in the cortex, similar to those in Pick's disease. Abnormal accumulations of neurotubule (tau) protein are found in both neurons and glia. This tau protein has some similarities to the tau protein found in some other neurodegenerative disorders, notably progressive supranuclear palsy and Pick's disease.

From: *Parkinson's Disease and Movement Disorders:*
Diagnosis and Treatment Guidelines for the Practicing Physician
Edited by: C. H. Adler and J. E. Ahlskog © Mayo Foundation
for Medical Education and Research, Rochester, MN

Table 19-1
Core Clinical Features
of the Corticobasal Degeneration Syndrome

PARA syndrome
* <u>P</u>rogressive course
* <u>A</u>symmetrical occurrence (typically starts in
  one limb)
* <u>R</u>igidity
* <u>A</u>praxia (disordered motor programming)

CLINICAL FEATURES

As in many other neurodegenerative disorders, symptoms begin insidiously in the sixth to eighth decade and gradually progress over 5 to 15 years until death. Men and women are equally affected. A relatively high frequency of coexisting autoimmune diseases was noted in one series, although we have not been impressed with that relationship in our patients. The disorder is typically sporadic, but there are rare reports of a similarly affected relative.

Progressive Asymmetrical Rigidity and Apraxia

The most consistent and distinctive manifestations of this condition are progressive asymmetrical rigidity and apraxia, which we have termed the "PARA syndrome" (Table 19-1). Symptoms typically begin in one limb (no predilection for the right or left side has been observed). Patients describe their limb difficulties as "clumsiness," "incoordination," or "stiffness." Many of our patients can recall the first instance when rigidity or apraxia became evident. One remembered being unable to turn a doorknob while carrying objects with the unaffected upper limb; another, who had onset of symptoms in a lower extremity, recalled frustration while unsuccessfully attempting to pedal a bicycle.

On examination, the limb is mildly to severely rigid and sometimes adopts a dystonic posture. In examining numerous patients, we have been struck by the features of both rigidity (i.e., velocity-independent increased tone) and spasticity (i.e., velocity-dependent increased tone) in the affected limbs. Alternating motion rates are markedly reduced. The affected limb is profoundly apractic[1]—unable to correctly perform simple activities (e.g., combing hair, bathing, feeding, handwriting) or gestures (e.g., waving good-bye, saluting like a soldier, throwing a ball, pounding a hammer). When asked to perform these activities or gestures with an involved hand, patients characteristically glare at the limb, as if perplexed by how to make the hand function properly. There is often a delay as they visibly struggle, which invariably elicits frustration. With time, the limb becomes completely useless, and other limbs become similarly affected.

---

[1]A generally accepted definition of apraxia is the inability to perform learned motor movements in the absence of other neurologic deficits, such as weakness, sensory loss, and rigidity. Although every patient with CBD has other coexisting neurologic deficits contributing to the dysfunctional limb or limbs, the term "apraxia" is still applied because it most aptly denotes this feature.

The spread of limb involvement and rate of progression vary. The most common course is PARA in one of the upper limbs for at least 2 years and then involvement of either the ipsilateral lower limb or contralateral upper limb, eventually leading to severe generalized disability several years later. Less commonly, a lower limb is affected first, or progression occurs rapidly over many months. With the patient bedbound, death typically results from aspiration pneumonia or urosepsis.

The pathologic substrate for the rigidity presumably results from nigral or striatal damage. Apraxia presumably is due to contralateral parietal or mesial frontal cortical degeneration.

Dystonia

Dystonic posturing of a limb is a common early manifestation, usually affecting one upper limb. One of our patients colorfully described this presentation as "Napoléon's arm," in which the shoulder is adducted and the elbow and wrist are flexed. This characteristic posture, reminiscent of a stroke, may erroneously suggest a cerebrovascular cause despite the insidious (rather than acute) onset and absence of an infarct on imaging studies.

As noted above, many patients have elements of rigidity and spasticity in affected limbs. Less commonly, dystonic posturing is evident only during walking or reaching (i.e., action-induced). In patients with symptoms beginning in a lower limb, the foot is often tonically inverted, and ambulation is severely limited. Pain is a fairly common complaint. Some have argued that dystonia results from striatal damage, although the position of the completely dystonic arm (which resembles the posture with middle cerebral artery infarcts) may represent severe degeneration of the motor (including premotor) cortex.

Myoclonus

The lightning-like jerks of myoclonus usually are low in amplitude and usually do not become a problem unless proximal muscle involvement causes the limb to strike the body or other objects. The frequency and amplitude of myoclonic jerks typically increase with tactile stimuli (i.e., stimulus-sensitive myoclonus) and action (i.e., action myoclonus). Recent electrophysiologic studies suggest that the myoclonus in this disorder results from enhanced direct sensory input to cortical motor areas.

Tremor

Tremor, another common presenting feature, coexists with rigidity and thus can lead to the incorrect diagnosis of Parkinson's disease. Unlike the tremor of Parkinson's disease, which is most prominent at rest and dampens with action, this tremor is amplified with activity and minimal at rest. Patients typically describe the affected extremity as "jerky," and often this quality reflects the combined effect of myoclonus and tremor.

Lack of Levodopa Response

Although some persons note a very mild improvement in overall well-being, significant clinical reversal of motor deficits does not occur with levodopa therapy. Lack of improvement with levodopa therapy can be regarded as a diagnostic feature of this disorder (with the recognition that other parkinsonism-plus syndromes may also not respond). We typically escalate the carbidopa-levodopa dosage to 1,000 mg of levodopa daily (divided doses, on an empty stomach) to confirm nonresponsiveness.

Alien Limb Phenomenon

The alien limb phenomenon is the most intriguing feature of CBD, characterized by elevation of an extremity, involuntary grasping of objects (i.e., utilization behavior), or overt oppositional behavior. Patients describe their affected limb as "alien," "uncontrollable," or "having a mind of its own" and often label the limb as "it" when describing the limb's behavior. When severe, this phenomenon can cause much embarrassment and frustration. One of our patients was reluctant to stand in crowds after "it" (his hand) had grabbed other persons. The movements are spontaneous and minimally affected by mental effort, sometimes requiring restraint by the contralateral limb. The alien behavior is often more prominent with the patient's eyes closed. This phenomenon is typically transitory, lasting months to a few years before progressive rigidity or dystonia supersedes it.

Cortical Sensory Loss

Cortical sensory loss, another common early manifestation, is sometimes described as "numbness" or "tingling." Impaired joint position sense, impaired two-point discrimination, agraphesthesia, astereognosis, and extinction to double simultaneous stimulation with intact primary sensory modalities are all evidence of cortical sensory loss. Cortical sensory loss per se may result in involuntary movements; these slow choreoathetoid-like movements due to sensory disconnection are termed "pseudoathetoid." We have seen several patients with PARA and "numbness" in one hand who had carpal tunnel release despite clinical and electrophysiologic findings inconsistent with median neuropathy at the wrist; not surprisingly, they had no relief of symptoms after surgery. We have also seen (rarely) persons with pain and paresthesias in the same distribution and other findings suggestive of central pain (see below). The pathologic substrate for cortical sensory loss is the contralateral parietal (somatosensory) cortex or thalamus.

Mirror Movements

A patient has mirror movements if the opposite limb inappropriately performs the same activity as the one being examined. For example, a person asked to perform rapid alternating movements with the left upper extremity often unknowingly makes similar spontaneous movements in the right upper limb. Mirror movements are often suppressible, but when the person is distracted and rechallenged with the same maneuver several minutes later, they recur. This phenomenon most likely results from parietal damage.

Dementia

Prominent dementia is not a typical early finding, but more subtle cognitive impairment may be present. Learning and memory are typically minimally affected until the terminal stage of the disease (when all four limbs are involved) in most patients. This finding is not surprising, because the hippocampi and other mesial temporal structures are relatively preserved when examined histologically. Neuropsychometric testing shows impairment in attention and concentration, verbal fluency, executive functions, and visuospatial functioning—all of which are often overshadowed by the predominant PARA features. If memory impairment is severe and interferes with daily functioning, atypical Alzheimer's disease should be considered as an alternative diagnosis.

Table 19-2
Clinical Features of Corticobasal Degeneration

Core findings (see Table 19-1)

Other common findings
- Focal or asymmetrical appendicular dystonia (especially upper limb)
- Focal or asymmetrical appendicular myoclonus (often stimulus-sensitive, especially upper limb)
- Focal or asymmetrical appendicular postural, or action, tremor (especially upper limb)
- Alien limb phenomenon
- Cortical sensory loss
- Mirror movements
- Lack of levodopa response

Infrequent, relatively mild, or late findings
- Aphasia
- Frontal lobe signs
- Appendicular ataxia
- Other involuntary movements (chorea, blepharospasm)
- Dysphagia

Variable findings
- Dysarthria or apraxia of speech
- Oculomotor impairment or apraxia of eyelid movements
- Corticospinal tract signs
- Postural instability

Other Findings

Several less specific clinical manifestations may also be seen. If aphasia occurs, it is usually a late development, as are frontal release signs (e.g., snout reflex, glabellar sign, palmomental reflex, grasp reflex). Hypokinetic (parkinsonian) dysarthria is not uncommon. Apraxia of speech is quite common; in one published case, in fact, speech apraxia was the primary presenting problem and "typical" features did not develop until at least 5 years later.

Oculomotor impairment occurs to some degree in almost all patients. This includes saccadic pursuit (brief interruptions in the normally smooth movements of eyes following a target) or ocular apraxia (difficulty initiating eye movements to command). Supranuclear gaze paresis is also occasionally present, sometimes indistinguishable from that in progressive supranuclear palsy. Apraxia of eyelid movements has also been noted.

Asymmetrical hyperreflexia or extensor toe responses (or both) occur frequently. Postural instability is very common, whereas appendicular ataxia, chorea, and blepharospasm are infrequent manifestations. Dysphagia begins insidiously in the later stages of this disorder and often leads to aspiration pneumonia, which may be fatal in advanced disease.

Summary of Clinical Findings

CBD is primarily a clinical diagnosis. The core findings are those of the PARA syndrome, but other clinical features support the diagnosis (Table 19-2).

LABORATORY FINDINGS

The syndrome of CBD is diagnosed clinically, and no laboratory test is diagnostic. However, certain results on several ancillary studies have been reported to support the clinical diagnosis. These studies may also demonstrate or suggest other diseases (see Differential Diagnosis, below).

Blood and Cerebrospinal Fluid

Results of routine blood tests (e.g., complete blood cell count, sedimentation rate, antinuclear antibody determination) and cerebrospinal fluid analyses are typically normal.

Neuropsychometrics

Neuropsychometric testing typically shows impairment in domains subserved by frontal, frontostriatal, and parietal cognitive networks: attention and concentration, verbal fluency, executive functions, praxis, and visuospatial functioning. Performance on tests of learning and memory, although not normal, tends to be mildly impaired. Alternative diagnoses, particularly Alzheimer's disease, should be considered if performance on delayed recall measures is markedly abnormal.

Neuroimaging

The purpose of a computed tomography scan or magnetic resonance imaging of the brain is to exclude a structural lesion, such as tumor, abscess, hematoma, or infarct. If these lesions are absent, certain findings can support the diagnosis of CBD, such as asymmetrical cortical atrophy, especially frontoparietal, with the more prominent atrophy existing contralateral to the side most severely affected clinically. The lateral ventricle in the maximally affected cerebral hemisphere can also be slightly larger than the opposite one. These findings are often subtle, and their presence or absence is of only minor diagnostic significance.

Other Studies

Certain findings on single-photon emission computed tomography, positron emission tomography, electroencephalography, and recording of somatosensory evoked potentials are reported to support the clinical diagnosis. These studies should not be considered critically necessary.

Brain Biopsy

Brain biopsy may be unrevealing because of sampling error. Since this procedure carries some risk and the results can be negative, plus the fact that CBD and alternative diagnoses have no effective treatment, it is seldom performed.

NEUROPATHOLOGY

Macroscopic

Asymmetrical frontoparietal cortical atrophy and pallor of the substantia nigra are the typical macroscopic pathologic findings. Some patients with typical clinical and microscopic findings do not have appreciable cortical atrophy, however.

Microscopic

Before the development of immunostains, the histologic hallmarks of this disorder were ballooned achromatic (nonstaining) neurons in cortical and subcortical structures. Recently developed immunostains have permitted visualization of three other findings: phosphorylated neurofilament protein-positive achromatic neurons, tau-positive astrocytic plaques, and tau-positive oligodendroglial coiled bodies. Although similar findings can be present in Pick's disease and progressive supranuclear palsy, CBD has a characteristic distribution and frequency. We have seen several instances in which classic CBD pathologic features were missed because phosphorylated neurofilament and tau immunostains were not used at autopsy. Use of these immunostains is critical for the pathologic diagnosis of CBD.

DIFFERENTIAL DIAGNOSIS

Pick's Disease

Although the classic features of Pick's disease are personality change and dementia, several reports describe patients with findings very similar to those of CBD. One patient had asymmetrical dystonia, myoclonus, cortical sensory loss, and, later, apraxia. Asymmetrical cortical atrophy and Pick bodies were observed pathologically. In a recently published report, classic Pick bodies were also present in a patient who had PARA, myoclonus, tremor, alien limb phenomenon, and cortical sensory loss. Thus, the clinical syndrome of CBD may occasionally occur with other pathologies (see below for further discussion).

Parkinson's Disease

The bradykinesia and asymmetrical rigidity of Parkinson's disease appear quite similar to early manifestations of CBD. In idiopathic Parkinson's disease, the tremor, rigidity, and bradykinesia can be reversed with levodopa therapy, which may require 750 mg of levodopa daily (i.e., 2 1/2 tablets of standard carbidopa-levodopa 25/100 taken three times a day on an empty stomach or 2 tablets of controlled-release carbidopa-levodopa 50/200 taken two times a day on an empty stomach). We therefore typically give carbidopa-levodopa to patients with the PARA syndrome and titrate upward as tolerated to at least 1,000 mg of levodopa daily in divided doses. The following features help differentiate CBD from Parkinson's disease: (1) apraxia, alien limb phenomenon, cortical sensory loss, and mirror movements, (2) "jerky" postural or action tremor rather than pill-rolling rest tremor, and (3) lack of response to 1,000 mg of levodopa daily.

Alzheimer's Disease

Alzheimer's disease typically begins as early and severe anterograde memory impairment, which is followed by progressive impairment in most if not all other cognitive domains until severe and generalized dementia results. In the vast majority of patients with Alzheimer's disease, motor signs, including extrapyramidal deficits, are relatively mild early in the course. Thus, in typical cases there is little diagnostic confusion. There are several clinicopathologic case reports in the literature, however, of atypical presentations of Alzheimer's disease erroneously diagnosed in life as CBD. In these patients, the PARA syndrome developed early in the course of Alzheimer's disease. The correct diagnosis may be difficult to make in such patients; fortunately for the clinician, these cases are rare.

Progressive Supranuclear Palsy

Rigidity, supranuclear gaze palsy, and postural instability are common to both CBD and progressive supranuclear palsy (PSP). However, the rigidity in PSP is typically mostly axial and symmetrical rather than primarily appendicular and asymmetrical. Supranuclear gaze palsy is common in PSP and less common in CBD. In patients with CBD who have initial upper limb presentations, falling is uncommon early in the course, whereas early falls are common in PSP. Several findings are distinctly uncommon in PSP: limb apraxia, myoclonus, alien limb phenomenon, cortical sensory loss, and mirror movements. Conversely, the wide-eyed, unblinking, severely masked face of PSP is not typically seen in CBD.

Other Considerations

Multiple system atrophy has clinical features quite different from those of typical CBD: various degrees of ataxia, parkinsonism, corticospinal tract signs, and dysautonomia. Cortical abnormalities (apraxia, agnosia, aphasia) are typically absent in multiple system atrophy, and dementia is usually nil or a late manifestation. Rare instances of Creutzfeldt-Jakob disease have exhibited rigidity and myoclonus but with very rapid progression (over months), early dementia, and characteristic periodic sharp wave complexes on electroencephalograms; these features are almost diagnostic of Creutzfeldt-Jakob disease. Extrapyramidal dysfunction and early dementia are typical of diffuse Lewy body disease; however, the prominent asymmetry of motor signs and apraxia typical of CBD are not expected in this disorder. Early hallucinations and psychosis are frequent in diffuse Lewy body disease but distinctly uncommon in CBD.

CLINICOPATHOLOGIC HETEROGENEITY

The clinical and pathologic findings described in the preceding text apply to the "typical" occurrences of CBD. Although the literature gives the impression that CBD is a distinct clinicopathologic entity, the evolving experience at several academic centers suggests that such a conclusion is not quite so simple. There are reports in which typical CBD clinical findings were present in patients with the pathologic hallmarks of Alzheimer's disease, Pick's disease, PSP, and nonspecific neurodegenerative changes on postmortem examination. There are also reports of pathologically confirmed CBD in which patients presented not with the PARA syndrome but rather with progressive aphasia or frontal lobe dementia. Although the practical consequences are not substantial when one is making distinctions among the various histopathologic patterns that can cause the PARA syndrome, the effect on research is substantial, because no biologic markers identified to date differentiate CBD from the mimicking disorders.

MANAGEMENT

Since there is no cure for CBD, management must be tailored to address the disabling symptoms of each patient. Unfortunately, pharmacotherapy provides minimal improvement. Levodopa, dopamine agonists, and baclofen have little effect on rigidity, bradykinesia, or tremor. However, levodopa should be titrated upward, as described above, to provide an adequate trial. Anticholinergic agents rarely relieve dystonia, and their use is limited by side effects.

Fortunately, central pain is rare in this disorder, but when present, it can be very difficult to treat. We have seen some patients with CBD and classic central pain in whom the usual analgesic strategies failed but who benefited from gabapentin. Botulinum toxin injections are occasionally effective in alleviating pain secondary to focal dystonia, such as that in an upper limb.

Tremor may respond initially to propranolol or primidone, but the effects wane with progression of the disorder. Clonazepam has reduced myoclonus in some cases.

Since the achromatic neurons of CBD resemble those in pellagra, we and others have tried niacin, but there has been no appreciable improvement. Vitamin E and other antioxidant agents have been tried in hopes of delaying progression of the disorder, but there is no supportive clinical evidence for their use.

Because of the poor response to pharmacotherapy, the mainstays of management are physical, occupational, and speech therapies. Passive range of motion exercises minimize development of contractures. Ambulation typically becomes impaired at some point, hence the need for gait assistance devices. Apraxia is often the most debilitating feature of the disorder. A home assessment by an occupational therapist can aid in determining which changes could be made to facilitate functional independence (e.g., replacing rotating doorknobs with handles, attaching specially formed pads around the handles of eating utensils and toothbrushes, and purchasing clothing with fabric closures [Velcro] instead of buttons or laces). Speech therapy and communication devices can optimize communication if the patient has dysarthria, apraxia of speech, or aphasia. Therapists also counsel patients and families on swallowing maneuvers and food additives to minimize aspiration when dysphagia occurs. Feeding gastrostomy may be necessary in some patients.

Other treatable comorbid illnesses must also be considered, most notably infections (e.g., pneumonia and urinary tract infections), psychiatric disorders (e.g., depression), and sleep disorders (e.g., obstructive sleep apnea syndrome, restless legs syndrome, and periodic limb movement disorder). Patients and caregivers eventually require assistance in maintaining optimal care, and this can be obtained either through home health care or in a skilled nursing facility.

SELECTED READING

Boeve BF, Maraganore DM, Parisi JE, Ahlskog JE, Graff-Radford N, Caselli RJ, Dickson DW, Kokmen E, Peterson RC. Pathologic heterogeneity in clinically diagnosed corticobasal degeneration. *Neurology* 1999; 53; 795-800.

Feany MB, Dickson DW. Widespread cytoskeletal pathology characterizes corticobasal degeneration. *Am J Pathol* 1995; 146: 1388-1396.

Kompoliti K, Goetz CG, Boeve BF, Maraganore DM, Ahlskog JE, Marsden CD, Bhatia KP, Greene PE, Przedborski S, Seal EC, Burns RS, Hauser RA, Gauger LL, Factor SA, Molho ES, Riley DE. Clinical presentation and pharmacological therapy in corticobasal degeneration. *Arch Neurol* 1998; 55: 957-961.

Pillon B, Blin J, Vidailhet M, Deweer B, Sirigu A, Dubois B, Agid Y. The neuropsychological pattern of corticobasal degeneration: comparison with progressive supranuclear palsy and Alzheimer's disease. *Neurology* 1995; 45: 1477-1483.

Rebeiz JJ, Kolodny EH, Richardson EP Jr. Corticodentatonigral degeneration with neuronal achromasia: a progressive disorder of late adult life. *Trans Am Neurol Assoc* 1967; 92: 23-26.

Riley DE, Lang AE, Lewis A, Resch L, Ashby P, Hornykiewicz O, Black S. Cortical-basal ganglionic degeneration. *Neurology* 1990; 40: 1203-1212.

Rinne JO, Lee MS, Thompson PD, Marsden CD. Corticobasal degeneration. A clinical study of 36 cases. *Brain* 1994; 117: 1183-1196.

20 Parkinsonism in Primary Degenerative Dementia

Richard J. Caselli, MD

CONTENTS

CORTICAL DEMENTIA
SUBCORTICAL DEMENTIA
MIXED CORTICAL-SUBCORTICAL DEMENTIA
SELECTED READING

"Dementia" is a term often used in reference to Alzheimer's disease (AD), but AD is only one of the many causes of dementia. Dementia is more than memory loss alone. If a tally were taken of the different cognitive skills potentially affected by dementia, one would find that language, spatial skills, mental speed, judgment, and a host of cognitive domains can be impaired in addition to memory. In fact, dementia is defined as a disabling level of impairment of multiple cognitive domains. Because many diseases can cause dementia, not all affect each cognitive domain to the same degree. For example, AD causes profound memory loss, but corticobasal ganglionic degeneration affects coordinated movement (praxis) more severely and progressive aphasia affects language skills more severely.

In an attempt to simplify the complexity of the multiple diseases and the multiple cognitive domains, a classification scheme is presented that divides dementia syndromes into three broad categories: cortical, subcortical, and mixed (Table 20-1).

"Cortical dementia" refers to diseases that resemble AD and not Parkinson's disease (PD). The cognitive domains most affected by cortical dementia are usually memory, language, praxis, and perception. Subcortical dementia shares certain physical signs of PD and affects mental speed in particular (sometimes termed "bradyphrenia"). Mixed cortical and subcortical features can also occur in relatively equal balance. Ideally, the anatomical distribution of degenerative disease should reflect these clinical features, so that cortical dementia would result from degeneration of cerebral cortex, subcortical dementia from degeneration of basal ganglia and other subcortical structures, and mixed dementia

From: *Parkinson's Disease and Movement Disorders:*
Diagnosis and Treatment Guidelines for the Practicing Physician
Edited by: C. H. Adler and J. E. Ahlskog © Mayo Foundation
for Medical Education and Research, Rochester, MN

Table 20-1
Primary Categories of Dementia Syndromes

Category	Example
Cortical	Alzheimer's disease
Subcortical	Parkinson's disease with dementia
	Progressive supranuclear palsy
Mixed	Corticobasal degeneration
	Diffuse Lewy body disease

from both. In truth, all dementing illnesses are mixed pathologically despite the predominance of clinical features favoring a cortical, subcortical, or mixed clinical diagnosis.

Now that we have described the ideal black-and-white dividing lines, they can be blurred into reality. Several general principles must be considered before any clinically meaningful distinctions can be made among different diseases in the real world. First, although parkinsonism can occur in at least some of the member diseases belonging to each of the three dementia categories, it is a defining feature of subcortical dementia. Second, neurodegenerative illnesses progress relentlessly over time, and in their terminal stages, it can be difficult or impossible to clinically differentiate one from another. In all patients with dementia who live until the terminal stages, therefore, some parkinsonian features can be expected. However, during the ambulatory phase, and especially during mild to moderate stages, clinical differences are readily apparent. Third, neuroleptic drug therapy, which is used in many patients with dementia at some point in the course of the disease, frequently causes parkinsonism. The remaining discussion applies to unmedicated patients in mild to moderate stages of dementia.

CORTICAL DEMENTIA

Generally, patients with cortical dementia have little parkinsonism and normal mental (psychomotor) speed. Instead, they have prominent memory loss (amnesia), difficulty coordinating their limbs in functionally meaningful ways (apraxia), impaired language skills like naming and comprehension (aphasia), impaired perceptual skills that, for example, may result in failure to recognize a familiar face (agnosia), or a variety of other complex cognitive problems referable to the frontal lobes. The exact combination of symptoms and signs is determined by the particular cortical topographic distribution of the neurodegenerative process.

Alzheimer's and Diffuse Lewy Body Diseases

AD is the most common cause of dementia, accounting for more than half of all cases. It increases in frequency with advancing age. The prevalence of severe dementia after age 60 is estimated at 5% and after age 85 between 20% and 50%. The lifetime risk of AD developing is estimated to be between 12% and 17%. After age, apolipoprotein E status is the most important risk factor. In both familial late onset and sporadic cases, the apolipoprotein E ε4 allele increases risk and the ε2 allele decreases risk.

AD is characterized pathologically by generalized cerebral cortical atrophy with widespread cortical neuritic (or senile) plaques and neurofibrillary tangles. Severe and early

Table 20-2
Dementing Disorders Associated With Lewy Bodies

Disorder	Substantia nigra Lewy bodies	Frequent cortical Lewy bodies	Prominent Alzheimer's disease pathology
Parkinson's disease	+	0	0
Diffuse Lewy body disease	+	+	0
Lewy body variant of Alzheimer's disease	+	+	+

involvement of the cholinergic basal forebrain is found, but similar involvement occurs in patients who have PD with dementia. Substantia nigra involvement, however, is far less in AD than in PD.

The pathologic hallmark of PD, the Lewy body, is also found in association with dementing illnesses. In a pathologic variant of AD, termed "Lewy body variant" by Katzman, there is involvement of the cortex as well as substantia nigra with Lewy bodies. Between 20% and 30% of brains of patients with a degenerative dementia have concomitant AD and PD changes. A much smaller percentage have diffuse Lewy bodies without AD changes. Pathologic change related to Lewy bodies is the second most common histopathologic finding behind AD. Thus, the dementing disorders associated with Lewy bodies can be categorized into three groups (Table 20-2): PD without cortical Lewy bodies[1] or AD changes (known simply as PD), PD with substantial numbers of cortical Lewy bodies but not concomitant AD changes (referred to as "diffuse Lewy body disease," or DLBD), and PD with cortical Lewy bodies and AD changes (also referred to as the "Lewy body variant of Alzheimer's disease," or LBV). The LBV has been observed in several genetically linked forms of AD (e.g., in families with presenilin 1 mutations, in rare Down's syndrome cases, and occasionally in apolipoprotein E ε4 cases), so that it is probably best considered within the domain of AD rather than DLBD.

Extrapyramidal findings, particularly mild rigidity and bradykinesia, occur with increasing frequency (in nearly 30%) as AD progresses, but how often the occurrence reflects disease related to Lewy bodies is uncertain. In an autopsy series of AD, 15% had some type of pathologic change related to Lewy bodies. Cognitively, the clinical features of AD and LBV are similar, although there is probably more mental slowing and less naming impairment in LBV than in AD.

Although parkinsonism that accompanies AD can be levodopa-responsive, levodopa may be more likely to result in psychosis in these patients, similar to that in DLBD (though not all studies support this clinical association). A greater clinical concern may be the use of neuroleptic medications in patients with AD who have parkinsonism, since the frequency and severity of parkinsonian side effects are much higher. In such patients, more selective agents, such as quetiapine or olanzapine, should be considered when neuroleptic therapy is believed to be necessary. Acetylcholinesterase inhibitors, such as tacrine and donepezil, which are used to treat cognitive deficits in these patients, might

---

[1]Rare cortical Lewy bodies are found in most or all patients with PD, but the frequency is minimal compared with that in diffuse Lewy body disease or the Lewy body variant of AD.

theoretically be thought to worsen the extrapyramidal symptoms, but their motor effects are generally minimal.

Asymmetrical Cortical Degeneration Syndromes

A large minority of patients who present with a cortical dementia pattern have an atypical form of dementia that can be divided into four broadly defined categories collectively referred to as "asymmetrical cortical degeneration syndromes": frontotemporal syndromes, progressive aphasia, perceptual-motor syndromes, and bitemporal syndromes. The frequency of extrapyramidal symptoms differs dramatically between these subgroups. Several histopathologic patterns have been found in association with these various syndromes, including the pathologies of Pick's disease, AD, and corticobasal ganglionic degeneration as well as nonspecific degeneration.

FRONTOTEMPORAL SYNDROMES

In patients with frontal lobe, frontotemporal, or bitemporal syndromes, the underlying pathology is that of Pick's disease, with involvement of the frontal and anterior temporal lobes in varying proportion and laterality. The most common clinical features are neuropsychiatric, including abulia, apathy, inertia, disinhibition, memory loss, and naming difficulties. Because the frontal lobes and amygdala (part of the anterior temporal lobe) are neuroanatomically part of the dorsal striatal circuit (i.e., the part of the basal ganglia affected by PD), patients often lack spontaneity (abulia) and can have a slow and somewhat shuffling gait, though rarely are they frankly parkinsonian. Neuroleptic medications do not appear to have the same severity of extrapyramidal side effects in these patients as in LBV. Recent genetic studies of certain kindreds with frontotemporal dementia and overlap syndromes of frontotemporal dementia, parkinsonism, and amyotrophic lateral sclerosis have shown mutations in the gene coding for cytoskeletal protein, tau.

PROGRESSIVE APHASIA

As the name implies, patients experience progressive language dysfunction, largely with sparing of other aspects of cognition. Generally, motor signs apart from speech are a rare occurrence in patients with primary progressive aphasia. If they occur, they do so late in the course or are mild.

PERCEPTUAL-MOTOR SYNDROMES

Asymmetrical apraxia is the clinical hallmark of the perceptual-motor syndromes. Apraxia implies loss of the ability to spatially and temporally sequence a series of elementary motor movements to make a more complex programmed movement. The apraxia is usually associated with rigidity, and the features of the asymmetrical rigidity and apraxia together form the primary clinical presentation of the corticobasal degeneration syndrome (see Chapter 19). The apraxia represents cortical dysfunction, whereas rigidity reflects subcortical (basal ganglia) pathology. Hence, the fully developed corticobasal degeneration syndrome is a mixed cortical-subcortical disorder (see below).

BITEMPORAL SYNDROMES

Amnesia and other neuropsychiatric features predominate in these patients, who have little or no parkinsonism.

Amyotrophic Lateral Sclerosis and Dementia Complex

Dementia rarely occurs in patients with amyotrophic lateral sclerosis, although known associations include the parkinsonism, amyotrophic lateral sclerosis, and dementia complex of Guam as well as familial and sporadic cases from around the world, including the United States. Clinical descriptions to date have varied, but the most prevalent cognitive syndromes are frontal lobe dementia and aphasic dementia. Bulbar symptoms have been consistently prominent. True parkinsonian signs are typically absent in sporadic cases of amyotrophic lateral sclerosis and dementia complex occurring outside of Guam, even though the frequency of nigral involvement at autopsy is roughly 50%.

SUBCORTICAL DEMENTIA

The core of the subcortical dementia syndrome, apart from memory loss, is parkinsonism and psychomotor slowing. By definition, therefore, parkinsonism is universally present in this subset of dementing diseases. Depression is very common, and to a lesser degree (though perhaps more characteristic of this group), hallucinations occur.

Progressive Supranuclear Palsy

Progressive supranuclear palsy was the original model for subcortical dementia. The concept was intended to differentiate subcortical dementia from the cortical dementia of AD. Subcortical memory impairment related to dementia is benefited more by cuing than is the memory disturbance of cortical dementia. Patients have equal or greater difficulty learning new material but have relatively less difficulty retrieving it, possibly a reflection of disturbed frontal or frontostriatal function involved in attending to and organizing the memory task. Subcortical dementia is not accompanied by aphasia, apraxia, or agnosia (at least not to the same degree as in cortical dementia). However, it is accompanied by psychomotor slowing, an early feature of Parkinson's dementia. Patients perform disproportionately poorly on timed tests whether motor responses (such as the digit symbol substitution subtest of the Wechsler Adult Intelligence Scale–Revised) or pure cognition (such as the controlled oral word association test) is used. Progressive supranuclear palsy is described in more detail in Chapter 16.

Parkinson's Dementia

Estimates of dementia prevalence in PD vary considerably, with most within the 20% to 30% range. The underlying pathologic features are variable. About half of demented parkinsonian patients have coincident AD pathologic changes. Degeneration of the cholinergic basal forebrain (implicated in memory) is more pronounced in patients who have PD with dementia than in those without but is insufficient to explain the cognitive impairment. Nondegenerative "quasiparkinsonian" syndromes with cognitive impairment have many causes, including multi-infarct dementia (perhaps the most common cause for subcortical dementia), normal pressure hydrocephalus, and postencephalitic parkinsonism.

Huntington's Disease

One widely cited series found that 90% of patients with Huntington's disease were demented by a mean age of 48.3 years and that dementia preceded the onset of chorea in 24%. Cognitive decline is correlated pathologically and radiologically with degree of caudate atrophy and hypometabolism, but it is not certain whether caudate disease itself is a critical determinant for dementia. Three particular types of difficulty that characterize the dementia of Huntington's disease are neuropsychiatric deficits related to the frontal lobe or frontostriatal region, declarative memory impairment, and motor learning difficulties. Patients with frontostriatal neuropsychiatric deficits may initially have either primarily psychiatric or primarily cognitive disturbances, although both may occur together. Psychomotor slowing (bradyphrenia) is a characteristic early feature, and memory impairment, although present early on, does not deteriorate as rapidly or correlate as well with caudate atrophy as does psychomotor speed. Patients with Huntington's disease have greater perseveration and less initiative than patients with AD and greater problems with concentration and sustained attention. The overall pattern of deficits related to the frontal lobe nonetheless appears qualitatively distinct from purely frontal cortical damage and has instead been attributed to frontostriatal dysfunction.

Huntington's disease causes memory loss, but it appears to be less severe than that in AD at matched stages of dementia severity (although this has been more consistent at milder stages of dementia). Most studies have found that although patients with Huntington's disease have difficulty learning a task (such as a list of words), they are more greatly benefited by cued recall or recognition in contrast to free or unassisted recall. The many similarities in memory disturbance between Huntington's disease and PD suggest similarities between the "subcortical dementia" illnesses, although the overlap is incomplete.

In contrast to the mild impairment of verbal learning, motor learning is severely impaired in Huntington's disease. There is a "double dissociation" between patients with AD and those with Huntington's disease in performances on tests of these two memory systems: patients with AD do poorly on verbal and well on motor learning, and patients with Huntington's disease do well on verbal and poor on motor learning. Demented patients with PD perform poorly on both.

MIXED CORTICAL-SUBCORTICAL DEMENTIA

In this category, contributions from cortical and subcortical structures are equally prominent, with components of both dementia and parkinsonism.

Corticobasal Degeneration

The vast majority of patients with corticobasal degeneration (see Chapter 19) have had a progressive perceptual-motor syndrome, with both apraxia and rigidity; hence the term "progressive asymmetrical rigidity and apraxia syndrome" has been proposed. Because of these two major clinical characteristics, both slowly progressive, this disorder is distinctive and easily recognized. A pronounced asymmetry of these features is the rule, and they are associated with other signs, such as hemisensory impairment (usually in the form of astereognosis), alien limb phenomenon, and myoclonic jerks. The apraxia includes limb apraxia (inability to use the limb in a meaningful way, such as to use a comb), gestural apraxia (inability to pantomime or imitate symbolic movements), dressing apraxia, constructional apraxia, and apraxic agraphia (inability to write). If the lateralized limb defects reflect dominant hemispheric dysfunction, mild aphasia is commonly present. If the predominantly involved limb reflects nondominant hemisphere dysfunction, however, it is distinctly uncommon to see hemineglect, though it can occur. Impaired arithmetic skills (acalculia) and, less frequently, other parietal signs can occasionally be seen. Most patients are slow on tests of psychomotor speed, such as the controlled oral word association test, in which subjects must generate within a minute as many words as possible beginning with a certain letter. Eye movements may become affected, perhaps related to involvement of the frontal eye fields in the cortex. Although patients with severe apraxia, psychomotor slowing, and mild aphasia may appear demented by neuropsychologic testing, they are readily found to be rational, insightful, and quite upset by their condition. In late stages, patients can barely move or speak despite preserved consciousness. Involvement of contiguous visual association cortices can occur early or late in the course of corticobasal degeneration, commonly leading to visuoperceptual disorders as well.

Diffuse Lewy Body Disease

The relationship between LBV and DLBD provides a clinicopathologic bridge between two otherwise seemingly disparate neurodegenerative diseases, AD and PD. As with many neurodegenerative conditions, the dividing lines between neighboring disorders are relative rather than absolute. For our purposes, DLBD is defined by its greater preponderance of cortical Lewy bodies and greater nigral degeneration than LBV, although this distinction may eventually be proved to be artificial, reflecting severity rather than nosology. Autopsy series suggest that the mean age at onset of DLBD is 57 years, that men are affected more often than women (male-to-female ratio of 1.7:1), and that death ensues after roughly 10 to 15 years. The prevalence of DLBD is, very roughly, 30% that of PD (although clinically diagnosed cases are less common than that).

Pathologically, the substantia nigra is pale. Brain stem Lewy bodies are present in the substantia nigra, locus ceruleus, and raphe nuclei. The locations of cortical Lewy bodies include the cingulate gyrus, insular cortex, and parahippocampal gyrus. The cholinergic neurons in the nucleus basalis are severely reduced in number.

Parkinsonian and cognitive symptoms, including both dementia and psychosis, occur early and are the primary presenting components. In general, the clinical picture of a mild parkinsonian syndrome with a more severe dementia and prominent visual hallucinations and paranoid delusions should prompt consideration of DLBD. Although the parkinsonian syndrome can be levodopa-responsive, levodopa may exacerbate the psychosis, and in early to middle stages, the neuropsychiatric disturbance is generally the more disabling feature. As the disease progresses, however, both components worsen, and the patient may eventually become severely parkinsonian and wheelchair-bound. Rest tremor is less frequent (about 29% of patients) than in PD, but bradykinesia is a predictable feature.

Pharmacotherapy for symptoms is exceptionally difficult because treatment of the parkinsonian syndrome may exacerbate the neuropsychiatric disorder and vice versa. Therefore, one must first decide whether treatment is necessary at all, and if so, what symptoms warrant treatment. Generally, the neuropsychiatric syndrome is the most disabling aspect in mild to moderate stages, and selective neuroleptic therapy with the newer atypical neuroleptic agents with less potential for exacerbating parkinsonism, such as quetiapine or olanzapine, should be considered the first choice. Treatment of the parkinsonian syndrome is similar to that of PD, generally beginning with carbidopa-levodopa at a very low dose and continuing with gradual titration of the dose to symptoms and side effects. Centrally acting acetylcholinesterase inhibitors, such as donepezil, which have become nearly standard for treatment of AD, can be used with similar success in DLBD.

SELECTED READING

Cortical Dementia

Caselli RJ. Focal and asymmetric cortical degeneration syndromes. *The Neurologist* 1995; 1: 1-19.
Terry R, Katzman R. Alzheimer disease and cognitive loss, in *Principles of Geriatric Neurology* (Katzman R, Rowe JW eds), F. A. Davis Company, Philadelphia, 1992, pp. 207-265.

Subcortical Dementia

Albert ML, Feldman RG, Willis AL. The 'subcortical dementia' of progressive supranuclear palsy. *J Neurol Neurosurg Psychiatry* 1974; 37: 121-130.
Heindel WC, Salmon DP, Shults CW, Walicke PA, Butters N. Neuropsychological evidence for multiple implicit memory systems: a comparison of Alzheimer's, Huntington's, and Parkinson's disease patients. *J Neurosci* 1989; 9: 582-587.
Pillon B, Dubois B, Ploska A, Agid Y. Severity and specificity of cognitive impairment in Alzheimer's, Huntington's, and Parkinson's diseases and progressive supranuclear palsy. *Neurology* 1991; 41: 634-643.
Stern Y, Richards M, Sano M, Mayeux R. Comparison of cognitive changes in patients with Alzheimer's and Parkinson's disease. *Arch Neurol* 1993; 50: 1040-1045.
Yoshimura M. Pathological basis for dementia in elderly patients with idiopathic Parkinson's disease. *Eur Neurol* 1988; 28 Suppl 1: 29-35.

Mixed

Beck BJ. Neuropsychiatric manifestations of diffuse Lewy body disease. *J Geriatr Psychiatry Neurol* 1995; 8: 189-196.
Byrne EJ, Lennox G, Lowe J, Godwin-Austen RB. Diffuse Lewy body disease: clinical features in 15 cases. *J Neurol Neurosurg Psychiatry* 1989; 52: 709-717.
Gibb WR, Luthert PJ, Marsden CD. Corticobasal degeneration. *Brain* 1989; 112: 1171-1192.
Kuzuhara S, Yoshimura M. Clinical and neuropathological aspects of diffuse Lewy body disease in the elderly. *Adv Neurol* 1993; 60: 464-469.
Pillon B, Blin J, Vidailhet M, Deweer B, Sirigu A, Dubois B, Agid Y. The neuropsychological pattern of corticobasal degeneration: comparison with progressive supranuclear palsy and Alzheimer's disease. *Neurology* 1995; 45: 1477-1483.

D

Movement Disorders Characterized by Excessive Movement (Hyperkinetic)

Simply stated, hyperkinetic movement disorders are disorders characterized by an excess of movement. Categorizing the disorders is primarily based on observation by the clinician, some key historical facts, and, if available, results of surface electromyographic studies. Obtaining a good medication history is critical, since almost all hyperkinetic movement disorders can be drug-induced. The implication in tardive disorders is that the movement disorder persists despite withdrawal of the offending medication. Similarly, a good family history is required because of the hereditary occurrences of tremor, chorea, tics, dystonia, and restless legs syndrome.

We begin this section with chapters on tremor disorders because they are the most common movement disorders in most clinical practices. *Tremor* is a rhythmic movement disorder that may occur at rest or with action and may involve any body part. *Dystonia* is characterized by a sustained posturing during the movement and is usually described by the location involved (e.g., eyes, blepharospasm; neck, torticollis). *Chorea* is manifested as a flowing movement of one or more body segments that is not sustained (in contrast to dystonia) or stereotyped (in contrast to most other hyperkinetic disorders). *Ballism* describes violent, flinging motions of a limb, which might be thought of as a severe form of chorea. *Myoclonus* is a very rapid, lightning-like movement that may appear, at one extreme, as a simple twitch of a finger or, at the other, as a sudden massive jerk of the entire body. Occasionally, simple forms of myoclonus may be difficult to differentiate from tics. *Tics* are also brief movements, but they can be simple or complex, motor or sensory, and are associated with an urge to move and the ability to partially suppress the movements for brief periods.

Quite commonly, these hyperkinetic movements overlap. For example, it is common for dystonia to occur with chorea in some patients and with tremor in others.

Some disorders are simply called *spasms,* and these are discussed in detail. *Restless legs syndrome is* also described. Like tics, this disorder has a strong sensory component along with involuntary leg movements, usually associated with prolonged sitting or being supine, and exacerbation in the evening.

The differential diagnosis is extensive for most of the hyperkinetic movement disorders. All the hyperkinetic disorders can result from a metabolic abnormality; all are worsened by stress and anxiety, and the possibility of a drug-induced cause should always be entertained. Vascular events or other structural lesions involving the basal ganglia, subthalamus, and related structures can cause a hyperkinetic movement disorder; hence, in selected cases, neuroimaging should be considered.

The treatment of hyperkinetic movement disorders may not always be satisfactory. For some of these disorders, oral medications given by a "trial-and-error" approach may be necessary. Botulinum toxin injections are the treatment of choice for many of these disorders. Surgical intervention may also be beneficial. Often, one must balance the likelihood of improvement against the consequences of side effects for all these treatment options.

21

Tremor Disorders: Overview

Joseph Y. Matsumoto, MD

CONTENTS

Tremor is the most common movement disorder. As a symptom, tremor may occur either as an isolated disorder, as in essential tremor, or as part of a movement disorder syndrome, as in Parkinson's disease or cerebellar disorders. Because it is so visible, tremor is often the patient's chief concern, even when other sources of disability are present. For these reasons, the practitioner needs an organized approach to the patient complaining of tremor.

Tremor may be defined as an involuntary, rhythmic movement. Thus, the fundamental appearance of any tremor disorder is oscillation—a sinusoidal movement of a body part that tends to be relatively constant in frequency and direction. Tremors are most often intermittent, waxing and waning through the patient's waking hours. Although the amplitude of tremor determines the degree of disability, it is not important in establishing the cause or diagnosis of a tremor disorder. Virtually all tremors worsen with emotion, anger, or anxiety and disappear with sleep. Such facts, when reported, should never be taken as evidence of psychiatric or feigned illness.

Some mild tremor disorders may be due to transmitted forces or may reflect the consequences of normal physiology. For instance, cardioballistic tremor is due to the force of a strong heartbeat resonating through the arm, and physiologic tremor results from the normal gaps in motor unit firing. In contrast, pathologic tremors are always the result of abnormal rhythmic grouped bursting of motor units. These rhythmic bursts of muscle activity are driven by similar bursts of activity within the central nervous system. However, exactly how and where this abnormal activity develops within the brain and spinal cord are not yet fully understood for any pathologic tremor disorder. Some believe

From: *Parkinson's Disease and Movement Disorders:*
Diagnosis and Treatment Guidelines for the Practicing Physician
Edited by: C. H. Adler and J. E. Ahlskog © Mayo Foundation
for Medical Education and Research, Rochester, MN

that groups of localized neurons act as rhythmic tremor generators. Others think that tremor reflects properties of systems of neurons that oscillate at resonant frequencies, much as a car might vibrate on a bumpy road.

The diagnosis and treatment of tremor disorders are based almost entirely on clinical history and examination. Bedside observations by the clinician allow classification into one of four major tremor types (rest, action-postural, intention, task-specific). Further details from history, examination, and laboratory testing allow a more specific etiologic diagnosis. With this organized approach, correct diagnosis and proper therapy may be accomplished for many patients.

CLASSIFICATION OF TREMOR

The first task is to determine that the movement observed is truly tremor. The frequency of the movement can be estimated by counting "one thousand one" and estimating the cycles of movement during this interval. With practice, tremor frequency can be ascertained with excellent accuracy, and repeating the measure gives an idea of the constancy of the tremor. The movement should be relatively constant in frequency and direction to be classified as tremor.

Once tremor is diagnosed, classification into one of the following four broad categories guides diagnosis and therapy.

Rest Tremor

Before diagnosing a rest tremor, the examiner must ensure that all muscular tension has been removed from the body part of interest. Placing the limb in a fully supported position and coaching the patient to relax may be necessary to ensure a completely resting state. The arm hanging at the side or the leg dangling over the examination table may be considered to be in the resting position. The head is resting when it is comfortably supported with the patient supine. By definition, one cannot observe resting tremor of the vocal structures at the bedside. However, resting tremor may be seen in the muscles of the chin or tongue.

The classic appearance of a resting tremor is pronation-supination of the forearm or a "pill-rolling" movement of the thumb and fingers. Typically, patients with resting tremor are more bothered by embarrassment than disability. Many patients report that they sit on their hands or place them in their pockets to conceal a resting tremor. In resting tremor, the initiation or even, at times, the thought of action of the involved limb causes the tremor to immediately diminish. Often, patients with severe resting tremor of the arm are still able to perform fine motor skills.

By far, the most common disease causing resting tremor is idiopathic Parkinson's disease. However, other conditions with resting tremor include other causes of parkinsonism (such as progressive supranuclear palsy), drug-induced parkinsonism (neuroleptic agents, metoclopramide), and hepatocerebral degenerative disorders such as Wilson's disease (Table 21-1).

Action-Postural Tremor

Tremors of the action-postural type become apparent with the activation of muscle contraction within the affected body part. These tremors are best observed when the body part is held in a sustained posture. Holding the arms elevated, especially with the fingers in front of the nose, augments a postural arm tremor. The legs must be held outstretched

Table 21-1
Differential Diagnosis of Rest Tremor

Parkinson's disease
Parkinsonism (neurodegenerative, infarct)
Drug-induced parkinsonism
    Neuroleptic agents
    Metoclopramide (Reglan)
    Prochlorperazine (Compazine)
    Catecholamine depletors
Hepatocerebral degeneration
    Hereditary (Wilson's disease)
    Acquired
Severe essential tremor

to reveal a postural leg tremor. Anytime the patient is holding the head erect, a postural tremor may be seen as either a back-and-forth ("yes-yes") or a side-to-side ("no-no") oscillation. Asking the patient to hold a prolonged vowel sound reveals a voice tremor as vibrato intonation. Maneuvers that induce muscle relaxation cause an action-postural tremor to disappear.

Action-postural tremors are also evident with movement of a limb. When present, this tremor occurs with the same magnitude throughout the range of movement. Handwriting and drawing are actions that allow observation and recording of action-postural arm tremors.

Typically, tremors of the action-postural type are extremely sensitive to the slightest anxiety. Creating mildly stressful circumstances in the clinic by having the patient recite the months of the year backwards, hold a piece of paper or cup of water outstretched, or draw between fine parallel lines may bring a mild action-postural tremor into prominence.

Action-postural tremors, when mild, are mainly a source of embarrassment. As they become more severe, however, significant disability develops: handwriting becomes illegible, eating with utensils or holding a cup becomes challenging, and shaving or putting on makeup becomes potentially hazardous.

Most patients with action-postural tremor have essential tremor. However, exaggerated physiologic tremor, toxic-metabolic tremors, and neuropathic tremor also are in this category (Table 21-2).

Intention Tremor

This unique type of tremor occurs almost exclusively with disease of the cerebellum or its connections. The fundamental feature is a marked exacerbation with action, especially the final or terminal portion of a visually guided movement. This phenomenon, termed "intention tremor" or "terminal tremor," differentiates this tremor from action-postural tremor. In early cerebellar disease, only terminal tremor may be present, but as the disease progresses, a severe cerebellar tremor syndrome affecting posture and all phases of movement develops. This syndrome has been termed "rubral tremor" and "cerebellar outflow tremor," but because the exact anatomical structures causing this tremor are unknown, these terms are best avoided. In this severe form of cerebellar tremor, proximal muscle groups are often involved, causing wild oscillations of large amplitude. Afflicted patients may be at risk for self-injury because of the violence of these

Table 21-2
Differential Diagnosis of Action-Postural Tremor

Physiologic tremor
Exaggerated physiologic tremor
Essential tremor
Drug-induced tremor
Neuropathic tremor
Parkinson's disease and parkinsonism
Metabolic disorders

movements. In general, the arms are afflicted by cerebellar tremor. However, the head may also oscillate in a movement termed "titubation." Because patients with the severe form of cerebellar tremor often have diseases causing paraparesis or gait dysfunction, leg tremor of this type is often not a problem.

At the bedside, cerebellar tremors are elicited by target-directed actions, such as finger-to-nose pointing. Oscillations markedly increase as the finger approaches the target. At times, it is difficult to differentiate terminal tremor from repetitive overshoot or undershoot movements that result from ataxia ("serial dysmetria"). The examiner is aided by returning to basic principles and determining whether the movements have a regular oscillatory frequency and consistent vector, as in tremor. The movements of dysmetria are irregular in frequency, amplitude, and direction.

Any form of cerebellar disease can potentially cause this form of tremor, but disorders affecting the white matter tracts of the superior cerebellar peduncle, such as multiple sclerosis and head trauma, are most common.

Tremor Occurring in Specific Circumstances

Tremor occurring exclusively during handwriting should suggest a diagnosis of primary writing tremor. Another unusual tremor disorder, "orthostatic tremor," occurs only when patients stand.

OTHER DIAGNOSTIC FEATURES OF TREMOR

Frequency

Because tremors are oscillations, one would expect that they could be characterized by their frequencies. Unfortunately, specific frequencies do not exist for the various tremor disorders. In general, intention tremors oscillate the slowest, at 2 to 4 Hz, parkinsonian rest tremors oscillate between 4 and 6 Hz, and essential tremor typically has a rate of 6 to 9 Hz. However, as patients with essential tremor become older and more severely affected, the rate of tremor tends to slow and overlap with frequencies in the parkinsonian range. These overlaps limit the usefulness of frequency in tremor diagnosis.

Accompanying Neurologic Signs

Any patient with tremor should have a complete neurologic examination. The presence or absence of accompanying clinical signs aids in more specific tremor diagnosis. In essential tremor, no other neurologic abnormalities are found. Clearly, a resting

Table 21-3
Causes of Exaggerated Physiologic Tremor

Mental state	Anxiety, stress, fatigue
Metabolic	Thyrotoxicosis, hypoglycemia, fever, pheochromocytoma
Drug-induced	Beta-adrenergic agonists, tricyclic antidepressants, anticonvulsants, lithium, thyroxine, steroids, neuroleptics
Toxins	Heavy metals, alcohol withdrawal
Diet	Xanthines (caffeine)

tremor accompanied by masked facies, rigidity, bradykinesia, micrographia, lack of arm swing, or shuffling gait suggests parkinsonism. An intention tremor seldom exists without accompanying evidence of more widespread cerebellar dysfunction, such as ataxic eye movements, nystagmus, slurred speech, clumsy limb movements, or a wide-based, unsteady gait.

Laboratory Evaluation

If a tremor disorder can be clearly diagnosed on clinical grounds, few laboratory tests are necessary. Thyroid function tests should be checked in patients with action-postural tremor to rule out possible hyperthyroidism. A serum ceruloplasmin study and slit-lamp examination are indicated, to rule out Wilson's disease, in patients younger than 50 with any tremor of unclear cause. Computed tomography or magnetic resonance imaging (MRI) of the brain is generally not necessary if the tremor is bilateral but may help in the diagnosis of structural disease, especially if the tremor is unilateral. MRI is recommended in the evaluation of intention tremor to search for demyelinating disease, stroke, cerebellar degeneration, or structural lesions.

Surface electromyographic (EMG) studies are invaluable in the diagnosis of orthostatic tremor, yielding a pathognomonic pattern of a 14- to 18-Hz tremor of the legs on standing. These studies are also somewhat useful in characterizing tremor frequency and exacerbating postures or actions. Psychogenic tremors are often distractible, and surface EMG studies may best detect this phenomenon. In such cases, surface EMG evaluation serves as an extension of the clinical examination.

SPECIFIC TREMOR DISORDERS

Exaggerated Physiologic Tremor

Normal physiologic tremor may become noticeable and bothersome as a result of excessive anxiety, thyrotoxicosis or other metabolic changes, drugs, toxins, or dietary factors (Table 21-3). The tremor always affects the hands and has a rapid, fine appearance, with frequencies of 8 to 12 Hz. Other signs of anxiety, such as sweaty palms or agitation, often accompany the tremor. Exaggerated physiologic tremor is generally self-limited, and treatment is aimed at the primary condition. When the period of anxiety can be predicted, as, for example, in the performing arts, propranolol (Inderal), 10 to 40 mg, may be prescribed for prophylaxis.

Tremor as a Manifestation
of Toxic-Metabolic Disease

Action-postural tremor may arise in a variety of diffuse disturbances of brain dysfunction. Delirium tremens and other drug withdrawal states are the most common examples, yet virtually any other encephalopathy can have a similar presentation. For example, tremor may be a severe complication of pulmonary or hepatic failure. In all such states, clouding of consciousness and myoclonus or asterixis may be accompanying signs. A variety of drugs may induce action-postural tremor. Prominent among these are valproic acid, lithium, and theophylline derivatives.

Essential Tremor

Essential tremor is covered in detail in Chapter 22. Several features that assist in diagnosis need mention here. Essential tremor is the most common tremor disorder, with an estimated prevalence of at least 300 cases per 100,000 population. The age at onset of essential tremor varies from the teenaged years to late life. Essential tremor is the prototypical action-postural tremor that may start in the head, voice, or arms. Leg tremor is an infrequent area of involvement. Involvement of the arms is generally bilateral, and a tremor that involves one limb for over a decade is unlikely to be essential tremor. A family history of tremor is obtained in up to 96% of patients with essential tremor, and an autosomal dominant pattern of inheritance is recognized.

Another distinguishing feature that may aid in the diagnosis of essential tremor is its frequent profound sensitivity to alcohol. Some patients with severe tremor may find themselves unintentionally dependent on alcohol's therapeutic effects. Unfortunately, the relation to alcohol is not diagnostic, since dystonia and other types of tremor may also be alcohol-responsive.

Parkinsonian Tremor

Rest tremor is a cardinal symptom of Parkinson's disease. In most cases, tremor starts asymmetrically, often affecting a single upper extremity for many years before spreading to the contralateral arm or either leg. On examination, the tremor is usually seen with the hand at rest in the lap. The examiner should also watch for the rest tremor when the patient performs rapid movements with the contralateral limb or when the patient walks and the hand is down at the side.

In most cases, other parkinsonian symptoms (rigidity and slowness of movement) are found in the tremulous limbs. In a variant of the disease, "benign tremulous Parkinson's disease," severe rest tremor may be the only manifestation for the patient's lifetime. Such patients have a relatively good prognosis because they are spared the disabling akinesia and postural instability of the more generalized Parkinson's disease. However, treatment of severe rest tremor may be difficult.

When extremely severe, parkinsonian tremor invariably begins to involve postural activities. In addition, approximately 15% of patients with Parkinson's disease have prominent action-postural tremor early in the course of the disease. Such patients pose a diagnostic problem that requires careful history-taking and examination to reveal the correct diagnosis.

Therapy for parkinsonian tremor reflects the same principles as the treatment of Parkinson's disease in general. Pharmacologic therapy is considered when the patient experiences functional or social disability. Levodopa (Sinemet) is the most effective agent, with significant tremor reduction experienced in approximately 50% of patients. Anticholinergic medications, such as trihexyphenidyl (Artane), have an antitremor potency equivalent to that of levodopa, but side effects such as memory impairment and urinary retention prohibit use in the elderly. Dopamine agonists and amantadine have a less reliable effect on tremor. The postural tremor of Parkinson's disease seldom is affected by levodopa and in fact is better treated with a medication such as propranolol or primidone (Mysoline). When tremor is severe and disabling, thalamotomy or thalamic stimulation (see Chapter 22) provides highly effective suppression. Both rest and action-postural tremor are suppressed by these neurosurgical interventions. Pallidotomy (see Chapter 13), on the other hand, markedly reduces resting tremor but has little effect on postural tremor.

Severe Cerebellar Tremor

Severe cerebellar tremor is often the most disabling of tremors. The wild oscillations brought on by action frequently render the limb useless and hazardous. Multiple sclerosis ranks as the commonest cause of this form of tremor. Head trauma, presumably producing shearing injury to midbrain white matter tracts, is the next most frequent cause. Instances of midbrain tumors, strokes, or other structural lesions causing this tremor syndrome are rare. If the cause of this tremor syndrome is not readily apparent, MRI of the brain is indicated.

Despite a variety of clinical trials, no medical therapy has proven even mildly effective in the treatment of this devastating movement disorder. Carbamazepine (Tegretol), isoniazid, tetracannabinol, beta-blockers, primidone, and benzodiazepines provide a litany of treatment failures. Until drug therapy is discovered, thalamotomy or thalamic stimulation offers the only hope for the treatment of severe cerebellar tremor. Treatment success has been estimated to be between 30% and 70% but is less predictable than that when the same procedure is applied to parkinsonian or essential tremor. One factor mitigating success is the persistent ataxia that may continue to disable patients even if the tremor is abolished.

Tremor of Wilson's Disease

Wilson's disease is a systemic disorder of copper metabolism (see Chapter 32) that causes a wide variety of movement disorders, including tremor, parkinsonism, and dystonia. The tremor of Wilson's disease can be of any type, from rest tremor to severe cerebellar tremor. Combinations of tremor types are often seen. One form of tremor attributed to Wilson's disease is a wing-beating or flapping movement of the arms when the elbows are held abducted. In general, a tremor in Wilson's disease is accompanied by dysarthria, dystonia, or other neuropsychiatric manifestations. This is not always the case, however, and any patient under the age of 50 with tremor of unclear origin should be screened for Wilson's disease, as the illness is treatable (see Chapter 32). This screening should include a careful search for Kayser-Fleischer rings (corneal copper deposition) by slit-lamp examination, measurement of the serum ceruloplasmin level (low in Wilson's disease), and a 24-hour urinary copper determination (high in Wilson's disease).

Orthostatic Tremor

Orthostatic tremor is a unique, situation-specific tremor occurring in the elderly. Afflicted patients give a history of feeling unsteady and experiencing leg tremor immediately on standing. The unsteadiness is so severe that patients become fearful of standing in open places and frequently lean dependently on walls or other means of support. While sitting, patients are entirely normal, and walking generally lessens the tremor and unsteadiness to some degree.

Examination of these patients when standing may reveal leg jerks and slight irregular contractions of leg muscles, but often there is no observable tremor. The examiner should look at the patient's pants or skirt to see if tremor appears on standing. Walking is generally wide-based and somewhat bizarre in appearance. Often, the panicked, lurching movements of these patients wrongly suggest a psychiatric illness. Only when surface EMG electrodes are placed on leg muscles is the nature of the tremor conclusively revealed. High-amplitude, 14- to 18-Hz EMG bursts are recorded immediately as the patient stands. These bursts occur too rapidly to be visible as tremor. Electroencephalographic waves of a similar frequency recorded at the top of the head may accompany the bursts.

Some believe that orthostatic tremor is an unusual form of essential tremor. However, the rapid frequency, absence of family history, and lack of alcohol effect argue against this. Also, orthostatic tremor is unusually responsive to clonazepam, with approximately 50% of patients having substantial relief. Beginning at 0.25 mg at bedtime, the dose can be increased to a maximum of 1 mg four times a day. If clonazepam is ineffective, there is little hope for effective therapy, although valproic acid (500 to 2,000 mg/day in divided doses), phenobarbital (30 to 120 mg/day), and gabapentin (100 to 2,400 mg/day in divided doses) have had some reported success.

Primary Writing Tremor

Primary writing tremor is a task-specific disorder that exclusively affects handwriting, sparing all other tasks using the involved hand. Task specificity is a feature shared with focal arm dystonia, leading some to equate the two disorders. The absence of abnormal posturing and an occasional response to propranolol suggest, however, that the writing disorder is more akin to essential tremor. Poor response to medication is the rule for this disorder; propranolol, clonazepam, and primidone are most often prescribed. Many patients benefit from learning to write with the nondominant hand or to rely on keyboarding. Extremely disabling cases may benefit from thalamotomy.

Tremor Associated With Peripheral Neuropathy

The association of tremor and peripheral neuropathy has long been recognized, but the pathophysiology has never been clearly understood. Presumably, disease of peripheral nerves can upset the balance of feedback systems and induce oscillation or tremor. Supporting this idea is the fact that tremor is observed most often in demyelinating neuropathies such as Guillain-Barré syndrome, chronic inflammatory demyelinating polyradiculoneuropathy, and paraproteinemic neuropathies. Roussy-Lévy syndrome is a rare hereditary form of peripheral neuropathy with associated action-postural tremor. Because there is no specific therapy for this type of tremor, treatment is directed toward the underlying neuropathy.

Palatal Tremor

Although previously referred to as "palatal myoclonus," this tremor is now known as "palatal tremor." Palatal tremor is slow in frequency (1 to 3 Hz), often is present during sleep, and can be associated with synchronous movements of the tongue, face, larynx, and neck. This disorder may be idiopathic or symptomatic. Idiopathic cases are characterized by ear clicking, and examination reveals rhythmic soft palate movements that are often visible by watching the hyoid "Adam's apple."

Symptomatic palatal tremor is characterized by brain stem or cerebellar signs on examination, frequently associated with eye movement abnormalities. Causes include brain stem infarct, multiple sclerosis, trauma, and degenerative diseases. MRI of the brain reveals hypertrophy of the inferior olive in patients with symptoms.

Treatment of palatal tremor includes benzodiazepines and anticonvulsants. Botulinum toxin injections in the tensor veli palatini muscle may benefit patients with ear clicking.

DIFFERENTIAL DIAGNOSIS OF TREMOR

Three movement disorders may masquerade as pathologic tremor. Myoclonus may occur continuously, mimicking tremor. This condition is variously termed "polyminimyoclonus" or "epilepsia partialis continua." The hand is most often affected, and close observation reveals that the apparent tremor is really an irregular series of very rapid muscle twitches.

Body parts affected by dystonia often display repetitive phasic contractions that may resemble tremor. Generally, the examiner can differentiate phasic dystonia from tremor by coexisting sustained abnormal posturing, a slower and often irregular frequency of the movements, and the occurrence of the phasic movements in a specific posture or with a specific task. The differentiation is made more difficult by the association of dystonia with essential tremor.

One final condition to be differentiated from organic tremor is psychogenic tremor. Wild, atypical movements that vary in frequency and diminish with distraction suggest a psychiatric source. Psychogenic tremors often change in direction or vector and are distractible. A bizarre gait disorder often accompanies psychogenic tremor. However, this is an extremely difficult diagnosis to make, and a specialist in movement disorders should evaluate such patients. See Chapter 35 for a detailed discussion.

SELECTED READING

Bain PG, Findley LJ, Thompson PD, Gresty MA, Rothwell JC, Harding AE, Marsden CD. A study of hereditary essential tremor. *Brain* 1994; 117: 805-824.

Deuschl G, Bain P, Brin M. Consensus statement of the Movement Disorder Society on Tremor. Ad Hoc Scientific Committee. *Mov Disord* 1998; 13 Suppl 3:2-23.

Deuschl G, Toro C, Valls-Sole J, Zeffiro T, Zee DS, Hallett M. Symptomatic and essential palatal tremor. I. Clinical, physiological and MRI analysis. *Brain* 1994; 117: 775-788.

Elble RJ, Koller WC. *Tremor.* Johns Hopkins University Press, Baltimore, 1990.

Findley LJ, Koller WC. *Handbook of Tremor Disorders.* M. Dekker, New York, 1995.

Koller WC, Vetere-Overfield B. Acute and chronic effects of propranolol and primidone in essential tremor. *Neurology* 1989; 39: 1587-1588.

22

Essential Tremor: Diagnosis and Treatment

Jean Pintar Hubble, MD

CONTENTS

PREVALENCE

Essential tremor (ET) has been recognized as a distinct condition for over 100 years. In the medical literature, it has been described as "common," "not uncommon," and even "rare." Reports of ET have been issued from far-flung locales, suggesting a global occurrence. Prevalence estimates vary widely, from 0.0005 to 5.55. Despite differences in methods and results, more recent epidemiologic studies clearly prove that ET is a frequently encountered disorder, especially among the elderly. In a Canadian community-based study, 14% of persons 65 years old or older had ET in contrast to 3% with Parkinson's disease. Applying the prevalence rate of ET derived from a door-to-door Finnish study to the U.S. Census Bureau population statistics (1988), one can estimate that over 5 million Americans may be affected. ET is now recognized to be the most common movement disorder.

GENETICS AND ETIOLOGY

The tendency of ET to occur in families has been recognized for many years. In the American medical literature, Dana reported on three families with tremor in 1887. In the ensuing years, authors reviewing ET have reaffirmed its heritable nature and published

From: *Parkinson's Disease and Movement Disorders:*
Diagnosis and Treatment Guidelines for the Practicing Physician
Edited by: C. H. Adler and J. E. Ahlskog © Mayo Foundation
for Medical Education and Research, Rochester, MN

pedigrees to support an autosomal dominant mode of inheritance. Positive family histories vary by report from 17% to 96%. The variability in reported figures is probably due to different methods of ascertainment.

Recent technologic advances in molecular genetics may ultimately provide the means to identify the genetic substrate for ET. Gene markers for other heritable movement disorders are being tested in ET. Because of the putative association of ET with dystonia, the gene marker for idiopathic torsion dystonia (9q32-34) was examined in families with ET, and no association was found. More recently, genetic loci on chromosome 2p22-p25 were linked to the occurrence of ET in a large family of Czech descent; repeat expansion detection analysis suggested that expanded CAG trinucleotide sequences were a factor in determining tremor in this family. In a separate report, ET was mapped to a locus on chromosome 3q13 on the basis of genomic screening of 16 Icelandic families with 75 affected members. The relevance of these results of chromosomal linkage to other families with ET or to sporadic cases has not been determined.

The cause of sporadic ET is unclear. Many of these cases may, indeed, have a familial basis, because family history can be unreliable or unattainable. Busenbark et al. found that 68% of patients with ET reported a positive family history during an initial clinic visit but that direct questioning of immediate family members revealed the family history to be positive in 96%.

CLINICAL PRESENTATION

Tremor is the sole clinical manifestation of ET. The diagnosis is made either incidentally or when the patient is examined because of the mechanical or social disability resulting from the tremor. Mild abnormalities of tone are occasionally reported. Tandem gait abnormalities occur in approximately 50% of patients with ET in contrast to approximately 25% of age-matched controls. Other neurologic signs or symptoms are typically absent. ET is usually postural, that is, the tremor is best seen when the limb is maintained in a fixed posture, such as arms held outstretched. It may be accentuated by goal-directed movement of the limbs (kinetic tremor), and in some instances of severe tremor, it may even be noticed at rest.

Body Region Affected by Essential Tremor

ET appears most frequently in the hands. An adduction-abduction movement of the fingers and a flexion-extension movement of the hand are typical; less often, a pronation-supination movement is seen, similar to the tremor of Parkinson's disease. Frequently, the tremor is unilateral at onset, but both sides become involved with time. Handwriting may become tremulous, and rounded letters take on a sharp angularity. The handwriting in ET does not become micrographic, differentiating it from the script in Parkinson's disease. Other fine motor activities may also be impaired, including pouring liquids, holding a utensil or cup, and sewing.

The next most frequently affected body part is the cranial musculature. Although the tongue, head, or voice may be affected in isolation (see "Clinical Variants of Essential Tremor," below), it is most common for tremor in these regions to occur in association with hand tremor. In advanced cases, tremor of the palate, voice, and tongue may result in dysarthric speech, but this infrequently occurs before age 65 years. The legs and trunk become affected infrequently and usually in the later stages of the illness. The frequency of involvement of body parts or regions in ET is summarized in Table 22-1.

Table 22-1
Body Parts Affected in Essential Tremor[a]

No. of subjects	Percent of subjects			
	Hands	Head	Voice	Legs
604	96	32	14	9

[a]Summary data from seven clinical series reported from 1960 to 1988.

Disease Onset and Progression

ET may begin at any age, but its incidence rises with advancing years. It has been reported that the onset of ET appears to be earlier in the familial form than in the sporadic form. It is rarely reported in infancy and childhood. Categorizing ET by age at onset has not proved useful because it does not assist in predicting clinical outcome or therapeutic response.

ET is generally considered to be a slowly progressive disorder, but variability in its clinical course has been noted. The progression of this disease can be defined as an increase in tremor amplitude or extension of tremor to previously unaffected body parts. Increased amplitude impairs the patient's voluntary movements and results in disability. On the basis of their detailed study of familial ET, Larsson and Sjögren characterized ET as typically appearing before age 50 as a fine, rapid hand tremor of little consequence; by age 65 years, the tremor is moderate, and at 70 years, it is quite pronounced (increased in amplitude and diminished in frequency). Despite its progressive nature, ET has long been recognized not to be associated with increased mortality.

Factors Influencing Essential Tremor

Tremor can be affected by a host of factors. Tremor frequency declines and amplitude increases with age. Handedness may be relevant to the expression and severity of ET. Biary and Koller found a higher incidence of left-handedness among patients with ET than among controls and a direct relationship between left-hand dominance and tremor severity. It is not unusual for patients to associate the onset of tremor with a specific incident or circumstance; however, a causal relationship is doubtful. Tremor resembling ET has been reported after head trauma; this is unlikely to be a form of ET, because of the abrupt onset, temporal link to trauma, and relative lack of response to drug therapy. A number of elements can have a short-term effect on tremor: stress, anxiety, fatigue, extremes of temperature, emotional upset, sexual arousal, central nervous system stimulants, and diurnal fluctuations in catecholamine levels. Caffeine, commonly considered to be a precipitant of tremor, produced no discernible effect in a formal study evaluating 50 patients with ET.

Alcohol may have a remarkably ameliorative effect on ET. In many patients, tremor is lessened by ingestion of even a small amount of an alcoholic beverage. Intravenous infusion of alcohol, but not local intra-arterial injection, decreases ET and suggests that the effect of alcohol is centrally mediated. Although the "antitremor" effect of alcohol is most typical in ET, it may also occur in other forms of tremor. As is true of most involuntary movements, ET typically remits during sleep; there are, however, rare reports of its persistence during slumber. In summary, the appearance of tremor in a given patient may vary, not only with the passage of years but even over the course of a single day.

Table 22-2
Common Functional Disabilities
in Essential Tremor

Handwriting
Drinking liquids
Eating
Dressing
Fine manipulations
Speaking
Embarrassment

Disability

The patient with ET may be handicapped by the physical limitations the tremor may create or by the embarrassment and resulting social isolation it may cause (Table 22-2). The physical disability is directly related to tremor amplitude; with aging or disease progression, amplitude increases and the ability to execute fine, discrete movements may be impaired. Occupational skills may be compromised; approximately 15% of patients with ET referred to a university clinic were pensioned or disabled as a result of tremor. One cannot estimate to what extent ET alters the quality of life because the social limitations that patients may impose on themselves because of embarrassment resulting from tremor vary greatly. Although ET is sometimes preceded by the term "benign," the condition can be disabling.

Diagnostic Pitfalls

ET can be misdiagnosed, most commonly as incipient Parkinson's disease (PD). The tremor of PD is usually most evident when the hands or arms are supported in a resting position. In contrast, the hand tremor of ET is usually accentuated with sustained posture, for example, arms outstretched. Other signs of PD that differentiate it from ET are muscle rigidity, bradykinesia (slowness in movement), and postural instability. The correct diagnosis of ET can also be clouded by the consideration of other conditions that may have tremor as a clinical manifestation, including neurologic disorders (e.g., Wilson's disease) and systemic illnesses (e.g., hyperthyroidism). In an effort to minimize diagnostic errors, clinical criteria for ET have been suggested (Table 22-3). Such guidelines not only assist in individual case assessment but also assure validity in the clinical investigation of ET.

Clinical Variants of Essential Tremor

A variety of atypical tremor disorders appear to be related to ET (Table 22-4). An association of these conditions with ET is suggested by the high occurrence of a family history of ET, frequent presence of a mild postural tremor, and tremor reduction with alcohol ingestion.

Some variants are of doubtful relationship to ET and are difficult to differentiate from other neurologic disorders. Primary writing tremor needs to be differentiated from a segmental dystonia of the hands, writer's cramp. Dystonic spasms are observed in writer's cramp rather than tremor. However, tremor may occur secondary to attempts to control the dystonia, and some families have been reported with different members exhibiting writer's cramp, ET, and primary writing tremor. Orthostatic tremor is a unique variant. In orthostatic tremor, the legs begin to shake after the patient stands for several seconds;

Table 22-3
Proposed Diagnostic Criteria for Essential Tremor

Intermittent or constant tremor of the hands, head, or voice

Tremor is postural or kinetic, or both

No other neurologic abnormalities related to systemic or
neurologic disease

No exposure to drugs known to cause tremor

Positive family history of tremor supports the diagnosis

Reduction of tremor with alcoholic beverages supports the
diagnosis

Table 22-4
Possible Clinical Variants of Essential Tremor

Kinetic predominant hand tremor
Combined resting-postural tremor
Primary writing tremor
Isolated voice, chin, or tongue tremor
Orthostatic tremor

the tremor increases with time and can lead to falling. There is little or no tremor during sitting, walking, or leaning against a firm support.

Association of Essential Tremor and Parkinson's Disease

The relationship of ET and PD has been controversial since James Parkinson first commented on the distinction between paralysis agitans and senile (essential) tremor in 1817. Because both disorders are relatively common, chance may account for ET or PD in family members and instances of concurrent ET and PD. In recent years, the issue of an association between ET and PD has been examined many times with disparate results. A clear link between these two conditions has yet to be discovered. Nevertheless, the controversy over the relationship between PD and ET persists. It is fueled by lack of diagnostic certainty and the etiologic ambiguity of both these disorders. It can be difficult, even impossible, to differentiate ET from PD early in the clinical course or in the very aged. Parkinsonian signs such as cogwheeling and rest tremor may be seen in ET, and postural tremor is not uncommon in PD. Table 22-5 lists historical and clinical features that may help in distinguishing ET and PD.

PATHOPHYSIOLOGY

Unfortunately, our knowledge of the anatomical localization of neural abnormalities in ET is minimal, and likewise there is almost no understanding of possible pathophysiologic mechanisms. Is the pathogenesis of ET related to dysfunction of the peripheral or central nervous system? A central site for the abnormality in ET is supported by the beneficial effect of thalamotomy and centrally acting antitremor drugs. A central mechanism is also suggested by reports of stroke alleviating ET. Several brain areas are candi-

Table 22-5
Comparison of Parkinson's Disease and Essential Tremor

	Parkinson's disease	Essential tremor
Characteristic		
Family history	Usually negative	Positive in > 50%
Alcohol	Usually no effect	Marked tremor reduction
Medical attention sought	Early in course	Often late in course
Age at onset	Usually > 60 years	May begin in childhood or adolescence
Tremor type	Resting	Postural, kinetic
Body part affected	Hands, legs	Hands, head, voice
Disease course	Progressive	Slowly progressive, static periods
Bradykinesia, rigidity, or postural instability	Present	Usually absent
Treatment		
Levodopa	Effective	No effect
Propranolol	May decrease tremor	Effective
Primidone	No effect	Effective

dates for the exaggerated oscillations responsible for ET. Areas with inherent rhythmicity in the frequency range of ET are the inferior olivary and thalamic nuclei. Hypermetabolism of the inferior olive can be observed on positron emission tomography during tremor activation. A tremor disorder that resembles ET can be caused by vascular or traumatic insults to the brain stem, particularly around the area of the red nucleus. Recent positron emission tomographic studies support the hypothesis that the cerebellum and its circuitry are integrally involved in the production of ET; abnormal activation of the cerebellum, red nucleus, and thalamus has been demonstrated.

Because of the relative lack of human data, animal models of ET would be important. Tremors in animals can be induced by drugs such as harmaline and oxotremorine and by anatomical lesions. However, the relevance of these conditions to ET is uncertain. It would appear that ET may be due to an abnormal oscillation of a central nervous system "pacemaker," the location of which is currently unknown. There have been few reported postmortem brain studies in patients with ET. These investigations have reported only nonspecific changes without evident selective neurodegeneration as in PD. Detailed studies of neurochemical changes in ET brains have not been performed to date. It has been theorized that ET results from alterations in γ-aminobutyric acid or serotonin in the neurotransmitter system. However, drug trials with the GABAergic and serotoninergic agonists progabide and trazodone, respectively, have yielded negative findings in ET, offering no substantiation of this notion.

It can be concluded that ET is probably produced by a central oscillator that can be enhanced or suppressed by reflex pathways. Further analysis of postmortem material from ET brains is needed if we are to understand the pathophysiology of ET.

TREATMENT

Despite recent advances in treatment, a uniformly safe and effective tremor-specific remedy has yet to be devised. What follows is a review of ET treatments, starting with the oldest (alcohol) and concluding with the newest (thalamic stimulation).

Pharmacologic Treatment

ALCOHOL

Alcohol temporarily causes a dramatic reduction in tremor in most patients with ET. Alcohol's mechanism of action is unknown. A better understanding of its action could lead to the development of other pharmacologic agents for ET. The famous neurologist Critchley warned that use of alcohol in ET "appeared only too often to have served as an excuse for habits of intemperance." Results from a retrospective chart survey supported the notion that alcohol use in ET could lead to addiction. However, a prospective study in the U.S. demonstrated that the prevalence of pathologic drinking in ET did not differ from that in other tremor disorders or in chronic neurologic disease without tremor. Likewise, similar surveys in Finland and Sweden found that patients with ET neither used more alcohol nor had a higher prevalence of alcoholism than the general population. Chronic alcoholism may itself result in a persistent postural tremor lasting up to 1 year during abstinence. It can be concluded that the occasional use of alcohol in ET is not contraindicated and that the risk of alcoholism is probably no greater than that in the general population.

BETA-ADRENERGIC BLOCKERS

The effectiveness of β-adrenergic blockers in reducing tremor has been recognized for over 2 decades. Most investigations, using both subjective and objective (accelerometer recordings) evaluation, have confirmed the efficacy of propranolol (Inderal) in reducing postural hand tremor. Tremor amplitude is decreased but tremor frequency is unchanged. Thus, propranolol became the drug of choice for ET. However, the clinical response to propranolol is variable and often incomplete. It is generally estimated that 50% to 70% of patients have some symptomatic relief. Dramatic improvement occurs in a much smaller percentage, and some patients have no response. Average tremor reduction of 50% to 60% is suggested by some studies. No well-defined factors predict therapeutic responsiveness. In a dose-response study of propranolol, 240 to 320 mg/day was found to be the optimal dose range. Doses above 320 mg/day conferred no additional benefit. A sustained-release preparation of propranolol (propranolol long-acting) designed for once-daily dosage has been reported to provide tremor reduction similar to or, in some cases, greater than that with divided doses. Administration should begin with 40 mg of long-acting propranolol per day, with an increase in dose up to 320 mg/day as needed or tolerated.

Relative contraindications for propranolol use are (1) heart failure, especially if poorly controlled, (2) second- or third-degree atrioventricular block, (3) asthma or other bronchospastic disease, and (4) insulin-dependent diabetes, in which propranolol may block the adrenergic manifestations of hypoglycemia. Most side effects of propranolol are related to β-blockade. The pulse rate is lowered in most patients. A pulse of 60 beats per minute is usually well tolerated. Less common adverse reactions are fatigue, weight gain, nausea, diarrhea, rash, impotency, and alteration of mental status (e.g., depression). It has been suggested that tolerance may develop to the effect of propranolol. In at least one study, however, most patients had not lost any effect of propranolol after 1 year of therapy.

The mechanism of action of propranolol in ET is unknown. A central action site has been proposed because of the lack of effect of local intravenous or intra-arterial administration of propranolol and a delay in the effect of oral therapy. However, several controlled studies have shown that propranolol causes an immediate and sustained reduction in tremor. A peripheral site of action has also been proposed, since β-blockers, which enter the central nervous system with difficulty, can reduce ET. Specific β_2-adrenergic antagonists (ICI 118551 and LI 32-468), which act predominantly peripherally, are effective in further decreasing tremor, supporting a peripheral β_2-blocking mechanism of action.

Other orally active β-adrenergic blocking drugs are available, such as metoprolol succinate (Toprol XL), nadolol (Corgard), atenolol (Tenormin), timolol (Blocadren), and pindolol (Visken). The β-blockers are classified according to whether or not they have β_1-adrenergic selectivity, intrinsic sympathomimetic activity (partial agonists), and membrane-stabilizing properties. Differences in potency, metabolism, half-life, protein binding, and excretion are recognized. Despite the therapeutic similarity of β-blocking drugs, differences in their pharmacodynamic properties may be clinically important.

The effectiveness of metoprolol in ET has been demonstrated in multiple case reports and a controlled study. Metoprolol differs from propranolol in preferentially antagonizing β_1-adrenergic receptors. The selectivity for β_1 receptors is, however, only relative. With higher doses, β_2-adrenergic receptors are also blocked. Some degree of β_2-blockage appears in patients at daily doses above 100 mg. Metoprolol is beneficial in ET at divided doses of 100 to 200 mg/day. Patients in whom propranolol had no effect also did not have a response to metoprolol. Because of its relative lack of β_2-blocking properties, metoprolol should theoretically be better tolerated than the nonspecific β-blocker propranolol in patients with bronchospastic disease. Several patients with asthma and ET were reported who could tolerate metoprolol but not propranolol. Metoprolol can, however, cause respiratory distress and should be used with caution in this patient population.

Nadolol, administered once daily at doses of 120 and 240 mg/day in a controlled study, significantly decreased ET in patients who also had a response to propranolol. Nadolol has a 24-hour half-life and can be taken once daily, avoiding inconveniences and compliance problems of multiple daily drug doses. Atenolol was reported to have little or no effect on tremor, timolol was found to decrease tremor, and pindolol was found to be ineffective. Because of its partial agonist activity, pindolol may actually produce tremors. The peripherally acting β-adrenergic blocker arotinolol, given at a dose of 30 mg/day, reportedly reduced ET in 15 patients. This suggests that peripherally acting drugs may be beneficial in ET. As an alternative explanation, arotinolol may have some central actions at the tested dose.

In summary, the usefulness of other β-adrenergic blockers is not much different from that of propranolol. Selective β-blockers, like metoprolol, can be used at low doses in bronchospastic disease. If side effects such as impotence occur with propranolol, changing to a different β-blocker may eliminate the adverse effect without reducing the beneficial action.

PRIMIDONE AND PHENOBARBITAL

It was first anecdotally noted that primidone (Mysoline) given to a patient with epilepsy and ET reduced tremor. The drug was then given to 20 other patients, with the dose starting at 125 mg and increasing to 750 mg/day. Twelve patients had a good clinical response, but six patients could not tolerate the drug. Other early studies of primidone in ET demonstrated effectiveness compromised by poor drug tolerance. Then, Koller and Royse, using objective recording techniques, found that primidone (50 to 1,000 mg/day) significantly reduced the amplitude of essential hand tremor in both untreated and propranolol-treated patients. Low doses (e.g., 250 mg/day) appeared to be as effective as high doses, with no correlation between serum levels and therapeutic response. Comparative drug studies demonstrated that both propranolol and primidone significantly reduce tremor, with no significant difference between the drugs. As with other drugs for tremor, tolerance to primidone may eventually develop. Some patients cannot tolerate the drug because of acute reactions to initial doses (nausea, sedation, lightheadedness) or side effects with long-term use (ataxia, confusion). In an effort to avoid or minimize side effects, administration of primidone should begin with a 50-mg tablet, one half tablet at bedtime, and then the dose can be slowly escalated as needed or tolerated up to 250 mg three times a day.

The mechanism of action of the antitremor effect of primidone is unknown. Primidone is converted to two active metabolites: phenylethylmalonamide, with a half-life of approximately 30 hours, and phenobarbital, with a half-life of approximately 10 days. The administration of high doses of phenylethylmalonamide has no effect on tremor. Similarly, phenobarbital has minimal antitremor action. Primidone itself or an unrecognized metabolite appears to be the responsible agent.

BENZODIAZEPINES

Benzodiazepines and other sedative-hypnotic drugs have been used to treat ET. Interestingly, no ET clinical trial has been done with the most commonly prescribed drug of this class, diazepam (Valium). It is generally thought that benzodiazepines possess only limited efficacy, probably related to a reduction in tension and anxiety, which can enhance tremor. Sedation is a frequent side effect with all benzodiazepine compounds. Clonazepam (Klonopin) may be started at a dose of 0.25 mg at bedtime and increased up to a total target dose of 1 to 3 mg/day. It is often suggested that clonazepam is most effective in the treatment of kinetic predominant hand tremor and in orthostatic tremor. Alprazolam (Xanax), a triazole analogue of the benzodiazepines, was assessed in a double-blind, placebo-controlled study of 24 patients with ET. Significant improvement occurred, but transient, mild fatigue or sedation occurred in half the patients.

CARBONIC ANHYDRASE INHIBITORS

The carbonic anhydrase inhibitor methazolamide (Neptazane) appeared to be effective in ET in an open-label trial, but its benefits were indistinguishable from those of placebo in a subsequent double-blind study. Methazolamide can be tried for patients in whom other pharmacotherapies have failed. Common adverse reactions are sedation, nausea, epigastric distress, anorexia, numbness, and paresthesias. Administration should start with a 25-mg tablet, one half tablet twice a day, with slow titration up to a maximum of 50 mg three times a day.

OTHER DRUGS

Patients with ET are sometimes given antiparkinsonian drugs, but administration is usually discontinued because of lack of benefit and side effects. Amantadine (Symmetrel) has been variably reported to reduce, enhance, or have no effect on tremor in ET. It would appear that amantadine is not generally useful in ET, but an occasional patient may have a response to the drug. Administration of amantadine begins at 100 mg/day, with a maximum of 100 mg three times a day.

On the basis of an open-label trial, clonidine (Catapres) was suggested as an agent to treat ET. The ability of clonidine to stimulate central α-adrenergic receptors was the proposed mechanism of action. However, no benefit was found in a double-blind placebo-controlled investigation. Similarly, the α-adrenergic blocking drug thymoxamine did not provide significant benefit in tremor reduction when examined in a controlled trial. There is no evidence that α-adrenergic mechanisms are involved in the pathogenesis of ET, and α-adrenergic drugs, agonists or antagonists, appear to have no role in the treatment of ET.

The anticonvulsant gabapentin (Neurontin) reduced tremor in ET in open-label trials and one controlled trial. Administration can start at 100 to 300 mg/day and increase to a maximum of 600 mg four times a day. Potential side effects are nausea, sedation, and inability to concentrate. Other agents tested in placebo-controlled trials that did not produce benefit in ET are trazodone (Desyrel), ritanserin, and progabide. The atypical neuroleptic agent clozapine (Clozaril) decreased ET in 9 of 12 patients in an open trial. Sedation was a major side effect and may have been related to the observed reduction in tremor. Clozapine may cause agranulocytosis in a fraction of exposed persons; therefore, weekly to bimonthly blood cell counts are mandatory. It is not considered a practical remedy for ET by most authorities. Clozapine doses range from 12.5 to 150 mg at bedtime. The anticonvulsant felbamate (Felbatol) has been anecdotally reported to reduce ET, but its propensity to cause blood dyscrasia and liver abnormalities severely limits its use.

Mephenesin, a centrally acting muscle relaxant, was said to have a limited effect in reducing ET; however, the drug never gained widespread use. Two calcium channel blockers, nifedipine (Procardia) and verapamil (Calan), have no apparent benefit in ET; nifedipine may even increase the intensity of tremor. Flunarizine, a calcium channel blocking drug not available in the U.S., has had variable reported effects on the tremor of ET. Flunarizine has caused intolerable side effects, including dystonia, parkinsonism, weight gain, and depression, in more than 25% of patients with ET studied. The calcium channel blocker nicardipine (Cardene) may transiently lessen tremor in ET. Nimodipine (Nimotop) also reduces tremor, but it has not gained widespread use for this indication, primarily because of cost.

BOTULINUM TOXIN INJECTION

Botulinum toxin type A (Botox) represents a novel approach to the treatment of ET. Intramuscular injection of minute amounts of the toxin causes muscle paresis by acting at peripheral nerve endings to block the release of acetylcholine. This agent has been shown to be effective in a variety of focal dystonias, and its administration is the treatment of choice for these disorders (see Chapter 31). In open-label studies, botulinum toxin type A intramuscular injections have effectively reduced hand tremor. Many questions remain about the optimal dose, the muscles to be injected, and the time frame for reinjection. Transient muscle weakness is the anticipated effect of this therapy, and weakness of finger and hand muscles may be poorly tolerated. Botulinum toxin type A is usually well tolerated when injected into the neck muscles or vocal cords to help control head tremor and voice tremor. Patients with tremor who derive benefit from botulinum toxin type A require reinjections, because the effects of the toxin typically last for 2 to 4 months.

INITIATING DRUG THERAPIES

Propranolol and other β-adrenergic blockers and primidone are the only drugs shown to be clearly effective in suppressing ET. It is unclear which should be the drug of first choice. As many as 20% of patients have side effects for several days after the first dose of primidone. If the patients are warned of these potential adverse reactions and encouraged to continue to take primidone, only a few will stop taking it. Side effects with long-term therapy are uncommon with primidone but are of much more concern with propranolol. Many elderly patients cannot take β-blockers. Some patients may require therapy with both propranolol and primidone.

The following treatment schedule can be tried.

1. Start with primidone, 25 mg at night (warn patient of possible side effects but recommend continuation of drug if tolerable side effects occur).
2. Increase dose of primidone slowly to 250 mg at night, if necessary.
3. Add or substitute propranolol, 40 mg in the morning.
4. Increase propranolol dose to 40 mg twice a day, if necessary.
5. Increase propranolol dose to 80 mg twice a day, if necessary.
6. Increase propranolol dose to 80 mg three times a day, if necessary.

Use long-acting propranolol if preferred. If there is no response to either primidone or propranolol, taper the dose and discontinue administration. Next consider alprazolam, gabapentin, clonazepam, or the other alternative agents noted above. Surgical interventions can be considered if tremor is disabling and refractory to medication.

TREATMENT OF CLINICAL VARIANTS

Propranolol, primidone, and other drugs used to treat typical ET have been tried with various degrees of success in the treatment of clinical variants, such as head, voice, and orthostatic tremors. For example, 5 of 10 subjects had reduced head tremor after treatment with botulinum toxin type A compared with 1 of 10 subjects treated with placebo. Head tremor does not seem to respond well to long-term β-blocker therapy. The carbonic anhydrase inhibitor methazolamide has been suggested to be helpful in alleviating ET, particularly tremor affecting the head and voice; however, a controlled study of its use did not support the initial findings. Orthostatic trunk and leg tremor may respond to clonazepam, gabapentin, or valproic acid (Depakote) therapy.

Surgical Treatment

THALAMOTOMY

Stereotactic thalamotomy is an effective procedure in the treatment of parkinsonian, cerebellar, and essential tremors. The technical aspects of the procedure have improved greatly in the past decade. Advances in the neurophysiologic confirmation of the site of the lesion now allow for more accurate discrete lesion placement. The site of the lesion selected is often the ventral intermediate (Vim) nucleus of the thalamus. Thalamotomy can be performed with the use of mild sedation and local anesthesia. The neurologic status of the patient can be monitored because the patient is awake during surgery. Lesion coordinates are generated by use of brain imaging and computerized programs. A microelectrode with a recording tip can be inserted into the brain, and the neuronal response further defines the anatomical site for the lesion. Heating of the electrode tip creates a small lesion. Benefit (tremor reduction) is achieved in more than 75% of patients.

Thalamotomy currently has a mortality rate of less than 0.3%. Temporary intellectual deficits and transitory hemiparesis may occur. Other uncommon adverse reactions are seizures, involuntary movements, and cerebellar signs. A transient deterioration of speech may occur occasionally after unilateral thalamotomy. Bilateral thalamotomies, however, can result in severe persistent dysarthria and permanent neuropsychologic deficits. Therefore, bilateral operations are rarely recommended.

Thalamotomy appears to be a neglected therapeutic option, but it should be reserved for patients with (1) severe unilateral or asymmetrical tremor, (2) marked functional disability, and (3) tremor that is unresponsive to maximally tolerated doses of propranolol and primidone.

THALAMIC STIMULATION

Clinical researchers from France initially reported their results on the use of electrical stimulation of the thalamus as treatment for tremor disorders. Applying high-frequency electrical stimulation to the Vim thalamic nucleus, they reported control of disabling tremor in ET and PD. The device has now been widely tested and approved for use in the United States. An electrical lead is implanted stereotactically into the Vim thalamic nucleus and is connected by an extension wire to a pulse generator implanted subcutaneously below the clavicle. The device is analogous to a cardiac pacemaker. The lead, extension, and pulse generator are entirely implanted; there are no visible external wires or hardware. The stimulation parameters (including electrical voltage and stimulation frequency) can be adjusted externally as needed by the physician or trained staff to provide the greatest degree of relief of tremor with the fewest side effects.

The most common side effects of thalamic stimulation are transient paresthesias and paresis. Risks or complications for the surgical implantation are similar to those with other stereotactic brain operations. This technique presumably causes a neuronal electrophysiologic blockade, with effects similar to those of the lesion surgical procedure, thalamotomy.

When the device is turned off, the tremor returns, so that thalamic stimulation appears to have little or no permanent effect. The patient can turn the device on and off with a handheld magnet. It is recommended that the stimulator be turned off during nighttime sleep to conserve pulse generator power. The pulse generator must be replaced every 3 to 5 years.

In clinical trials, patients with ET had significant improvements in dressing, drinking, eating, bathing, and handwriting with thalamic stimulation. In 1997, the Food and Drug Administration in the U.S. approved unilateral thalamic stimulation (Activa) for the treatment of disabling medication-refractory hand tremor in ET and parkinsonian tremor. A direct comparison of the effects and safety of thalamic stimulation to those of thalamotomy has not been undertaken, but stimulation appears to offer the same or similar benefits without permanent neuronal damage.

SUMMARY

ET is the most common movement disorder, affecting millions worldwide. It almost always involves the hands and can be disabling. The primary drugs for therapy are β-blocking agents (propranolol) and primidone. The drugs are not uniformly effective, and side effects are common, especially in the elderly. Because the cause and pathophysiology of ET are not understood, efforts to develop more effective remedies are greatly hindered. Surgical remedies, including thalamotomy and thalamic stimulation, are reserved for persons with severe, disabling, medication-refractory tremor.

SELECTED READING

Biary N, Koller W. Handedness and essential tremor. *Arch Neurol* 1985; 42: 1082-1083.

Busenbark KL, Nash J, Nash S, Hubble JP, Koller WC. Is essential tremor benign? *Neurology* 1991; 41: 1982-1983.

Hubble JP, Busenbark KL, Pahwa R, Lyons K, Koller WC. Clinical expression of essential tremor: effects of gender and age. *Mov Disord* 1997; 12: 969-972.

Hubble JP, Busenbark KL, Wilkinson S, Penn RD, Lyons K, Koller WC. Deep brain stimulation for essential tremor. *Neurology* 1996; 46: 1150-1153.

Janokovic J, Cardoso F, Grossman RG, Hamilton WJ. Outcome after stereotactic thalamotomy for parkinsonian, essential, and other types of tremor. *Neurosurgery* 1995; 37: 680-686.

Koller WC. Long-acting propranolol in essential tremor. *Neurology* 1985; 35: 108-110.

Koller WC, Hubble JP, Busenbark KL. Essential tremor, in *Neurodegenerative Disorders* (Calne DB ed), W. B. Saunders, Philadelphia, 1994, pp. 717-742.

Koller WC, Royse V. Efficacy of primidone in essential tremor. *Neurology* 1986; 36: 121-124.

Larsson T, Sjögren T. Essential tremor: a clinical and genetic population study. *Acta Psychiat Scand* 1960; 36(Suppl 144): 1-176.

Pahwa R, Busenbark K, Swanson-Hyland EF, Dubinsky RM, Hubble JP, Gray C, Koller WC. Botulinum toxin treatment of essential head tremor. *Neurology* 1995; 45: 822-824.

Papa SM, Gershanik OS. Orthostatic tremor: an essential tremor variant? *Mov Disord* 1988; 3: 97-108.

23 Dystonia

Daniel Tarsy, MD

CONTENTS

OVERVIEW
PRIMARY DYSTONIA
SECONDARY DYSTONIA
EVALUATION OF THE PATIENT WITH DYSTONIA
MEDICAL TREATMENT
BOTULINUM TOXIN
SURGICAL TREATMENT
SELECTED READING

OVERVIEW

Dystonia is a syndrome of sustained muscle contractions that produce involuntary twisting and repetitive movements and abnormal postures of the trunk, neck, face, and extremities. Somewhat confusingly, the term "dystonia" has been used to designate a specific type of abnormal movement or posture, a syndrome that occurs secondary to a number of underlying neurologic disorders, and a primary disorder known as "idiopathic torsion dystonia." Dystonic movements can be either slow or rapid and are usually more stereotyped and patterned than movements in chorea (random and unpredictable), myoclonus (brief and shocklike), and tremor (rhythmical and not associated with alterations in posture). Dystonia is characterized by failure of normal reciprocal inhibition, which causes co-contraction of agonist and antagonist muscles accompanied by excessive contraction of other regional muscles. This chapter is limited to conditions characterized by continuous rather than intermittent dystonia and therefore does not include discussion of the paroxysmal dystonias. Psychogenic dystonia is discussed in Chapter 35 together with other functional movement disorders.

Dystonia is usually more prominent during the execution of voluntary movements. Action dystonia is present only during a voluntary movement and absent when the affected body part is at rest. Some dystonias are task-specific, such as the occupational hand cramp disorders; jaw, tongue, or vocal cord spasms during speaking; throat contractions during swallowing; and foot or toe dystonias during walking. As dystonia progresses, it may become activated by movements in body parts other than the involved area. In advanced cases, the dystonia progresses to produce posturing at rest. It differs from contractures,

From: *Parkinson's Disease and Movement Disorders:*
Diagnosis and Treatment Guidelines for the Practicing Physician
Edited by: C. H. Adler and J. E. Ahlskog © Mayo Foundation
for Medical Education and Research, Rochester, MN

Table 23-1
Classification of Dystonia by Age at Onset

Onset	Description
Childhood	Focal onset in leg or arm, usually progressing to the trunk and other limbs
Adult	Focal onset with limited if any progression to adjacent regions

since voluntary or passive movements can still be achieved. Dystonia often increases with stress, anxiety, or fatigue and decreases after rest or sleep. Many patients discover sensory tricks (geste antagoniste), such as touching or holding the affected body part, that partially or completely suppress the dystonia.

Dystonia may be classified by age at onset, topographic distribution, and etiology. Age classification (Table 23-1) is broadly divided into childhood-onset dystonia and adult-onset dystonia. There is a relationship among age at onset, the initially affected body part, and prognosis. Patients with onset in one leg typically are younger than 10 years, whereas those with initial arm or cervical involvement are more likely to be in the early teenage years or older. Evolution into generalized dystonia, with a more rapid rate of progression, occurs in approximately one-half of patients with early-onset leg dystonia but in only 15% of patients with later-onset arm or axial dystonia. By contrast, only about 2% of adult-onset focal dystonias become generalized.

Topographic classification (Table 23-2) includes focal dystonia, involving one limb or body part; segmental dystonia, involving two or more adjacent body parts; multifocal dystonia, involving scattered body parts; hemidystonia, involving the ipsilateral arm and leg; and generalized dystonia, involving trunk, legs, and one other body part. We also classify dystonia simply into primary (inherited and sporadic) and secondary (structural, heredodegenerative, metabolic, and drug-induced) types.

PRIMARY DYSTONIA

Primary dystonia is clinically characterized by the absence of any other associated neurologic abnormalities, with the exception of tremor and occasionally myoclonus. A genetic basis for some primary dystonias has already been discovered and is strongly suspected for others in this group of dystonias. The primary dystonias can be subdivided into generalized and focal dystonias. Childhood-onset dystonia accounts for 10% to 15% and adult-onset dystonia for 85% to 90% of all primary dystonias.

Primary Generalized Dystonia

Also known as "idiopathic torsion dystonia" and formerly as "dystonia musculorum deformans," this disorder begins in childhood or young adulthood. It typically is focal in onset, involving the foot, hand, or axial muscles, and spreads over months or years to involve other body regions. Symptoms in childhood usually appear in one foot, with inversion of the ankle and plantar flexion of the toes during walking, and are absent with the patient seated or lying down. Curiously, the foot dystonia is often alleviated by running, dancing, or walking backwards. Because of lower extremity and truncal involvement, abnormal gait is the most striking manifestation of generalized dystonia. A variety of

Table 23-2
Classification of Dystonia by Distribution

Type of dystonia	Part of body affected
Focal	One body region
Segmental	Adjacent body regions
Multifocal	More than one nonadjacent body region
Hemidystonia	More than one ipsilateral body region
Generalized	Both legs or one leg and trunk plus another body region

abnormal postures may be seen, such as exaggerated abduction of the hip, excessive elevation or hyperextension of the knee, and inversion of the foot. Trunk involvement often produces a characteristic "dromedary" appearance, with excessive flexion at the hips and hyperextension of the neck or upper trunk.

Despite the persistent spasmodic contraction of muscles, pain is surprisingly uncommon in most body regions, with the exception of cervical dystonia, in which pain is prominent in 75% of patients. As the disease progresses, walking becomes increasingly impossible, and patients are sometimes forced to crawl to get around. Over 5 to 10 years, action dystonia may progress to more constant dystonic postures and produce fixed axial, cranial, and extremity deformities that may confine the patient to wheelchair or bed. Nevertheless, except for tremor, the only neurologic signs remain dystonic postures and movements. Tremor is often a rhythmical manifestation of dystonia due to co-contraction of agonist and antagonist muscles. Tremor may also appear intermittently when the patient attempts to move a body part in a direction opposite to the direction of the dystonia. Weakness, spasticity, abnormal reflexes, cerebellar ataxia, eye findings, dementia, or seizures indicate the possible existence of a secondary form of dystonia and should be investigated (see below).

Early-onset primary generalized dystonia is inherited as an autosomal dominant trait in Jewish Ashkenazi families of east European origin as well as in non-Jewish families. A series of genetic linkage studies has mapped the gene for primary torsion dystonia, known as *DYT1*, to chromosome 9. The *DYT1* gene mutation is a guanine-adenine-guanine deletion that results in a glutamate deletion in the protein Torsin A, which resembles the heat shock family of proteins. Genetic testing of potentially affected persons is now possible, even without a positive family history. The mechanism by which this abnormal protein produces dystonia is unknown. Clinical features of early extremity involvement with progression to axial muscles and sparing of cranial muscles are common to all families identified with the *DYT1* gene mutation. This mutation accounts for virtually all cases of primary generalized dystonia in Jewish Ashkenazi families and most cases in non-Jewish families. There is a reduced penetrance rate of approximately 30% to 40% in affected families (thus, having the gene mutation does not guarantee that dystonia will occur), and various degrees of clinical expression are common. The factors accounting for variable phenotypic expression in these families are unknown, although the potential influence of environmental events, such as trauma, and other unknown factors have been suggested.

Primary Focal Dystonia

The clinical expression of primary dystonia in adults is usually above the waist in the cranial muscles, neck, and arm. Primary focal dystonia in adults is estimated to be 10 times more common than early-onset generalized dystonia, with an estimated prevalence of 30 per 100,000, probably an underestimate. Although focal dystonia has long been considered to be sporadic, there is increasing evidence that it may also be genetically determined. Several family studies have found that as many as 25% of examined parents and siblings exhibit focal dystonia not evident from a routine family history. These study findings are consistent with autosomal dominant inheritance with reduced penetrance. Studies of two large non-Jewish families with cervical-cranial dystonia excluded a *DYT1* locus and produced mapping to loci on chromosomes 8 *(DYT6)* in one and 18 *(DYT7)* in the other. At the same time, sporadic cases of cervical-cranial dystonia usually could not be mapped to any of these sites. Thus, genetic heterogeneity is suggested among both individuals and families with adult-onset focal dystonia.

As a group, focal dystonias tend to appear in middle life or later, and with the exception of writer's cramp, focal dystonia is more common in women than in men, shows gradual progression in the first 2 years followed by a static course, and sometimes spreads to involve adjacent body regions. Similar to generalized dystonia, focal dystonias are often activated by voluntary movements, as in writer's cramp, and may be relieved by sensory tricks. As they progress, they become less task-specific and occur during a variety of other motor activities as well as at rest.

CERVICAL DYSTONIA

Cervical dystonia, or spasmodic torticollis, is the most common focal dystonia that comes to medical attention. Although most cases are primary in origin, secondary forms may result from a variety of orthopedic, neurologic, and infectious disturbances of the craniocervical junction, most commonly in children or young adults. Cervical dystonia was once thought to be psychogenic in origin, but psychologic studies and general clinical experience do not support this conclusion. Onset is usually between ages 30 and 50, and women are affected more often than men. Symptoms usually begin with mild neck stiffness and subtle postural deviations of the head. Patients are often treated for more common diagnoses, such as cervical strain, disk disease, and osteoarthritis, before dystonia is considered. As the disorder progresses, posterior cervical and shoulder pain become prominent in 75% of patients. Pain is much more common than in the other focal dystonias, and although the pain is most likely to be musculoskeletal, spondylosis and radiculopathy may complicate long-standing cervical dystonia.

Several abnormal head postures may occur, consisting of various combinations of rotation (torticollis), lateral tilt (laterocollis), head extension (retrocollis), and forward flexion (anterocollis) (Fig. 23-1). Torticollis and laterocollis or combinations of the two are the most common pattern of cervical dystonia. The most common head position is rotation to one side, with upward deviation of the chin, lateral tilt of the head to the opposite side, and ipsilateral shoulder elevation. Pure retrocollis is uncommon in primary cervical dystonia and more common in neuroleptic-induced tardive dystonia. Isolated antecollis is the least common form of cervical dystonia but can be a hallmark of multiple system atrophy, in which other extrapyramidal signs should be obvious. Symptoms are often absent first thing in the morning (the "honeymoon" period) and increase late in the day in association with fatigue. Sensory tricks are often helpful early in the course and may include a light touch on the chin or a hand on the back or side of the head. Head support during reclining or lying down is often helpful, but some patients experience persistent symptoms even while reclining. Most patients exhibit tonic head devia-

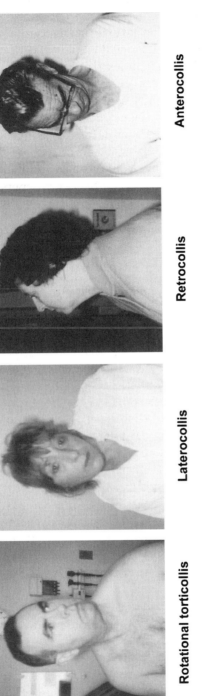

Rotational torticollis   **Laterocollis**   **Retrocollis**   **Anterocollis**

Fig. 23-1. Abnormal head postures in cervical dystonia (left to right): rotation, lateral tilt, head extension, and forward flexion.

tion, but in some cases clonic jerks are present. Head tremor is common and is either a dystonic tremor due to attempts to control the head movements or a tremor related to underlying essential tremor, which may accompany cervical dystonia and may also be evident in the hands.

CRANIAL DYSTONIA

Cranial dystonia comprises focal dystonias of the eyelids, face, jaw, tongue, pharynx, platysma, and vocal cords. Involvement of the eyelids, face, and jaw together is referred to as "Meige's syndrome." As in cervical dystonia, women are affected more often than men, but age at onset is somewhat later, between 40 and 60 years.

Essential blepharospasm is the most common manifestation of cranial dystonia, producing various combinations of increased eyeblink frequency and forced closure of the eyelids. Blepharospasm is bilateral, although it may begin asymmetrically. Unilateral blepharospasm is nearly always due to hemifacial spasm, which is a nondystonic irritative disorder of the facial nerve characterized by twitching of ipsilateral upper and lower facial muscles (see Chapter 24). Contractions of the orbicularis oculi may be brief, causing repetitive blinking, or more persistent, causing prolonged and forceful closure of the eyes or persistent narrowing of the palpebral fissures. Symptoms are often aggravated by driving, bright lights, watching television, or reading and can be severe enough to cause functional blindness. Patients may display either more or less blepharospasm while relating their history than while sitting passively during the examination. Sensory tricks, such as touching the upper lids, voluntary movements of the lower face, coughing, humming, and chewing gum, are often used to overcome blepharospasm. Elevation of the eyebrows due to frontalis contraction and ticlike movements of the lower face and mouth are common. These usually represent abortive attempts at controlling the blepharospasm and do not indicate a diagnosis of Meige's syndrome.

Differential diagnosis should include secondary forms of blepharospasm due to an ocular abnormality, tardive dyskinesia, Parkinson's disease, progressive supranuclear palsy, and rare brain stem lesions. Early in the disorder, patients may report eye irritation or dryness, but primary eyelid or lacrimal gland disorders do not cause chronic blepharospasm. Some patients have apparent inhibition or apraxia of eyelid opening without orbicularis oculi spasm. In contrast to patients with orbicularis oculi spasm, in which the eyebrows are often displaced downward, patients with eyelid opening apraxia have frontalis contraction with prominent elevation of the eyebrows.

Oromandibular dystonia is the second most common manifestation of cranial dystonia. It may occur in isolation but is frequently associated with other cranial dystonias. Jaw muscles may be involved bilaterally, asymmetrically, or unilaterally and produce various combinations of involuntary jaw opening, jaw closing, jaw protrusion, or jaw deviation. Sensory tricks are often used, such as chewing gum; holding a toothpick, pipe, or pencil in the mouth; speaking; and holding one hand beneath the jaw. Associated dyskinesias are common, including platysma contractions, pursing lip movements, tongue protrusion or dyskinesia, spasmodic contractions of the mouth and pharynx, and antecollic neck movements. In early stages, jaw movements may be triggered by speaking or chewing, but eventually they become continuous. Oromandibular dystonia is cosmetically disfiguring, and severe forms may produce major disability, such as jaw pain, dysarthria, difficulty chewing and swallowing, and trauma to oral and dental structures. Differential diagnosis includes tardive dystonia, edentulous jaw movements of the elderly, bruxism, hemimasticatory spasm (a variant of hemifacial spasm), and temporomandibular joint disorders.

Spasmodic dysphonia, or laryngeal dystonia, is an action dystonia in which involuntary adduction or abduction of the vocal cords is precipitated by speech. Again, women are affected more often than men, with onset most commonly between ages 30 and 50 years. Adductor spasmodic dysphonia accounts for 90% of cases and results from approximation of the vocal cords due to excessive contraction of the thyroarytenoid muscles during speech. This produces an effortful and strained voice pattern with frequent voice breaks. There may be an associated voice tremor. Singing may be less impaired than speaking. Having the patient hold a prolonged vowel sound elicits the characteristic voice pattern. In abductor spasmodic dysphonia, separation of the vocal cords results from excessive contraction of the posterior cricoarytenoid muscles and produces a breathy voice pattern. Some patients with adductor dysphonia compensate by voluntarily whispering, and this may somewhat confuse separation of the two forms of dysphonia. Spasmodic dysphonia is often misdiagnosed as psychogenic in origin by physicians unfamiliar with the disorder, but it can be differentiated from psychogenic voice disorders by otolaryngologic examination and formal voice evaluation techniques, such as videolaryngostroboscopy. Differential diagnosis includes primary voice tremor (a manifestation of essential tremor); pharyngeal, jaw, or facial dystonias affecting voice production; and structural or inflammatory disorders of the vocal cords.

Miscellaneous cranial dystonias include rare and isolated dystonias of the pharynx, tongue, and ears. Pharyngeal dystonia causes severe dysphagia and is frequently associated with oromandibular or cervical dystonia. Isolated lingual dystonia is also very rare and is usually associated with oromandibular dystonia.

LIMB DYSTONIA

Limb dystonia, the least common of the primary adult-onset focal dystonias, produces involuntary twisting and other repetitive movements or postures of the upper or lower extremity. When limited to distal portions of the limb, it produces flexion or extension posturing of the hand or fingers and inversion or flexion postures of the foot. If entirely involved, the arm may extend, abduct, elevate, or reach behind the back while the patient is walking. Limb dystonia is often present as part of a segmental dystonia that also involves the neck or trunk. Many upper extremity limb dystonias are action dystonias precipitated by a wide variety of highly skilled motor tasks, such as writing, playing a musical instrument, typing, dart throwing, and holding a golf club, and are referred to as "occupational cramp disorders." In some cases, these remain task-specific dystonias, and in others, they progress to affect other voluntary uses of the hand.

Writer's cramp is the most common task-specific action dystonia. Originally described in scribes in the 19th century, it is not limited to persons who do a lot of writing. Patients present with complaints of slow and awkward handwriting without necessarily noticing excessive muscle contraction. Observation of writing may disclose, individually or in various combinations, excessive squeezing of the pen with the thumb and index finger, involuntary flexion of other digits, extensor posturing of individual fingers away from the pen, involuntary flexion or extension of the wrist, and elevation of the arm off the table. Writing is slow, effortful, and jerky and may be associated with aching discomfort in the forearm. Some persons are unable to hold a pen long enough to write more than a few words. Writing larger on a blackboard by using more proximal muscles is more easily done but impractical for writing on paper. In patients who learn to write with the opposite hand, similar symptoms sometimes appear on that side as well. Hand tremor is sometimes present and, similar to other focal dystonias, may either be a dystonic tremor limited to certain specific hand postures or resemble essential tremor.

Other task-specific dystonias are more commonly occupationally related and occur on a background of repetitive or highly skilled learned movements. Examples among musicians are finger flexion dystonias in pianists, cellists, and guitarists and finger extension dystonias in wind musicians. In musicians, these dystonias may be associated with orthopedic overuse syndromes and entrapment neuropathies; unfortunately, they do not respond well to simple rest. Arm dystonia may progress to interfere with other actions, such as shaving and using tools or utensils.

Unlike in children, primary focal dystonia of the lower extremity is very uncommon in adults and should raise the possibility of a secondary dystonia. The foot and ankle are primarily involved by a combination of inversion and plantar flexion at the ankle and curling of the toes. In early stages, the dystonia may only occur during walking, but often it progresses to a fixed dystonic posture. More proximal involvement may produce hyperextension of the knee and abduction or internal rotation of the hip. Focal limb dystonia is common as a presenting sign of young-onset Parkinson's disease, usually involving the foot. Progressive supranuclear palsy and corticobasal ganglionic degeneration may begin with focal upper extremity dystonia. Differential diagnosis also includes focal lesions of the basal ganglia, such as stroke, tumor, abscess, or other structural lesions capable of producing focal dystonia or hemidystonia.

TRUNCAL DYSTONIAS

This is a group of unusual appearing, uncommon, and relatively isolated dystonias of the trunk, pelvis, abdominal wall, back, and shoulders. As a group, these appear as a complex of repetitive, stereotyped, and sometimes spasmodic movements of one body part with mixed features, which may include rhythmical oscillations, muscle jerks, and torsional movements. These disorders have been differentiated from myoclonus, chorea, and tics and are classified as focal dystonias on the basis of clinical appearance and electromyographic characteristics.

SECONDARY DYSTONIA

There is no current evidence that the *DYT1* gene for primary dystonia is a risk factor for any of the secondary dystonias. It remains uncertain whether peripheral trauma can directly cause dystonia. Trauma to the neck, limb, and vocal cords anecdotally has been related to focal dystonias of those regions. Whether this represents true cause and effect is unclear. These dystonias often occur rapidly shortly after trauma, are present at rest from the outset, and lack a sensory trick. Fixed postures and causalgia are common, and response to treatment is usually poor.

Acquired Structural Lesions

Acquired structural disorders of the brain typically produce a pattern of focal dystonia or hemidystonia, with neuroimaging yielding the structural cause. Patients with hemidystonia often have other neurologic findings, such as hemiparesis, reflex abnormalities, gait disturbance, cerebellar signs, and speech disturbance. Early occurrence of fixed dystonia of an extremity at rest and persistent dystonia in sleep are much more common in secondary than in primary dystonia.

Any structural lesion of the basal ganglia is capable of producing contralateral hemidystonia. The vast majority of these involve the putamen or thalamus, and most are

Table 23-3
Some Heredodegenerative Diseases
That Cause Dystonia

Metabolic disorders
     Wilson's disease
     Aminoacidurias
     Metachromatic leukodystrophy
     Ceroid lipofuscinosis
     Juvenile dystonic lipidosis
     Gangliosidoses
     Lesch-Nyhan syndrome
     Leigh's disease
     Mitochondrial disorders
     Familial basal ganglia calcification

Degenerative disorders
     Parkinson's disease
     Multiple system atrophy
     Huntington's disease
     Progressive supranuclear palsy
     Corticobasal ganglionic degeneration
     Lubag
     Pallidal degenerations
     Hallervorden-Spatz disease
     Neuroacanthocytosis
     Ataxia telangiectasia
     Rapid-onset dystonia-parkinsonism

Data from Weiner WJ, Lang AE. *Movement Disorders: A Comprehensive Survey.* Futura Publishing Company, Mount Kisco, New York, 1989.

readily identified by neuroimaging. More common causes are infarcts or hemorrhage, traumatic brain injury, perinatal brain injury, arteriovenous malformation, anoxia, carbon monoxide poisoning, multiple sclerosis, brain tumor, brain cyst, and focal infectious processes. Miscellaneous toxic insults are more likely to produce axial or bilateral dystonia and often do not produce other neurologic abnormalities. These include manganese toxicity, methanol poisoning, and wasp sting encephalopathy.

Heredodegenerative Diseases

The heredodegenerative diseases include a long list of progressive degenerative or metabolic diseases that produce dystonia in addition to other neurologic abnormalities. They are characterized by pathologic brain abnormalities, many of which directly involve the basal ganglia and produce a variety of other extrapyramidal signs, especially parkinsonism. These are not discussed in detail here but are listed in Table 23-3. They can be broadly divided into (1) hereditary disorders of childhood or young adulthood due to known or unknown metabolic defects and (2) progressive degenerative disorders of unknown cause, most of which occur in adulthood and are characterized by distinctive neuropathology. All these disorders have phenotypes that include dystonia as just one characteristic among other extrapyramidal signs.

Dopa-Responsive Dystonia

Although rare, dopa-responsive dystonia (DRD) is important because of its well-understood genetics, pathophysiology, and marked response to replacement therapy with small amounts of levodopa. Onset is nearly always as a gait disorder in early childhood, but it may appear at any age. The appearance of initial symptoms at about 4 to 8 years of age after normal early development differentiates it from cerebral palsy, with which it is commonly confused. Clinical features usually include a prominent stiff-legged gait associated with dystonic postures of the feet, legs, trunk, arms, or neck. One of its major hallmarks is the increase of symptoms toward the end of the day, known as diurnal fluctuation. Other names for DRD are "Segawa's disease" and "hereditary progressive dystonia." Hyperreflexia, extensor plantar responses, and dystonic postures of the large toes are sometimes present, furthur confusing the disorder with spastic forms of cerebral palsy. Unusual presentations occasionally occur, such as isolated kyphoscoliosis, focal or segmental dystonia, and clinical features suggesting primary muscle disease. If the disorder is left untreated, neurologic signs progress slowly over many years, although more rapid progression has also been described. Some children display parkinsonian features, such as facial masking, bradykinesia, rigidity, and postural disturbance. Parkinsonism may be the only manifestation in the less common adult-onset form of the disease. Both dystonia and parkinsonian signs respond within days to weeks to relatively small doses of carbidopa-levodopa (Sinemet) (25/100 mg, 1 to 6 tablets/day).

DRD is inherited as an autosomal dominant trait and is due to a mutation in the gene for guanosine triphosphate cyclohydrolase I on chromosome 14q. This is the rate-limiting enzyme for synthesis of tetrahydrobiopterin (a cofactor for tyrosine hydroxylase). Multiple point mutations in this gene have been discovered in different families with DRD. Deficiency of this cofactor may account for reduced dopamine synthesis, although it remains possible that biopterin deficiency is a marker for, rather than a cause of, dopamine deficiency. Fluorodopa positron emission tomography (PET), dopamine transporter imaging by single-photon emission computed tomography (SPECT), and neuropathologic studies all show no abnormality of nigrostriatal dopaminergic cells in patients with DRD. DRD is therefore a metabolic disorder due to defective dopamine synthesis, which accounts for its exquisite sensitivity to dopamine replacement therapy even many years after onset of symptoms.

DRD must be differentiated from juvenile Parkinson's disease, which is due to nigrostriatal dopamine cell damage and is associated with abnormal results of PET and SPECT dopamine imaging and a gradually declining response to levodopa, with motor fluctuations. Primary childhood-onset generalized dystonia is differentiated from DRD by the lack of therapeutic response to levodopa and the absence of parkinsonian features. Late adult-onset DRD associated with parkinsonism is suggested by a family history of dystonia, an unusually dramatic and persistent response to low doses of levodopa, and normal results of brain dopamine imaging.

Drug-Induced Dystonia

Drug-induced dystonia can occur after short exposure or long-term therapy. Acute dystonic reactions, including torticollis, opisthotonos, and ocular, oromandibular, or lingual dystonia, frequently occur with dopamine receptor blocking agents, such as haloperidol (Haldol), pimozide (Orap), metoclopramide (Reglan), and prochlorperazine (Compazine). Intravenous administration of benztropine (Cogentin), 1 to 2 mg, or diphenhydramine (Benadryl), 50 to 100 mg, is usually effective for acute dystonic reactions. Long-term treatment with neuroleptic agents may result in tardive dystonia (Chapter 26). Dopaminergic agents, such as levodopa and the dopamine agonists, often cause dystonia in patients with Parkinson's disease.

EVALUATION OF THE PATIENT WITH DYSTONIA

Age at onset, distribution of dystonia, pattern of progression, and presence or absence of other neurologic findings are all important in the evaluation of dystonia. Childhood forms usually begin in the lower extremity and become generalized regardless of whether primary or secondary. Delayed-onset dystonia may occur after perinatal brain injury due to anoxia, trauma, or kernicterus. When evaluating torticollis in a child, one should always check for orthopedic or cervicomedullary junction abnormalities as well as trochlear nerve palsy. Children with hemidystonia should have imaging studies for structural lesions. Generalized dystonia requires a much more extensive evaluation to search for Wilson's disease, inborn errors of metabolism, and structural abnormality. Genetic testing for the *DYT1* gene mutation can also be done.

By contrast, adult forms of primary dystonia nearly always begin above the waist and are usually nonprogressive while remaining focal or segmental. These rarely have an underlying cause, and laboratory and radiologic investigations are rarely indicated. Only rarely does neuroimaging of the brain or spine disclose a structural lesion. If onset is before age 50, serum ceruloplasmin, 24-hour urine sample for copper, and slit-lamp examination for Kayser-Fleischer rings should be pursued to rule out Wilson's disease (Chapter 32). Hemidystonia in a child or adult is strongly indicative of a contralateral basal ganglia abnormality and warrants neuroimaging. Action dystonia is more common in primary dystonia, whereas the early appearance of a fixed dystonic posture usually indicates a secondary dystonia.

Taking a careful history is critical. Diurnal variability in a child should suggest DRD. Birth, developmental, medication, toxin, trauma, and family histories are all very important in identifying causes of secondary dystonia. The medication history may need to include examination of pharmacy records if drug-induced dystonia appears likely. Examination of patients with dystonia should emphasize careful observation under a variety of circumstances, especially while they perform tasks that precipitate dystonia. Variability of symptoms under different circumstances is a hallmark of primary dystonia and does not necessarily indicate a psychogenic cause. If other abnormal neurologic findings are present, a search should be made for an underlying structural or metabolic cause.

MEDICAL TREATMENT

If a specific therapy is available for the underlying cause of dystonia, treatment is straightforward. The two best current examples of this are DRD and Wilson's disease. Because of the importance of making the diagnosis of DRD and the positive response to levodopa treatment, all patients with segmental and generalized dystonia in whom an alternative cause has not been identified should have a therapeutic trial of carbidopa-levodopa. This is especially true since the phenotypic expression of this disorder appears to be so highly variable. These patients have a response to low doses of carbidopa-levodopa (25/100 mg, 1 to 6 tablets/day) given over only a few weeks. Wilson's disease is treated with chelating agents that deplete body stores of copper or inhibit its absorption (Chapter 32). Drug-induced or toxic dystonias are treated by removing the offending agent. In tardive dyskinesia, this step is not immediately effective and furthur therapeutic strategies may be required (Chapter 26).

Unfortunately, in most patients with either primary or secondary dystonia, the pharmacology of the disorder is unknown and treatment is nonspecific and symptomatic. A large number of drugs with a variety of pharmacologic actions have been reported to ameliorate dystonia (Table 23-4). However, most of these reports are anecdotal, and results are much less consistent when the drugs are administered to larger numbers of patients. The key to use of oral medications is to start at a low dose, slowly titrate upward to minimize side effects, use the lowest effective dose, and warn recipients of the need to be patient while awaiting a response. There is no "quick fix" with oral medications. All drugs should be given in divided doses throughout the day.

Anticholinergic drugs are the most common systemic agents used in the treatment of dystonia, but benefit is achieved in only 40% to 50% of patients. Children have more frequent and dramatic improvement with fewer side effects than adults. Peripheral antimuscarinic side effects, such as dry mouth, blurred vision, constipation, and urinary retention, may at least in part be managed by coadministration of peripheral cholinergic agents, such as pilocarpine eyedrops and oral pyridostigmine (Mestinon). Central nervous system side effects, such as memory impairment, confusion, and hallucinations, cannot be prevented in this way and require a lowering of dose. The mechanism of action of anticholinergic drugs in dystonia is unknown.

Benzodiazepines, particularly clonazepam (Klonopin), 1 to 4 mg daily, are widely used in the treatment of dystonia but have not been subjected to formal study. They appear to be effective in only about 15% of patients. Their mechanism of action is unknown, but it is likely to be nonspecific by virtue of their effects on anxiety, pain, and muscle spasm. Oral baclofen (Lioresal), 30 to 120 mg/day, in uncontrolled studies produced marked therapeutic benefit in a small proportion of patients with focal and generalized dystonias. However, large doses are usually required, which may cause sedation and effects of drug withdrawal if administration is discontinued. With an effect similar to that in spasticity, intrathecal baclofen has been used successfully in a small number of patients with dystonia, and further trials of this approach are in progress. Treatment begins with a test dose of 25 to 250 μg intrathecally. If there is a convincing clinical response, a permanent pump can be implanted for continuous intrathecal infusion of baclofen.

Dopamine receptor blocking drugs, such as the phenothiazines, haloperidol, and pimozide, are of variable efficacy in dystonia and in recent years have fallen out of favor because of concern for tardive dyskinesia. Tetrabenazine (Nitoman), 50 to 200 mg daily, is a dopamine-depleting agent available in Europe and Canada that has not been reported to cause tardive dyskinesia and is sometimes used in place of a neuroleptic agent. However, it also has dopamine receptor blocking properties and may cause sedation and depression. Unfortunately, this drug is not available in the United States. It is probably most effective in the treatment of tardive dyskinesia. Metyrosine (Demser) also depletes catecholamines and may be of benefit but with potential side effects of hypotension, depression, and sedation. Carbamazepine (Tegretol), gabapentin (Neurontin), and mexiletine (Mexitil) have also been used in the treatment of focal dystonia but with generally poor results.

The order of introducing drugs in a given patient should be individualized, but on the basis of experience in large dystonia clinics, a trial of levodopa should be followed serially by an anticholinergic agent, baclofen, and a benzodiazepine. In some patients, combination therapy with two or more of these drugs may be considered.

Table 23-4
Oral Medications for Dystonia

	Start, mg/day	Escalate, mg/day/wk	Daily dose, mg/day
Dopaminergic			
Carbidopa-levodopa (Sinemet)	25/100	25/100	25/100 to 150/600
Anticholinergic			
Trihexyphenidyl (Artane)	1	1	6 to 80
Benztropine (Cogentin)	0.5	0.5	4 to 8
Benzodiazepine			
Clonazepam (Klonopin)	0.25	0.25	1 to 4
Diazepam (Valium)	2	2	2 to 40
Baclofen (Lioresal)	10	10	30 to 120
Dopamine depletor			
Metyrosine (Demser)	250	250	1,000
Anticonvulsants			
Carbamazepine (Tegretol)	200	200	600 to 1,200
Dopamine antagonists	Depends on drug used		

BOTULINUM TOXIN

The use of botulinum toxin (BTX) has revolutionized the management of focal dystonia. In brief, BTX is a neurotoxin that inhibits neuromuscular transmission. It is injected into dystonic muscles, causing mild, temporary muscle weakness that reduces the dystonia. The effect lasts a few months, and then reinjection is necessary. The most significant, but relatively uncommon, side effect is overweakening of the injected or adjacent muscles, leading to focal dysfunction (Chapter 31). Administration of BTX has become the treatment of first choice for most adult-onset focal dystonias because of the limited efficacy of other medical treatments, desire to avoid systemic side effects, and the good results achieved. Its direct effect by injection into symptomatic muscles without producing systemic side effects is inherently appealing to patients and accounts in large part for its success.

BTX is used in torticollis, blepharospasm, and spasmodic dysphonia for first-choice treatment, with 75% to 95% of patients having a response. It is also very effective for jaw-closing and jaw-deviation forms of oromandibular dystonia but is less effective for jaw-opening forms of this disorder. BTX is increasingly used for patients with limb dystonia, including children with dystonic or spastic forms of cerebral palsy with gait disturbances. It is often useful for relatively isolated task-specific limb dystonias, such as writer's or musician's cramp, although the success rate in this condition is less than that for other focal dystonias. The limiting factor is overweakening of the hand muscles, which may be more distressing to the patient than the dystonia.

SURGICAL TREATMENT

In the 1950s, following earlier successes in the treatment of Parkinson's disease, thalamotomy was introduced for the treatment of dystonia. A variety of targets in the thalamus were selected for creation of lesions, the most common site being the posterior ventrolateral nucleus. Variable results of surgery among different centers, especially in patients with axial dystonia; poor documentation of outcomes; and a disturbingly high incidence of neurologic side effects led to waning use of these procedures. Recently, several factors led to renewed interest in stereotactic procedures for treatment of dystonia. These include increased understanding of functional relationships within the basal ganglia, improved localization because of new neuroimaging and microelectrode recording technology, reversal of levodopa-induced dystonia and dyskinesia by pallidotomy in Parkinson's disease, and the advent of high-frequency, deep-brain stimulation techniques. As a result, several centers are beginning to carefully reexamine the effects of thalamotomy, pallidotomy, and high-frequency thalamic and pallidal deep-brain stimulation in dystonia. Currently, patients with hemidystonia unresponsive to oral medication and BTX are being considered for these procedures.

Peripheral denervation procedures include bilateral anterior cervical rhizotomy, selective peripheral denervation (posterior ramisectomy), microvascular decompression of the accessory nerve for treatment of cervical dystonia, lysis of peripheral branches of the facial nerve or myectomy for treatment of blepharospasm, and laryngeal nerve section for treatment of spasmodic dysphonia. Since the introduction of BTX, the last three operations have been largely abandoned and rhizotomy is currently done infrequently because of side effects of excessive weakness, pain, and dysphagia. Selective peripheral denervation of the posterior rami is designed to denervate involved muscles while sparing others to preserve as much normal cervical mobility as possible. With experienced surgeons, this procedure has had a high rate of success with minimal morbidity. However, it is laborious and requires considerable experience, careful preoperative clinical and electromyographic evaluation, and extensive postoperative physical therapy. Currently, the procedure is nearly always reserved for patients in whom treatment with BTX and oral medications has failed.

SELECTED READING

Bressman SB, de Leon D, Kramer PL, Ozelius LJ, Brin MF, Greene PE, Fahn S, Breakefield XO, Risch NJ. Dystonia in Ashkenazi Jews: clinical characterization of a founder mutation. *Ann Neurol* 1994; 36: 771-777.

Chan J, Brin MF, Fahn S. Idiopathic cervical dystonia: clinical characteristics. *Mov Disord* 1991; 6: 119-126.

Fahn S. Concept and classification of dystonia. *Adv Neurol* 1988; 50: 1-8.

Furukawa Y, Shimadzu M, Rajput AH, Shimizu Y, Tagawa T, Mori H, Yokochi M, Narabayashi H, Hornykiewicz O, Mizuno Y, Kish SJ. GTP-cyclohydrolase I gene mutations in hereditary progressive and dopa-responsive dystonia. *Ann Neurol* 1996; 39: 609-617.

Greene P, Shale H, Fahn S. Analysis of open-label trials in torsion dystonia using high dosages of anticholinergics and other drugs. *Mov Disord* 1988; 3: 46-60.

Hallett M. The neurophysiology of dystonia. *Arch Neurol* 1998; 55: 601-603.

Jankovic J, Brin MF. Therapeutic uses of botulinum toxin. *N Engl J Med* 1991; 324: 1186-1194.

Jankovic J, Ford J. Blepharospasm and orofacial-cervical dystonia: clinical and pharmacological findings in 100 patients. *Ann Neurol* 1983; 13: 402-411.

Lang AE. Dopamine agonists and antagonists in the treatment of idiopathic dystonia. *Adv Neurol* 1988; 50: 561-570.

Lang AE, Sheehy MP, Marsden CD. Anticholinergics in adult-onset focal dystonia. *Can J Neurol Sci* 1982; 9: 313-319.

Marsden CD. The focal dystonias. *Clin Neuropharmacol* 1986; 9 Suppl 2: S49-S60.

Marsden CD, Obeso JA, Zarranz JJ, Lang AE. The anatomical basis of symptomatic hemidystonia. *Brain* 1985; 108: 463-483.

Marsden CD, Quinn NP. The dystonias. *BMJ* 1990; 300: 139-144.

Nygaard TG, Marsden CD, Fahn S. Dopa-responsive dystonia: long-term treatment response and prognosis. *Neurology* 1991; 41: 174-181.

Rivest J, Quinn N, Marsden CD. Dystonia in Parkinson's disease, multiple system atrophy, and progressive supranuclear palsy. *Neurology* 1990; 40: 1571-1578.

Tsui JKC, Calne DB. *Handbook of Dystonia*. M. Dekker, New York, 1995.

Waddy HM, Fletcher NA, Harding AE, Marsden CD. A genetic study of idiopathic focal dystonias. *Ann Neurol* 1991; 29: 320-324.

Weiner WJ, Lang AE. *Movement Disorders: A Comprehensive Survey*. Futura Publishing Company, Mount Kisco, New York, 1989.

24

Hemifacial Spasm

Mark F. Lew, MD

CONTENTS

GENERAL DESCRIPTION

Hemifacial spasm (HFS) is one of the most common craniofacial movement disorders. It differs from most other movement disorders because the underlying cause is routinely identifiable. Localization is generally to a lower motor neuron abnormality of the peripheral nervous system. In most cases, a blood vessel is compressing the seventh cranial nerve as it exits the brain stem. This peripherally induced facial spasm may also be considered to be segmental myoclonus.

HFS has been described in the past as "tic convulsif" (see Differential Diagnosis), hemispasm, reflex facial spasm, and the seventh disease. The syndrome is characterized by an involuntary, episodic, synchronous contraction of the muscles innervated by the facial nerve on one-half of the face. The disorder generally begins with intermittent spasms of the eyelids (orbicularis oculi). The main complaint is usually eye spasms that cause the eye to close and can interfere with activities like reading, driving, and watching television. The spasms may spread over time to include the lower facial musculature and can cause mouth and cheek pulling and twitching. Some patients notice a pulse of tinnitus with the facial spasm, caused by contraction of the stapedius muscle. Others may have spasms of the platysma muscle in the neck. Both of these muscles are controlled by the facial nerve.

From: *Parkinson's Disease and Movement Disorders:*
Diagnosis and Treatment Guidelines for the Practicing Physician
Edited by: C. H. Adler and J. E. Ahlskog © Mayo Foundation
for Medical Education and Research, Rochester, MN

The spasm may begin as a fleeting clonic burst and worsen to become tonic, lasting from several seconds to longer than 1 minute. This involuntary facial dyskinesia worsens with fatigue, anxiety, or stressful situations. The jerking spasms are irregular and intermittent. Hours or days may pass with patients noticing little to no spasm. Rare instances of spontaneous remission lasting weeks to years have been described. In general, the condition is chronic and lifelong, necessitating continuing therapy. In long-standing occurrences, facial weakness can be appreciated in approximately one-half of those afflicted.

HFS rarely occurs bilaterally, with the second side becoming affected months to more than a decade later. Bilateral spasm is never synchronous. Familial cases exist in the literature, but the vast majority of occurrences are spontaneous.

HFS may be present during sleep but is decreased. Some patients state that they are clearly awakened by the facial spasm. Any voluntary facial expression or movement, such as chewing, eating, smiling, or laughing, may provoke attacks. At presentation, many patients have mild facial weakness or facial contracture related to tonic muscle activity. Involvement of a cranial nerve other than the facial nerve, such as the fifth, eighth, ninth, or tenth cranial nerve, suggests a space-occupying lesion. HFS causes patients functional compromise, cosmetic deformity, and social embarrassment. Nevertheless, coexisting psychopathology and illness-related life changes are relatively lacking.

HFS occurs in adults and has only rarely been seen in persons younger than 20 years (see Differential Diagnosis). The average age at onset is between the middle fifth and the early sixth decades, and the range is between the third and the eighth decades. The female-to-male ratio is 3:2.

PATHOPHYSIOLOGY

HFS can be caused rarely by central nervous system lesions, such as multiple sclerosis, or other intraparenchymal brain stem abnormalities. Electrophysiologic monitoring of HFS reveals that F waves are abnormally enhanced. F waves, which result from the backfiring of antidromically activated motor neurons of the facial motor nucleus, are indices of excitability in this nucleus. Measurement of blink reflexes and abnormal muscle responses (lateral spread), characteristic signs in HFS, has been used to investigate the pathophysiologic mechanism of HFS.

Surgical studies have evaluated electromyographic responses evoked by the facial nerve, blink reflexes, and abnormal muscle responses both preoperatively and postoperatively. F waves and blink reflexes were enhanced before surgery, and all patients had lateral spread on the spasm side but not on the normal side. Postoperatively, hyperexcitability of the facial nerve nucleus did not resolve immediately and in some patients with clinically resolved spasm, did not occur for months afterward (up to 2 years). Therefore, repeat surgical procedures are not recommended for at least this length of time.

Two principal theories exist on the irregular nerve transmission and muscle spasm. The first implicates ephaptic transmission of the hyperactive and abnormal signals originating from a small area of demyelination of the nerve trunk secondary to the vascular impingement. This causes "short-circuiting" of adjacent nerve fibers. The second theory states that the automatic or abnormal signals originate in the facial nerve nucleus itself and are related to vascular compression. A better explanation may combine

Table 24-1
Differential Diagnosis of Hemifacial Spasm (HFS)

	Central nervous system	Peripheral nervous system	Present during sleep	Unilateral	Bilateral	Other cranial nerves
HFS	- (rare)	+	+	+	-	-
Tardive dyskinesia	+	-	-	- (rare)	+	+
Tics	+	-	+	+	+	+
Dystonia	+	-	-	-	+	+
Epilepsy	+	-	+	+	+	+
Chorea	+	-	-	-	+	+

both theories, assuming that a partial lesion of the facial nerve reorganizes the neuronal machinery within the facial nerve central motonucleus.

DIFFERENTIAL DIAGNOSIS

Although originally thought to be idiopathic, most instances of HFS are due to a blood vessel compressing cranial nerve (CN) VII at the root entry zone into the brain stem. Usually, the vertebral, anterior inferior cerebellar, or posterior inferior cerebellar artery is the cause. The differential diagnosis of HFS (Table 24-1) also involves any space-occupying lesion of the facial nerve or cerebellopontine (CP) angle. Included are posterior fossa aneurysms, CN VII schwanommas, temporal bone cholesteatomas, eighth nerve tumors (such as an acoustic neuroma), lipomas, meningiomas of the petrous ridge, CP angle arachnoiditis or astrocytomas, and parotid gland masses or inflammation. Bony abnormalities, such as Paget's disease and basilar impression, are less frequent etiologic factors. HFS has rarely occurred as a result of benign intracranial hypertension, subarachnoid cysticercal cysts, and intrauterine facial nerve damage (one case of congenital HFS). HFS in a patient with Lyme disease and HFS caused by a supratentorial occipital falcine meningioma have been reported. Facial synkinesia related to previous facial trauma, surgery, or idiopathic seventh nerve palsy (Bell's palsy) can closely resemble HFS. The manifestations are eye spasms when the mouth moves (chewing, smiling) or mouth and cheek spasms when the eye blinks or closes. Synkinesis is due to aberrant regeneration of the damaged facial nerve.

Focal epilepsy localized to the facial representation of the motor strip needs to be considered. These movements, however, are routinely very regular, and other findings on examination might confirm a diagnosis of seizure. An unusual epileptic syndrome of cerebellar origin related to a cerebellar ganglioglioma has been described as hemifacial seizures in infants. The child has unilateral facial contraction, head and eye deviation, nystagmus, autonomic dysfunction, and retained consciousness. Fewer than 1% of patients with HFS are younger than 30. In this population, a compressing vascular structure is frequently not the cause. Thickening of the surrounding arachnoid and focal compression commonly exist.

A rare syndrome called "painful tic convulsif" describes the simultaneous occurrence of HFS and trigeminal neuralgia related to compression of the facial and trigeminal nerves by a CP epidermoid tumor. Rarely, CP angle epithelial cysts have been reported, as have cholesterol granulomas at the petrous apex. Contralateral space-occupying masses can cause HFS, which appears as a false localizing sign.

HFS should be clinically distinguishable from cranial dystonias. Blepharospasm (eye or upper face dystonia) and Meige's syndrome (upper and lower facial dystonia) are bilateral and rarely profoundly asymmetrical. The lower face dystonias usually result in jaw or tongue movements, sites not involved in HFS. Dystonia completely remits with sleep. Other movement disorders that may be confused with HFS but that substantially differ from it are facial tics (can be voluntarily suppressed, are stereotypical, and involve a nerve other than CN VII), facial myokymia (facial quivering with brain stem lesions or multiple sclerosis), tardive dyskinesias (rarely unilateral and mostly dystonic or choreiform), and hemimasticatory spasm (a fifth nerve abnormality).

DIAGNOSTIC TESTING

Routine evaluation for HFS should include magnetic resonance imaging (MRI) to rule out mass lesions in the posterior cranial fossa, which are present in approximately 5% of patients. MRI also rules out intra-axial brain stem lesions, such as multiple sclerosis. Magnetic resonance angiography may help find a blood vessel in the CP angle. Magnetic resonance angiography with maximum intensity projection technique has been shown to reliably allow recognition and characterization of the neurovascular conflict. Magnetic resonance tomographic angiography, a technique that uses computer reconstruction of magnetic resonance tomographic images, has been found to be more sensitive and specific than magnetic resonance angiography in identifying vascular compression in HFS. Because a small percentage of control patients have vascular compression radiographically, other factors may contribute to HFS.

Blink reflex and facial electromyographic studies may identify damage to the facial nerve. In some instances, an electroencephalogram may be useful to rule out focal motor epilepsy.

MEDICAL THERAPY

Pharmacologic therapy for HFS offers modest benefit at best. Empirical treatment with carbamazepine, phenytoin, clonazepam, other benzodiazepines, baclofen, and gabapentin often provides temporary but not significant long-term benefit (Table 24-2). Little literature exists on double-blind, placebo-controlled trials for any of these medications. Unlike botulinum toxin injections (see below), oral medications must be taken several times a day and have many potential side effects.

BOTULINUM TOXIN AND OTHER INJECTABLES

Intramuscular injection of botulinum toxin type A (BTX-A) has become the treatment of choice for HFS for more than a decade. Local injections of BTX-A into active facial musculature offer a useful alternative to surgery. BTX-A is a bacterial exotoxin that binds to the presynaptic membrane of the neuromuscular junction and inhibits the release of acetylcholine. The result is chemical denervation and weakness of muscles injected. Most patients receive substantial benefit within 48 to 72 hours after an injection. Benefit peaks by 2 to 3 weeks and wanes, on average, 3 to 4 months later, with 6 months

Table 24-2
Treatment of Hemifacial Spasm

Medical

	Starting dose, mg	Maximum dose, mg	Side effects
Carbamazepine (Tegretol)	100 q.d.	200 t.i.d.	Dizziness, nausea, drowsiness
Phenytoin (Dilantin)	200 h.s.	400 h.s.	Nystagmus, ataxia
Clonazepam (Klonopin)	0.25 h.s.	0.5-1 t.i.d.	Sedation, ataxia
Baclofen (Lioresal)	10 q.d.	20 q.i.d.	Nausea, sedation
Gabapentin (Neurontin)	100-300 q.d.	600 q.i.d.	Somnolence, dizziness

Surgical

    Microvascular decompression

    Endoscopic microvascular decompression

Injection therapy

    Botulinum toxin type A (Botox)

    Doxorubicin (experimental)

as the greatest extent. Most patients describe a 75% to 95% reduction in symptoms. Complete remission after injection with BTX-A has been reported rarely.

Side effects of injections for HFS are generally related to injections into the orbicularis oculi. Complications of treatment include weakening of the levator muscle, resulting in ptosis, and lagophthalmos, or weakness of lid closure (orbicularis oculi). The complications of the latter include possible exposure keratitis, dry eyes, blurred vision, and hypersecretion epiphora (excessive tearing). Complications are minimal and transient, lasting days to weeks, with complete resolution in time. Other side effects are double vision (rare) and drooping of the lower face or mouth. Long-term use of BTX-A (more than a decade) has been documented to have continuing efficacy, with rare development of tolerance or tachyphylaxis.

Recently, use of the chemotherapeutic agent doxorubicin was reported for the treatment of HFS. In a nonrandomized study, six of nine patients tolerated doxorubicin chemomyectomy. The main cause of withdrawal from the study was skin inflammation. Five of the six patients required no further treatment after a course of injections several months apart. One patient required intermittent BTX-A injections. Further work is under way to evaluate cotreatments to improve skin tolerance and long-term efficacy of this treatment.

SURGICAL THERAPY

Historically, many different extracranial procedures have been performed for HFS, including sectioning of the entire seventh nerve root or trunk, root or distal neurotomy, and thermolysis or chemolysis (with alcohol or phenol) of the root or nerve. Results were variable at best, and morbidity was significant.

Currently favored surgical intervention for the treatment of HFS involves microvascular decompression of the seventh cranial nerve at the entry or exit zone of the nerve. Polytef (Teflon) microsponges are often placed between the offending vessel and the facial nerve. Proline, polyethylene terephthalate (Dacron), and sequestered blood products are used less commonly. The most common offending vascular loops are the anterior inferior cerebellar artery, the posterior inferior cerebellar artery, the vertebral artery, and combinations including the vertebral artery. Tortuosity, elongation, or dilatation of the artery or arterial variant is the cause of compression. Narrowing of the CP angle cistern has recently been reported as a contributory factor in HFS. Rare complications of the surgical use of polytef include the formation of a polytef-induced giant cell granuloma compressing the eighth nerve and causing unilateral, progressive asymmetrical loss of hearing.

Intraoperative electrophysiologic monitoring of both the seventh and the eighth nerves has been helpful in identifying the specific causative vessel and also in decreasing the surgical morbidity. From 65% to 75% of patients have complete cessation of facial twitching, 25% to 35% have partial improvement, and 5% experience surgical failure. Approximately 10% of patients operated on who have immediate cessation of movement have some recurrence of facial spasm. This generally occurs within the first 2 years after the procedure. A second microvascular decompression may help these patients. Only 1% of patients who remain free of spasms after 2 years have recurrent HFS.

The most common surgical morbidities are both temporary and permanent facial nerve damage and hearing impairment. Less commonly, dysequilibrium, sensory changes over the scalp, anosmia, meningitis (both bacterial and aseptic), cerebrospinal fluid rhinorrhea, otitis media, and wound infection can occur. Rarely, brain stem strokes or even death has been reported. Approximately 25% of patients have transient complications, and fewer than 10% have permanent postoperative sequelae. Oculostapedial synkinesis (ear clicking with eyelid closure) has been described as a rare morbidity of surgery. More experimental techniques have been described, including a retrosigmoid endoscopic approach to microvascular decompression, neurointerventional catheterization to identify the compressive artery, and three-dimensional MRI to guide surgical exploration. Transtympanic needling of the facial nerve has been recommended more recently.

SUMMARY

Hemifacial spasm is characterized by unilateral spasms of the upper and lower facial muscles innervated by CN VII. Most occurrences are caused by a blood vessel compressing CN VII as it exits the brain stem. MRI can rule out other causes of compression. The treatments of choice are botulinum toxin injections and microvascular decompressive surgery, although some patients obtain benefit from oral medications.

SELECTED READING

Adler CH, Zimmerman RA, Savino PJ, Bernardi B, Bosley TM, Sergott RC. Hemifacial spasm: evaluation by magnetic resonance imaging and magnetic resonance tomographic angiography. *Ann Neurol* 1992; 32: 502-506.

Alexander GE, Moses H III. Carbamazepine for hemifacial spasm. *Neurology* 1982; 32: 286-287.

Brin MF, Fahn S, Moskowitz C, Friedman A, Shale HM, Greene PE, Blitzer A, List T, Lange D, Lovelace RE, McMahon D. Localized injections of botulinum toxin for the treatment of focal dystonia and hemifacial spasm. *Adv Neurol* 1988; 50: 599-608.

Evidente VG, Adler CH. Hemifacial spasm and other craniofacial movement disorders. *Mayo Clin Proc* 1998; 73: 67-71.

Ishikawa M, Ohira T, Namiki J, Gotoh K, Takase M, Toya S. Electrophysiological investigation of hemifacial spasm: F-waves of the facial muscles. *Acta Neurochir (Wien)* 1996; 138: 24-32.

Janetta PJ. Cranial rhizopathies, in *Neurological Surgery: A Comprehensive Reference Guide to the Diagnosis and Management of Neurosurgical Problems* (Youmans JR ed), 3rd ed, W. B. Saunders, Philadelphia, 1990, pp. 4169-4182.

Jankovic J, Schwartz K, Donovan DT. Botulinum toxin treatment of cranial-cervical dystonia, spasmodic dysphonia, other focal dystonias and hemifacial spasm. *J Neurol Neurosurg Psychiatry* 1990; 53: 633-639.

Tsui JK. Botulinum toxin as a therapeutic agent. *Pharmacol Ther* 1996; 72: 13-24.

Wilkins RH. Hemifacial spasm: a review. *Surg Neurol* 1991; 36: 251-277.

Wirtschafter JD, McLoon LK. Long-term efficacy of local doxorubicin chemomyectomy in patients with blepharospasm and hemifacial spasm. *Ophthalmology* 1998; 105: 342-346.

25

Huntington's Disease
and Other Choreas

John N. Caviness, MD

CONTENTS

Chorea consists of continuous, unsustained, nonstereotyped movements of variably changing speed and direction that seemingly flow from one muscle group to another, thus giving the appearance of "dancing." The differential diagnosis of chorea is large, consisting of many diseases and disparate causes (Table 25-1). Most causes of chorea are hypothesized to involve basal ganglia dysfunction. Physiologically, hypoactivity in the subthalamic nucleus and medial pallidum is thought to correlate with chorea.

NEURODEGENERATIVE DISORDERS

Huntington's Disease

Huntington's disease is a progressive disorder characterized by chorea, mental status changes, and autosomal dominant inheritance. The symptoms of chorea and mental status changes may occur simultaneously or in succession. The average age at onset is 35 to 45 years, but a wide range of onset ages is common (childhood to more than 80 years). Because onset typically is chronic, precise age at onset often is difficult to ascertain. The medical history may reveal mildly excessive movements or personality changes months to years before medical attention is sought. Family and acquaintances may consider such subtle manifestations to be "normal" for that person. Duration of symptoms also varies widely, with an average period of 10 to 20 years. Onset at younger ages often is associated with a more rapidly progressive course.

From: *Parkinson's Disease and Movement Disorders:*
Diagnosis and Treatment Guidelines for the Practicing Physician
Edited by: C. H. Adler and J. E. Ahlskog © Mayo Foundation
for Medical Education and Research, Rochester, MN

Table 25-1
Common Causes of Chorea

Neurodegenerative disorders
    Huntington's disease
    Dentatorubropallidoluysian atrophy
    Cerebellar system degenerations

Lesions of the basal ganglia
    Vascular (stroke)
    Neoplastic
    Infectious
    Inflammatory

Drugs
    Tardive syndromes (dopamine antagonists)
    Antiparkinsonian agents
    Stimulants
    Opiates
    Antiepileptic agents
    Exogenous hormones (estrogens)

Metabolic conditions
    Wilson's disease
    Hyperthyroidism
    Hyperglycemia
    Hypoglycemia
    Electrolyte disorders

Other systemic disorders
    Sydenham's chorea
    Lupus erythematosus
    Polycythemia vera
    Neuroacanthocytosis
    Chorea gravidarum
    Acquired hepatocerebral degeneration

Essential chorea syndromes
    Benign familial chorea
    Senile chorea

Paroxysmal chorea
    Paroxysmal kinesigenic choreoathetosis
    Paroxysmal dystonic choreoathetosis

CHOREA

Initially, chorea is mild and may involve only the face or hands. Some persons at first appear fidgety or restless. The movements may be merged into intentional gestures, seeming to be semipurposeful or unusual mannerisms. One must be careful not to overestimate the significance of subtle movements, since nervousness and fear in the doctor's office may cause a patient to appear fidgety or restless. As time passes, the amplitude of the movements becomes more obviously abnormal. Subsequently, most of the musculature is usually involved.

It is useful to observe for chorea when interviewing the patient. When the examination begins, the patient, with eyes closed, should be observed for at least 1 minute, with and without mental activation. Chorea can cause an inability to maintain a constant level of postural tone when the tongue is protruded ("flycatcher's tongue"), the examiner's hand is gripped ("milkmaid's grip"), or the hands are held out supine ("dishing"). Deep tendon reflexes may exhibit delayed relaxation ("hang up"). Chorea can give the gait a hyperkinetic character, with random limb and trunk movements as well as lurching. Although the gait can appear quite clumsy, the person with hyperkinesis can maintain balance quite well in the early stages of the disease. Hyperkinetic characteristics also can affect speech, producing abnormal articulatory movements and irregular variations in voice loudness. Chorea in other disorders can manifest itself in a similar manner.

In addition to chorea, other movement disorders and related phenomena can be seen. Juvenile patients (onset before age 20) often have a variable combination of parkinsonism, dystonia, seizures, myoclonus, and ataxia. Tics have been noted in some patients. As Huntington's disease progresses, hyperkinetic movements often give way to a hypokinetic state of parkinsonism and fixed dystonic postures. Balance problems may be due to chorea, parkinsonism, dystonia, or impaired postural reflexes. Likewise, repetitive movements of the fingers, hands, and feet may show disruption due to hyperkinetic or hypokinetic abnormalities or both types. Swallowing dysfunction is one of the end-stage characteristics of the disease. Eye movement abnormalities consist of the slowing and limited excursion of both voluntary pursuits and saccades.

MENTAL STATUS

Personality changes, memory loss, and attention deficits are signs of evolving dementia. In many patients, depression is prominent. Other mental status problems are psychotic behavior, anxiety, impulsiveness, severe mood swings, irritability, obsessive behavior, and aggression. Suicidal ideation is not uncommon in Huntington's disease. The dementia is progressive and can lead to an inability to handle personal affairs. Remarkably, many patients, even those with end-stage disease, can recognize people and things quite well.

AUTOSOMAL DOMINANT INHERITANCE

Autosomal dominant inheritance implies a 50% risk of passage from an affected person to a child. The disease is fully penetrant and does not skip generations. If a generation is unaffected by the disease, the disease will not show up in subsequent generations unless the mutation is reintroduced into the family. The gene for Huntington's disease has been identified on chromosome 4. It codes for a protein that has been designated "huntingtin." Although its function is not yet known, the gene product seems to be widely expressed in the human body, including non-neural tissues. The mutation is characterized by an expanded CAG trinucleotide repeat within the gene product. When males transmit the mutation, there can be greater repeat length expansion, and symptoms may develop in the child at an earlier age ("anticipation"). The expanded stretch of CAG repeats results in an increased polyglutamine string that is believed to create a toxic property for the huntingtin protein.

PATHOPHYSIOLOGY

The basal ganglia are important for the scaling of amplitude and velocity of movements. In Parkinson's disease, the lack of dopaminergic stimulation causes a reduction in the ratio of excitatory to inhibitory activity in the basal ganglia loops, which in turn prevents the proper facilitation of movement. The use of levodopa and other antiparkinsonian agents restores this ratio closer to normal. Anticholinergic medication has a similar effect. The existence of excitatory and inhibitory nuclei in the basal ganglia also provides the background for the usefulness of pallidotomy in Parkinson's disease.

Huntington's disease is the prototypical condition for a situation in the basal ganglia that is exactly opposite to that in Parkinson's disease. There is selective vulnerability for which neurons undergo degeneration in the basal ganglia. This causes an increase in the ratio of excitatory to inhibitory activity in the basal ganglia loops, which in turn creates an exaggerated facilitation of movement that is apparent in patients with the chorea of Huntington's disease. As one would expect, dopaminergic and anticholinergic agents can worsen the chorea in Huntington's disease. Some stimulants can have a similar effect. In later stages of the disease, all basal ganglia loops are affected, and the result is a clinical state more similar to parkinsonism. Much less is known about the pathophysiology of the mental status changes in Huntington's disease. The cortex is involved in Huntington's disease, and pathologic effects there may combine with basal ganglia dysfunction to produce the mental status changes.

DIAGNOSIS

The diagnosis of Huntington's disease has been simplified somewhat by the commercial availability of a direct genetic test. With genetic testing, however, new problems and dilemmas have appeared. Although a CAG repeat number of >40 in the huntingtin gene is diagnostic for Huntington's disease, several important issues need to be considered before genetic testing is performed. First, does the person receiving genetic testing have symptoms? Asymptomatic testing is done in persons genetically at risk for Huntington's disease. If one parent had the disease, each child, through autosomal dominant inheritance, is at 50% risk for carrying the gene and being considered presymptomatic for the disease. The person at risk may or may not have definite symptoms. If not, the person should first receive genetic counseling before any genetic testing is done. Such counseling should conform to established guidelines. Neurologic history, physical and psychologic examinations, and counseling that deals with the effect of genetic testing on family relationships, other social relationships, employment, insurance, and self-image should be included. If the person at risk has definite symptoms (usually chorea) that can be considered to be indicative of symptomatic Huntington's disease, the testing is confirmatory rather than asymptomatic. Counseling should be considered in those instances as well, but the clinical situation may dictate a need for a definitive diagnosis. If it is uncertain whether a person has signs of Huntington's disease, a neurologist experienced in dealing with the disease should be consulted.

TREATMENT

At present, no treatment prevents the progression of Huntington's disease. Controlled clinical trials are under way to test drugs with the goal of slowing the process that kills neurons in Huntington's disease. Until those drugs are proven to be useful, only treatment of symptoms will be available for Huntington's disease. Such treatment can be both pharmacologic and nonpharmacologic. Pharmacologic treatment can be directed at chorea, depression, mood swings, and other problems. Pharmacologic treatment of symptoms in Huntington's disease is problematic because of side effects, so that such treatment should be undertaken only if a symptom is causing disability and cannot be treated nonpharmacologically alone. Medication should begin at a low dosage and be increased gradually, as tolerated. Since the signs and symptoms of Huntington's disease are always evolving, it is extremely important to continually follow up the therapeutic effects and side effects produced by the drug, as they will change. The decision to continue the treatment is based on the relative magnitude of these effects.

Treatment of the chorea in Huntington's disease is generally discouraged unless the movements impair balance or activities of daily living. One should never treat the movements in Huntington's disease simply to make them "disappear," because that practice usually brings more side effects than benefit. No agent is approved for symptomatic treatment of chorea, so that all drugs are used "off label." Clonazepam (Klonopin), dopamine antagonists, and dopamine depleting agents are most commonly used. Clonazepam, started at 0.5 mg/day, can be useful in total daily dosages of 1 to 4 mg. Side effects include, but are not limited to, sedation, mental status changes, and balance problems. Dopamine antagonists are more effective in treating chorea but also are associated with more severe side effects; thus, these agents are tried after clonazepam fails. Several dopamine antagonists are available; among these I most frequently use risperidone (Risperdal) (2 to 6 mg/day), fluphenazine (Prolixin) (0.5 to 5.0 mg/day), and haloperidol (Haldol) (0.5 to 5.0 mg/day). These drugs are antipsychotic agents and are most often used to treat schizophrenia. Reserpine (Serapasil) and tetrabenazine (Nitoman) are dopamine-depleting agents that have been used to treat chorea. These agents are difficult to obtain, since the production of reserpine has been discontinued and tetrabenazine has not been approved by the Food and Drug Administration for any use in the United States. Although effective, these agents can cause severe depression, sedation, orthostatic hypotension, and parkinsonism, so they should be used with caution. Reserpine is most commonly given in the range of 0.2 to 0.6 mg/day, and tetrabenazine is used in the range of 50 to 200 mg/day.

It is very important to assess the degree of disability produced by the side effects of antichorea treatment. Parkinsonism, depression, and sedation are common reasons to decrease the dose or discontinue administration. If necessary, the same drugs can be used to treat the symptoms in other causes of chorea, discussed below.

Mental status problems are difficult to treat in Huntington's disease. Increasing the awareness of family members and caretakers and counseling them make coping with these problems easier for all concerned. Depression is common and may respond to conventional tricyclic antidepressants or selective serotonin reuptake inhibitors. A 6-week trial is necessary to evaluate the effectiveness of each agent. Inquiries about suicidal ideation should be sought routinely, and if this mental concept is found, the appropriate social, psychologic, and medical measures should be undertaken. Buspirone (BuSpar) and benzodiazepines can be used to treat anxiety, but overuse of these agents can lead to tolerance, dependence, and detrimental side effects. Psychotic behavior, such as delusions, paranoia, and hallucinations, often requires the antipsychotic agents mentioned above for the treatment of chorea. Severe mood swings, irritability, and aggressive behavior may be helped by carbamazepine (Tegretol), selective serotonin reuptake inhibitors, or clonazepam. Unfortunately, controlled trials of selected agents for mental status problems in Huntington's disease are rare, and there is not much evidence on which to base a drug preference for a particular problem. Drug dosages and side effect profiles for the drugs mentioned above are the same as those for the corresponding problems in persons without Huntington's disease.

The possible advantages of nonpharmacologic treatment should be emphasized. Topics to consider are speech, swallowing, physical therapy, adaptation strategies, and social counseling. Speech pathologists can address any communication problem that arises from the cognitive and speech production deficits. In addition, because choking is a common cause of death for persons with Huntington's disease, it is often important to obtain a swallowing evaluation from the speech pathologist, who can assess the dysphagia and give suggestions for decreasing the risk of choking and aspiration. Physical therapy can be used to delay debilitation and may simply consist of stretching, instruction on how to preserve balance, or exercises for endurance and strength. Often, adapting the home for physical disability can make things easier for both the patients with Huntington's disease and their caretakers. Social issues need constant vigilance. Huntington's disease can place an overwhelming stress on emotional and financial reserves. Physicians should probe for these problems and refer to social services if necessary. The fact that this disease is passed on to children introduces additional stress and concern for the family. Caretakers can become exhausted and, at the same time, have feelings of guilt and helplessness. Breaks from the constant responsibility of caring for the person with Huntington's disease are beneficial. Counseling can address the grief and stress that this illness brings to patients, families, and caretakers.

Other Neurodegenerative Diseases

Dentatorubropallidoluysian atrophy (DRPLA) can manifest chorea among other phenotypes and has an autosomal dominant inheritance pattern. The genetic mutation is located on chromosome 12, and the genetic test is commercially available. Like Huntington's disease, DRPLA involves extra CAG repeat sequences. DRPLA can also appear with parkinsonism, ataxia, myoclonus, and seizures. Other cerebellar system degenerations may include chorea as a manifestation; some have had mutations defined, and for them genetic testing is commercially available.

LESIONS OF THE BASAL GANGLIA

Focal lesions have long been known to produce chorea. The area most often involved is the subthalamic nucleus, but other regions, such as the striatum and thalamus, may produce chorea when they are sites of lesions. In these patients, chorea is usually in a hemidistribution, often associated with ballismus, and most commonly caused by a vascular process. The chorea often is sudden in onset and spontaneously resolves in vascular lesions, but atypical temporal courses occur. The hyperkinesia usually is relieved by benzodiazepine or antidopaminergic treatment. Neoplastic, infectious, and inflammatory lesions can also produce chorea. Thus, neuroimaging in patients with new-onset chorea is recommended.

DRUGS

Tardive syndromes can cause chorea (see Chapter 26). Choreiform movements can be a complication of antiparkinsonian agents (see Chapter 7). Stimulants and other drugs that can affect biogenic amine systems in the brain have the potential to create chorea. More rarely, narcotics and antiepileptic agents induce choreiform movements.

Chorea induced by oral contraceptives is a rare complication of those agents. Most patients appear to have a history of rheumatic fever or Sydenham's chorea. The chorea most often has a hemidistribution, usually appears a few months after the estrogen preparation is first given, and resolves a few weeks after administration of the oral contraceptive is discontinued. No particular high- or low-dose estrogen preparation is more likely to produce chorea than another. Most authors believe that the chorea represents an estrogen-induced hypersensitivity of dopaminergic systems in a predisposed person who has acquired a basal ganglia abnormality through rheumatic fever.

METABOLIC CONDITIONS

Hyperthyroidism

The reported incidence for chorea occurring during hyperthyroidism has ranged from 0.1% to 2%. The chorea is generalized, occurs in the 2nd through 4th decades, and is associated often with tremor and sometimes with myopathy. The chorea rarely is paroxysmal, and many reports comment on the exacerbating effect of excitement on the chorea. Some reports propose that the chorea results from the hyperadrenergic state, but most authors agree that the chorea parallels the level of thyroid hormone. The most common syndrome in which it occurs is Graves' disease, and when euthyroidism is established by propylthiouracil, the chorea usually abates. Propranolol and neuroleptics have also been used for treatment. Evidence suggests that the hyperthyroid state induces increased sensitivity of striatal dopaminergic receptors. Autopsy and imaging data have found no abnormalities.

Other Metabolic Causes

Wilson's disease (see Chapter 32) is an important cause of chorea to recognize in young persons. Serum ceruloplasmin and urinary copper determinations, slit-lamp eye examination, and liver biopsy can be used to confirm the diagnosis. Glucose and electrolyte disorders have the potential to produce chorea, but the mechanism is not clear.

OTHER SYSTEMIC DISORDERS

Sydenham's Chorea

Sydenham's chorea is a major manifestation of rheumatic fever, and it appears as a consequence of α- or β-hemolytic streptococcal infection after 1 to 6 months. Studies on the incidence and prevalence of rheumatic fever have shown its decline during this century. Sydenham's chorea has likewise declined both in overall incidence and in percentage incidence of rheumatic fever. Nevertheless, the disorder still occurs and warrants attention and study. It develops in 10% to 30% of patients with rheumatic fever, has a gradual onset, is generalized in 80% and hemichoreic in 20%, lasts 5 to 15 weeks, recurs in 20%, and can be associated with mental status changes. Most instances of Sydenham's chorea are unassociated with the other manifestations of rheumatic fever. Most cases occur between ages 5 and 15 years, with a female predisposition only in the later teenage years. Antineuronal antibodies have been found in the serum of patients and may have pathogenic significance. The chorea responds to antidopaminergic agents, and valproic acid is known to be effective. Although it is usually assumed that Sydenham's chorea resolves without ill effects, motor coordination difficulties can persist.

Lupus Erythematosus

Chorea occurs in fewer than 1% of patients with lupus erythematosus and accounts for 2% of neurologic complications in that disorder. The chorea is usually generalized, is often accompanied by other neurologic manifestations, and can occur before, during, or after the diagnosis is made. It occurs at a median age of 20 years, which is younger than that in the general population with lupus. Most cases last less than 3 months but some last up to 3 years. Chorea in lupus can be recurrent or episodic or can take the form of chorea gravidarum. In contrast to idiopathic lupus, drug-induced lupus rarely causes central nervous system complications, and chorea has never been reported. Treatment with steroids can afford improvement gradually over several weeks, and the chorea also responds to dopamine antagonists. Although vascular disease has been reported in the basal ganglia, it is poorly correlated with the occurrence of chorea. The pathophysiology of the basal ganglia dysfunction has also been associated with cytotoxic antibodies. Chorea has been associated with the lupus anticoagulant without other serologic and clinical features of lupus.

Polycythemia Vera

Polycythemic chorea occurs in 1% to 2.5% of patients with polycythemia vera. Chorea in these patients has a greater incidence in females, occurs at a median age of 61, is associated with clear mentation, and is generalized in distribution but occurs mainly in the head, neck, and shoulders. Although many instances of chorea coincide with polycythemia and diminish with treatment of polycythemia, no such correlation exists in a significant number of cases. This observation has prompted some authors to favor a biochemical effect for the causation of the chorea rather than vascular effects.

Neuroacanthocytosis

Chorea is often part of the rare syndrome named "neuroacanthocytosis," a multisystem neurologic disease associated with acanthocytes on the peripheral blood smear. This syndrome can also include other movement disorders, cognitive deficits, psychiatric manifestations, seizures, neuropathy, dysarthria, dysphagia, and diffuse neuropathologic changes, including involvement of the basal ganglia.

Chorea Gravidarum

Chorea gravidarum usually occurs in the first half of the first pregnancy, often has a hemidistribution, persists until after delivery, and can recur in a later pregnancy. Sixty percent of pregnant patients with a history of chorea induced by oral contraceptives experience chorea gravidarum. The incidence of chorea gravidarum has decreased during the last half of this century, presumably because rheumatic fever has also decreased. These observations have led to the hypothesis that increased female hormone levels during pregnancy cause the chorea to become apparent in a predisposed person, similar to the circumstance in chorea induced by oral contraceptives mentioned above. Because of the pregnancy, treatment is usually conservative, but sometimes phenobarbital is used. Some experts believe that lupus erythematosus and Huntington's disease are now responsible for most cases of chorea gravidarum occurring since the decline of rheumatic fever.

Acquired Hepatocerebral Degeneration

Various causes of liver dysfunction that occur either chronically or episodically can be associated with movement disorders, including chorea. The temporal profile may be static or progressive. Brain magnetic resonance imaging may show hyperintensity of the basal ganglia on T1-weighted images.

ESSENTIAL CHOREA SYNDROMES

The existence of essential chorea syndromes (senile chorea and benign nonprogressive chorea) is surrounded by controversy because of arguments about the extent to which "essential" cases may actually represent other, undiagnosed entities. Despite these difficulties, benign hereditary chorea is believed to be a distinct entity, and more than 100 cases have been reported in various kindreds. In most patients, generalized chorea begins before age 5 years, inheritance is autosomal dominant, the course is nonprogressive, mentation is normal, and usually there is little or no associated disability. Normal CAG repeat length has been reported in cases labeled senile chorea, but this essential senile syndrome is probably etiologically diverse.

PAROXYSMAL CHOREA

Paroxysmal dyskinesias of which chorea may be a part occur, but a paroxysmal movement disorder consisting of pure chorea is exceedingly rare. Usually, such paroxysmal dyskinesias resemble dystonic movements. These paroxysmal dyskinesias are usually divided into paroxysmal kinesigenic choreoathetosis (PKC), paroxysmal dystonic choreoathetosis (PDC), and an "intermediate" form. PKC is exacerbated by quick movement and usually lasts seconds or minutes. In contrast, PDC is exacerbated by stress, exercise, caffeine, and alcohol and lasts minutes to hours. These paroxysmal dyskinesias are often familial and may be associated with ataxia. PKC is the most common and best responds to anticonvulsants, whereas benzodiazepine therapy has a mild to moderate effect on PDC. The pathophysiology of paroxysmal dyskinesias is unknown.

BALLISMUS

Ballismus refers to ballistic, large-amplitude, proximal limb movements that have the appearance of wild flailing motion. Many experts believe that ballismus resembles severe proximal chorea. Physiologically, subthalamic lesions are often responsible for ballismus, adding to the similarity with chorea. As ballismus abates, the movements may change to chorea. Ballismus most often occurs in a hemidistribution. Strokes are the most common lesion causing hemiballismus, and the lesion occurs contralateral to the movements. Neuroimaging is recommended. Because the movements are severe, orthopedic injury, exhaustion, pneumonia, and heart failure have been known to occur in these patients. For this reason, symptomatic treatment with a benzodiazepine, haloperidol (or other dopamine antagonist), or a dopamine depletor drug is recommended. Many patients experience remission, and therapy should be withdrawn when appropriate. In rare cases, brain surgery in the contralateral basal ganglia loop is necessary.

SELECTED READING

Caviness JN, Muenter MD. An unusual cause of recurrent chorea. *Mov Disord* 1991; 6: 355-357.

Hardie RJ, Pullon HWH, Harding AE, Owen JS, Pires M, Daniels GL, Imai Y, Misra VP, King RHM, Jacobs JM, Tippet P, Duchen LW, Thomas PK, Marsden CD. Neuroacanthocytosis. A clinical, haematological and pathological study of 19 cases. *Brain* 1991; 114; 13-49.

Hayden MR, Bloch M, Wiggins S. Psychological effects of predictive testing for Huntington's disease, in *Behavioral Neurology of Movement Disorders* (Weiner WJ, Lang AE, eds), Raven Press, New York, 1995, pp. 201-210.

Lance JW. Familial paroxysmal dystonic choreoathetosis and its differentiation from related syndromes. *Ann Neurol* 1977; 2: 285-293.

Lee MS, Marsden CD. Movement disorders following lesions of the thalamus or subthalamic region. *Mov Disord* 1994; 9: 493-507.

Paulson GW, Prior TW. Issues related to DNA testing for Huntington's disease in symptomatic patients. *Semin Neurol* 1997; 17: 235-238.

Pulsinelli WA, Hamill RW. Chorea complicating oral contraceptive therapy. Case report and review of the literature. *Am J Med* 1978; 65: 557-559.

Shoulson I. On chorea. *Clin Neuropharmacol* 1986; 9 Suppl 2: S85-99.

Weiner WJ, Lang AE. *Movement Disorders: A Comprehensive Survey.* Futura, Mount Kisco, New York, 1989, pp. 293-346, 569-597.

26 Tardive Dyskinesias

Kapil D. Sethi, MD, FRCP

Contents

"Tardive dyskinesia" is a generic term used to describe all persistent and sometimes irreversible abnormal involuntary movements that occur with prolonged neuroleptic therapy. This disorder came to light after the advent of chlorpromazine (Thorazine), which was first developed in France in 1952 and introduced in the United States in 1954. Acute reactions, such as akathisia, drug-induced parkinsonism, and dystonia, were soon recognized. A more persistent dyskinesia, that is, tardive dyskinesia (TD), was first recognized in the late 1950s. Subsequently, TD variants were described, including tardive dystonia, tardive akathisia, myoclonus, tics, and, possibly, tardive tremor (Table 26-1).

Classic TD is easy to recognize, but the variants may be misdiagnosed. The movement disorder is usually "choreic" in type, and since it usually involves the mouth and the tongue, it is called "classic orobuccolinguomasticatory dyskinesia."

EPIDEMIOLOGY

Movement disorders were seen in psychotic patients before the neuroleptic era. Descriptions of schizophrenia dating to the beginning of this century include mention of abnormal movements similar to those of TD. In a recent study, 3 of 22 (14%) schizophrenic patients who had never received medication met the research diagnostic criteria for probable spontaneous dyskinesia. However, epidemiologic evidence suggests that the prevalence of TD in medicated schizophrenic patients is far higher than in the pre-neuroleptic era, suggesting that TD is associated with neuroleptic therapy. All classes of dopamine blocking agents (DBAs) have been implicated in the development of TD. These include phenothiazines, such as chlorpromazine and thioridazine (Mellaril), and

From: *Parkinson's Disease and Movement Disorders:*
Diagnosis and Treatment Guidelines for the Practicing Physician
Edited by: C. H. Adler and J. E. Ahlskog © Mayo Foundation
for Medical Education and Research, Rochester, MN

Table 26-1
Classification of Tardive Dyskinesias

Tardive dyskinesia
    Classic orobuccolingual dyskinesia
Tardive dyskinesia variants
    Tardive dystonia
    Tardive akathisia
    Tardive myoclonus
    Tardive tics
    Tardive tremor

butyrophenones, such as haloperidol (Haldol). DBAs like metoclopramide (Reglan) and prochlorperazine (Compazine) given for gastrointestinal disorders are also associated with the development of TD. A list of drugs that may cause TD appears in Table 26-2.

Because spontaneous dyskinesia may resemble TD, current estimates of TD may be inflated. The overall prevalence of TD among patients with schizophrenia receiving long-term treatment is 24%, and the incidence in young adults is 5% per year. Aging is a major risk factor for development of TD, and other risk factors are female sex, mood disorders, "organic" brain dysfunction, and early extrapyramidal side effects. There are conflicting reports on brain atrophy seen on neuroimaging and susceptibility to the development of TD.

VARIOUS TARDIVE DISORDERS

Classic Orobuccolinguomasticatory Syndrome

Classic orobuccolingual dyskinesia (OBLD) was the first type of TD described. It consists of stereotyped oral and facial movements, including twisting and protrusion of the tongue, lip smacking and puckering, and chewing. One of the early signs of OBLD is slow, writhing movements of the tongue in the floor of the mouth. Movements are usually confined to the orofacial area but sometimes spread to involve the extremities. OBLD is often more disturbing to the physician and the family than to the patient, but it may cause dry mouth, dysarthria, and dysphagia. The upper face is often uninvolved in TD, but blepharospasm may occur with tardive dystonia. Occasionally, diaphragmatic dyskinesia can result in dyspnea and hypoxia. OBLD is frequently associated with tardive akathisia, tardive dystonia, and parkinsonism.

TD may begin while a person is taking a DBA, but often it makes its first appearance when DBA administration is discontinued or the dosage is reduced. This form of dyskinesia has been called "withdrawal dyskinesia" or "covert dyskinesia." Withdrawal dyskinesia disappears within 3 months after administration of the drug has stopped, whereas covert dyskinesia becomes apparent on reduction of neuroleptic therapy and persists for longer periods. However, this distinction is arbitrary, because some of the covert dyskinesias disappear after prolonged follow-up.

OBLD is usually associated with long-term DBA therapy but may occur in other situations. The known causes of OBLDs are listed in Table 26-3.

Table 26-2
Drugs That May Cause Tardive Dyskinesia

Generic name	Trade name	Generic name	Trade name
Acetophenazine maleate	Tindal	Perphenazine	Trilafon
Amoxapine	Asendin	Perphenazine-	Triavil
Butaperazine maleate	Repoise maleate	amitriptyline	
Carphenazine maleate	Proketazine	Piperacetazine	Quide
Chlorpromazine	Thorazine	Prochlorperazine	Compazine
Chlorprothixene	Taractan	Promazine	Sparine
Fluphenazine	Prolixin	Promethazine	Phenergan
Haloperidol	Haldol	Risperidone	Risperdal
Loxapine	Loxitane	Thiethylperazine	Torecan
Mesoridazine	Serentil	Thioridazine	Mellaril
Metoclopramide	Reglan	Thiothixene	Navane
Molidone	Moban	Trifluoperazine	Stelazine
Olanzapine	Zyprexa	Triflupromazine	Vesprin
		Trimeprazine	Temaril

Table 26-3
Differential Diagnosis
of Orobuccolingual Dyskinesia

Spontaneous dyskinesia
    of elderly(usually dystonic)
Hereditary choreas
Basal ganglia strokes
Systemic lupus erythematosus
Edentulous dyskinesia
Other drugs causing dyskinesias
       Levodopa
       Amphetamines
       Cocaine
       Tricyclic antidepressants
       Cimetidine
       Flunarizine
       Antihistamines
       Phenytoin

Tardive Akathisia

Akathisia is characterized by an inability to sit still accompanied by an inner sense of restlessness. Clinically, the patients are fidgety, may march in place, and may show complex and stereotyped movements. They can often suppress these movements for brief periods. The legs are most frequently affected, as in acute akathisia, and truncal movements may also occur. Some patients moan constantly. Persistent akathisia may occur as a subtype of TD as opposed to acute akathisia, which is apparent within a short period of the beginning of DBA therapy. Persistent akathisia is defined as an occurrence present for at least 1 month when the patient is receiving a constant DBA dose. If both subjective and objective criteria are used to define akathisia, it is apparent that tardive akathisia is quite common. It occurs in 20% to 40% of DBA-treated patients with schizophrenia. No case-control studies have been done to ascertain whether akathisia is more prevalent in schizophrenic patients treated with neuroleptic agents. However, other than in restless leg syndrome and Parkinson's disease, persistent akathisia is very uncommon, and it appears that neuroleptic agents contribute to the genesis of tardive akathisia.

As in classic TD, several types of tardive akathisia have been described, for example, covert and withdrawal akathisias. In one report, the mean age of patients with tardive akathisia was 58 years and women outnumbered men by 2 to 1. Almost all classes of DBAs have been responsible.

Tardive Dystonia

In 1982, Burke and associates reported on 42 patients with tardive dystonia seen in three movement disorders centers. However, scattered reports have been available since 1962. This variant is not uncommon, and recent studies have found a prevalence of 9% to 13% in schizophrenic patients with long-term medication.

Multiple reasons exist to differentiate classic TD from tardive dystonia. The abnormal movements are distinct from those of classic TD. Whereas classic dyskinetic movements are rapid and stereotyped, dystonic movements are slower and twisting. Whereas TD seems to occur more commonly in elderly women, tardive dystonia seems to be more common in younger patients and to have no predilection for either sex. Moreover, anticholinergic drugs tend to worsen classic TD but are beneficial in tardive dystonia. Tardive dystonia tends to be more persistent than classic TD.

The dystonic movements of patients with tardive dystonia are indistinguishable from those of idiopathic torsion dystonia. However, other types of involuntary movements, such as OBLD, may coexist. In general, multiple movement disorders in a patient should alert the physician to the possibility of a drug-induced movement disorder.

The dystonia may be focal or segmental, rarely generalized. Most patients have segmental dystonia. If dystonia involves the neck, the usual result is retrocollis. Similarly, if it involves the trunk, the usual result is truncal extension. However, flexion dystonia may occur in tardive dystonia.

A rare variant is "reverse obstructive sleep apnea syndrome." Patients with this type have complete upper airway obstruction during the day, but as soon as they sleep, the tardive dystonia disappears and breathing is normal during sleep. This sequence is exactly the opposite of obstructive sleep apnea syndrome, in which sleep actually brings on the abnormal breathing pattern. This effect could be life-threatening and may require tracheostomy. Another uncommon variant is recurrent oculogyric crises accompanied by obsessional thoughts and hallucinations.

Tardive Tics or Tardive Tourettism

Rarely, motor and vocal tics appear for the first time in patients receiving long-term antipsychotic therapy. The tics may be accompanied by coprolalia.

Tardive Myoclonus

Tardive myoclonus has been described as a late complication of prolonged neuroleptic therapy. Usually, it is a postural myoclonus of the upper extremities, and associated movement disorders are common.

Tardive Tremor

Tardive tremor has only rarely been reported. It is said to be more postural and kinetic than the rest tremor in drug-induced parkinsonism and usually is not associated with other parkinsonian signs. This tremor responds poorly to the dopaminergic drugs used in Parkinson's disease.

MECHANISM

The pathophysiology of TD and its variants is thought to be an increased number and affinity of postsynaptic dopamine D_2 receptors due to long-term dopamine receptor blockade. This belief is based on rodent models of TD in which, after 2 weeks of exposure to conventional DBAs, affinity and numbers of dopamine D_2 receptors are increased. It is also based on the dyskinesias produced by levodopa and dopamine agonists in patients with Parkinson's disease. However, the effect is universal in animals, whereas TD occurs in only about 20% of patients exposed to DBAs. Moreover, the role of other neurotransmitters, such as γ-aminobutyric acid (GABA) and norepinephrine, may be important. In a postmortem study of patients with TD who were free of neuroleptic drugs for 1 year before death, the dopamine D_2 receptor density was diminished in the striatum but increased in the pallidum. A positron emission tomography (PET) study using N-11c-methyl-spiperone did not show a difference in D_2 receptor density in patients with or without TD. Another PET study showed that metabolic rates were increased in the motor cortex and the globus pallidus of patients with TD, a suggestion of overactivity of these regions.

PROGNOSIS

TD was thought to be persistent in most cases, but several studies have shown that up to 40% of patients improved after cessation of neuroleptic therapy. This response is more likely in younger patients and in patients with a shorter history of TD. Tardive dystonia tends to remit less often than classic TD.

MANAGEMENT

DBAs should not be prescribed for anxiety and other neuroses, because of the danger that TD will develop and because of the availability of other drugs to treat these disorders. The only acceptable indication for long-term neuroleptic therapy is chronic schizophrenia. The need for neuroleptic therapy should be reviewed periodically, even in patients without signs of TD. Some patients with chronic psychosis may not require the high doses of neuroleptic agents that were prescribed initially. Metoclopramide (Reglan) should be used only for short-term therapy, that is, less than 6 months, and only if clearly indicated. The combination antidepressants like Triavil (perphenazine and amitriptyline), which combines a DBA with an antidepressant, should not be used for depression. Instead, a tricyclic antidepressant or a selective serotonin reuptake inhibitor antidepressant should be used alone.

Any patient requiring DBA treatment should be examined every 3 to 6 months for early signs of TD. If TD is present, switching to clozapine (Clozaril) or possibly olanzapine (Zyprexa) or quetiapine (Seroquel) may prevent worsening of TD. Controlled studies to test this approach are lacking.

Paradoxically, TD movements may lessen with an increase in the dose of the offending agent. Tempting as it may appear, this strategy is counterproductive. In most cases, after a variable delay, the movements break through and necessitate further increases in the neuroleptic dose. Because a vicious cycle results, increasing the dose of neuroleptic medication to treat TD is not recommended.

A variety of drugs have been used to treat the symptoms of TD, but the most important issue is to *stop* administering the neuroleptic drugs if possible. With all drugs used for TD, the dose should be low at first and then be gradually titrated upward. The mainstay of treatment is a dopamine-depleting agent like reserpine (Serpasil), metyrosine (Demser), or tetrabenazine (Nitoman). The usual dose of reserpine is 0.25 mg/day, increased gradually to 3 to 5 mg/day. Metyrosine is started at a dose of 250 mg/day, which is gradually increased to a maximum of 250 mg four times a day. Tetrabenazine is given at 25 mg/day and gradually increased up to 150 mg/day. This drug is not available in the United States and has to be obtained from Canada or Europe. The side effects of all dopamine depletors include sedation, hypotension, and depression.

In general, anticholinergic drugs exacerbate orobuccolingual TD and should be withdrawn. A number of cholinergic drugs, such as choline chloride, have been used with variable results. Some patients with TD have been treated with low doses (i.e., 2.5 to 10 mg/day) of bromocriptine (Parlodel). Baclofen (Lioresal) has been used because of its GABA agonist properties with some success. The starting dose is 10 mg/day, to be increased to 20 mg four times a day. This drug may cause sedation and precipitate seizures in susceptible persons. Clonazepam (Klonopin), 0.25 mg at bedtime up to 1 mg four times a day, may be useful in some cases. Calcium channel blockers have produced varying results, with nifedipine (Adalat) and verapamil (Isoptin) somewhat efficacious and diltiazem (Cardizem) ineffective.

Although there has been much interest in vitamin E, at a beginning dose of 400 IU twice a day, it produces a modest response at best. Risperidone (Risperdal), 4 to 12 mg/day, was reported in a retrospective review to relieve TD, but it also can cause TD and drug-induced parkinsonism. Clozapine has been used in patients with TD and tardive dystonia (see below). A summary of drugs used to treat TD appears in Table 26-4.

Tardive Akathisia

Unlike acute akathisia, tardive akathisia does not disappear when administration of neuroleptic agents is stopped. The treatment of tardive akathisia is very similar to that of TD. If possible, neuroleptic drugs should be withdrawn. If this is not possible, treatment of symptoms with the dopamine-depleting agents should be attempted. As with classic TD, reserpine or tetrabenazine may be used. Tardive akathisia does not respond to opiates, such as propoxyphene (Darvocet), or to beta blockers, such as propranolol (Inderal), as is true with acute akathisia.

Tardive Dystonia

When tardive dystonia appears in patients receiving neuroleptic therapy, one should rule out other causes, like Wilson's disease and focal lesions in the basal ganglia. Family history is important to rule out inherited dystonias. The DYT1 gene is not present in patients with tardive dystonia (see Chapter 23). The treatment of tardive dystonia differs from that of TD because anticholinergic drugs seem to benefit patients with tardive dystonia but worsen the condition of those with classic TD. However, dopamine-depleting

Table 26-4
Drugs Used to Treat Tardive Dyskinesia and Variants

Drug	Starting dose	Maximum dose	Side effects
Clonazepam (Klonopin)	0.25 mg q.h.s.	2 mg q.i.d.	Sedation, ataxia
Baclofen (Lioresal)	10 mg q.d.	20 mg q.i.d.	Nausea, sedation
Reserpine (Serpasil)	0.25 mg q.d.	1 mg q.i.d.	Sedation, depression, hypotension, worsening parkinsonism
Tetrabenazine (Nitoman)	25 mg q.d.	50 mg t.i.d.	Sedation, worsening parkinsonism
Vitamin E	400 IU b.i.d.	1,000 IU b.i.d.	
Trihexyphenidyl (Artane)	2 mg q.d.	20 mg q.i.d.	Dry mouth, constipation, urinary retention
Botulinum toxin type A (Botox)	Variable	Variable	Weakness of injected muscles

agents are useful in both. It may also be worth trying clonazepam or baclofen (Table 26-4). Botulinum toxin can be used in focal or segmental tardive dystonia (see Chapter 31).

Tardive Tics, Tremor, and Myoclonus

The treatment of tics is not clear, but withdrawing the offending drug and giving dopamine depletors might be helpful. Myoclonus may be helped by clonazepam, and tremor may respond to tetrabenazine.

CLOZAPINE AND TARDIVE DYSKINESIA

Clozapine, a dibenzodiazepine, is an atypical neuroleptic drug that is currently approved for the treatment of neuroleptic-resistant schizophrenia. It has an extremely low propensity to cause extrapyramidal side effects. In fact, no convincing cases of TD secondary to clozapine monotherapy have been reported. Several uncontrolled observations have suggested that clozapine benefits about 40% of patients with TD, particularly those with tardive dystonia. Dosages in the range of 500 to 900 mg/day usually are required, and the improvement may occur within days or may take months. The mechanism of this beneficial effect is unclear. Clozapine may have a passive effect by removal of the offending agent, allowing the underlying pathophysiologic processes to reverse. It may only suppress TD like the neuroleptic drugs; however, this action appears unlikely. Lastly, it may possess a specific antidyskinetic action, which may be related to its blockade of both D_1 and D_2 receptors. The newer atypical antipsychotic agents (risperidone, olanzapine, quetiapine) have a lower incidence of TD than haloperidol, but they have not been systematically tested in the treatment of tardive syndromes.

Clozapine has been reported to induce neuroleptic malignant syndrome and may cause positive and negative myoclonus. Another risk of clozapine treatment is agranulocytosis, so that a complete blood cell count must be monitored weekly during the first 6 months of treatment and then every other week during the entire treatment period.

SUMMARY

TD is an iatrogenically induced movement disorder caused by dopaminergic blocking drugs. These include drugs used to treat schizophrenia and some gastrointestinal disorders. The movements are usually dyskinesias of the mouth, tongue, and jaw, but can also include other body regions and be more dystonic in phenomenology. Akathisia can occur as well. The key to treatment is to *stop* administering the offending drug in expectation that the movements will resolve spontaneously. If they do not, trials of various oral medications can be undertaken (Table 26-4).

SELECTED READING

Bharucha KJ, Sethi KD. Tardive tourettism after exposure to neuroleptic therapy. *Mov Disord* 1995; 10: 791-793.

Burke RE, Kang UJ, Jankovic J, Miller LG, Fahn S. Tardive akathisia: an analysis of clinical features and response to open therapeutic trials. *Mov Disord* 1989; 4: 157-175.

Factor SA, Friedman JH. The emerging role of clozapine in the treatment of movement disorders. *Mov Disord* 1997; 12: 483-496.

Gardos G, Cole JO. Overview: public health issues in tardive dyskinesia. *Am J Psychiatry* 1980; 137: 776-781.

Jeste DV, Caligiuri MP. Tardive dyskinesia. *Schizophr Bull* 1993; 19: 303-315.

Kang UJ, Burke RE, Fahn S. Natural history and treatment of tardive dystonia. *Mov Disord* 1986; 1: 193-208.

Marsden CD, Jenner P. The pathophysiology of extrapyramidal side-effects of neuroleptic drugs. *Psychol Med* 1980; 10: 55-72.

Tarsy D, Kaufman D, Sethi KD, Rivner MH, Molho E, Factor S. An open-label study of botulinum toxin A for treatment of tardive dystonia. *Clin Neuropharmacol* 1997; 20: 90-93.

Weiner WJ, Goetz CG, Nausieda PA, Klawans HL. Respiratory dyskinesias: extrapyramidal dysfunction and dyspnea. *Ann Intern Med* 1978; 88: 327-331.

27 Myoclonus

John N. Caviness, MD

CONTENTS

CLINICAL DIAGNOSTIC APPROACH
DIAGNOSTIC TESTING
TREATMENT
SELECTED READING

Myoclonus is defined as sudden, brief, shocklike involuntary movements caused by muscular contractions (positive myoclonus) or inhibitions (negative myoclonus). "Myoclonus" should be considered a descriptive term that refers to a *symptom* or *sign*. As with many neurologic symptoms and signs, myoclonus does not constitute a diagnosis, and it is nonspecific relative to its neuroanatomical source, its pathogenesis, and the cause of the syndrome in which it occurs. Myoclonus often has strong implications for the diagnosis, prognosis, and treatment of the underlying disorder.

Other abnormal movements should be differentiated from myoclonus. Tics may be distinguished by their complexity, ability to be suppressed, and build up of "psychic tension" that is not relieved until the movement is performed. However, a quick, simple tic can be virtually indistinguishable from myoclonus on the basis of appearance alone. Chorea also may produce muscle jerking. Its continuous and seemingly random change of direction, its nonstereotyped appearance, and the smooth flow of movement from one body part to another differentiate chorea from myoclonus. Rhythmic myoclonus may be differentiated from tremor by its lack of a sinusoidal pattern. Its "square wave" temporal pattern contains a distinguishable interval between each myoclonic movement. Quick movements can occur in the syndrome of dystonia, which is characterized by sustained muscle contractions and abnormal postures, but only rarely do these movements appear quick enough to warrant the term "myoclonus." Some experts consider exaggerated startle reflex to be an example of a generalized myoclonic movement, whereas others make a separate designation. Although myoclonus may be triggered by startle, an exaggerated startle reflex is, by definition, triggered from a startle stimulus. A fasciculation is produced by a spontaneous discharge of a motor unit in lower

From: *Parkinson's Disease and Movement Disorders:
Diagnosis and Treatment Guidelines for the Practicing Physician*
Edited by: C. H. Adler and J. E. Ahlskog © Mayo Foundation
for Medical Education and Research, Rochester, MN

motor neuron disease (e.g., peripheral nerve lesion) and usually, unlike myoclonus, does not result in the movement of a whole muscle.

CLINICAL DIAGNOSTIC APPROACH

The best strategy for using the symptom of myoclonus in the diagnosis of its underlying syndrome is to view it within the context of the total clinical picture. Because of the large differential diagnosis of myoclonus, a comprehensive history and physical examination should be performed. Important information obtained from the history should include the temporal course of onset, drug or toxin exposure, history of seizures, past or current medical problems, mental status changes, and family history.

The basic parts of the myoclonus examination include distribution, temporal profile, and activation characteristics of the movement. The distribution can be focal, multifocal, segmental, or generalized. A multifocal myoclonus distribution may have bilaterally synchronous movements as well. The temporal profile can be continuous or intermittent as well as rhythmic or irregular. If intermittent, myoclonus can occur sporadically or in trains. The activation of myoclonus may be at rest (spontaneous), induced by various stimuli (reflex myoclonus), induced by voluntary movement (action myoclonus), or the result of some combination of these. Patients may have more than one pattern of myoclonus, and all distributions and temporal patterns should be described. All activation characteristics should be noted as absent or present.

The rest of the neurologic examination is no less important. Certain myoclonic syndromes commonly have ataxia, mental status changes, parkinsonism, or other movement disorders associated with them. Other neurologic findings (positive or negative) are necessary for the diagnosis of the underlying syndrome.

A diagnostic approach to myoclonus can be organized according to the etiologic classification scheme designed by Marsden and associates. An updated version of this scheme is in Table 27-1. The major categories of myoclonus in this scheme are as follows: physiologic, essential, epileptic, and symptomatic (secondary). Each of these major categories is associated with different clinical circumstances. Most instances of myoclonus are in the symptomatic category, followed by epileptic and essential. The first task is to determine which major category reflects the clinical circumstances of the patient. The second task is to match the patient's specific clinical characteristics and results from diagnostic studies with those of a specific diagnosis under the corresponding major category.

Physiologic Myoclonus

Physiologic myoclonus occurs in neurologically normal persons. There is minimal or no associated disability, and the physical examination discloses no relevant abnormalities. The ability to identify the jerks associated with physiologic myoclonus on the basis of the history enables the clinician to reassure the patient and family that no treatment is necessary. Sleep is the most common circumstance of physiologic myoclonus. The spouse, who complains of being kicked in bed, often takes more notice of the jerks during sleep than the patient does. Various types of sudden movement occur during sleep or sleep transitions. Nocturnal myoclonus consists of stereotyped, repetitive dorsiflexion of the toes and foot and, occasionally, flexion of the knee and hip. Nocturnal myoclonus is the same type of movement that occurs in the restless legs syndrome, in which the movements can disrupt sleep.

Essential Myoclonus

Essential myoclonus is clinically significant and is usually the most prominent or only clinical finding. Thus, the myoclonus is an isolated or "essential" phenomenon, but the patient usually experiences some disability. Essential myoclonus is idiopathic and progresses slowly or not at all. Hereditary and sporadic forms exist. Hereditary essential myoclonus is characterized by onset before age 20 years, dominant inheritance with variable severity, a benign course compatible with an active life and normal life span, lack of other neurologic deficits, and normal findings on an electroencephalogram (EEG). The myoclonus usually is distributed throughout the upper body, exacerbated by muscle activation, and responsive to alcohol. Sporadic essential myoclonus is more clinically heterogeneous than is hereditary essential myoclonus. This "entity" has the nonspecific inclusion criteria of any idiopathic case that cannot be included in any other myoclonic category. Thus, the term "sporadic essential myoclonus" is, in reality, a miscellaneous list. This term, as it has been used, most likely represents various heterogeneous and yet undiscovered causes and false-negative findings on family histories.

Epileptic Myoclonus

Epileptic myoclonus refers to myoclonus in patients with epilepsy—that is, a chronic seizure disorder. Myoclonus can occur as only one component of a seizure, the only seizure manifestation, or one of multiple seizure types within an epileptic syndrome. Seizures usually dominate the clinical picture in epileptic myoclonus, and the disorder is often idiopathic. Associated seizure manifestations, such as lapses in consciousness and epileptiform abnormalities on an EEG, can help to clarify the underlying epileptic syndrome. The myoclonus in epileptic syndromes is presumed to be of cortical origin. Myoclonus may occur as a relatively minor observation within the epileptic syndrome or be described as the only manifestation. A generalized tonic-clonic seizure can be preceded by generalized myoclonic jerks, and the clonic phase is typified by a series of myoclonic jerks that usually signal the end of the seizure. Myoclonus can accompany absence seizures as movements of the eyelid or head or as more extensive body jerks, and the generalized ictal EEG pattern has a frequency of approximately 3 Hz, which is typical for absence syndromes. Photosensitive myoclonus can be demonstrated by intermittent photic stimulation that induces myoclonus associated with a generalized EEG discharge. This phenomenon commonly is observed in persons with generalized epilepsy and also can occur as an isolated event. Epilepsia partialis continua often resembles spontaneous myoclonus; it occurs irregularly or regularly at intervals no longer than 10 seconds, is confined to one part of the body, and continues for hours, days, or weeks. The EEG usually, but not always, shows a focal abnormality appropriate for the affected region.

Myoclonic seizures are epileptic seizures in which the motor manifestation is myoclonus. The myoclonus is accompanied by a generalized ictal epileptiform EEG discharge. Infantile spasms and Lennox-Gastaut syndrome are well-known childhood epileptic syndromes that exhibit myoclonic seizures and are associated with other severe neurologic dysfunction. Interictal EEG abnormalities help define these syndromes. Juvenile myoclonic epilepsy (awakening myoclonus of Janz) is the classic idiopathic syndrome in which myoclonic seizures may occur in conjunction with generalized tonic-clonic or absence seizures (or both) but without other neurologic disability in the patient. It is characterized by onset during adolescence or young adulthood and by generalized myoclonus exacerbated by sudden awakening, sleep deprivation, or photic stimulation (or some combination of these entities). The ictal EEG shows a 4- to 6-Hz polyspike-and-wave pattern. The interictal EEG may demonstrate the same pattern or normal findings. Valproic acid can be effective in controlling all seizure types associated with this syndrome.

Symptomatic Myoclonus

Symptomatic (secondary) myoclonus appears in patients with an identifiable underlying disorder, neurologic or non-neurologic. These symptomatic syndromes are the most common cause of myoclonus. Often, there is clinical or pathologic evidence of diffuse nervous system involvement. Clinical manifestations are usually multiple, and some may be more prominent than the symptom of myoclonus. Chronic clinical progression suggests symptomatic myoclonus. Although seizures with appearances other than myoclonic may be present, they are not the only prominent clinical manifestation. Mental status abnormalities and ataxia are common clinical associations in symptomatic myoclonic syndromes, and most often the cortex is believed to be the myoclonic source in these cases. The subcategories of causes discussed below are in Table 27-1.

STORAGE DISEASES

Various storage diseases have been classified under the clinical syndrome of progressive myoclonic epilepsy. A chronic, progressive neurologic syndrome that contains some combination of myoclonus, seizures, ataxia, and dementia characterizes progressive myoclonic epilepsy. These disorders usually affect persons younger than 30 years and are often fatal. Differences in age at onset, rate of progression, details of clinical expression, other clinical manifestations, and pattern of stimulus sensitivity exist among individual storage diseases. The neuropathologic changes in the brain are widespread. Tissue biopsy and enzyme activity measurements are useful for diagnosis (Table 27-2). A few years ago, no specific treatment was available for the storage diseases. Currently, enzyme replacement therapy is feasible in some instances.

NEURODEGENERATIVE SYNDROMES

Neurodegenerative syndromes have widely distributed pathologic features with variable involvement of cerebellar pathways, basal ganglia, and cerebral cortex. Action myoclonus is often characteristic in cerebellar myoclonic syndromes. Myoclonus as a finding in Parkinson's disease is best known as a consequence of levodopa therapy, although rare instances of occurrence without medication have been reported. Multiple system atrophy occurs as various degrees of parkinsonism, ataxia, and autonomic dysfunction, and a stimulus-sensitive distal limb myoclonus was found in 31% of patients in one series. Myoclonic jerks, both action and stimulus sensitive, commonly occur in corticobasal degeneration. The distribution of the myoclonus in corticobasal degeneration is either asymmetrical or focal and is similar to that of the other clinico-pathologic manifestations of the disease. Myoclonus has been documented in a few patients with Huntington's disease, although the various reports have implicated different mechanisms for the myoclonus. The myoclonus in Creutzfeldt-Jakob disease, a clinical hallmark of the disorder, can occur at rest or is exacerbated by action or a stimulus. The usual presentation of myoclonus in Alzheimer's disease is small, multifocal distal jerking. In the dementia syndrome known as diffuse Lewy body disease, which often has parkinsonian features, myoclonus was noted in 15% of the total patients in one series. A cortical source for the myoclonus is hypothesized for many of these degenerative syndromes.

INFECTIOUS AND POSTINFECTIOUS SOURCES

Myoclonus has been known for decades to occur with a wide variety of infectious agents. Acute and "slow" viral agents, prion protein diseases, and bacterial, fungal, and parasitic infections have all been implicated. Cortical and segmental physiologic types have been reported with these causes of myoclonus. Numerous case reports have recently shown that human immunodeficiency virus infection is a primary cause of myoclonus and a secondary cause through its many complications. Criteria for a "postinfectious" diagnosis remain nonspecific. In many such instances, the infectious source is not documented but is inferred from nonspecific constitutional symptoms. These syndromes usually have a subacute presentation, from days to weeks, but acute and chronic presentations are seen as well.

DRUGS, TOXINS, AND METABOLIC CONDITIONS

The toxic and drug-induced causes of myoclonus are numerous (Table 27-3). In metabolic conditions, myoclonus often occurs in the hospital, frequently with mental status changes. The myoclonus may be multifocal and subtle or generalized and almost constant, as in the entity "myoclonic status epilepticus." Hypoxia is the most common cause of myoclonic status epilepticus, and this myoclonus is associated with generalized epileptiform activity on an EEG. Prognosis in such cases depends on the severity and reversibility of the underlying process. Asterixis, which is known as "negative myoclonus," is a well-known accompaniment of toxic and metabolic encephalopathies. It is characterized by brief lapses in postural tone and is particularly common in kidney and liver failure. Deficiency states and mitochondria disorders are recently described causes of myoclonus. Most of the drugs, toxins, and metabolic conditions that cause myoclonus also can produce mental status changes and tremor. Usually, the effects are lessened or abolished when use of the agent is discontinued or the metabolic condition is reversed.

HYPOXIA

In 1963, Lance and Adams reported myoclonus *after* recovery from severe hypoxic episodes. All patients experience hypoxic coma for several hours to days, and spontaneous myoclonus or seizures (or both) may or may not be present during the coma. When the patient regains consciousness, myoclonus is present or subsequently develops. Action myoclonus is characteristic, and ataxia as well as mental status changes are frequent parts of the syndrome. A cortical origin for the myoclonus is established, and the disorder has been named "cortical reflex myoclonus."

Posthypoxic myoclonus may respond to oral administration of 5-hydroxytryptophan, which is a precursor of the neurotransmitter serotonin. Cerebrospinal fluid serotonin metabolites are often decreased in these patients, and this finding has led to the hypothesis that depression of serotonin system activity is important in the pathophysiology of this syndrome. Conversion of 5-hydroxytryptophan to serotonin in the central nervous system increases serotonergic activity, which somehow suppresses the myoclonus. Post-traumatic action myoclonus has a similar biochemical pattern.

Table 27-1
Classification of Myoclonus

I. *Physiologic myoclonus* (normal subjects)
>Sleep jerks (hypnic jerks)
>Anxiety-induced
>Exercise-induced
>Hiccup (singultus)
>Benign infantile myoclonus with feeding

II. *Essential myoclonus* (no known cause and no other gross neurologic deficit)
>A. Hereditary (autosomal dominant)
>B. Sporadic

III. *Epileptic myoclonus* (seizures dominate and no encephalopathy, at least initially)
>A. Fragments of epilepsy
>>Isolated epileptic myoclonic jerks
>>Epilepsia partialis continua
>>Idiopathic stimulus-sensitive myoclonus
>>Photosensitive myoclonus
>>Myoclonic absences in petit mal epilepsy
>B. Childhood myoclonic epilepsy
>>Infantile spasms
>>Myoclonic astatic epilepsy (Lennox-Gastaut)
>>Cryptogenic myoclonus epilepsy (Aicardi)
>>Awakening myoclonus epilepsy of Janz
>C. Benign familial myoclonic epilepsy (Rabot)
>D. Progressive myoclonus epilepsy: Baltic myoclonus
>>(Unverricht-Lundborg)

IV. *Symptomatic myoclonus* (progressive or static encephalopathy dominates)
>A. Storage disease
>>Lafora's body disease
>>GM_2 gangliosidosis (late infantile, juvenile)
>>Tay-Sachs disease
>>Gaucher's disease (noninfantile neuronopathic form)
>>Krabbe's leukodystrophy
>>Ceroid lipofuscinosis (Batten)
>>Sialidosis (cherry-red spot) (types I and II)
>B. Spinocerebellar degenerations
>>Ramsay Hunt syndrome
>>Friedreich's ataxia
>>Ataxia-telangiectasia
>>Other spinocerebellar degenerations
>C. Basal ganglia degenerations
>>Wilson's disease
>>Torsion dystonia
>>Hallervorden-Spatz disease
>>Progressive supranuclear palsy
>>Huntington's disease
>>Parkinson's disease
>>Multisystem atrophy

(Table 27-1 continued)

(Symptomatic myoclonus cont.)
                Corticobasal degeneration
                Dentatorubropallidoluysian atrophy
    D. Dementias
                Creutzfeldt-Jakob disease
                Alzheimer's disease
                Lewy body disease
    E. Infections and postinfectious causes
                Subacute sclerosing panencephalitis
                Encephalitis lethargica
                Arbovirus encephalitis
                Herpes simplex encephalitis
                Human immunodeficiency virus (HIV)
                Postinfectious encephalitis
                Malaria
                Cryptococcosis
    F. Metabolic disorders
                Hyperthyroidism
                Hepatic failure
                Renal failure
                Dialysis syndrome
                Hyponatremia
                Hypoglycemia
                Nonketotic hyperglycemia
                Multiple carboxylase deficiency
                Biotin deficiency
                Mitochondrial dysfunction
    G. Toxic and drug-induced syndromes (Table 27-3)
    H. Physical encephalopathies
                Posthypoxia (Lance-Adams)
                Post-traumatic
                Heat stroke
                Electric shock
                Decompression injury
    I. Focal central nervous system damage
                Poststroke
                Post-thalamotomy
                Tumor
                Trauma
                Inflammation (e.g., multiple sclerosis)
                Olivodendate lesions (palatal myoclonus)
                Peripheral nerve lesions
    J. Malabsorbtion
                Celiac disease
                Whipple's disease
    K. Eosinophilia-myalgia syndrome
    L. Paraneoplastic encephalopathies
    M. Opsoclonus-myoclonus syndrome
                Idiopathic
                Paraneoplastic

(Table 27-1 continued)

(Table 27-1 continued)
(Symptomatic myoclonus cont.)

           Brain stem lymphoma
           Sarcoidosis
           Cocaine
           Toluene
           Nonketotic hyperosmolar coma
           Stroke
N.   Exaggerated startle syndromes
           Hereditary
           Sporadic

Modified from Fahn S, Marsden CD, Van Woert MH. Definition and classification of myoclonus. *Adv Neurol* 1986; 43: 1-5. By permission of Raven Press.

Table 27-2
Advanced Testing for Myoclonus

Condition or disorder	Helpful diagnostic studies
Segmental lesion	Magnetic resonance imaging of segment
Paraneoplastic syndrome	Paraneoplastic antibodies, body imaging for occult cancer
Ataxia-telangiectasia	Serum alpha-fetoprotein level (high), cytogenetic analysis, radiosensitivity of DNA synthesis
Wilson's disease	Serum ceruloplasmin level (low), 24-hour urine copper level (high), slit-lamp examination for Kayser-Fleischer rings
Hallervorden-Spatz disease	Magnetic resonance imaging
Lafora's body disease	Skin biopsy: characteristic inclusion bodies in eccrine duct cells and peripheral nerve
Ceroid lipofuscinosis	Skin biopsy or lymphocytes: characteristic inclusions
Gaucher's disease	Beta-glucocerebroside activity in leukocytes (decreased)
Sialidosis type I	Neuraminidase activity in lymphocytes and fibroblasts (decreased)
Sialidosis type II	Neuraminidase activity in lymphocytes and fibroblasts (decreased)
Juvenile GM_2 gangliosidosis	Hexosaminidase A activity in lymphocytes and fibroblasts (decreased)
Tay-Sachs disease (infantile GM_2 gangliosidosis)	Hexosaminidase A activity in lymphocytes and fibroblasts (decreased)
Krabbe's disease	Galactocerebroside β-galactosidase activity in leukocytes (decreased)
Multiple carboxylase deficiency	Serum biotinidase activity (decreased)

Focal and Segmental Causes

Lesions placed at different locations along the neuraxis may produce myoclonus. Focal lesions in the cortex, from a variety of causes, can infrequently cause focal myoclonus. Even more rarely, myoclonus has occurred after a lesion in the thalamus.

Brain stem lesions involving the dentate–inferior olive pathway are believed to be important in the generation of palatal myoclonus. This myoclonus is the most common example that arises in a focal or segmental distribution. The movement is rhythmic and

Table 27-3
Toxins and Drugs Associated With Myoclonus

I. Toxins
  Bismuth
  Aluminum
  Inorganic mercury
  Organic mercury
  Tetraethyl lead
  Methyl bromide
  Dichloroethane (dry cleaning
    fluid)
  Oven cleaner
  Rapeseed oil (anilines)
  Chloralose
  Strychnine
  Tetanus toxoid
  Marijuana
II. Drugs
  A. Psychiatric medications
    Cyclic antidepressants:
      imipramine, desipramine,
      amitriptyline, doxepin,
      trazodone, nortriptyline,
      maprotiline
    Selective serotonin uptake
      inhibitors: fluoxetine,
      paroxetine, sertraline,
      fluvoxamine,
      clomipramine
    Monoamine oxidase
      inhibitors: phenelzine,
      clorigyline
    Lithium
    L-tryptophan
    Buspirone
    Methaqualone
    Bromisovalum
    Clozapine
    Tardive syndrome
      (antipsychotic use)
      Diazepam withdrawal
  B. Anti-infectious agents
    Penicillin
    Carbenicillin
    Ticarcillin
    Cefmetazole

  Monolactam
  Pefloxacin
  Isoniazid
  Piperazine
  Acyclovir
  Vidarabine
  C. Narcotics
    Morphine
    Hydromorphone
    Meperidine
    Diamorphine
    Fentanyl
    Sufentanil
  D. Anticonvulsants
    Phenytoin
    Hydantoin
    Valproic acid
    Carbamazepine
    Vigabatrin
  E. Anesthetics
    Etomidate
    Enflurane
    Isoflurane
    Tetracaine
    Midazolam
  F. Contrast media
  G. Cardiac medications
    Calcium channel blockers: diltiazem,
      nifedipine, verapamil
    Antiarrhythmic agents: flecainide,
      propafenone
  H. Antihistamines
    Pseudoephedrine
    Tripelennamine
  I. Antineoplastics
    Chlorambucil
    Prednimustine
  J. Miscellaneous medications
    Levodopa
    Bromocriptine
    Metoclopramide
    Physostigmine
    Tumor necrosis factor

From Caviness JN. Myoclonus. *Mayo Clinic Proc* 1996; 71: 679. By permission of Mayo Foundation for Medical Education and Research.

usually bilateral, with a rate between 1 and 4 Hz, other rates being less common. Because of the rate, rhythmic property, and relative slowness of the palatal movement in comparison with other examples of myoclonus, many clinicians believe that it more closely resembles a tremor phenotype than myoclonus. Palatal myoclonus may be associated with similar movements in other body segments. In many patients, no brain stem lesion is found, and these instances have been termed "essential palatal myoclonus." In contrast to "symptomatic palatal myoclonus," in which an identifiable lesion exists, the essential type occurs in younger patients, has slower rates, is more likely to be associated with an audible ear click as the chief complaint, and is less likely to involve other muscles. In essential palatal myoclonus, patients frequently seek medical attention because of the audible ear clicks. In symptomatic palatal myoclonus, patients are more likely to seek medical attention because of other neurologic problems associated with the lesion (e.g., stroke or trauma) than because of the palatal myoclonus per se.

The medullary reticular formation is believed to be the origin of "reticular reflex myoclonus" in humans. This rare type of myoclonus consists of generalized reflex-sensitive axial and proximal limb jerks, and occurrence may be a result of hypoxia or uremia. It has been suggested that urea can inhibit the binding of glycine to its receptor in the brain stem. Thus, this may be a possible mechanism for urea-induced reticular reflex myoclonus.

Tumor is the most commonly reported cause of spinal cord myoclonus, and viral infection, trauma, and ischemia are also known causes. The myoclonus originating from the spinal cord is usually rhythmic, is slow (less than 4 Hz), and involves only the musculature from a few spinal levels (spinal segmental myoclonus). A newly described type of myoclonus arising from the spinal cord, termed "axial propriospinal myoclonus," consists of axial jerks that seem to start from a single spinal level and spread both rostrally and caudally. These jerks are reflex sensitive and can involve several spinal levels as well as limb muscles. Investigators believe that the neuronal hyperactivity spreads through propriospinal pathways.

Peripheral nervous system lesions may cause myoclonus; however, it is unclear if the pathophysiology is wholly peripheral. Possibly, in these peripheral cases, the peripheral lesion affects central nervous system excitability, which in turn causes myoclonus.

OTHER MEDICAL CONDITIONS

A variety of medical illnesses can be uncommonly associated with myoclonus. These include celiac disease, Whipple's disease, eosinophilia-myalgia syndrome, and paraneoplastic syndromes. Other medical illnesses also are listed in Table 27-1 under the heading "opsoclonus-myoclonus syndrome" (see below).

OPSOCLONUS-MYOCLONUS SYNDROME

This syndrome can have a variety of symptomatic causes, of which idiopathic and paraneoplastic are the most common. Opsoclonus is the occurrence of involuntary, repetitive, rapid conjugate eye movements (ocular saccades) that are irregular in amplitude and frequency and occur in all directions without an intersaccadic interval. There is often a prodrome of systemic complaints, including nausea, vomiting, vertigo, oscillopsia (jerks in vision), and gait difficulties. The onset can be gradual or abrupt. When onset is gradual, during weeks to months, mild ataxia can occur before the onset of opsoclonus and myoclonus. More often, onset of the condition is rapid, and ataxia, opsoclonus, myoclonus, and other movement disorders occur together during a period of days to a few weeks. Multifocal action myoclonus is the most common form of myoclonus. There are clinical and etiologic differences between adult and pediatric presentations.

EXAGGERATED STARTLE SYNDROMES

Sporadic and hereditary forms of hyperekplexia (exaggerated startle reflex) have been described. Included in the clinical syndrome, besides a nonfatigable startle reflex, are hypertonia in infancy, hyperreflexia, falling episodes with retained consciousness, and tonic spasms. Autosomal dominant inheritance exists in the hereditary form, and a genetic mutation has been found in the glycine receptor.

DIAGNOSTIC TESTING

If a diagnosis is not apparent after a thorough history and physical examination, diagnostic testing is justified to arrive confidently at a diagnosis. Such testing is usually necessary because certain diagnoses need to have certain pathologic conditions ruled out for diagnostic certainty. Both the myoclonus and the other aspects of the clinical presentation determine what type of testing should be done. For example, if an infectious or inflammatory syndrome is present, the cerebrospinal fluid should be examined. Knowledge of the various diagnostic entities in Table 27-1 facilitates the proper diagnostic confirmation. The following minimal testing should be done in all unexplained cases of myoclonus:

Electrolytes
Glucose
Renal function tests
Hepatic function tests
Drug and toxin screen (if prompted by history)
Brain imaging
Electroencephalography

Electrolyte, glucose, kidney, and liver abnormalities can be particularly relevant in hospitalized patients. The most effective way to determine exposure to drugs and toxins is by directly questioning the patient or by reviewing medical records. However, if the patient is unconscious, if unknown poisoning is a possibility, or if abuse is suspected, a drug and toxin screen is useful. Imaging can show structural lesions and atrophic abnormalities of the brain. Magnetic resonance imaging is preferred over computed tomography for better visualization of the brain stem. An EEG provides information on specific and nonspecific patterns, both encephalopathic and epileptic. Examples include the spike-and-wave pattern in juvenile myoclonic epilepsy and absence seizure syndromes as well as the periodic discharges of subacute sclerosing panencephalitis and Creutzfeldt-Jakob disease.

If the combination of historical and physical information and basic testing does not reveal the diagnosis, more advanced testing (Table 27-2) should be considered. There are several hereditary causes of myoclonus. Inheritance can be dominant, recessive, or maternal. Genetic testing is commercially available for a few disorders (spinocerebellar ataxias, Huntington's disease, mitochondrial disorders, and others). Before genetic testing is done, the patient should be fully aware of the implications for both positive and negative results. If appropriate, genetic counseling is recommended.

Myoclonus pathophysiology can be broadly classified by its source as cortical, subcortical, or segmental. Of these, cortical is the most common. Short duration (less than 50 msec) surface electromyographic discharges, enlarged somatosensory evoked potentials, and exaggerated long latency electromyographic responses are electrophysiologic findings that support a cortical source.

TREATMENT

The ideal strategy for the treatment of myoclonus is, of course, treatment of the underlying disorder. Some causes of myoclonus can be reversed partially or totally, such as an acquired abnormal metabolic state, a removable medication or toxin, or an excisable lesion. In most instances of myoclonus, however, treatment of the underlying disorder is not possible.

Valproic acid and clonazepam (Klonopin) are the two most useful agents in the symptomatic treatment of myoclonus. Valproic acid is given at 1,200 to 2,000 mg/day in divided doses, and common side effects are nausea, vomiting, diarrhea, sedation, and liver toxicity. Clonazepam is used at 2 to 6 mg/day in divided doses, and common side effects are sedation, imbalance, ataxia, and behavioral changes. For both agents, it is useful to start with low doses and increase the medications slowly. This can result in a lower incidence of side effects.

Several other agents are reported to be particularly useful in certain myoclonic syndromes. 5-Hydroxytryptophan has been used with success in the treatment of posthypoxic (Lance-Adams) myoclonus, although many patients do not have a response. This drug is not currently approved by the Food and Drug Administration. A wide dosage range of 100 to 3,000 mg/day has been used. 5-Hydroxytryptophan is given with carbidopa to prevent peripheral metabolism. Anorexia, nausea, diarrhea, and mood changes often limit the use of 5-hydroxytryptophan. Drugs that act to block the reuptake of serotonin at nerve terminals can be used to lower the required dosage of 5-hydroxytryptophan and carbidopa and to decrease side effects. 5-Hydroxytryptophan has been used to treat other kinds of myoclonus as well, although the beneficial effect is not as great.

Piracetam, which is not available in the United States, has been reported to have antimyoclonic activity, particularly in combination with other agents. Piracetam is used at 2,400 to 16,800 mg and is usually well tolerated, although nausea and flatulence may occur. This drug is indicated only for cortical myoclonus. Anticholinergic agents, such as trihexyphenidyl (Artane) and benztropine (Cogentin), may be useful for treating rhythmic myoclonus, such as palatal myoclonus, and for treating essential myoclonus. The anticonvulsants phenytoin (Dilantin), phenobarbital, and carbamazepine (Tegretol) are used for rhythmic myoclonus; success varies. Benzodiazepines can be helpful in an attempt to treat acute myoclonic seizures after a severe hypoxic insult. Tetrabenazine is another drug that is considered experimental, but it has been used successfully in types of rhythmic myoclonus, such as palatal myoclonus and spinal myoclonus. Opiates and dopaminergic agents have been used to treat abnormal myoclonus during sleep (see Chapter 30).

Myoclonus treatment is often unsuccessful. High expectations for a marked decrease in or total elimination of the myoclonus are usually unrealistic. For example, rhythmic segmental myoclonus commonly can be refractory to all treatment trials. Multiple drug trials are sometimes necessary to find the best drug for a certain patient. Combination treatment regimens may be beneficial, particularly in cortical myoclonus. Sometimes the treatment of myoclonus may seem modest yet produce a significant decrease in disability, such as enabling a patient to more easily stand or walk. Our growing knowledge about myoclonus will serve to produce better treatment options.

SELECTED READING

Brown P. Myoclonus: a practical guide to drug therapy. *CNS Drugs* 1995; 3: 22-29.
Caviness JN. Myoclonus. *Mayo Clin Proc* 1996; 71: 679-688.
Marsden CD, Hallett M, Fahn S. The nosology and pathophysiology of myoclonus, in *Movement Disorders 1 and 2* (Marsden CD, Fahn S eds), Edition reissue, Butterworth-Heinemann, Oxford, 1995, pp. 196-248.
Pranzatelli MR, Snodgrass SR. The pharmacology of myoclonus. *Clin Neuropharmacol* 1985; 8: 99-130.

28 Spasms and Stiff-man Syndrome

C. Michel Harper, Jr., MD

CONTENTS

CENTRAL NERVOUS SYSTEM DISORDERS
PERIPHERAL NERVE DISORDERS
MUSCLE DISORDERS
SELECTED READING

A variety of central and peripheral nervous system diseases have muscle spasms or stiffness as the primary complaint (Table 28-1). Many of these disorders are discussed in other chapters. This chapter focuses on disorders in which muscle spasms or stiffness is produced by hyperexcitability of the upper motor neuron in the central nervous system, the motor axon in the peripheral nervous system, the neuromuscular junction, or the muscle. The alpha motor neuron receives segmental input from the muscle spindle, Golgi tendon organ, cutaneous and visceral afferents, Renshaw interneurons, and gamma motor neurons (Fig. 28-1). Interneurons inhibit the alpha motor neuron presynaptically through γ-aminobutyric acid (GABA) and postsynaptically through the release of glycine. Excitatory postsynaptic potentials are produced in alpha motor neurons by glutamate, aspartate, peptides, and catecholamines. Suprasegmental control of lower motor neuron excitability originates at the level of the cerebral cortex and the brain stem. Cortical influence is mediated primarily by the corticospinal tract but also indirectly through feedback loops connecting the cortex with the basal ganglia, cerebellum, and a number of brainstem nuclei.

CENTRAL NERVOUS SYSTEM DISORDERS

Spasticity

Spasticity refers to the velocity-dependent increase in muscle tone and stiffness associated with suprasegmental lesions that interfere with the descending motor pathways in the motor cortex (corticospinal tract), brain stem (rubrospinal and reticulospinal tracts), or spinal cord. The hallmarks of spasticity are the disinhibition of the tonic stretch reflex and the presence of the clasp-knife phenomenon. In addition to hyperactivity of deep tendon reflexes, disinhibition of the stretch reflex causes an increase in baseline muscle tone that is further enhanced by muscle stretch or other stimuli. Thus, cutaneous, prop-

From: *Parkinson's Disease and Movement Disorders:*
Diagnosis and Treatment Guidelines for the Practicing Physician
Edited by: C. H. Adler and J. E. Ahlskog © Mayo Foundation
for Medical Education and Research, Rochester, MN

Table 28-1
Causes of Muscle Stiffness or Spasms

Central nervous system disorders

Spasticity	Lesions of the corticospinal tract or other upper motor neuron pathways
Rigidity	Parkinsonism and other disorders of the basal ganglia
Dystonia	Idiopathic, acquired, focal, segmental, generalized
Anterior horn cell hyperexcitability	Stiff-man syndrome, progressive encephalomyelitis with rigidity, tetanus, strychnine poisoning, hereditary hyperekplexia

Peripheral nervous system disorders

Motor axon	Cramps, symptomatic tetany, cramp-fasciculation syndrome, focal or generalized myokymia, Isaacs' syndrome (neuromyotonia)
Neuromuscular junction	Acetylcholinesterase inhibitor intoxication
Muscle	Myotonic disorders (muscle channelopathies), rippling muscle disease, metabolic myopathies

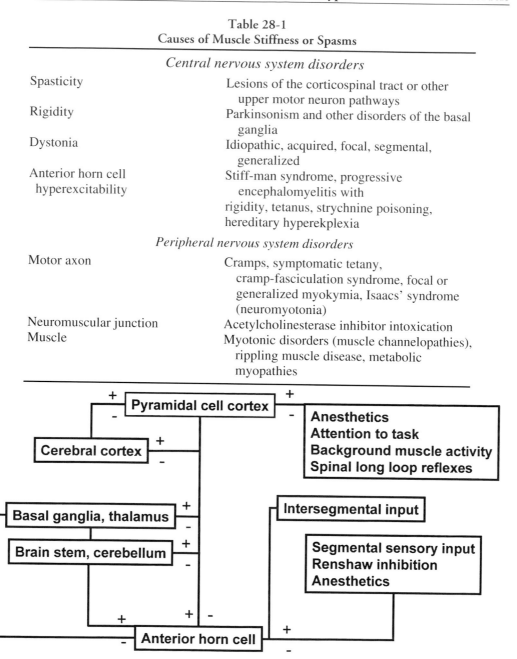

Fig. 28-1. Schematic of various excitatory (+) and inhibitory (-) inputs into the anterior horn cell.

rioceptive, or visceral stimulation can lead to painful transient "spasms" in spastic limbs. The clasp-knife phenomenon refers to a sudden decrease in tone that immediately follows the increased tone elicited by muscle stretch. This is an inhibitory reflex mediated by the Golgi tendon organ. Spasticity is typically associated with other "upper motor neuron" signs, such as extensor toe responses, muscle weakness, loss of dexterity, and dysfunction of the autonomic nervous system. If left untreated, spasticity eventually leads to permanent muscle shortening due to the formation of contractures.

Stiff-man Syndrome
and Progressive Encephalomyelitis With Rigidity

Stiff-man syndrome is a rare disorder of the central nervous system associated with hyperexcitability of alpha motor neurons. Moersch and Woltman initially described the disease in 1956. The diagnosis is made on clinical grounds assisted by electrophysiologic and serologic studies. The clinical picture is fairly characteristic, yet the disease is often misdiagnosed or attributed to psychosomatic factors.

The disorder typically begins with constant but fluctuating stiffness of axial muscles that spreads over weeks and months to involve proximal muscles of the limbs. Constant contraction of the paraspinal muscles produces characteristic accentuation of the lumbar lordosis and palpable tightness of the affected muscles. The gait becomes slow and labored by constant rigidity of paraspinal and lower extremity muscles. Painful spasms superimposed on underlying muscular rigidity develop in most patients. These spasms occur spontaneously or in response to emotional, auditory, cutaneous, or proprioceptive stimuli. They can produce enough force to fracture limbs and, along with constant muscle stiffness, eventually lead to complete immobilization of the patient if left untreated. Rarely, the stiffness and superimposed spasms are confined to one or both lower limbs. Cranial muscles are rarely if ever involved in stiff-man syndrome. This lack of involvement helps differentiate it from tetanus, with which it shares many other clinical features. The autonomic nervous system is affected in some patients, who have diaphoresis, tachycardia, and fluctuations in blood pressure. Sudden death, presumably related to autonomic instability, has been reported. Deep tendon reflexes are hyperactive, but other signs of spasticity are absent, and muscle strength is preserved.

An autoimmune pathogenesis for stiff-man syndrome is supported by the increased incidence of other autoimmune diseases in affected patients, the beneficial response to immunomodulating therapy, the existence of oligoclonal bands in the cerebrospinal fluid, and the detection of antibodies to glutamic acid decarboxylase in the spinal fluid or serum of 40% to 60% of patients with this disorder. These antibodies serve as a marker for the disease and probably contribute to the pathogenesis by reducing levels of GABA, an important inhibitory neurotransmitter, in the spinal cord.

Stiff-man syndrome can occur as a paraneoplastic disorder associated with thymoma, lymphoma, small cell lung cancer, and breast carcinoma. The term "progressive encephalomyelitis with rigidity" is used in this condition because there is often pathologic, clinical, and serologic evidence of immune-mediated neuronal degeneration in the spinal cord and other areas of the central nervous system. Thus, in addition to stiffness, rigidity, and painful intermittent spasms, these patients often have signs of motor neuron degeneration, sensory neuronopathy, autonomic neuropathy, myelopathy, limbic or brain stem encephalitis, or subacute cerebellar degeneration. Antibodies to amphiphysin, a vesicle-associated protein with important functions in the regulation of uptake and release of various neurotransmitters, are frequently detected in the serum of patients with stiff-man syndrome or progressive encephalomyelitis with rigidity associated with breast carcinoma.

The diagnosis of stiff-man syndrome is based primarily on clinical grounds. Key features include continuous stiffness and rigidity beginning in axial muscles and spreading over time to the extremities, an exaggerated lumbar lordosis that persists during flexion, painful transient spasms induced by movement or by auditory or tactile stimulation, a positive personal or family history of autoimmune disease, and clinical responsiveness to benzodiazepines or other GABA agonist medications. Anti–glutamic acid decarboxylase antibodies in the serum or cerebrospinal fluid confirm the diagnosis in the appropriate clinical situation. Electrophysiologic studies provide information in support of the diagnosis of stiff-man syndrome and help exclude conditions that may mimic the disorder. Needle electromyography (EMG) of rigid muscles demonstrates constant activity of otherwise normal motor unit potentials (MUPs) that cannot be suppressed by attempts to relax the muscle. Results of routine nerve conduction studies are normal except for high-amplitude H-reflexes. Abnormalities of the blink reflex and the acoustic startle reflex have been reported in some patients, suggesting brain stem involvement, although the effect is less severe than that in tetanus. Also, in contrast to tetanus, the silent period in both cranial and limb muscles, measured after either mechanical or electrical stimuli, is normal in stiff-man syndrome. Important features that help differentiate stiff-man syndrome from tetanus, strychnine poisoning, and progressive encephalomyelitis with rigidity are listed in Table 28-2.

Diazepam (Valium) is the mainstay of treatment in stiff-man syndrome. Characteristically, patients tolerate the very high doses of diazepam (40 to 100 mg/day) required to control the painful spasms and reduce the underlying constant contraction of axial muscles. Other drugs that enhance activity in GABA-containing neurons, such as baclofen (Lioresal), 30 to 120 mg/day; clonazepam (Klonopin), 1 to 4 mg/day; valproic acid (Depakote), 750 to 2,000 mg/day; and vigabatrin (not yet available in the U.S.), may also reduce stiffness and relieve spasms. Plasmapheresis, intravenous immunoglobulin, prednisone, azathioprine, and other immunosuppressive drugs have been used to suppress the autoimmune response with variable success. Intrathecal baclofen or intermittent injection of botulinum toxin into paraspinal muscles has been reported to benefit patients who are refractory to standard therapy.

Tetanus

Tetanus is produced by a 150-kd neurotoxin, tetanospasmin, released from *Clostridium tetani*. This organism is an anaerobic, gram-positive bacillus that is widespread in the environment, especially in soil contaminated with organic material. The disease is rare in the United States because of widespread immunization of the population. Most reported cases occur in neonates, in older adults who never received a primary immunization, and in intravenous drug users.

After the organism enters the skin, a strict anaerobic environment is required for growth. Thus, puncture wounds and injuries associated with soil contamination, foreign body introduction, or necrotic tissue favor growth of the organism and local production of toxin. Once produced, tetanospasmin is taken up by local nerve endings and conveyed by retrograde axonal transport to the central nervous system. Usually, tetanospasmin is transported by motor axons. Tetanospasmin can induce neuromuscular blockade and axonal degeneration in some cases, but the major site of action is in the central nervous system, where it inhibits the release of both GABA and glycine, an effect leading to disinhibition of alpha and gamma motor neurons. The exact mechanism by which tetanospasmin inhibits neurotransmitter release is unknown, but recent evidence indicates that the toxin interferes with synaptobrevin and other peptides that mediate vesicle docking and fusion to the membrane of the nerve terminal.

In mild cases, hyperactivity of lower motor neurons is confined to a single segment of the spinal cord (local tetanus) or brain stem (cephalic tetanus). More commonly, the toxin

Table 28-2
Differential Diagnosis of Stiff-man Syndrome

	SMS	PERM	Tetanus	Strychnine
Clinical manifestations	Axial and proximal limb rigidity with intermittent painful spasms, legs > arms, proximal > distal, exaggerated lordosis, often dramatic initial response to benzodiazepines	Similar to SMS but associated with cancer and one or more additional signs: motor neuron degeneration, sensory neuronopathy, myelopathy, limbic or brain stem encephalitis, cerebellar degeneration	Prominent cranial involvement, severe autonomic instability in generalized tetanus	Seizures, myoclonus, and encephalopathy in addition to rigidity and spasms of cranial, axial, and limb muscles
Temporal course	Subacute to chronic, progressive	Subacute, progressive (more rapid than SMS)	Acute, fulminant	Acute, fulminant
Laboratory studies	Anti-GAD antibodies	Other paraneoplastic antibodies, including anti-amphiphysin	May culture *Clostridium tetani* from wound	
Electrophysiologic studies	Normal EMG except failure to relax MUPs, normal silent period and startle reflexes	Abnormal NCS and EMG in patient with motor neuron or sensory neuron degeneration, silent period normal or abnormal, abnormal startle reflexes	Normal EMG except failure to relax MUPs, abnormal silent period and startle reflexes	Normal EMG except failure to relax MUPs, abnormal silent period and startle reflexes, abnormal EEG

EEG, electroencephalogram; EMG, electromyogram; GAD, glutamic acid decarboxylase; MUP, motor unit potential; NCS, nerve conduction studies; PERM, progressive encephalomyelitis with rigidity; SMS, stiff-man syndrome.

causes diffuse disinhibition of lower motor neurons innervating both skeletal and smooth muscle. Involvement of smooth muscle results in the classic clinical features of generalized tetanus, including trismus (continuous contraction of muscles of mastication), risus sardonicus (facial grimace caused by contraction of facial muscles), and opisthotonos (extension of spine and boardlike rigidity of the limbs). As in stiff-man syndrome, constant generalized muscle contraction occurs with superimposed painful focal or generalized tonic spasms. These spasms occur spontaneously and are triggered by any type of sensory stimulus, voluntary movement, or emotional distress.

In untreated patients, death occurs from tonic spasms of pharyngeal, laryngeal, and respiratory muscles or from autonomic instability related to disinhibition of autonomic motor neurons in the spinal cord and brain stem. Cardiac arrhythmias, labile blood pressure, diaphoresis, and fever often persist and continue to pose a major risk even when the muscle spasms and stiffness are controlled with neuromuscular blocking agents. Generalized seizures can occur in severe cases, but in general, the neurons of the cerebrum are less likely to be affected by tetanospasmin than the lower motor neurons. Despite aggressive supportive treatment, the case-fatality rate remains between 20% and 50% in generalized tetanus.

The diagnosis of tetanus is primarily clinical. Acute onset of symptoms, involvement of cranial muscles, and history of exposure in an incompletely immunized person are key features in the diagnosis. Rarely, *Clostridium tetani* can be cultured from the point of entry. Electrophysiologic study findings are relatively nonspecific. On needle EMG, spontaneous activity is difficult to assess because of excess MUP activity. MUP size and morphology are normal, but in generalized tetanus, even minimal stimulation or attempts at voluntary movement trigger painful, prolonged generalized tonic spasms associated with a full interference pattern on EMG recordings. Results of standard nerve conduction studies are usually normal except for large H-reflexes. The acoustic startle reflex and other brain stem reflexes are greatly increased. The silent period is absent in clinically involved muscles.

The treatment of generalized tetanus is limited to supportive measures. The point of entry should be identified, debrided, and cleansed. Human tetanus immunoglobulin and intravenous antibiotics help prevent the formation of additional toxin, but there is no way to eliminate the toxin that has already gained access to the nervous system. Mild cases can be treated with high doses of benzodiazepines, but most patients with generalized tetanus require complete neuromuscular blockade and mechanical ventilation. Manifestations of autonomic instability should be treated and can often be minimized by heavily sedating the patient and limiting unnecessary verbal or cutaneous stimulation. Neuromuscular blocking agents and sedation may have to be continued for several weeks or even months in severely affected patients. This extended duration emphasizes the importance of good general medical and nursing care to prevent complications such as deep venous thrombosis, decubiti, and secondary infections. Patients who survive the acute complications of the illness usually recover completely over several months.

Strychnine Poisoning

Strychnine poisoning is very similar to tetanus clinically. Strychnine inhibits the release of glycine in the central nervous system. Reduction of glycine-mediated inhibition at the spinal, brain stem, and cortical levels leads to generalized muscle stiffness, tonic muscle spasms, and generalized tonic-clonic seizures. Trismus and generalized muscle rigidity can lead to death due to asphyxiation, aspiration, rhabdomyolysis, and lactic acidosis. Because the toxin has a very short half-life, the symptoms are of much shorter duration than with tetanus (several days). Neurophysiologic studies have not been reported. Treatment is supportive and similar to that of tetanus. If muscle rigidity, autonomic instability, and seizures are treated aggressively, many patients recover completely.

Hereditary Hyperekplexia

Hereditary hyperekplexia is caused by a point mutation in the gene for the α_1 subunit of the glycine receptor. The mutation reduces the affinity of the receptor for glycine, leading to a reduction in glycine-mediated inhibition in the brain stem and spinal cord. Most patients have symptoms in infancy, with generalized increased muscle tone producing a flexed posture of the trunk and limbs. Muscle stiffness becomes less severe over time into adulthood, producing a slow, laborious gait and mild generalized rigidity. The most prominent symptom in this disorder is an exaggerated generalized myoclonic jerk or spasm in response to an unexpected startle. The startle response in this condition is severe enough to produce injury secondary to generalized stiffness that produces falls and abolishes protective reflexes. The startle response can be measured electrically by recording surface EMG signals over cranial, axial, and limb muscles after a brief, unexpected auditory stimulus. Clonazepam, 1 to 4 mg/day, has been shown to diminish the startle response and reduce muscle stiffness in most patients.

PERIPHERAL NERVE DISORDERS

Disorders associated with hyperexcitability of the motor unit along its course from the anterior horn cell to the motor axon and nerve terminal can produce a wide variety of clinical disorders characterized by constant or intermittent involuntary muscle contraction. The involuntary activity can be focal or diffuse and at times is associated with myalgia or hypertrophy (or both) of involved muscles. Clinical syndromes associated with hyperexcitability of the motor axon are listed in Table 28-3; included are common cramps, symptomatic tetany, and the syndrome of continuous muscle fiber activity. The continuous muscle fiber activity syndrome includes cramp-fasciculation syndrome, focal or generalized myokymia, and Isaacs' syndrome. Each of the continuous muscle fiber activity syndromes can be inherited or acquired and can occur with or without a peripheral neuropathy. Characteristic abnormalities on nerve conduction studies and needle EMG help classify and differentiate disorders of peripheral nerve hyperexcitability.

Common Cramps and Symptomatic Tetany

The common muscle cramp is characterized by the sudden onset of painful muscle contraction that is often precipitated by muscle shortening and relieved by muscle stretch. On EMG, these symptoms are associated with a high-frequency discharge of otherwise normal MUPs that characteristically builds in a crescendo pattern. Both clinically and by EMG, cramps are usually preceded and followed by a flurry of fasciculations, which by definition represent random depolarization of individual motor units within the muscle. Muscle cramps with or without fasciculation potentials frequently occur in normal persons, especially after exercise, with use of caffeine or other stimulants, or in association with metabolic disturbances. They usually are confined to lower extremity muscles but

Table 28-3
Clinical Syndromes Associated
With Hyperexcitability of the Motor Axon

Common cramps
Symptomatic tetany
Continuous muscle fiber activity syndrome
    Cramp-fasciculation syndrome (normocalcemic or
        idiopathic tetany)
    Focal or generalized myokymia
    Isaacs' syndrome (neuromyotonia)

occasionally involve upper limb, axial, or cranial muscles. The term "benign cramps or fasciculations" is used when there are no other symptoms or signs of neurologic disease and the cramps respond to simple measures, that is, replacement of fluid and electrolytes, correction of other metabolic disturbances, regular stretching exercises, avoidance of caffeine and other stimulants, and treatment with calcium, magnesium, or quinine. Cramps can also occur during intoxication with acetylcholinesterase inhibitors (e.g., organophosphates, pyridostigmine) or in patients with diseases that damage or cause degeneration of the anterior horn cell or peripheral motor axon (Table 28-4). Other manifestations of damage to the motor unit (e.g., atrophy, weakness, reduced reflexes) are present in these disorders.

Symptomatic tetany occurs in association with electrolyte disturbances and is characterized by frequent and disabling cramps and fasciculations that are widespread in distribution, worsened by hyperventilation, and often triggered by percussion of a nerve (e.g., Chvostek's sign) or partial limb ischemia (e.g., Trousseau's carpopedal spasm). Although nonspecific, the diagnosis is confirmed when these manifestations occur with hypocalcemia, hypomagnesemia, or another electrolyte disturbance and resolve when the metabolic disturbance is corrected.

Cramp-Fasciculation Syndrome

The cramp-fasciculation syndrome (CFS) has many features in common with tetany, although no underlying metabolic disorder can be identified. Clinically, patients complain of constant but fluctuating fasciculations and frequent muscle cramps that often affect unusual muscles (e.g., hands, abdominal wall, paraspinal, thigh) and occur with sufficient frequency and severity to be disabling. Patients may also complain of migratory paresthesias, which may represent hyperexcitability of peripheral sensory neurons. There is clinical or electrophysiologic evidence of a peripheral neuropathy in some patients, but in most, no other peripheral nerve pathologic condition is identified. Electrophysiologic studies reveal indirect evidence of peripheral nerve hyperexcitability. Electrical stimulation of the peripheral nerve may produce repetitive firing of F waves, and repetitive stimulation may induce persistent MUP activity or cramp discharges. Needle EMG reveals frequent fasciculation potentials, MUPs that fire in doublets or triplets, and frequent cramp discharges. The cramp discharges may occur spontaneously or be triggered by either hyperventilation or partial limb ischemia. Electrolytes and laboratory values of renal, liver, and endocrine function are normal.

Table 28-4
Disorders Associated With Fasciculations and Muscle Cramps

Idiopathic
    Benign cramps or fasciculations, or both (often associated with
       exercise)
Metabolic disorders
    Fluid and electrolyte imbalance, uremia, hemodialysis, hepatic
       insufficiency, pregnancy, hypothyroidism, corticosteroid excess
       or deficiency
Motor neuron disease
    Amyotrophic lateral sclerosis, spinal muscular atrophy,
       poliomyelitis (including old polio)
Disorders of the motor axon
    Radiculopathy, peripheral neuropathy, continuous muscle fiber
       activity (cramp-fasciculation syndrome, focal or generalized
       myokymia, Isaacs' syndrome)
Disorders of neuromuscular transmission
    Acetylcholinesterase inhibitor intoxication

There is growing evidence that CFS is an autoimmune disease in some patients. This possibility is supported by (1) a history of preceding viral or flulike illness in some patients; (2) occurrence of CFS in patients with a history of other autoimmune diseases (e.g., myasthenia gravis, connective tissue disease, autoimmune thyroid disease) or with occult malignant disease; (3) clinical and electrophysiologic similarity of CFS to Isaacs' syndrome, which is associated with antibodies to potassium channels; and (4) beneficial response to immunomodulating therapy in some patients with CFS.

Treatment of mild cases of CFS is the same as that of benign cramps. Quinine is often effective, and some patients are helped by calcium or magnesium supplements, even if serum electrolyte concentrations are normal. Patients with more severe CFS often have a response to anticonvulsants (carbamazepine, phenytoin, phenobarbital, valproic acid), GABA agonists (benzodiazepines, baclofen), or other membrane stabilizing medications (quinidine sulfate, procainamide, mexilitine). If these medications are unsuccessful and the cramps are disabling, corticosteroids can be tried; however, the risks of long-term steroid therapy should be carefully weighed against the severity of the symptoms in each patient.

Focal or Generalized Myokymia

Clinical myokymia refers to the characteristic "bag of worms" appearance of the skin caused by semirhythmic involuntary contractions of groups of motor units in the underlying muscle. Frequent fasciculations, cramps, and slight voluntary contraction of the muscle may produce an appearance similar to clinical myokymia. On needle EMG, myokymic discharges consist of groups of MUPs firing in a semirhythmic, involuntary pattern. Myokymic discharges are thought to be generated by ephaptic transmission of action potentials between adjacent axons. They can be seen in neuropathic lesions that are focal (e.g., carpal tunnel syndrome, radiculopathy, brain stem lesions affecting the intra-axial portion of the facial nerve), multifocal (e.g., radiation plexopathy), or generalized (acute or chronic inflammatory demyelinating neuropathy). Generalized myokymia can be caused by gold therapy and snake venom, and it can be observed in hereditary conditions such as Charcot-Marie-Tooth disease, familial episodic ataxia (linked to the potassium channel gene on chromosome 12p13), and the Schwartz-Jampel syndrome. The last-named, which is also referred to as "chondrodystrophic myotonia," is characterized by generalized myokymia, flexion contractures of limb muscles, dwarfism, kyphoscoliosis, and blepharophimosis. Myokymia is an important marker for the diseases outlined above but does not by itself require treatment.

Isaacs' Syndrome

As originally described by Isaacs in 1961, this syndrome is characterized by the insidious onset and progression of generalized muscle stiffness. The stiffness may be associated with fasciculations, clinical myokymia, and occasional cramps, although muscle pain is usually not a prominent feature. Eventually, the constant motor unit activity produces generalized muscle hypertrophy. In addition, symptoms of a hypermetabolic state with heat intolerance, sweating, and tachycardia may be noted. Electrophysiologic study findings are very similar to those in CFS, although characteristic high-frequency (200 to 400 Hz) neuromyotonic discharges (in addition to fasciculations, cramps, myokymic discharges, and doublets and triplets) are noted on needle EMG.

The similarities between CFS and Isaacs' syndrome suggest that these disorders are on opposite ends of a continuous spectrum. The continuous muscle fiber activity of Isaacs' syndrome (and, when studied, most instances of CFS as well) persists during sleep and after peripheral nerve block but is abolished by neuromuscular blockade. This observation suggests that the nerve terminal very likely is the site of origin of peripheral nerve hyperexcitability in these disorders. Antibodies directed against potassium channels of the nerve terminals have been detected in patients with Isaacs' syndrome. These antibodies block potassium channels and produce recurrent spontaneous nerve action potentials in vitro. Isaacs' original patients had a response to carbamazepine or phenytoin. If these agents are unsuccessful, the same treatment regimen outline for CFS above (including immunomodulating therapy) should be used.

MUSCLE DISORDERS

Myotonic Disorders (Muscle Channelopathies)

Muscle stiffness may be due to clinical myotonia, which is characterized by prolonged contraction and delayed relaxation of muscle. Myotonia may occur spontaneously or be triggered by voluntary activation or percussion of the muscle. In some disorders, changes in ambient temperature, diet, and exercise affect the frequency and severity of clinical myotonia. In conditions associated with frequent generalized myotonia, muscle hypertrophy or pain (or both) may occur. Myotonia results from electrical instability of the muscle membrane. The membrane is depolarized more easily than normal, and prolonged, repeated depolarizations may occur. In general, these disorders result from an alteration in the structure or function of ion channels within the muscle membrane. The underlying genetic defect responsible for most of the muscle channelopathies has been identified (Table 28-5). Clinical myotonia occurs in all inherited channelopathies except hypokalemic periodic paralysis. The clinical and electrophysiologic characteristics of the myotonic disorders are outlined in Table 28-6. On needle EMG, myotonic discharges always accompany clinical myotonia. These discharges have a characteristic "dive-bomber" sound because of

Table 28-5
Muscle Channelopathies

Disease	Genetic defect
Myotonic dystrophy	Protein kinase
Myotonia congenita	Chloride channel
Paramyotonia congenita	Sodium channel
Potassium-sensitive myotonia (myotonia fluctuans)	Sodium channel
Proximal myotonic myopathy	Sodium channel
Hyperkalemic periodic paralysis	Sodium channel
Hypokalemic periodic paralysis	Calcium channel

their waxing and waning frequency. Some conditions produce myotonic discharges without clinical myotonia (e.g., inflammatory myopathies, acid maltase deficiency, hypothyroidism, cholesterol-lowering medications). Clinical myotonia can be treated by avoiding precipating factors (e.g., cold; foods high in sodium, potassium, and carbohydrates; excessive exercise) and by a variety of membrane-stabilizing medications (e.g., phenytoin, carbamazepine, valproic acid, mexilitine, quinidine sulfate, procainamide).

Rippling Muscle Disease

Rippling muscle disease is a rare disorder signaled by subjective muscle stiffness and myalgias, usually after exercise. Characteristic "waves" or "ripples" are observed over the muscle surface and are typically triggered by stretch or percussion of the affected muscle. When the muscle is examined with needle EMG during the involuntary contractions, there may be electrical silence or involuntary discharge of MUPs similar in appearance to a voluntary contraction. Most cases associated with electrical silence are inherited, and the defect in at least one kindred has been linked to the long arm of chromosome 1. Acquired cases of rippling muscle disease may be autoimmune in origin. No well-documented forms of treatment have been reported.

Metabolic Myopathies

Most metabolic myopathies can produce proximal weakness, myalgias, and sometimes myoglobinuria. Muscle diseases associated with painful muscle episodic "spasms" (actually, electrically silent contractures) are confined to disorders of glycogen or lipid metabolism (Table 28-7). Disorders of glycolysis are more likely to produce muscle contracture under conditions of anaerobic exercise. The ischemic forearm exercise test shows a failure of serum lactate to increase under anaerobic conditions. Disorders of lipid metabolism are more likely to produce myalgias and subjective tightness than true muscle contractures, especially during endurance (aerobic) exercise, when free fatty acids form the main source of fuel for skeletal muscle. Mitochondrial disorders may produce fatigue, weakness, and myalgias but are not typically associated with muscle stiffness or spasms.

Table 28-6
Myotonic Disorders

Disease	Clinical manifestations	EMG findings
Myotonic dystrophy	Autosomal dominant with variable expression; myotonia, facial and distal weakness, cataracts, baldness, cardiac conduction defects, mild mental retardation, glucose intolerance, gastrointestinal dysmotility, testicular atrophy	Small MUPs (secondary to myopathy); myotonic discharges, increase with cold, decrease with exercize or warming
Proximal myotonic myopathy	Autosomal dominant; myotonia, proximal weakness, cataracts	Same as myotonic dystrophy
Myotonia congenita	Autosomal dominant (Thomsen's) and recessive (Becker's) forms; myotonia, muscle hypertrophy, rarely proximal weakness, rarely myalgia; mild increase in myotonia in the cold but rapid disappearance with exercise or warming	Rarely, small MUPs; myotonic discharges (shorter bursts and higher frequency than with myotonic dystrophy), increase with cold, decrease with exercise or warming
Paramyotonia congenita	Autosomal dominant with variable expression; myotonia with or without myalgia that increases dramatically in the cold or with exercise; prolonged exposure to cold or exercise may induce transient weakness or paralysis	Rarely, small MUPs; myotonic discharges similar to those of myotonia congenita, except that discharges initially increase and then decrease with cold and exercise, sometimes leading to transient paralysis
Hyperkalemic periodic paralysis	Autosomal dominant with variable expression; same as paramytonia except that episodic muscle weakness more prominent; episodes of weakness last 2 to 24 hours and are associated with hyperkalemia	Same as paramyotonia c ogenita; prolonged exercise test may be positive (decrease in amplitude of compound muscle action potential 30 to 40 minutes after 5 minutes of isometric exercise)
Potassium-sensitive myotonia	Autosomal dominant with variable expression; fluctuating myotonia unaffected by temperature but worse during period of rest after exercise or with exposure to potassium	Myotonic discharges in symptomatic muscles

EMG, electromyographic; MUP, motor unit potential.

Table 28-7
Metabolic Myopathies Associated With Muscle Stiffness and Spasms

Disorders of glycogen metabolism

Myophosphorylase deficiency (McArdle's disease)
 Autosomal recessive; myalgias; stiffness and muscle contractures
  with or without myoglobinuria induced by anaerobic exercise;
  proximal myopathy may develop; contractures associated with
  increased serum creatine kinase that are electrically silent on
  needle EMG
Phosphorylase *b* kinase deficiency
 Autosomal recessive; one of four syndromes; manifestations similar
  to those of McArdle's disease in childhood
Phosphofructokinase deficiency
 Autosomal recessive, male predominance; multiple clinical
  syndromes, but the most common is very similar to McArdle's
  disease
Phosphoglycerate kinase deficiency
 X-linked recessive; multiple syndromes; rarely, manifestations
  similar to those of McArdle's disease, especially in symptomatic
  heterozygotic females
Phosphoglycerate mutase deficiency
 Autosomal recessive; manifestations similar to those of McArdle's
  disease, although milder

Disorders of lipid metabolism

Carnitine palmityltransferase (CPT) deficiency
 Autosomal recessive; myalgias; muscle stiffness and spasms with or
  without myoglobinuria induced by prolonged aerobic exercise,
  fasting, exposure to cold, or general anesthesia
Acyl-CoA dehydrogenase deficiency
 Autosomal recessive; defect of one of multiple enzymes that
  perform beta oxidation of free fatty acids; multiple syndromes,
  including one similar to CPT deficiency

SELECTED READING

Auger RG. AAEM minimonograph #44: diseases associated with excess motor unit activity. *Muscle Nerve* 1994; 17: 1250-1263.

Harper CM Jr. Myopathies, in *Pediatric Clinical Electromyography* (Jones HR Jr, Bolton CF, Harper CM Jr eds), Lippincott-Raven, Philadelphia, 1996, pp. 387-443.

Isaacs H. A syndrome of continuous muscle-fibre activity. *J Neurol Neurosurg Psychiat* 1961; 24: 319-325.

McEvoy KM. Stiff-man syndrome. *Semin Neurol* 1991; 11: 197-205.

Moersch FP, Woltman HW. Progressive fluctuating muscular rigidity and spasm ("stiff-man" syndrome): report of a case and some observations in 13 other cases. *Proc Staff Meet Mayo Clin* 1956; 31: 421-427.

Montecucco C, Schiavo G. Tetanus and botulism neurotoxins: a new group of zinc proteases. *Trends Biochem Sci* 1993; 18: 324-327.

Nishiyama T, Nagase M. Strychnine poisoning: natural course of a nonfatal case. *Am J Emerg Med* 1995; 13: 172-173.

Pascuzzi RM. Schwartz-Jampel syndrome. *Semin Neurol* 1991; 11: 267-273.

Smith BA. Strychnine poisoning. *J Emerg Med* 1990; 8: 321-325.

Solimena M, Folli F, Aparisi R, Pozza G, De Camilli P. Autoantibodies to GABA-ergic neurons and pancreatic beta cells in stiff-man syndrome. *N Engl J Med* 1990; 322: 1555-1560.

Tahmoush AJ, Alonso RJ, Tahmoush GP, Heiman-Patterson TD. Cramp-fasciculation syndrome: a treatable hyperexcitable peripheral nerve disorder. Neurology 1991; 41: 1021-1024.

Weinstein L. Tetanus. N Engl J Med 1973; 289: 1293-1296.

29

Gilles de la Tourette's Syndrome and Tic Disorders

Kathleen A. Kujawa, MD, PhD
and Christopher G. Goetz, MD

CONTENTS

CLINICAL FEATURES

Tics can be characterized as repetitive, usually rapid and brief darting movements (motor tics) or sounds made through the nose, mouth, or throat (vocal tics). Unlike many other movement disorders, tics are not constantly present and naturally wax and wane. The disorder can be very severe for short periods and then abate and even transiently disappear. Exacerbation of tic activity can occur when the patient is anxious, excited, or sleep-deprived. However, tics tend to decrease when the patient is involved in a focused activity and become more prominent during relaxation. Whereas any body part can be affected, the eyes, face, and neck are the most common areas of tic involvement. An important characteristic of tics is voluntary suppression, and patients, especially adults, can often keep their tics suppressed for minutes at a time.

Motor and vocal tics can be simple or complex. Simple motor tics are sudden, brief movements restricted to a particular body part, such as eye blinking, head jerks, and lip pouting. Less commonly, they can be sustained twisting movements, which are then considered tonic or dystonic tics. These dystonic tics are bursts of movement that can cause a posturing of short duration, but they do not tend to be continuous. Simple motor tics also tend to express themselves in a series, such as several eye blinks or a run of

From: *Parkinson's Disease and Movement Disorders:*
Diagnosis and Treatment Guidelines for the Practicing Physician
Edited by: C. H. Adler and J. E. Ahlskog © Mayo Foundation
for Medical Education and Research, Rochester, MN

shoulder shrugs. Simple vocal tics can appear as single meaningless sounds, such as sniffing, grunting, barking, hissing, or clearing of the throat.

More complicated movements that involve several muscle groups are considered complex motor tics. These movements can appear purposeful, such as touching other people, throwing, jumping, clapping, gesturing obscenely (copropraxia), and repeating movements of other people (echopraxia), or nonpurposeful, such as kicking, gyrating, and performing sequenced facial movements, such as biting the mouth, rolling the eyes, and sticking out the tongue. Motor tics are repetitive but not necessarily identical each time they occur, and over time they can disappear from one part of the body only to reappear in another. Complex vocal tics consist of meaningful words, phrases, or sentences and can include echolalia (repeating words of others), palilalia (repeating one's own words), and coprolalia (using obscene words). The tics may occur at points of language transition, such as at the beginning of a sentence, or at transition points resulting in altered volume, emphasized word, or accented syllable.

Patients with tics describe an unpleasant urge to perform the movement, and performing such a movement transiently relieves the sensation. The patient can often voluntarily suppress the tic for various lengths of time. But when the tic is suppressed, the patient experiences an inner feeling of discomfort that can be relieved only by a rebound burst of tic activity. Often, the patient suppresses the tic while in the physician's office, so that the examination findings do not coincide with the history. If the patient is left alone in a room while being videotaped, the tic often becomes quite apparent.

Gilles de la Tourette's syndrome (GTS) is a disorder that begins before age 21 years and is characterized by fluctuating multiple motor and vocal tics lasting at least 1 year. Often, these tics occur in association with comorbid conditions, including attention-deficit hyperactivity disorder (ADHD) and obsessive-compulsive disorder (OCD). ADHD, characterized by difficulty with attention and concentration, impulsivity, and motor hyperactivity, can precede the development of tics. Obsessions (involuntary thoughts or impulses) and compulsions (behaviors in response to obsessions) can become apparent at the same time as tics but usually appear later in the clinical course. They can be difficult to differentiate from complex tics.

ETIOLOGY

Chronic tics and GTS are generally accepted to be genetic disorders. Despite familial studies suggesting that chronic tics and GTS may have an autosomal dominant inheritance pattern with variable expression and gender-specific penetrance, studies to localize the responsible gene have been unsuccessful.

An autoimmune mechanism causing or exacerbating tic disorders, and possibly OCD, has been suggested. Testing a hypothesis that infection with group A streptococci can trigger an autoimmune reaction, a study was performed in which immunologic treatments, such as plasmapheresis, intravenously administered immunoglobulin, and high doses of prednisone, were used in four patients with acute onset or worsening of tics or OCD. Two patients had evidence of a recent group A streptococcal infection, two had a history of a recent viral illness, and all had a clinically significant response to treatment. Although no specific association has been established, this area of research is worthy of physician attention. Some movement disorder specialists obtain a streptococcal throat culture whenever tic exacerbation occurs. Any patient with a positive throat culture is then treated with a full course of antibiotics.

PATHOPHYSIOLOGY

Abnormalities in neurotransmitter function, specifically dopamine, serotonin, and norepinephrine metabolism, may be related to the motor movements seen in GTS. Indeed, dopamine blocking agents, such as haloperidol (Haldol), can relieve the clinical symptoms in patients with GTS. The noradrenergic system has been implicated because of the therapeutic effects of clonidine (Catapres), a drug that reduces noradrenergic activity. In studies of biogenic amine metabolism in samples of cerebrospinal fluid, mean baseline levels of homovanillic acid, which correlate with dopamine levels in the brain, were significantly lower in children with GTS. Cerebrospinal fluid levels of 5-hydroxyindoleacetic acid, which reflect serotonin levels, were normal, yet serotonin turnover appears impaired in some patients.

A study of biogenic amine metabolism before and after haloperidol treatment revealed a significant increase in homovanillic acid levels toward the normal range. This suggests that patients with GTS have an increased sensitivity of the dopamine receptor, resulting in reduced release of dopamine through negative feedback to the presynaptic dopamine neuron. Haloperidol, by blocking the presynaptic dopamine receptors, reduces the negative feedback to the presynaptic neuron and therefore could allow more dopamine to be released into the synaptic cleft.

The cholinergic, noradrenergic, GABAergic, second-messenger cyclic adenosine monophosphate, and endogenous opioid systems have also been theorized to contribute to the pathophysiology of GTS, but all lack confirmation. Norepinephrine, determined by baseline levels of the metabolite 3-methoxy-4-hydroxyphenylethylene glycol, is normal in affected patients. In clinical practice, biogenic amine levels in cerebrospinal fluid are not useful for evaluating for tic disorders.

In addition to biochemical dysfunction, neuroimaging techniques have been used to assess structural and functional abnormalities in patients with tic disorders. Subtle structural changes in the basal ganglia, particularly the caudate nucleus, may contribute to the pathologic changes of GTS. A study evaluating magnetic resonance imaging (MRI) in monozygotic twin pairs with tic disorders revealed differences in the right caudate volume in the more severely affected twin. However, MRI findings in most subjects with tics appear normal. Functional neuroimaging studies, such as cerebral blood flow recorded by single-photon emission computed tomography, have also been used to characterize the pathologic characteristics of GTS. A relative increase in metabolic activity and regional cerebral blood flow to certain frontal cortical areas in OCD and decreased relative metabolic activity and regional cerebral blood flow to the basal ganglia of patients with GTS are consistent with suggestions that these are the regions affected in this disease process. However, normal striatal dopamine metabolism was found in patients with GTS and OCD when positron emission tomography was used to quantify dopamine receptors in the brain.

DIFFERENTIAL DIAGNOSIS

Inherited disorders that may feature tics include Huntington's disease, neuroacanthocytosis, Wilson's disease, and torsion dystonia (Table 29-1). Tics can develop as a consequence of medications, including antiepileptic agents and levodopa. Head trauma, stroke, carbon monoxide poisoning, and infections (encephalitis, prion diseases, Sydenham's chorea) have also been implicated.

Table 29-1
Differential Diagnosis of Tic Disorders

Inherited disorders
   Huntington's disease
   Neuroacanthocytosis
   Torsion dystonia
   Wilson's disease
Medication-related tics
   Antiepileptic agents
   Levodopa
   Long-term use of neuroleptic agents (tardive dyskinesia)
   Stimulants
Infection
   Encephalitis
   Prion disease
   Sydenham's chorea
Other
   Head trauma
   Stroke
   Carbon monoxide poisoning

Tics can be a manifestation of tardive dyskinesia associated with long-term use of neuroleptic agents. Tics have also been described in association with use of central nervous system stimulants, although the direct role of these drugs in their causation or exacerbation has not been fully studied. In patients with ADHD who require methylphenidate (Ritalin) or amphetamine products, judicious use is important for those who have a history or a family history of tic disorders, since some patients may experience an increase in their tics.

DIAGNOSTIC ASSESSMENT

The diagnosis of a tic disorder can be relatively straightforward after a thorough history and careful physical examination. In the clinician's office, results of neurologic examination may be normal, with no evidence of vocal or motor tics, because tics are transiently suppressed. It therefore may be useful to videotape the patient in a quiet environment, without medical personnel, to document the type and frequency of the tic disorder. The usefulness of videotaping patients was demonstrated in a study in which tic frequency in the examiner's presence was only 27% of that after the examiner left the room.

When the history and physical examination, with or without videotape, suggest a primary tic disorder, an extensive laboratory or neuroradiologic assessment is usually not necessary. A blood smear (to detect acanthocytes), thyroid function tests, and serum ceruloplasmin level (low in Wilson's disease) are usually adequate. A slit-lamp examination by an ophthalmologist to rule out Kayser-Fleischer rings (seen in Wilson's disease) may also be ordered. If the presentation is acute, without a family history of the disorder, a toxicology screen, antistreptolysin O titers, and imaging studies (MRI) may be indicated. Because ADHD and OCD are commonly encountered comorbid conditions, neuropsychologic testing may be indicated in evaluating for GTS.

Table 29-2
Treatment of Tics and Related Disorders

Indication	Medication	Usual initial dose	Usual maximum dose (per day)	Possible side effects
Tics	Clonidine (Catapres)	0.1 mg at bedtime	0.8 mg	Sedation, hypotension, insomnia, dry mouth
	Haloperidol (Haldol)	0.25 mg at bedtime	20 mg	Sedation, acute dystonia, parkinsonism, tardive dyskinesia
	Pimozide (Orap)	1 mg at bedtime	10 mg	As above, prolonged QT interval
ADHD	Clonidine (Catapres)	As above	60 mg (children)	Anxiety, agitation, insomnia, decreased seizure threshold, worsening of tics
	Methylphenidate (Ritalin)	5 mg/day	2 mg/kg (adults)	
	Pemoline (Cylert)	18.75 mg/day	112.5 mg	As above, hepatic toxicity
	Desipramine (Norpramin)	25 mg at bedtime	150 mg	Dry mouth, cardiotoxicity
OCD	Fluoxetine (Prozac)	10 mg/day	80 mg	Anxiety, insomnia, anorexia
	Clomipramine (Anafranil)	25 mg at bedtime	200 mg	Sedation, dry mouth, tremor, seizures

ADHD, attention-deficit hyperactivity disorder; OCD, obsessive-compulsive disorder.

TREATMENT

Many patients never need pharmacologic treatment. In such instances, education and regular follow-up evaluation are usually sufficient. Drug therapy should be considered if the tic disorder interferes with the patient's or family's lifestyle (Table 29-2). Because the most bothersome symptom may be related to ADHD, OCD, or other behavioral problems, therapy needs to be directed at the most disruptive manifestation of the disorder.

In patients with mild tics, especially those with behavior problems, clonidine, which has a low side-effect profile and is thought to reduce central noradrenergic activity, can be given. The dose can begin at 0.1 mg at bedtime and be increased by 0.1 mg every 3 days until adequate control is achieved or side effects, such as sedation and lightheadedness secondary to hypotension, occur. Because the drug has a short half-life, three or four doses a day may be necessary. If administration is to be discontinued, tapering over 2 weeks may prevent symptoms of acute withdrawal, such as agitation and manic behaviors. Clonidine patches can be considered, but local irritation and the inability to maintain the patch on active children often limit their usefulness.

If tics are not adequately controlled by clonidine, a neuroleptic agent can be added, or if there is no response, clonidine can be replaced with a neuroleptic drug. Haloperidol is probably the most commonly prescribed medication for GTS; however, pimozide (Orap) is the only neuroleptic drug specifically marketed in the United States for this disorder. Pimozide has fewer side effects, such as sedation and extrapyramidal symptoms, than haloperidol and has the advantage of once- or twice-a-day dosage. Treatment can be initiated at 1 mg every evening and the dose increased by this amount every 7 days until a maintenance dose that controls the tics (usually 1 to 4 mg) is achieved. Haloperidol can be started at 0.25 mg at bedtime, and a reasonable response can be expected at 2 mg/day in two or three divided doses. Fluphenazine (Prolixin) and risperidone (Risperdal) have also been shown to reduce the severity and frequency of tics in children and adolescents with chronic tic disorders. Fluphenazine can be given as an intramuscular injection once every 3 to 4 weeks.

Less commonly used agents for tic disorders are calcium channel blockers, such as nifedipine (Procardia), 10 mg three times a day, and tetrabenazine, which is not available in the United States. Tetrabenazine reduces dopaminergic transmission by depleting the presynaptic storage of dopamine. The benzodiazepine clonazepam (Klonopin) has also been used to treat chronic tic disorders. For tic disorders refractory to more conventional drug therapies, a trial of botulinum toxin injections can be considered for tics affecting specific muscle groups. Botulinum toxin inhibits the release of acetylcholine at the neuromuscular junction, resulting in decreased muscle activity (see Chapter 31).

Although the preceding discussion focused on the pharmacologic treatment of tics and GTS, ADHD or OCD may be the predominant source of disability. In addition to its use in treating tics, clonidine, as described previously, can be used in the treatment of ADHD. Although its efficacy has not been clearly demonstrated for ADHD, clonidine can be a safer alternative to central nervous system stimulants, such as methylphenidate (Ritalin), pemoline (Cylert), and dextroamphetamine (Dexedrine), which can aggravate tics in some patients. However, because stimulants remain one of the most effective medications for the treatment of primary ADHD, they are used in many persons with GTS whose main disability is ADHD. Methylphenidate, 5 mg in the morning, increased by 5 mg each week, with a goal of 15 to 20 mg/day, should be given two or three times a day to be effective. Pemoline, 18.75 mg in the morning, increased by 18.75 mg every 7 days to a maximum of 112.5 mg/day, has the advantage of once-a-day dosage, but the patient must be monitored for hepatic toxicity.

If the patient does not receive adequate relief with clonidine therapy and if stimulants exacerbate the tic disorder, a tricyclic antidepressant, such as desipramine (Norpramin), 25 mg at bedtime, increased every 7 days to a total dose of 100 mg in a single dose, can also be considered. However, there have been reports of sudden death in children treated with desipramine because of the presumed cardiotoxicity of the drug (electrocardiograms should be monitored for QT prolongation). Bupropion (Wellbutrin), fluoxetine (Prozac), sertraline (Zoloft), and imipramine (Tofranil) have also been used in the treatment of ADHD.

The treatment of OCD, another associated comorbid behavioral disorder in GTS, usually involves serotonin reuptake inhibitors. Fluoxetine, 10 mg every morning to a maximum of 80 mg/day in divided doses, or tricyclic antidepressants, such as clomipramine (Anafranil), 25 mg at bedtime, increased every 7 days to a total dose of 75 to 100 mg in a single dose, may be of benefit.

In addition to pharmacologic therapy, physicians and health care workers need to educate the families of patients about tic disorders and related behavioral problems. The Tourette Syndrome Association and local support groups can provide valuable information and counseling for families and individuals suffering from these disorders. A team approach with referrals to psychiatrists or behavioral psychologists for school difficulties may be necessary in the total care of these patients.

PROGNOSIS

Tics begin in childhood, peak in early to middle adolescence, and decrease in severity in the late teenage years. It is unusual for tics to increase in severity during adulthood. In a study of 58 adult patients with GTS diagnosed in childhood, the severity of tics during childhood did not correlate with poor outcome. Therefore, parents of affected children with moderate or severe tics can be counseled that the disorder does not necessarily lead to impaired adult function. Indeed, 98% of patients in the study graduated from high school and 90% were full-time students or employees.

SELECTED READING

Bruun RD, Budman CL. The course and prognosis of Tourette syndrome. *Neurol Clin* 1997; 15: 291-298.

Butler IJ, Koslow SH, Seifert WE Jr, Caprioli RM, Singer HS. Biogenic amine metabolism in Tourette syndrome. *Ann Neurol* 1979; 6: 37-39.

Chappell PB, Scahill LD, Leckman JF. Future therapies of Tourette syndrome. *Neurol Clin* 1997; 15: 429-450.

Coffey BJ, Miguel EC, Savage CR, Rauch SL. Tourette's disorder and related problems: a review and update. *Harv Rev Psychiatry* 1994; 2: 121-132.

Devor EJ. Linkage studies in 16 St. Louis families. Present status and pursuit of an adjunct strategy. *Adv Neurol* 1992; 58: 181-187.

Goetz CG, Tanner CM, Stebbins GT, Leipzig G, Carr WC. Adult tics in Gilles de la Tourette's syndrome: description and risk factors. *Neurology* 1992; 42: 784-788.

Hyde TM, Stacey ME, Coppola R, Handel SF, Rickler KC, Weinberger DR. Cerebral morphometric abnormalities in Tourette's syndrome: a quantitative MRI study of monozygotic twins. *Neurology* 1995; 45: 1176-1182.

Kurlan R, Trinidad KS. Treatment of tics, in *Treatment of Movement Disorders* (Kurlan R ed), J B Lippincott, Philadelphia, 1995, pp. 365-406.

Lombroso PJ, Scahill L, King RA, Lynch KA, Chappell PB, Peterson BS, McDougle CJ, Leckman JF. Risperidone treatment of children and adolescents with chronic tic disorders: a preliminary report. *J Am Acad Child Adolesc Psychiatry* 1995; 34: 1147-1152.

Pauls DL. Issues in genetic linkage studies of Tourette syndrome. Phenotypic spectrum and genetic model parameters. *Adv Neurol* 1992; 58: 151-157.

Pauls DL, Leckman JF. The inheritance of Gilles de la Tourette's syndrome and associated behaviors: evidence for autosomal dominant transmission. *N Engl J Med* 1986; 315: 993-997.

Singer HS, Tune LE, Butler IJ, Zaczek R, Coyle JT. Clinical symptomatology, CSF neurotransmitter metabolites, and serum haloperidol levels in Tourette syndrome. *Adv Neurol* 1982; 35: 177-183.

Swedo SE, Leonard HL, Garvey M, Mittleman B, Allen AJ, Perlmutter S, Lougee L, Dow S, Zamkoff J, Dubbert BK. Pediatric autoimmune neuropsychiatric disorders associated with streptococcal infections: clinical description of the first 50 cases. *Am J Psychiatry* 1998; 155: 264-271.

The Tourette Syndrome Classification Study Group. Definitions and classification of tic disorders. *Arch Neurol* 1993; 50: 1013-1016.

Wilkie PJ, Ahmann PA, Hardacre J, LaPlant RJ, Hiner BC, Weber JL. Application of microsatellite DNA polymorphisms to linkage mapping of Tourette syndrome gene(s). *Adv Neurol* 1992; 58: 173-180.

30 Restless Legs Syndrome

Virgilio Gerald H. Evidente, MD
and Charles H. Adler, MD, PHD

CONTENTS

Restless legs syndrome (RLS) is a disorder that was first described by Thomas Willis in 1672. Much later, authors referred to this condition variably as "anxietas tibiarum" and "leg jitters." On the basis of a large series of patients, Ekbom, a Swedish physician, coined the term "restless legs" in 1944 and gave the first full clinical description of RLS. Thus, RLS is sometimes referred to as "Ekbom's syndrome." He found that about 5% of normal subjects report occasional crawling sensations in their legs and that familial clustering occurs in some cases. He also described secondary forms due to anemia or pregnancy. Ekbom's son later described instances of RLS after gastric surgery, presumably due to iron deficiency.

Current data suggest that RLS afflicts up to 10% or even 15% of the population. A questionnaire survey of 515 patients seen in an outpatient clinic revealed that about 29% report symptoms of RLS. Despite its common occurrence, RLS is often unrecognized and misdiagnosed, leading to significant physical and emotional disability. One study revealed that, on average, patients in whom symptoms develop before age 20 years do not seek medical advice until after age 32 and do not receive the correct diagnosis until after age 50.

DEFINITION AND CLINICAL FEATURES

The symptoms of RLS include deep paresthesias and crawling and drawing sensations in the calves and legs that occur exclusively during inactive, seated, or recumbent wakefulness or rest. RLS is usually idiopathic (primary RLS), although occasional secondary

From: *Parkinson's Disease and Movement Disorders:*
Diagnosis and Treatment Guidelines for the Practicing Physician
Edited by: C. H. Adler and J. E. Ahlskog © Mayo Foundation
for Medical Education and Research, Rochester, MN

Table 30-1
Clinical Criteria Suggested by the International
Restless Legs Syndrome Study Group
for the Diagnosis of Restless Legs Syndrome

Paresthesias or dysesthesias, with the desire to move the limbs
Appearance or exacerbation of symptoms during rest or inactivity, with at least partial or temporary relief by activity
Motor restlessness
Nocturnal exacerbation of symptoms

From Walters AS. Mov Disord 1995; 10: 634-642.

causes may be identified. It is often familial. In 1992, the International Restless Legs Syndrome Study Group proposed criteria in establishing the diagnosis of RLS (Table 30-1). The condition may be acute in onset but is usually chronic (more than 3 months in duration). Commonly, symptoms fluctuate in severity and frequency. Other features of RLS are periodic limb movements in sleep, dyskinesias while awake, and sleep disturbance.

Paresthesias or Dysesthesias, With a Desire to Move the Limbs

Patients often describe abnormal, unpleasant sensations in the limbs (dysesthesias or paresthesias), most commonly in the calves and occasionally in the thighs, feet, or upper limbs. Most patients simply relate a vague, indescribable, nonpainful discomfort in the limbs. Terms used to describe such unpleasant sensations include "creeping," "crawling," "burning," "jittery," "tingling," "itching," "deep-seated," and "aching." The symptoms are usually, but not universally, bilateral.

Appearance or Exacerbation of Symptoms During Rest or Inactivity

The unpleasant limb sensations occur especially during rest or inactivity or on lying down or sitting still. Prolonged periods of inactivity, such as taking an airplane ride or a long car trip or watching a play or movie in the theater, are particularly distressing to the patient. The discomfort is relieved, at least partially, by motor activity.

Nocturnal Exacerbation

Symptoms of RLS are characteristically worse in the evening, usually on going to bed or in the middle of the night (commonly around 3 a.m.). Less often, and usually in more chronic cases, symptoms begin after dinner or a few hours before sleeping. In severe cases, symptoms may occur during the day. The nocturnal worsening of RLS is not only related to increased inactivity or rest at night but also most likely due to an independent circadian rhythm. The appearance of symptoms seems to coincide with the falling phase of the circadian body temperature cycle and their disappearance with the rise of body temperature in the morning.

Motor Restlessness

To relieve the limb discomfort, patients with RLS commonly resort to a repertoire of movements, including pacing the floor, tossing and turning in bed, stretching or shaking the legs, exercising, and running. They are compelled to perform such motor activities, with voluntary suppression resulting in increased discomfort and a build-up of inner tension or even in involuntary limb jerking. The limb movements in RLS are partly voluntary, because the patient chooses to move to relieve the discomfort, and partly involuntary, since the person is compelled to move.

Periodic Limb Movements in Sleep

About 80% of patients with RLS have unilateral or bilateral periodic limb movements in sleep (PLMS). These are stereotyped, repetitive flexion movements of the limbs (legs alone or legs more than arms) during stage 1 or 2 sleep. Other terms for PLMS are "periodic leg movements," "periodic movements in sleep," and "nocturnal myoclonus." In the lower limbs, repetitive dorsiflexion of the big toe with fanning of the small toes is seen, along with flexion of the ankles, knees, and thighs. PLMS occur semirhythmically at intervals of 5 to 90 seconds, with electromyographic (EMG) activity lasting 0.5 to 5 seconds. The movements are not truly myoclonic (which, strictly speaking, implies EMG bursts of 0.1-second duration or less), instead resembling the slower triple flexion or Babinski response. PLMS commonly occur in elderly persons (50 years old and older) without RLS or sleep disruption. Severe PLMS can disrupt the sleep of both patient and sleeping partner.

Dyskinesias While Awake

Dyskinesias while awake (DWA), also referred to as "periodic limb movements while awake," are involuntary limb movements that are phenomenologically identical to PLMS but are less periodic and more myoclonic. DWA occur in 30% to 50% of patients with RLS. They are associated with the typical sensory symptoms of RLS, disappear with voluntary movement, and can be suppressed by the patient to some extent.

Sleep Disturbance

Insomnia, due to either disturbed sleep onset (prolonged sleep latency) or disturbed sleep maintenance (frequent nocturnal awakenings), is experienced to varying degrees by most patients with RLS. The cause is largely limb discomfort but can also be severe PLMS. Insomnia can lead to excessive daytime sleepiness and fatigability, though not to the same degree as that in obstructive sleep apnea or narcolepsy.

Mean Age, Age at Onset, and Course

Most patients with RLS are middle-aged or older. The mean age reported in the 1960s of patients seen at the Mayo Clinic was 53 years. In 1996, Ondo and Jankovic reported a mean age of 63 years, with an average age at onset of 34. The condition may begin at any age, even as early as infancy, childhood, or adolescence. The more severely affected persons are usually middle-aged or older. Although the condition may remain static, more than two-thirds of patients report progression with time. At least 16% experience remission of symptoms for a month or more.

Family History

Primary RLS has been reported to be familial in 25% to 92% of cases, often with an autosomal dominant pattern of transmission. The exact gene locus remains unidentified. Some patients with PLMS without RLS symptoms are believed to have a forme fruste of autosomal dominant RLS. The phenomenon of "anticipation" (progressive decrease in age at onset with subsequent generations) is often seen in familial RLS. Features that may differentiate familial RLS from sporadic RLS are younger age at onset, slower progression, and greater frequency of myoclonus.

Neurologic Examination

The neurologic examination findings in idiopathic, or primary, RLS are normal. Signs of peripheral neuropathy (e.g., hypoactive or absent muscle stretch reflexes and decreased pain, light touch, or temperature sensation) are seen in "neuropathic" RLS. The International Classification of Sleep Disorders stipulates that peripheral neuropathy be ruled out by medical history and clinical examination before the diagnosis of primary RLS is made. Despite normal results of neurologic examination, detailed neurophysiologic studies may reveal signs of axonal neuropathy in patients with presumed primary RLS. Thus, nerve conduction studies and a needle EMG examination may be indicated.

ETIOLOGY

Although idiopathic in most cases, RLS has been reported to be associated with certain conditions or disorders (Table 30-2). Despite the long list of associated factors, only iron deficiency, uremia, pregnancy, and certain types of peripheral neuropathy are convincingly linked to RLS. The other associated conditions are, in general, common in the general population (especially the elderly), so that their precise relationship to RLS is difficult to establish. Some of these associations, if not most, may be merely coincidental and should be interpreted with considerable caution.

The association of iron deficiency anemia and RLS was first appreciated by Ekbom. RLS can, in fact, be the initial manifestation of iron deficiency anemia. Serum iron levels, however, are often unreliable in the diagnosis of iron deficiency because of the multiple factors that affect iron transport. A more stable indicator of iron stores is the serum ferritin level. In a study by O'Keefe and associates (1994) of 18 patients with RLS and 18 normal elderly persons who had similar serum iron levels, those with RLS had significantly lower serum ferritin levels. The severity of symptoms was inversely related to the ferritin levels. Treatment with ferrous sulfate resulted in lessening of symptoms in those with reduced ferritin, especially serum ferritin levels of 18 μg/L or less. Thus, depletion of iron stores without overt iron deficiency can lead to RLS.

RLS occurs in as many as one-third of pregnant women, especially in the second half of pregnancy, and among those with folate deficiency. Although symptoms usually resolve post partum, they may persist and evolve.

"Neuropathic" RLS develops in about 5% of patients with clinical and electrical evidence of polyneuropathy, especially among patients with sensory neuropathy. Treatment of the polyneuropathy can lead to reduction or resolution of symptoms. It is estimated that RLS develops in 17% to 40% of patients with uremia or chronic renal

Table 30-2
Factors or Conditions Possibly
Associated With Restless Legs Syndrome

Iron deficiency anemia

Pregnancy, especially with folate deficiency

Polyneuropathy due to uremia, rheumatoid arthritis, diabetes mellitus, alcohol abuse, amyloidosis, Sjögren's syndrome, lumbosacral radiculopathy, idiopathic polyneuropathy, monoclonal gammopathy of undetermined significance, Lyme disease, or B_{12} deficiency

Parkinson's disease

Gastric surgery

Chronic obstructive pulmonary disease

Carcinoma

Chronic venous insufficiency or varicose veins

Myelopathy or myelitis

Hypothyroidism or hyperthyroidism

Acute intermittent porphyria

Drugs: lithium, neuroleptics, beta-blockers, H_2 antagonists, caffeine, alcohol, antidepressants (e.g., paroxetine, amitriptyline, and mianserin, a tetracyclic antidepressant), terodiline (anticholinergic drug for urinary incontinence), and anticonvulsants (e.g., methsuximide and phenytoin)

Withdrawal from vasodilators, sedatives, or imipramine

Fibromyalgia syndrome

Arborizing telangiectasia of the lower limbs

Peripheral cholesterol microemboli

Cigarette smoking

failure, 7% to 17% of patients with diabetes, and 25% to 30% of patients with rheumatoid arthritis. In uremic peripheral neuropathy, RLS is most severe in the early stage and may actually disappear as the neuropathy progresses.

PATHOGENESIS

The pathogenesis of the periodic limb movements and sensory symptoms in RLS remains unclear. Ekbom originally proposed that RLS is due mainly to vascular dysfunction causing accumulation of metabolites in the legs. This possibility is supported by the reduction of symptoms with motor activity and by the association of RLS with chronic venous insufficiency, varicose veins, and pregnancy. Similar vascular mechanisms have been thought to cause PLMS. This theory may not adequately explain the symptoms in idiopathic RLS.

Peripheral nerve abnormalities also appear to cause RLS. Although many patients have sensory and motor nerve conduction abnormalities with neuropathic RLS, those with primary or secondary non-neuropathic RLS have no structural changes in nerve endings.

Most experts favor a central origin for both PLMS and RLS. The periodicity of PLMS and DWA suggests origin from a subcortical or brain stem (reticular) oscillator. Since the face is spared in RLS, such an oscillator is likely to be below the level of the facial nucleus. On electroencephalographic back-averaging, no cortical myoclonic activity or movement-related cortical potential (Bereitschaftspotential) is detected in relation to the myoclonus or dyskinesias in RLS, implying that such movements are of subcortical or spinal origin. Some believe that PLMS result from sleep-related disruption of the descending inhibitory reticulospinal pathways that are normally active at the brain stem or spinal cord level. Furthermore, the disinhibition of the reticulospinal tract may be modulated by abnormal peripheral sensory input (as in peripheral neuropathy). This hypothesis is supported by electrophysiologic data localizing the abnormality to the pons or rostral to it and by the association of PLMS with myelopathy or spinal anesthesia. Structures above the brain stem may also be involved, because functional magnetic resonance imaging in idiopathic RLS may show bilateral cerebellar and contralateral thalamic activation associated with the sensory discomfort and show additional activation of the red nuclei and brain stem close to the reticular formation with PLMS.

On the basis of response to treatment, RLS may be deduced to involve abnormalities in the dopaminergic or opiate systems. Dopamine and opiate agonists relieve symptoms, an effect reversed by treatment with their antagonists. Since dopamine blockade reactivates symptoms in patients treated with opiates or dopaminergic agents, whereas opioid antagonism reverses symptoms only in opiate-treated persons, it is thought that the endogenous opiate system acts by influencing the dopaminergic pathways. Single-photon emission computed tomography scan data suggest a deficiency of dopamine D_2 receptors in RLS. D_2 receptor hypofunction has also been linked to iron deficiency in animal studies. Since iron is an integral part of the opiate μ and dopamine D_2 receptors, iron deficiency may produce RLS through dysfunction of these receptors.

The sympathetic, serotonin, and γ-aminobutyric acid (GABA) neurotransmitter systems are also implicated in RLS. Sympathetic (adrenergic or noradrenergic) hyperactivity in RLS is suggested by the following observations: Sympathetic vasoconstriction elicits PLMS either spontaneously or during spinal anesthesia, sympathetic nerve blockade relieves PLMS, tricyclic antidepressants (with sympatholytic effects) exacerbate PLMS, and α-adrenergic blockers reduce RLS symptoms. Studies on response to medications suggest possible underactivity of the serotonin and GABA neurotransmitter systems in RLS.

Various authors have noted a high prevalence of anxiety and depression, including suicide, among persons with RLS. It is uncertain, though, whether depression and anxiety are causes or results of RLS.

DIAGNOSIS

The diagnosis of RLS is established mainly on clinical grounds. In patients with recent worsening or abnormal findings on neurologic examination, serologic workup to rule out secondary causes of RLS should be done, including determination of blood urea nitrogen, creatinine, fasting blood glucose, glycosylated hemoglobin, complete blood cell count, ferritin, magnesium, thyroid-stimulating hormone, and folate

Table 30-3
Nonpharmacologic Treatment of Restless Legs Syndrome

General measures
  Avoidance of caffeine, smoking, and alcohol intake
  Avoidance of offending drugs (e.g., lithium, neuroleptics, beta-blockers, H_2 antagonists)
Physical measures
  Massage
  Warm or cold baths
  Transcutaneous electrical stimulation
  Vibration
  Acupuncture
  Pneumatic leg compressors
  Magnets
Treatment of secondary causes of restless legs syndrome, especially iron deficiency, folate deficiency during pregnancy, magnesium deficiency, and peripheral neuropathy

(especially during pregnancy). Needle EMG and nerve conduction studies should be performed if polyneuropathy is suspected. In most cases, polysomnography is not necessary but may be used to document PLMS, DWA, or sleep disturbance if questions about the diagnosis remain.

NONPHARMACOLOGIC MANAGEMENT

General Measures

Patients who have symptoms with caffeine should avoid this substance (Table 30-3). Reduction or cessation of smoking or alcohol intake may benefit some patients. Administration of medications such as lithium, neuroleptic agents, beta-blockers, and H_2 antagonists should be discontinued if the RLS symptoms are time-locked to the intake of these drugs. Patients with RLS report that a variety of physical measures may partially or temporarily alleviate symptoms, including a whirlpool hot bath (less effectively, a cold bath) before sleeping, massage, pneumatic leg compressors, elastic stockings, magnets, acupuncture, transcutaneous electrical nerve stimulation, and vibration.

Treatment of Underlying Conditions

Patients with reduced ferritin levels, with or without overt iron deficiency, may have a favorable response to iron supplementation without needing further pharmacotherapy. Correction of vitamin (e.g., folate) or electrolyte (e.g., magnesium) deficiency can lead to dramatic improvement. RLS symptoms associated with certain endocrine disorders (e.g., diabetes, hypothyroidism, and hyperthyroidism) abate with control of the associated condition. Persons with RLS and leg varicosities may be helped by intravenous injection of sclerosing agents. RLS in uremic patients may be relieved by kidney transplantation or by correction of the associated anemia with erythropoietin. Vitamin E and vitamin C have also been anecdotally reported to be effective in selected cases.

PHARMACOLOGIC MANAGEMENT

Pharmacologic treatment should be instituted if the symptoms are severe enough to disturb sleep (either at onset or during), cause emotional or physical distress, or interfere with activities of daily living. Drug therapy is largely directed at symptoms in idiopathic RLS, because cure is possible only in secondary RLS. Administration should begin at a low dose, which should be taken a couple of hours before sleep to allow sufficient absorption and action. If RLS awakens the patient in the middle of the night or if evening symptoms are severe enough, additional doses can be given. Severe RLS may require multiple daytime and nighttime doses. Some patients do well for prolonged periods (even many years) with a single drug. However, tolerance may develop, and effectiveness diminishes in many persons. Finding two or three effective medications and then rotating them monthly may help prevent the development of tolerance. Some advocate alternating chemically unrelated drugs weekly or biweekly. Since some patients with RLS experience remission of symptoms for a month or more, brief drug holidays may be attempted in those with well-controlled symptoms. The mainstays of treatment of RLS include the dopaminergic agents, opioids, and benzodiazepines (Table 30-4). If these drugs are ineffective, other options are anticonvulsants, clonidine, and beta-blockers.

Levodopa

Carbidopa-levodopa (Sinemet) diminishes the symptoms of RLS and reduces the frequency of PLMS. For symptoms that begin before sleep, one 25/100 carbidopa-levodopa tablet can be taken 1 to 2 hours before bedtime. If the symptoms occur later during the night, one 25/100 controlled-release carbidopa-levodopa (Sinemet CR) tablet can be used. In patients with symptoms before bedtime and during the night, a combination of the short-acting and controlled-release tablets can be given. Some patients may require higher or more frequent doses as well as daytime doses. Most patients report sufficient benefit with 100 to 500 mg of levodopa per day. In rare cases, doses of 1,000 to 1,500 mg/day are needed. To enhance absorption, carbidopa-levodopa should be taken on an empty stomach.

The most common side effect is nausea, especially with higher doses. Nausea may diminish with time or with intake of the pills with crackers or food. In more persistent cases, addition of carbidopa (Lodosyn), 25 mg with each dose of carbidopa-levodopa, may be beneficial.

The major drawback with levodopa for RLS is the development of "augmentation" in 80% of patients as early as a few months after initiation of treatment with the drug. Augmentation can be signaled by earlier onset of RLS symptoms during the evening, shorter latency to onset after assumption of a restful position, increased intensity, and extension of the symptoms to the upper body. With the usual evening or nighttime dose of levodopa, symptoms may occur later in the night, the following morning, or at an earlier time the following evening. The increase in severity of symptoms occurring in the morning is referred to as "rebound." Augmentation is more likely to develop if RLS symptoms are severe and if the total daily dose of levodopa exceeds 200 mg. Some advocate using levodopa five times a week and other agents the other two nights to prevent augmentation. Once augmentation or rebound develops, increasing the dose of levodopa only temporarily relieves RLS and eventually leads to reaugmentation and

Table 30-4
Pharmacologic Treatment
of Restless Legs Syndrome

Dopaminergic agents
    Levodopa (Sinemet)
    Dopamine agonists
        Pergolide (Permax)
        Bromocriptine (Parlodel)
        Pramipexole (Mirapex)
        Ropinirole (Requip)

Opioids
    Codeine
    Oxycodone (Percocet, Percodan)
    Propoxyphene (Darvocet, Darvon)
    Methadone
    Levorphanol

Benzodiazepines
    Clonazepam (Klonopin)
    Diazepam (Valium)
    Alprazolam (Xanax)
    Temazepam (Restoril)

Anticonvulsants
    Gabapentin (Neurontin)
    Carbamazepine (Tegretol)
    Valproic acid (Depakote)

Other agents
    Clonidine (Catapres)
    Baclofen (Lioresal)
    Tricyclic antidepressants
    Beta-blockers
    Tramadol (Ultram)
    Tryptophan

further worsening of symptoms. This is a Catch-22. In most cases, alternative treatment should be tried and levodopa withdrawn if possible.

Dopamine Agonists

Because of the augmentation and rebound associated with levodopa, many experts currently favor using a dopamine agonist for first-line therapy for RLS. Pergolide (Permax), a potent, long-acting dopamine D_1 and D_2 receptor agonist, has been shown to be effective in treating RLS, even among patients with severe sensory leg symptoms unresponsive to levodopa. However, patients with prominent PLMS have a better response to levodopa than to pergolide. The initial dose of pergolide is 0.05 mg about 2 hours before bedtime, with 0.05-mg increases in dose every 3 to 5 days until relief is obtained or side effects develop. Dosage can be split at dinner and bedtime and, if necessary, daytime as well. The effective dose ranges from as little as 0.05 mg to about 1.5 mg/day.

Bromocriptine (Parlodel), a dopamine D_2 receptor agonist, has also been shown to be effective in RLS. Administration should start at a low dose (1.25 mg/evening), with increases every few days of 1.25 mg until benefit is obtained or side effects develop. The effective dose ranges from 1.25 to 15 mg daily.

Pramipexole (Mirapex) and ropinirole (Requip) are newer dopamine D_2 receptor agonists that were approved by the Food and Drug Administration for Parkinson's disease in 1997. Pramipexole also has some D_3 agonist activity. Data suggest that both of these two agonists may be highly effective, even in refractory cases.

Pramipexole can be given at 0.125 mg before bedtime, and the dose is slowly increased at bedtime or dinnertime by 0.125-mg increments. Similarly, the dose of ropinirole can begin at 0.25 mg and be increased in the same fashion. Daytime doses may be required in more severe cases.

Side effects of dopamine agonists include nausea, lightheadedness, drowsiness, and postural hypotension. These can be minimized by lower doses and slower dose escalations. Dopamine agonists appear less likely to produce augmentation or rebound and can be of benefit to patients in whom these complications develop with levodopa.

Opioids

Opioids acting on the μ receptor have been found to be effective in RLS. Opiate antagonists, such as naloxone, reverse these effects. Low-potency opiates, such as codeine and propoxyphene, can benefit those with mild and intermittent symptoms, whereas high-potency agents, such as oxycodone, methadone, and levorphanol, may be used in refractory cases. Codeine (codeine sulfate or acetaminophen with codeine) can be effective in RLS at doses of 15 to 120 mg/day. Propoxyphene (Darvon), at 65 to 250 mg/day, has been shown to improve sleep efficiency and reduce frequency of arousals in patients with PLMS. Oxycodone (Percocet or Percodan), at doses of 5 to 15 mg/day, may have a beneficial effect on RLS symptoms, PLMS, sleep arousals, and sleep efficiency. Some physicians use methadone (5 to 30 mg/day) or levorphanol (2 to 10 mg/day) for refractory cases. Severe RLS has also been treated with epidural administration of morphine to help reduce the total opioid dosage and systemic effects.

The potential side effects of opioids include dizziness, sedation, nausea, vomiting, constipation, abdominal discomfort, weakness, euphoria, hallucinations, hypotension, and respiratory depression. Although some report continued long-term benefit from opioid treatment for up to 15 years with little risk of addiction, others have reported addiction after use of short-acting opioids in vulnerable patients. Most physicians are often hesitant to use opioids in RLS, but we recommend their use if other treatments are ineffective.

Benzodiazepines

Benzodiazepines, particularly clonazepam (Klonopin), were among the earliest drugs used to treat RLS. The initial dose of clonazepam is 0.25 mg at bedtime, and this can be increased by 0.25 mg every week to a maximum of 3 to 4 mg/day. Doses of 0.5 to 2.0 mg/night are especially effective in reducing PLMS. Other benzodiazepines have been anecdotally reported to be effective in RLS, particularly diazepam (Valium), 1 to 10 mg/day; alprazolam (Xanax), 0.25 to 2 mg/day; and temazepam (Restoril), 15 to 60 mg/day.

The major side effects of benzodiazepines are daytime drowsiness and confusion, unsteadiness, falls, and aggravation of sleep apnea (which can coexist with RLS). Dependency and tolerance are also of concern. Benzodiazepines, in general, are useful as monotherapy in patients with mild or intermittent symptoms and as additions to therapy with other oral drugs, especially dopaminergic agents, in more severe cases.

Anticonvulsants

Gabapentin (Neurontin) relieves PLMS and the sensory symptoms of RLS, even in refractory cases. The effective dose ranges from 300 to 2,400 mg daily. Gabapentin therapy can be initiated at a dose of 100 to 300 mg at bedtime, with increases of 100 to 300 mg every 3 days to a maximum of 2,400 mg/day. The drug is generally well tolerated but may cause transient or mild side effects such as somnolence, dizziness, ataxia, and fatigue.

Carbamazepine (Tegretol), at a dose of 200 to 400 mg daily, was shown to be effective in reducing RLS sensory symptoms in double-blind studies, especially among young patients with recent onset and severe symptoms. However, the drug has no apparent effect on PLMS, and subsequent clinical experience has been disappointing. Furthermore, side effects, including diplopia, drowsiness, confusion, and nausea, develop in about 40% of patients treated with carbamazepine and are severe enough to cause administration of the drug to be discontinued.

There are anecdotal reports of valproic acid (Depakote) benefiting RLS, especially if symptoms are mild. Doses of 250 to 1,000 mg/day can be used. There are no controlled trials investigating its efficacy. Potential side effects include hair loss, weight gain, thrombocytopenia, and liver toxicity.

Clonidine

Clonidine (Catapres), a presynaptic α_2-adrenergic agonist, can be effective in uremic and idiopathic RLS. Sensory and motor restlessness appear to respond, with no change in PLMS. The effective dose ranges from 0.1 to 1.0 mg/day, with an average of 0.5 mg/day. A double-blind, placebo-controlled study of six patients with uremic RLS showed moderate to marked diminishing of symptoms in all patients at a dose of 0.2 to 0.3 mg/day. Treatment with clonidine should be initiated at 0.1 mg at bedtime, with the dose increased every week by 0.1 mg to a maximum of 1 mg/day. The more common side effects are dry mouth, decreased cognition, lightheadedness, sleepiness, and constipation. Some patients actually report worsening of RLS symptoms with clonidine.

Other Medications

Baclofen (Lioresal), at a dose of 20 to 40 mg/night, was reported to reduce the sensory symptoms of RLS, improve sleep efficiency, and decrease the amplitude of PLMS. Although tricyclic antidepressants have been shown to exacerbate RLS, they may benefit some patients with RLS, especially those with associated depression. Imipramine (Tofranil) was also shown to eliminate PLMS in five patients at a dose of 25 mg before bedtime. Similarly, although beta-blockers have been associated with RLS symptoms, propranolol (Inderal), at a dose of 5 to 10 mg/night, may be useful in some patients. L-tryptophan, 5-hydroxytryptophan, and the analgesic tramadol (Ultram) have been anecdotally reported to be effective in some patients.

Treatment Summary

Oral medications for RLS and PLMS may be effective for symptoms but are not curative. Treatment is based on a trial-and-error approach, with dopamine agonists or gabapentin being our first choice to start. Often, combinations of these medications are required to minimize side effects. Rotating various effective medications may help deal with the issue of tolerance.

SUMMARY

1. RLS is characterized by abnormal sensory symptoms of the legs that worsen in the evenings and at rest and remit with leg movements.

2. RLS can be associated with PLMS, motor restlessness, and sleep disturbance.
3. Although most cases are idiopathic, RLS may be caused by iron deficiency anemia, uremia, pregnancy, certain medications, and neuropathy.
4. Multiple different medications may be effective in treating RLS and must be tried systematically.

The Restless Legs Syndrome Foundation, Inc., is a nonprofit organization that provides information, a newsletter, and local support groups devoted to RLS. It has recently begun funding research grants to advance knowledge related to this disorder. The organization can be reached at 819 Second Street, S.W., Rochester, MN 55902-2985. It has an Internet site at www.rls.org.

SELECTED READING

Allen RP, Earley CJ. Augmentation of the restless legs syndrome with carbidopa/levodopa. *Sleep* 1996; 19: 205-213.

Allen RP, Kaplan PW, Buchholz DW, Earley CJ, Walters JK. Double-blinded, placebo controlled comparison of high dose propoxyphene and moderate dose carbidopa/levodopa for treatment of periodic limb movements in sleep (abstract). *Sleep Res* 1992; 21: 166.

Becker PM, Jamieson AO, Brown WD. Dopaminergic agents in restless legs syndrome and periodic limb movements of sleep: response and complications of extended treatment in 49 cases. *Sleep* 1993; 16: 713-716.

Bucher SF, Seelos KC, Oertel WH, Reiser M, Trenkwalder C. Cerebral generators involved in the pathogenesis of the restless legs syndrome. *Ann Neurol* 1997; 41: 639-645.

Earley CJ, Allen RP. Pergolide and carbidopa/levodopa treatment of the restless legs syndrome and periodic leg movements in sleep in a consecutive series of patients. *Sleep* 1996; 19: 801-810.

Evidente VGH, Adler CH. How to help patients with restless legs syndrome: discerning the indescribable and relaxing the restless. *Postgrad Med* 1999; 105: 59-74.

Hening WA, Walters A, Kavey N, Gidro-Frank S, Cote L, Fahn S. Dyskinesias while awake and periodic movements in sleep in restless legs syndrome: treatment with opioids. *Neurology* 1986; 36: 1363-1366.

Iannaccone S, Zucconi M, Marchettini P, Ferini-Strambi L, Nemni R, Quattrini A, Palazzi S, Lacerenza M, Formaglio F, Smirne S. Evidence of peripheral axonal neuropathy in primary restless legs syndrome. *Mov Disord* 1995; 10: 2-9.

Krueger BR. Restless legs syndrome and periodic movements of sleep. *Mayo Clin Proc* 1990; 65: 999-1006.

Montplaisir J, Godbout R, Boghen D, DeChamplain J, Young SN, Lapierre G. Familial restless legs with periodic movements in sleep: electrophysiologic, biochemical, and pharmacologic study. *Neurology* 1985; 35: 130-134.

O'Keeffe ST, Gavin K, Lavan JN. Iron status and restless legs syndrome in the elderly. *Age Ageing* 1994; 23: 200-203.

O'Keeffe ST, Noel J, Lavan JN. Restless legs syndrome in the elderly. *Postgrad Med J* 1993; 69: 701-703.

Ondo W, Jankovic J. Restless legs syndrome: clinicoetiologic correlates. *Neurology* 1996; 47: 1435-1441.

Poceta JS. Restless legs syndrome and nocturnal myoclonus, in *Sleep Disorders: Diagnosis and Treatment* (Poceta JS, Mitler MM eds), Humana Press Inc., Totowa, 1998, pp. 75-93.

Rutkove SB, Matheson JK, Logigian EL. Restless legs syndrome in patients with polyneuropathy. *Muscle Nerve* 1996; 19: 670-672.

Silber MH. Restless legs syndrome. Mayo Clin Proc 1997; 72: 261-264.

Walters AS. Toward a better definition of the restless legs syndrome. The International Restless Legs Syndrome Study Group. *Mov Disord* 1995; 10: 634-642.

31

Botulinum Toxin Treatment of Movement Disorders

Charles H. Adler, MD, PhD

CONTENTS

GENERAL OVERVIEW

Botulinum toxins are the most deadly neurotoxins known and are produced by the anaerobic bacterium *Clostridium botulinum*. *C. botulinum* produces seven antigenically (immunologically) distinct neurotoxins: A, B, C₁, D, E, F, and G. These neurotoxins block neuromuscular transmission, resulting in both skeletal and smooth muscle paralysis. Clinically, botulism can occur after ingestion of contaminated food or from a wound infection. Signs of botulism can include limb paralysis, facial weakness, ophthalmoplegia, dysarthria, dysphagia, dyspnea progressing to respiratory arrest, constipation progressing to ileus, and urinary retention.

Botulinum toxin (BTX) was first used to treat human disorders involving excessive muscle spasms in 1981. In December 1989, the U.S. Food and Drug Administration approved the use of botulinum toxin type A (Botox, BTX-A) for use in strabismus, blepharospasm, hemifacial spasm, and other disorders of the seventh cranial nerve in patients 12 years of age or older. Currently, the only commercially available BTX is type A. Clinical research with types B, C, and F is continuing.

MECHANISM OF ACTION AND BASIC SCIENCE

All BTXs act by inhibiting the release of acetylcholine at the neuromuscular junction. Multiple proteins at the nerve terminal mediate the exocytotic release of acetylcholine. Each BTX acts by cleaving one of these proteins in a different location, resulting in

From: *Parkinson's Disease and Movement Disorders:*
Diagnosis and Treatment Guidelines for the Practicing Physician
Edited by: C. H. Adler and J. E. Ahlskog © Mayo Foundation
for Medical Education and Research, Rochester, MN

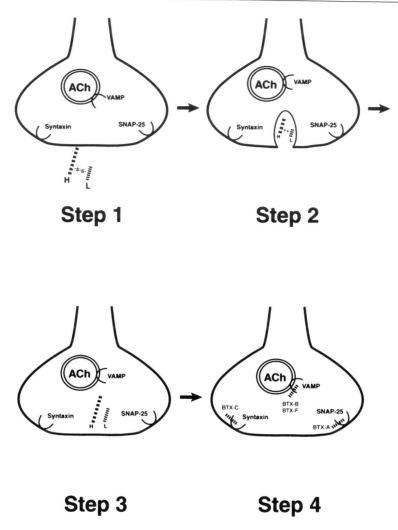

Fig. 31-1. Mechanism of botulinum toxin (BTX) inhibition of acetylcholine (ACh) release. All serotypes of BTX have a heavy (H) and light (L) chain joined by an S-S bond. *Step 1,* ACh is contained in vesicles that require multiple proteins (syntaxin, VAMP, SNAP-25) to mediate exocytosis. The H chain contains the binding site for BTX on the nerve terminal, whereas the L chain mediates cleavage of the appropriate protein. *Step 2,* It is theorized that the H and L chains enter the nerve terminal via endocytosis. *Step 3,* Once internalized, the H and L chains separate. Step 4, Each type of BTX (A, B, C, F) has a unique L chain. The L chain from BTX-A cleaves SNAP-25, the L chains from BTX-B and BTX-F cleave VAMP, and that from BTX-C cleaves syntaxin. The result is inhibition of ACh release by way of exocytosis.

inhibition of acetylcholine release, disruption of neuromuscular transmission, and, consequently, paralysis of the muscle (Fig. 31-1). BTX-A cleaves SNAP-25 and BTX-C cleaves syntaxin, both of which are cytosolic, presynaptic membrane–associated proteins, whereas BTX-B and BTX-F cleave synaptobrevin, or VAMP, an acetylcholine vesicle membrane protein.

BTX-A is produced by culturing *C. botulinum* and fermenting the bacteria, which results in lysis and toxin release. The toxin is then harvested, purified, crystallized with

Table 31-1
Botulinum Toxin

- Is the most potent neurotoxin known.
- Inhibits acetylcholine release at the
  neuromuscular junction, resulting in
  muscle weakness.
- Can be used, by injection of small doses, to
  weaken hyperactive muscles and reduce
  involuntary movements.

ammonium sulfate, diluted, and freeze-dried. Before injection, BTX-A must be diluted with preservative-free saline, and the preparation should be used within 4 hours of reconstitution.

The potency of BTX is expressed in mouse units, with 1 mouse unit equal to the median lethal dose (LD_{50}) for mice. BTX-A is packaged in 100-unit vials. It is estimated that the LD_{50} for humans is approximately 40 U/kg given intravenously, or 3,000 U in a 75-kg person. Most clinicians dilute the 100-U vial with 1 to 4 mL of saline, for a concentration of 2.5 to 10 U/0.1 mL, and use the toxin immediately. Doses depend on the part of the body to receive the injection and the muscle size; small muscles, such as the vocal cords, receive as little as 0.75 U, larger neck muscles may require 100 to 150 U, and upper leg muscles may need 200 to 300 U to weaken them.

The clinical use of BTX requires injection (intramuscular or subcutaneous) into or around a hyperactive muscle to partially or completely paralyze that muscle (Table 31-1). Once injected, the toxin spreads through the muscle and can spread to neighboring muscles. The goal is to weaken the muscles enough to relieve the involuntary contractions while maintaining some voluntary muscle control. After injection, BTX begins weakening the muscle in 24 to 72 hours. Maximal weakness occurs after about 14 days, and the effect begins wearing off over 2 to 6 months as the nerves begin to reinervate the muscle. Usually, an initial period of mild overweakening is followed by good control of the unwanted overactivity of the muscle and then a return of the hyperactivity or movement disorder. In small muscles, a single injection usually suffices, whereas in large muscles, BTX is usually injected into two or three sites to allow spread of the toxin. If a muscle fully atrophies, or is extremely weak, but the movement disorder continues, other muscles must be involved.

CLINICAL INDICATIONS

Overview

BTX-A is approved for use in strabismus, blepharospasm, hemifacial spasm, and other disorders of the seventh cranial nerve in patients 12 years of age or older (Table 31-2). Other indications have been found, but none has been officially approved. Movement disorders, such as focal dystonia (idiopathic or secondary forms) and hemifacial spasm, are the most commonly treated conditions. The goal of treatment with BTX is to reduce the involuntary movement disorder. The key is titrating the dose to allow the patient to continue using the muscles for voluntary activities. Various other disorders may also benefit, including spasticity due to multiple sclerosis, stroke, cerebral palsy, and brain or spinal cord injury; disorders with increased muscle spasms, such as facial synkinesis,

Table 31-2
Clinical Indications
for Botulinum Toxin Injections

Hemifacial spasm
Focal dystonia
　　Blepharospasm
　　Torticollis, cervical dystonia
　　Spasmodic dysphonia
　　Oromandibular dystonia
　　Limb dystonia
　　　　Writer's cramp
　　　　Musician's cramp
　　Truncal dystonia
Myoclonus
Tics
Tremors
Spasticity
Strabismus

bruxism, back spasms, tension headaches, stiff-man syndrome, and cricopharyngeal dysphagia; and hyperactive smooth muscle disorders, such as achalasia, detrusor sphincter dyssynergia, anismus, and rectal fissures. BTX-A has recently found a niche in the treatment of facial wrinkles and is being investigated for the treatment of hyperhidrosis.

All patients under consideration for BTX treatment must be carefully examined to determine the anatomy of the disorders. It is critical to determine which muscles are hyperactive and causing the involuntary movement and which may be acting in a compensatory fashion. The dose of BTX used should be based on muscle mass, muscle location, degree of hyperactivity, and potential adverse effects if the muscle (or adjacent muscles) is overly weakened. Small muscles need smaller doses of BTX than large muscles. For example, the vocal cords may require only 0.75 to 2.5 U of BTX-A to become fully weakened, whereas the quadriceps may require 300 to 400 U for only partial weakness. A good clinical dictum: You can always inject more BTX at a later date, but once the dose is injected, you cannot take it away. Therefore, dosage should be kept low initially and then be increased as needed.

The patient should be seen for follow-up, if possible, 2 to 4 weeks after the first injection. The patient can then be examined, side effects and benefit can be noted, and a plan for future injections can be created. Patients should be involved in formulating goals. It is recommended that patients not be given a "booster dose" 2 to 3 weeks after the initial injection because of the possibility of immunoresistance developing. Injection sessions should be at the longest possible intervals, certainly 2 to 3 months or longer. Patients and physicians must realize that frequent doses may lead to immunoresistance. Patients must be told that BTX-A is not a cure but rather a treatment that requires repeat injections approximately two to five times a year. It is important for patients and physicians to understand the limitations and expectations of this method of treatment. No laboratory testing is required, either before or after BTX-A injections.

Blepharospasm

BTX-A is the drug of choice for treating blepharospasm. Numerous trials demonstrate effectiveness in up to 95% of patients treated. Most clinicians use a tuberculin syringe and

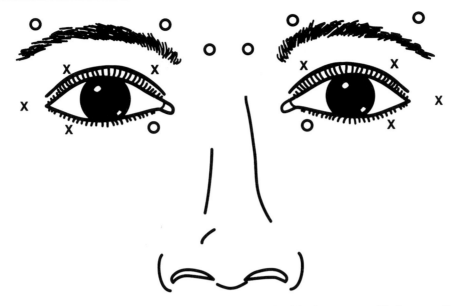

Fig. 31-2. Botulinum toxin injection sites for patients with blepharospasm. X, the most frequent sites of injection in the orbicularis oculi muscle are on the upper eyelid, on the lower eyelid, and lateral to the lateral canthus. O, the medial lower lid orbicularis oculi, frontalis muscle (above the eyebrows), and corrugator muscle (above the nose) are the sites of injection if spasms in these regions cause problems.

a 27- to 30-gauge needle to inject BTX-A. Injection techniques vary, but a starting dose of 1.25 to 5 U is injected subcutaneously (raising a bleb) into multiple locations of the orbicularis oculi (Fig. 31-2). This muscle is responsible for closing the upper and lower eyelids. Some patients with blepharospasm also have marked contractions of the frontalis or the corrugator muscles of the forehead, and injections can be made into these muscles if necessary (Fig. 31-2).

Injections for blepharospasm have potential side effects, including bruising, ptosis, blurred vision, diplopia, change in tearing, and lagophthalmos (difficulty completely closing the eye). All these side effects are transient, and most require no intervention. If lagophthalmos develops, the patient should use eyedrops or ointment during the day and tape the eye shut at night to avoid exposure keratitis and corneal abrasions. A lower dose of BTX-A should be injected at the next treatment session.

Torticollis or Cervical Dystonia

All forms of cervical dystonia may respond to BTX injections. The key to treatment is injecting the correct dose into the appropriate muscles. Starting doses are established for each muscle, and the doses are adjusted on the basis of response. Head turning is predominantly mediated by the contralateral sternocleidomastoid and the ipsilateral splenius capitis muscles. Tilting and shoulder elevation are mediated by the ipsilateral splenius capitis and levator scapulae. Retrocollis is the result of hyperactivity in the bilateral splenius capitii, trapezeii, and deep posterior cervical muscles, whereas anterocollis results from hyperactivity of the bilateral sternocleidomastoids, scalenes, and deep anterior cervical muscles. The injections must be intramuscular, and they can be done with or without electromyographic (EMG) guidance. Doses of BTX-A range from 20 to 150 U per muscle treated.

A good response to injections is achieved in 75% to 90% of patients with torticollis. However, patients and physicians must appreciate that a good response may not mean 100% improvement; instead, expectations should be for a significant reduction in movement and control of any pain. Injections for cervical dystonia can result in dysphagia, difficulty turning the head, difficulty keeping the head up (if the neck extensors are overweakened), or difficulty pulling the chin down (if the neck flexors are overweakened). Other potential side effects are fatigue, generalized weakness, pain at the injection site, and a paradoxical increase in muscle spasms. Brachial plexopathy has also been reported.

Oromandibular Dystonia

As in cervical dystonia, the critical issue in treating oromandibular dystonia is understanding the anatomy. Jaw-closing dystonia usually involves the masseter and temporalis muscles bilaterally; jaw-opening dystonia requires injections into the digastric muscles (anterior belly or posterior belly, or both) and in some cases the lateral pterygoids, geniohyoids, and mylohyoids. Jaw protrusion and lateral deviation involve the lateral pterygoids, and tongue protrusion involves the genioglossus and hyoglossus muscles. Benefit is obtained in most jaw closing and opening dystonias, with less success achieved with jaw deviation and lingual dystonias.

Initial doses should be conservative, since the main side effects are dysarthria, dysphagia, trouble chewing, and biting the inner cheek. With tongue injections, airway obstruction is a concern if the tongue is overweakened. These effects are all transient and usually require no intervention. If dysphagia is severe, patients must be extra cautious when eating to avoid aspiration.

Spasmodic Dysphonia or Laryngeal Dystonia

Adductor spasmodic dysphonia is characterized by a strained, choked voice. Enough BTX-A is injected into the true vocal cords, the thyroarytenoid muscles, to allow effortless speech without overweakening the cords and causing breathiness or dysphagia. Several injection techniques are in use. We inject percutaneously; a hollow, monopolar EMG needle is passed through the cricothyroid membrane into the thyroarytenoid muscle, the muscle is localized by monitoring EMG activity, and then BTX-A is injected through the EMG needle. The dose of BTX-A used in adductor spasmodic dysphonia varies. Initially, we usually inject 1.25 U bilaterally and then adjust the dose according to the response. Approximately 95% of patients with adductor spasmodic dysphonia have a good response.

Abductor spasmodic dysphonia is a rarer disorder that is more difficult to treat. The voice has a breathy, whispering quality due to hyperabduction of the vocal cords. The muscles, treated in these patients are the posterior cricoarytenoid muscles or the cricothyroid muscles, or both together. Injections can be done unilaterally, but if no benefit is achieved, bilateral injections may be needed. Bilateral injections are potentially dangerous, since they may result in stridor or airway obstruction. Only about 75% of patients with abductor spasmodic dysphonia have a good response to BTX-A.

It is critical that all patients with either type of spasmodic dysphonia be treated by physicians who can manage the potential acute complications of stridor and airway obstruction. If the thyroarytenoids swell acutely or if overweakening of the posterior cricoarytenoids results in an inability to abduct the vocal cords, either intubation or tracheostomy may be required. Therefore, most clinicians injecting BTX for spasmodic dysphonia are either otolaryngologists or neurologists working with otolaryngologists.

Limb Dystonia

For upper limb dystonia, it is critical to assess finger, wrist, and forearm posturing to determine the pattern of abnormal finger and wrist flexor or extensor activity. This is best done by examination with palpation and visualization. In some patients, further information may be obtained with either surface or needle EMG recordings. BTX-A is then injected into the target muscles. Doses of BTX-A are usually lower for the finger and wrist extensors (2.5 to 50 U) than for the flexor muscles (5 to 100 U).

For the leg, most patients have dystonic foot plantar extension, inturning, or toe movements. Injections must target the controlling muscle. These injections can be done by palpation (especially the gastrocnemius and soleus muscles) or with EMG guidance. Doses range from 50 to 400 U in each muscle group, since these are generally large, powerful muscles.

Limb injections can overweaken the hand and finger extensors and flexors and in the leg can result in footdrop or difficulty walking. Given the proximity of the various finger extensors and flexors to one another, injections to weaken one finger can result in weakness of neighboring fingers. The result may be difficulty performing voluntary activities with the limb, and in some cases the loss of this voluntary control may be a greater problem than the actual dystonia.

Hemifacial Spasm

Hemifacial spasm is a disorder characterized by unilateral muscle contractions of the upper face (forehead, eyelids) or lower face (mouth, cheek, and superficial neck) or both regions (see Chapter 24). This is not a form of dystonia but instead is usually secondary to irritation of the seventh cranial nerve by an artery compressing the nerve as it exits the brain stem ("root entry zone"). Injections of BTX-A need to be tailored to the facial muscles that are in spasm in patients with hemifacial spasm, and they differ from patient to patient. Initially, since the most troublesome symptom is usually involuntary eye closing, injections only into the orbicularis oculi may result in excellent benefit. These injections are done in a fashion similar to that described for blepharospasm, and the side effects are similar, only unilateral.

If relief of the eye-closing spasms is not sufficient to reduce forehead or lower facial spasms, BTX-A can be injected into other facial muscles, including the frontalis, risorius, zygomaticus major, orbicularis oris, and platysma. The treatment strategy may change with time, with lower facial injections avoided in some patients initially and then tried in the future. The main side effects of lower facial injections are facial weakness with facial asymmetry, face and mouth droop, loss of facial expression, drooling, and biting of the cheek. These can be quite undesirable and may resemble Bell's palsy. Forehead injections can result in brow ptosis or loss of eyebrow elevation.

Other Movement Disorder Indications

BTX-A can also be effective in treating dystonic tics. Treatments of eye closure tics, neck and shoulder tics, and vocal tics (injections into the vocal cords) have been successful. A number of recent studies have shown BTX-A to be partially effective in various tremor disorders, including head tremors, parkinsonian and essential hand tremors, and voice tremor. Myoclonus may also be responsive to BTX-A.

ADVERSE EFFECTS

Side effects of BTX-A injections are mainly localized to the region of injection and are discussed above. At the time of injection, patients may describe stinging, especially with injections of the eyelid and facial regions. Bruising at the site of injection may occur. Most adverse effects are local and are related to overweakening of the treated or surrounding musculature. These effects take a few days to develop and are transient.

Systemic side effects after BTX-A injections have also been reported. Some patients describe a flulike syndrome with fever, chills, generalized weakness, and fatigue. These symptoms are all transient and may last up to a few weeks. It is not difficult to manage the side effects, since most are all local phenomena. The only serious side effects are dysphagia and respiratory compromise, which can occur with injections into the neck, mouth region, tongue, or larynx.

Several studies have used EMG to investigate distant effects of BTX-A. There is evidence that injections in the neck or face can result in remote neuromuscular effects in the limbs. These remote effects were determined electrophysiologically with single-fiber EMG, and none of the patients had remote effects by clinical criteria.

In recent reports, two patients had a systemic botulism-like reaction to BTX-A injections. Severe generalized weakness, including bulbar weakness, developed in both and resolved over several weeks. One patient had only one series of injections for spasticity; the other had undergone injections for torticollis for a number of years.

Long-term side effects of BTX treatment have been minimal. One of the main effects may be a change in the phenomenology of the movement disorder. Some patients with torticollis treated on multiple occasions experience a shift in their disorder, whether new turning, tilting, flexion, or extension, compared with the baseline state. Most believe this happens because the central generator of the dystonia activates muscles not involved in the initial phenomenology.

SECONDARY NONRESPONSIVENESS

Unfortunately, a small number of patients may not have a response (primary non-responders), and others experience a loss of response to BTX-A (secondary nonresponders). Secondary nonresponse may occur after one or two treatment sessions or after many. The loss of response may be due to involvement of muscles not receiving injections, contractures, or a change in the patient's perception of response. If atrophy of muscles does not develop, even if high doses are injected, immunoresistance is suspected. Most instances of immunoresistance have occurred in patients with cervical dystonia, and up to 5% to 10% of those receiving injections may be affected. The development of antibodies appears to correlate with earlier age at onset of dystonia, higher mean BTX-A dose per visit, and higher cumulative BTX dose. Another factor thought to contribute to antibody development is giving doses too frequently (more than one series of injections less than 1 month apart).

Currently, in vivo tests for antibodies to BTX-A have been the only widely accepted methods for determining immunoresistance. A mouse neutralization assay is performed with the patient's blood. If mice survive a lethal dose of BTX-A after injection of the patient's serum, the patient most likely has circulating antibodies to BTX-A. An enzyme-linked immunosorbent assay technique has been developed, but results do not

Table 31-3
Relative Contraindications for Use of Botulinum Toxin

- Pregnancy
- Myasthenia gravis or Lambert-Eaton syndrome
- Motor neuron disease (e.g., amyotrophic lateral sclerosis)
- Concomitant use of aminoglycosides

correlate well with those of in vivo methods. A common technique that clinicians use is to inject the toxin unilaterally into the frontalis muscle (7.5 U into two sites) and check for weakness in elevation of the ipsilateral eyebrow 2 weeks later. If the forehead is weakened, the patient is not immune to BTX-A.

For patients in whom BTX-A becomes ineffective, treatment options include trials of oral medications or, it is hoped, treatment in the future with one of the other BTXs. Since the various BTXs are antigenically distinct, it is thought that patients will have a response to BTX type B, C, or F despite resistance to BTX-A. Reports of these toxins suggest that they are effective, but they are not yet commercially available. BTX-B has been shown to be effective in patients with torticollis who continue to respond to BTX-A and in those who have lost their response to BTX-A. BTX-B is awaiting approval by the Food and Drug Administration at this time.

CONTRAINDICATIONS

Currently, there are no known absolute contraindications to the use of BTX-A. The relative contraindications are pregnancy, lactation, disorders of neuromuscular transmission (myasthenia gravis and Lambert-Eaton syndrome), motor neuron disease, and concurrent use of aminoglycosides (Table 31-3). Therefore, the treating physician must weigh the apparent need for BTX treatment with the unknown potential complications. Anecdotally, BTX-A has been injected into pregnant women without any known adverse effects on the mother, the fetus, or the birth. Similarly, a few patients with myasthenia gravis have received BTX-A without problems.

Since there is a limit to the total dose a patient can receive (about 500 to 600 U at an injection session), patients with generalized dystonia (involving multiple body regions) cannot be treated completely. If the patient has a specific region that is most troublesome, because of disability or pain, that region could be selected for injection.

SUMMARY

BTX is a neurotoxin that inhibits acetylcholine release at the neuromuscular junction. Local injections of BTX into hyperactive muscles can cause muscle weakness and atrophy, thus reducing involuntary movements. Although BTX can be used to treat almost any movement disorder, its primary use is for focal dystonias and hemifacial spasm. The main adverse effects are those caused by overweakness of the muscles that received injections or weakness of adjacent muscles.

SELECTED READING

Blitzer A, Brin MF. Laryngeal dystonia: a series with botulinum toxin therapy. Ann Otol Rhinol Laryngol 1991; 100: 85-89.

Brin MF, Blitzer A, Herman S, Stewart C. Oromandibular dystonia: treatment of 96 patients with botulinum toxin type A, in Therapy With Botulinum Toxin (Jankovic J, Hallett M eds), Marcel Dekker, New York, 1994, pp. 429-435.

Brin MF, Lew MF, Adler CH, Comella CL, Jankovic J, O'Brien C, Murray JJ, Wallace JD, Willmer-Holme A, Koller M. Safety and efficacy of NeuroBloc (botulinum toxin type B) in type A-resistant cervical dystonia. Neurology 1999; 53: 1431-1438.

Greene P. Controlled trials of botulinum toxin for cervical dystonia: a critical review, in Therapy With Botulinum Toxin (Jankovic J, Hallett M eds), Marcel Dekker, New York, 1994, pp. 279-287.

Greene P, Fahn S, Diamond B. Development of resistance to botulinum toxin type A in patients with torticollis. Mov Disord 1994; 9: 213-217.

Jankovic J. Botulinum toxin in movement disorders. Curr Opin Neurol 1994; 7: 358-366.

Jankovic J, Schwartz K. Botulinum toxin treatment of tremors. Neurology 1991; 41: 1185-1188.

Ludlow CL, Rhew K, Nash EA. Botulinum toxin injection for adductor spasmodic dysphonia, in Therapy With Botulinum Toxin (Jankovic J, Hallett M eds), Marcel Dekker, New York, 1994, pp. 437-450.

Pahwa R, Busenbark K, Swanson-Hyland EF, Dubinsky RM, Hubble JP, Gray C, Koller WC. Botulinum toxin treatment of essential head tremor. Neurology 1995; 45: 822-824.

Price J, Farish S, Taylor H, O'Day J. Blepharospasm and hemifacial spasm. Randomized trial to determine the most appropriate location for botulinum toxin injections. Ophthalmology 1997; 104: 865-868.

Pulmann SL, Greene P, Fahn S, Pedersen SF. Approach to the treatment of limb disorders with botulinum toxin A. Experience with 187 patients. Arch Neurol 1996; 53: 617-624.

Schantz EJ, Johnson EA. Botulinum toxin: the story of its development for the treatment of human disease. Perspect Biol Med 1997; 40: 317-327.

Yoshimura DM, Aminoff MJ, Tami TA, Scott AB. Treatment of hemifacial spasm with botulinum toxin. Muscle Nerve 1992; 15: 1045-1049.

E

OTHER MOVEMENT DISORDERS

32   Wilson's Disease

Katrina A. Gwinn-Hardy, MD

CONTENTS

In 1912, S. A. Kinnear Wilson described hepatolenticular degeneration, which was later named "Wilson's disease." This autosomal recessive disease results in the deposition of excess copper in the body, especially in the brain and liver. Although our understanding of the clinical findings, laboratory abnormalities, molecular aspects, and treatment strategies is sophisticated, we still do not fully understand the pathogenesis of this fascinating disease. Inevitably fatal at the time of its description, Wilson's disease is now one of the most treatable hereditary neurologic disorders. Because it has a wide clinical spectrum and can be treated, Wilson's disease should be considered in the differential diagnostic evaluation of patients seen by neurologists, psychiatrists, gastroenterologists, and other physicians.

CLINICAL FEATURES

Wilson's disease (WD) is largely a disorder of children and young adults, although occurrences have been found as early as 3 years and as late as the seventh decade. Hepatic, neurologic, and psychiatric symptoms are common and may be seen alone or in combination. Renal, rheumatologic, and hematologic symptoms may also be seen, but less commonly.

Hepatic, extrapyramidal, and mood disorders should lead to consideration of the diagnosis of WD. Asymptomatic parents and siblings of affected persons should be evaluated for the disease, as should other relatives in consanguineous families. A family history of hepatic, neurologic, or psychiatric disease should further raise suspicion.

From: *Parkinson's Disease and Movement Disorders:*
Diagnosis and Treatment Guidelines for the Practicing Physician
Edited by: C. H. Adler and J. E. Ahlskog © Mayo Foundation
for Medical Education and Research, Rochester, MN

Because the presentation is so varied, it is better to be overly enthusiastic in the workup of WD than to miss a single case of this treatable but otherwise disabling and fatal illness.

Hepatic Manifestations

The most frequent presentation of WD, seen in up to 50% of patients, is that of hepatic symptoms and signs. Hepatic disease is most common in children, with the average age at onset of about 11 years. For unknown reasons, hepatic presentation is especially common in Asian persons with WD. The range of hepatic symptoms, like that of the neurologic and psychiatric symptoms, is broad. However, the most common hepatic manifestation of WD is progressive cirrhosis. The clinical picture is not unique to WD, resembling that of cirrhosis of other causes. Patients may have splenomegaly, ascites, esophageal varices, and hepatic encephalopathy. Other hepatic illnesses in WD are acute transient hepatitis, acute fulminant hepatitis, chronic active hepatitis, and asymptomatic enlargement of the liver or spleen. The hepatic disorders in WD can also lead to neurologic symptoms.

Neurologic Dysfunction

Neurologic dysfunction is likewise protean (Table 32-1). Whereas hepatic disease usually occurs in childhood, the neurologic manifestations of WD are most common in the teenage years, with an average age at onset of about 19 years. The age range is broad, however, and any patient younger than 50 years with a movement disorder could have WD and should undergo evaluation. If the disease is left untreated, all patients eventually acquire neurologic dysfunction.

Dysarthria and drooling are common neurologic signs. Uncontrolled drooling may occur in isolation. Speech can become severely strained, sometimes with dystonic features. Dystonia, especially of the face and neck, is often seen. However, dystonia may occur in almost any of its manifestations, including focal limb dystonia, task-specific dystonia, hemidystonia, and generalized dystonia. Gait can be ataxic or parkinsonian.

Tremor is frequent with neurologic WD, occurring in one-third to one-half of patients. The classic teaching is that the tremor associated with WD is "wing beating" (worse with arms bent at the elbows and held up at shoulder level in front of the chest, with hands in the midline); however, this type of tremor is only occasionally seen. Postural tremor and mixed tremor (resting and postural components) are probably the most common tremor manifestations. Tremor in WD may be task specific, may wax and wane in severity, and may not resemble any of the classic types of tremor. Parkinsonism, with rigidity, resting tremor, and slowness of movement, also occurs. Cerebellar findings, especially cerebellar tremor, occur in about one-fourth of patients with WD.

The disease can be asymmetrical or even focal in its manifestations, despite being a systemic illness. The neurologic symptoms can be quite bizarre, may be distractible, and have a waxing and waning quality. Many patients with WD have previously been informed that they have a psychogenic movement disorder. Thus, WD challenges the practitioner to maintain an open mind toward the cases that do not "make sense" according to the usual understanding of phenomenology. Some neurologic dysfunction may be permanent if treatment is not begun early enough to prevent permanent cell loss. Tremor seems to respond more quickly and completely to therapy than dystonia and dysarthria.

Seizures are rare in WD, occurring in fewer than 6% of patients, usually in children. Upper motor neuron signs (spasticity, hyperreflexia) are usually not seen. Cranial nerve

Table 32-1
Neurologic Manifestations in Wilson's Disease

Very frequent	Common	Rare or not seen
Dysarthria	Parkinsonism	Upper motor neuron signs[a]
Dystonia	Gait abnormalities	Lower motor neuron signs
Incoordination	Tremor	Seizures[a]
Sialorrhea	Eye movement	Other cranial nerve abnormalities
Rigidity	abnormalities	Dementia

[a]May occur with hepatic manifestations of the disease.

abnormalities suggest an alternative or additional diagnosis. Autonomic disorders are rare. Lower motor neuron findings do not occur directly as the result of WD but may be secondary to electrolyte abnormalities due to complications of the disease. Dementia, as noted above, is very rarely found.

Psychiatric Manifestations

Psychiatric manifestations are common and often overlooked, with up to 50% of patients having received psychiatric treatment, including hospitalization, before diagnosis. Almost all patients with WD have a psychiatric complaint at some point in their illness. As with the hepatic and neurologic manifestations noted above, no pathognomonic psychiatric features allow one to make the clinical diagnosis. The most frequent behavioral complaints of WD are personality changes (such as irritability and emotional lability) and depression (which can be severe, leading to suicide). Many individuals or their family members note subtle cognitive changes, including lapses in memory, trouble with concentration, difficulty with previously acquired skills (such as mathematics and vocabulary), and disinhibition. Almost all note some difficulty in functioning at school or work. Delusions may occur, but frank psychosis is rare.

Other Manifestations

OCULAR FINDINGS

The classic ocular finding in WD is the Kayser-Fleischer (KF) ring. Although reported, neurologic manifestations without KF rings are exceedingly rare. Originally noted by Kayser in 1902 and separately by Fleischer in 1903, the rings were not related to WD for another decade. KF rings in WD are due to the deposition of copper in Descemet's membrane of the cornea. They are usually bilateral. Typically brown, they can also be green or gold. They are more easily seen in blue eyes than in brown. The superior aspect of the cornea is involved initially, followed sequentially by the inferior, medial, and, ultimately, lateral aspects. Usually, slit-lamp examination by an experienced ophthalmologist or neuro-ophthalmologist is necessary to detect KF rings. Several other causes of corneal deposition indistinguishable from KF rings exist. These include the ocular findings of hepatic diseases such as cirrhosis of causes other than WD, multiple myeloma, and increased gamma globulin states.

Another ocular manifestation of WD is the sunflower cataract, which results from copper deposition in the lens. This is rare, occurring in fewer than 20% of patients.

Sunflower cataracts have a green, gold, gray, or brown appearance. Neither KF rings nor sunflower cataracts typically interfere with vision.

RHEUMATOLOGIC SYMPTOMS

Patients with WD often have bone or joint pain. Additionally, penicillamine, a common therapeutic agent (see below), can cause significant rheumatologic symptoms, which are sometimes difficult to differentiate from those due to WD. Tissue destruction by copper and iron deposition in joints may be responsible for the pain, although the cause is not yet known.

RENAL DISEASE

Renal disease may occur in WD, largely secondary to renal tubular necrosis due to excessive excretion of copper. This, in turn, can lead to secondary calcium and phosphate abnormalities.

HEMATOLOGIC DISORDERS

Hemolytic anemia (Coombs'-negative) is the most common hematologic occurrence and is the illness at presentation in 10% to 15% of patients. The hepatic manifestations may also lead to secondary hematologic abnormalities.

EPIDEMIOLOGY AND GENETICS

The disease is quite rare, with an estimated prevalence of 30 cases per 1 million population. The frequency of carriers is estimated at 1 per 90 persons. WD is less common in northern Europe and the U.S. than it is in Japan, the Middle East, and Sardinia, which have higher rates of consanguinity.

WD is autosomal recessive. Heterozygotes do not have symptoms and do not require treatment, although they may exhibit mild abnormalities on diagnostic testing of copper metabolism. In the mid-1980s, the gene was linked to chromosome 13q14.3. The gene has been found to be a copper-transporting P-type adenosinetriphosphatase (ATPase). At least 27 distinct mutations have been found. A given mutation type does not seem to correlate with phenotypic presentation or severity. Because of the large number of mutations that could be responsible, genetic testing as a clinical tool is unlikely to emerge in the near future. In a given family, however, linkage analysis may be possible.

PATHOPHYSIOLOGY

The normal body absorbs and excretes from 1 to 5 mg of copper per day. When the body's mechanisms for binding and eliminating excess copper are saturated, copper deposition occurs. This is the situation in WD. Most of the manifestations of the disease are due to accumulation of copper in the brain and liver, although other regions of the body, including the kidneys and joints, can be affected, as noted above. Ceruloplasmin, an α-globulin, is the major transporter of copper in the blood and is essential to the elimination of copper. Because ceruloplasmin is usually low in WD, ceruloplasmin deficiency was long suspected to be the cause of the disease. However, genetic evidence makes it clear that WD is a disorder not of ceruloplasmin but of abnormal ATPase activity. How the abnormal ATPase activity leads to impaired copper excretion is currently a subject of extensive research.

NEUROLOGIC DIFFERENTIAL DIAGNOSIS

With clinical manifestations as variable as those described above, the differential diagnosis is clearly quite broad. Although an exhaustive list of the differential diagnostic entities is not practical, several of the more common disorders that can resemble WD are briefly mentioned.

In patients with dysarthria, the differential diagnosis includes Parkinson's disease and other causes of parkinsonism, tardive dyskinesia or dystonia, Sydenham's chorea, and Huntington's disease. Degenerative diseases of the cerebellum, including the hereditary ataxias, are also within the differential diagnosis. Stroke may be manifested by dysarthria. The assistance of a skilled speech pathologist cannot be underestimated in the assessment of the features of speech, because a flaccid or purely spastic dysarthria is unlikely in WD.

Patients who have tremor also must be considered for the entities mentioned above as well as for essential, task-specific, and dystonic tremors. Enhanced physiologic tremor should also be considered. Tremor can be exacerbated or produced by a variety of agents, including anticonvulsant drugs, antiasthma therapy, some antidepressants, and lithium. Thyroid disease must also be ruled out.

Dystonia can be due to WD as well as to several of the causes listed above. Other hereditary causes of dystonia are dopa-responsive dystonia, paroxysmal dystonia, Hallervorden-Spatz disease, neuroacanthocytosis, X-linked dystonia parkinsonism, and Huntington's disease. Metabolic disorders that can cause dystonia, especially in children, include mitochondrial disorders such as Leigh's disease, vitamin E deficiency, and hexosaminidase A deficiency. A variety of environmental insults can cause parkinsonism or dystonia, including carbon monoxide poisoning, central nervous system tumor or trauma, and infections of the nervous system.

Discussion other than this brief summary is beyond the scope of this text. However, one must consider WD when evaluating movement disorders in most persons between 3 and 50 years old.

DIAGNOSTIC TESTS

The four most useful tests widely available for neurologic WD are serum ceruloplasmin determination, 24-hour urinary copper measurement, slit-lamp examination for KF rings by an experienced ophthalmologist, and liver biopsy for copper measurements. A brief discussion of each of these tests and of others is given below.

A summary of the important aspects of the evaluation for WD is as follows:

1. Initial screening includes a careful history and examination, liver function tests, serum ceruloplasmin level determination, 24-hour urinary copper content measurement, and slit-lamp examination.
2. In persons with hepatic disease primarily, a liver biopsy is usually necessary because they may not have KF rings and urinary copper and serum ceruloplasmin findings may be falsely negative.
3. In persons with neurologic or psychiatric symptoms, magnetic resonance imaging (MRI) is a useful adjunctive test.
4. If results of slit-lamp, 24-hour urinary copper, and ceruloplasmin tests are ambiguous or negative, a liver biopsy may be necessary.
5. Persons without symptoms who have affected siblings should have a thorough diagnostic workup.

Table 32-2
Causes of Abnormal Ceruloplasmin Levels

Low serum ceruloplasmin	High serum ceruloplasmin
Wilson's disease	Oral contraceptive pills
Heterozygotes for Wilson's disease	Postmenopausal hormone replacement
Menkes' syndrome	Acute phase reactant in systemic illness
Protein-losing enteropathy	
Nephrotic syndrome	
Hepatic failure	
Deficient protein intake	
Infancy	

Serum Ceruloplasmin

A normal ceruloplasmin level is 20 to 40 mg/dL in most laboratories. The concentration is abnormally low in about 80% of patients with WD (Table 32-2). It is important to have serum ceruloplasmin measured in a laboratory that is experienced in testing for this protein. Ceruloplasmin levels may be within or only slightly below the normal range in up to 20% of affected persons; thus, this test should not be used as the sole measurement. Heterozygotes may have low ceruloplasmin levels but do not require any treatment. Ceruloplasmin can be decreased in a number of disorders other than WD, including Menkes' syndrome (another hereditary disorder of copper metabolism), protein-losing enteropathy, nephrotic syndrome, hepatic failure, sprue, and deficiency in calorie and protein intake. Infants less that 6 months of age normally have low ceruloplasmin levels. An increased ceruloplasmin value is of no known clinical significance and often occurs in patients taking oral contraceptives or receiving female hormonal replacement therapy. Factors that normally increase ceruloplasmin may lead to false-negative values in patients with WD. Ceruloplasmin is an acute phase reactant and thus can be increased in patients who are otherwise ill.

Urinary Copper Measurements

When free copper is excessively high in the bloodstream, urinary copper excretion increases. Thus, urinary copper is increased in untreated WD. A urinary copper test should be done on a 24-hour collected specimen. It is essential that the laboratory doing the test supply the patient with copper-free jugs for the urine collection. The patient should be advised to be on a copper-free diet during the collection if at all possible (see below). In normal persons, copper excretion in the urine is less than 70 μg/day. Typically, urinary copper levels in symptomatic WD are 100 μg/day or greater. Persons with values between 60 and 100 μg/day are in an indeterminate range, and the 24-hour urine test for copper should be repeated or further testing, including liver biopsy, may be necessary. Urinary copper values may be in the normal range early in the manifestations of the disease, when free copper is accumulating in the liver, and in the early stages of disease, this test result may be falsely negative. False-positive results can occur with biliary cirrhosis due to other causes and with chronic active hepatitis. Increased copper in the diet can cause a falsely increased urinary copper result as well; thus, copper measurements should be made after advising the patient to maintain a copper-free diet for the 24 hours preceding and the 24 hours during collection of the specimen.

Slit-Lamp Examination for Kayser-Fleischer Rings

As noted above, an experienced ophthalmologist or neuro-ophthalmologist should perform the slit-lamp examination. The absence of KF rings in a patient with neurologic manifestations is highly suggestive of an alternative diagnosis. To date, only one person has been reported to have WD without KF rings. KF rings can sometimes be seen on bedside evaluation, but even experienced clinicians may not be able to detect them in that situation. In patients with hepatic disease only, KF rings may be absent, so that the lack of KF rings does not rule out WD.

Liver Biopsy for Copper Level Measurements

Liver biopsy for copper level measurements remains the most accurate test for WD that is widely available. Even patients with no hepatic symptoms have an increased value for copper in the liver. Sampling error may occur, because copper is not uniformly distributed in the liver in WD. The biopsy needle and containers must be free of copper contamination. A disposable steel biopsy needle should be used, and the plastic syringe for biopsy should contain 5% dextrose solution, not saline solution. A normal liver copper value is 15 µg/g of dry weight tissue, but in WD this is usually 200 µg or greater. Heterozygotes may have increased liver copper concentrations, but they are usually between 55 and 200 µg/g of dry weight. A strongly positive liver biopsy result can be misleading, because chronic active hepatitis and primary biliary cirrhosis also cause increased copper in the liver. A lower than expected liver copper value can be found in persons biopsied during treatment, but they usually have levels of about 100 µg/g of dry weight. Because liver biopsy carries a significant risk, it is usually done when results of other tests (ceruloplasmin, slit-lamp examination, urinary copper) are inconclusive or negative and suspicion remains high.

Imaging Studies

Results of imaging studies, including MRI of the brain, are occasionally normal in WD; however, MRI abnormalities are usually seen. On T2-weighted images, WD is suggested by atrophy, hyperintensity, or hypointensity bilaterally in the putamen, with abnormalities also found in the substantia nigra, periaqueductal gray matter, and thalamus. Hepatic abnormalities may also lead to increased T1 signal in the basal ganglia. Positron emission tomography scanning with deoxyglucose and fluorodopa may yield abnormal findings as well, but these testing modalities are not routinely used in clinical practice.

Radioactive Copper Studies

Measurement of the incorporation of radioactive copper (^{64}CU) into ceruloplasmin has been used to evaluate WD. However, this test is not widely available and remains largely a research tool at the time of this writing. In this test, if the person does not have WD, serum ^{64}CU initially increases after oral administration of the isotope as it enters the blood and binds to albumin and other serum components. Subsequently, the ^{64}CU level decreases as the agent is cleared by the liver and then increases as it is released into the bloodstream once again, this time bound to ceruloplasmin. In persons with WD, the second increase does not occur, even if the ceruloplasmin level is normal. The test result can be abnormal in patients with liver disease due to other causes. Radioactive copper studies have definite utility in the diagnosis of WD and may be more widely available in the future. The drawbacks include the use of a radioactive isotope.

Genetic Testing

As noted above, numerous mutations can lead to WD, so that development of commercial genetic tests for WD is difficult. In certain families, linkage analysis may be possible, although such testing is not widely available and requires the support of a research laboratory with facilities for such analyses.

Serum Copper Measurements

Measurement of copper in the serum is not usually useful in WD, because most laboratories measure both bound and unbound copper, and the values obtained may not be outside normal ranges in WD. The level of free serum copper would be expected to be high in WD, if the measurement is available.

Summary

The diagnosis is confirmed in the following ways.

1. Neuropsychiatric manifestations are found, and both KF rings and low ceruloplasmin level are present.
2. Neuropsychiatric manifestations and either KF rings or low ceruloplasmin levels are found, and the 24-hour urinary copper level is increased to 100 µg/day or greater.
3. Neuropsychiatric manifestations and only KF rings, increased urinary copper, or decreased ceruloplasmin is present, and the diagnosis is confirmed by increased copper content in the liver.
4. The liver biopsy finding is abnormal, with copper content greater than 200 µg/g of dry weight.
5. Studies of ^{64}CU uptake, if available, yield positive results.
6. Linkage analysis is possible and confirms the diagnosis.

TREATMENT

For treatment of WD, a negative copper balance must be achieved. Thus, one or all of the following strategies are used.

1. Reduction of copper intake.
2. Reduction of copper absorption.
3. Increase in copper elimination.
4. Liver transplantation.

Treatment must be lifelong. In addition to the treatment strategies outlined below and in Table 32-3, treatment of symptoms should be tailored to the manifestations of disease. For example, botulinum toxin may be useful for dystonia, antitremor medications can be used, and depressive symptoms should be treated, if indicated, by pharmacotherapy and counseling. Because most of the patients are children, teenagers, and young adults, involving the family in the treatment is essential, and genetic counseling should be made available.

Reduction of Copper Intake and Dietary Recommendations

The average American diet includes about 1 mg of copper per day. Foods high in copper content should be avoided, and foods moderate in copper content should be

Table 32-3
Comparison of Treatment Modalities in Wilson's Disease

Treatment	Dosage	Advantages	Disadvantages
Zinc (decreased absorption)	50 mg T.I.D.; can be increased up to 250 mg/day, taken in divided doses on an empty stomach	Low toxicity. Can be used in combination with chelation. Treatment of choice for maintenance therapy, for presymptomatic treatment, and in pregnancy.	Slow to act. May cause gastrointestinal irritation.
Ammonium tetrathiomolybdate (decreased absorption, decreased free serum copper)	20 mg with each meal and an additional 20 mg between meals	May be particularly useful in neurologic WD because of limit of free serum copper.	Little experience with this agent. May be difficult to monitor.
Penicillamine (increased cupruresis)	250 mg/day	Most experience with this agent.	Numerous adverse reactions, some life-threatening.
Trientine (increased cupruresis)	750 to 2,000 mg/day in three divided doses	Infrequent toxic reactions. Rapid results. Useful for initial therapy in combination with zinc.	Less experience than with penicillamine. Lupus nephritis, iron deficiency may occur.
Liver transplantation	Not applicable	May correct underlying defect, eliminating need for pharmacotherapy.	Mortality risk is up to 30%. Limited donor organ availability.

Table 32-4
Copper Content of Various Common Foods

	Low (< 0.1 mg)	*Moderate (0.1 to 0.2 mg)*	*High (≥ 0.2 mg)*
Meats and dairy products	Beef, cheese, chicken, eggs, light meat of turkey	Dark meat of turkey, peanut butter, fish	Pork, duck, lamb, squid, tofu, shellfish of all types
Fats and oils	Butter, mayonnaise, margarine	Olives, cream	Avocados
Starch	Refined-flour breads, oatmeal, pasta, rice	Most cereals, whole wheat bread	Bran breads and cereals, dried beans, wheat germ
Vegetables and fruits	Tomatoes, apples, bananas	Mangoes, pears, pineapples	Mushrooms, dried fruits, raisins, dates, prunes
Desserts	Jams, jellies, candies	Licorice	Chocolate

limited (Table 32-4). In some areas, tap water may contain substantial amounts of copper. It is best if patients with WD drink distilled water or water known to contain less than 100 μg of copper per liter. Domestic water softeners may add copper to the household water supply. Vitamins and health food products can contain significant amounts of copper and should not be used. Alcohol, although not a significant source of copper, is contraindicated because of its potentially damaging effects on the liver. Patients and their families should meet with a dietitian to further learn which foods are acceptable and how to interpret labeling of foods and other ingested products. Limitation of dietary intake alone is not adequate therapy and must be used in conjunction with other treatment.

Reduction of Copper Absorption

ZINC

Zinc can prevent absorption of copper through the gastrointestinal tract. The mediating protein is metallothionein, which has a high affinity for zinc but an even higher affinity for copper. When zinc is administered on an empty stomach, metallothionein formation is induced. This induction allows binding of the zinc, and as a secondary effect, dietary copper is also bound, thus decreasing the absorption of copper through gastrointestinal cells. The bound copper is excreted in the feces when these cells are shed naturally. Zinc also blocks the small amount of copper absorbed through saliva and gastric juices via the same mechanism. Zinc is generally well tolerated and has virtually no toxicity. Thus, the therapy is appealing for asymptomatic persons and those who are pregnant. The major drawback of zinc therapy is that action is delayed, with up to 2 weeks required for an effect on copper absorption, which may not be rapid enough for the patient with symptoms. Gastrointestinal upset can occur. Zinc use may lead to iron deficiency anemia. Once a patient with symptoms is in stable condition, switching from more potent agents to zinc for maintenance therapy is reasonable. Zinc may also be given along with faster acting agents at the beginning of treatment. It is taken as either zinc acetate or zinc sulfate. The usual dose is 50 mg orally three times a day.

AMMONIUM TETRATHIOMOLYBDATE

Ammonium tetrathiomolybdate (TM), like zinc, limits gastrointestinal absorption of copper, but it does so by forming a complex with copper and albumin in the gut lumen.

This prevents transport of the copper into the intestinal mucosal cells. Subsequently, the entire complex is excreted in the feces. Furthermore, TM is absorbed into the bloodstream, and the copper-TM-albumin complex forms there as well. This effect further prevents uptake of copper by the tissues and prevents end organ damage. Unlike that of zinc, the effect is prompt. Because TM forms a complex that is eliminated, it is nontoxic. Use requires titration to match the plasma molybdenum levels with the free serum copper levels, and this determination can be somewhat difficult. Unlike more aggressive agents, such as penicillamine (below), TM does not cause initial worsening of neurologic symptoms in affected persons. Rarely, the serious adverse side effect of bone marrow suppression occurs. Because animal studies have shown that TM may damage epiphyses, it should not be used in growing children. Experience with this agent remains limited, a current potential drawback.

The usual dose is 20 mg orally at mealtimes. If these three doses are tolerated, three between-meals doses of 20 to 60 mg are added. Thus, the usual final maximal dose is 20 mg with breakfast, 60 mg at midmorning, 20 mg with lunch, 60 mg in the midafternoon, 20 mg with dinner, and a final 60 mg in early evening (a total of six doses a day).

Increased Copper Elimination

As noted above, the average American diet results in a copper intake of about 1 mg. About 0.8 mg is lost in feces, and the rest is normally primarily excreted through the urine. Thus, if a 24-hour urinary copper excretion of 2 mg or greater can be achieved, the result is the elimination of copper from storage in tissues. The chelating agents described below exert their effect by increasing cupruresis.

PENICILLAMINE

The chelating agent in use the longest is penicillamine. It is believed that penicillamine exerts its primary beneficial effect through copper chelation and cupruresis; however, it may also produce some benefit from induction of metallothionein. The benefit is rapid in onset, so that the drug is useful in patients with severe symptoms. However, it worsens the neurologic symptoms initially during treatment, and patients and their family members must be warned about this phenomenon. This effect is believed to be due to immobilization of copper from the liver into the bloodstream, with secondary uptake into the nervous system, before excretion. Improvement in function overall may occur as early as within 2 weeks of treatment. Continued treatment may allow benefit for up to 2 years.

The adverse effects of penicillamine are many, and noncompliance is a problem in many patients. Sensitivity is common, occurring in up to 30% of those treated. Rash, fever, eosinophilia, thrombocytopenia, and leukopenia may all occur. These reactions usually appear in the first 2 weeks of treatment and may be dose-related. Lowering the dose may be adequate to allow the patient to tolerate treatment, especially if the drug is given concurrently with oral prednisone. After long-term administration, penicillamine can lead to nephrotic syndrome, a lupus-like syndrome, myasthenic syndrome, polyarthritis, and loss of taste. (Fortunately, loss of taste abates with zinc, which can be used for concurrent therapy.) Dermatologic problems are common in both short-term and long-term treatment with penicillamine. Rash, as noted above, occurs early and can be severe. Long-term treatment leads to brown discoloration of the skin from frequent incidental subcutaneous bleeding. The initial worsening of neurologic symptoms described above can also lead to noncompliance. Goodpasture's syndrome, with hemoptysis and hematuria, can be a fatal complication of this treatment, and it should lead to immediate discontinuation of administration of the drug.

Penicillamine is usually given on an empty stomach, between 1 and 2 g/day. Treatment should be initiated at 250 mg orally every day and the dose gradually increased by 250 mg every 3 to 5 days until copper excretion is thought to be adequate (see below). If sensitivity occurs, penicillamine should be withdrawn. Prednisone can be given after discontinuation for sensitivity, at 20 mg orally every day. After 3 days to a week of prednisone treatment, penicillamine can be readministered at a dose of 250 mg/day. If no sensitivity occurs over the next 2 to 3 weeks, the dose of prednisone should be gradually tapered.

TRIENTINE

Triethylenetetramine dihydrochloride (trientine) is also a copper chelating agent that results in increased cupruresis. It is less potent than penicillamine and thus is not used for first-line therapy. Rather, it is useful as an alternative for copper chelation if penicillamine is not tolerated. Even with trientine, however, a lupus-like nephritis and anemia can occur. The dosage is 750 to 2,000 mg/day in divided doses.

Liver Transplantation

Orthotopic liver transplantation is the treatment of choice in fulminant liver failure when death otherwise is the inevitable outcome. In addition to improvement in liver function, liver transplantation results in reduction in the neurologic and psychiatric symptoms, because the copper metabolism abnormality responsible for WD normalizes after liver transplantation. Transplantation is usually done when medical treatment has failed or is not possible.

MONITORING THERAPY

Noncompliance is a concern with all treatments of WD except liver transplantation. Reasons for noncompliance are that once the initial phase of treatment is over, the treatment benefits become less apparent, the regimen can be demanding and side effects unpleasant, the patients are usually children or teenagers, and the treatment must be lifelong. Regular long-term follow-up should take place with a neurologist, ophthalmologist, gastroenterologist, and psychiatrist.

Monitoring of therapy can vary somewhat among agents used, as noted below. During therapy, neuropsychiatric and hepatic improvement should occur. KF rings gradually recede in reverse order of their formation (see above). The findings on MRI may also decrease, although imaging is not an adequate method for following the progress of treatment.

Zinc

Monitoring is accomplished as follows:

1. Baseline values for serum ceruloplasmin and 24-hour urinary excretion of copper and zinc are obtained before treatment begins.
2. These measurements are checked after the first month of therapy and every 3 months thereafter. A gradual reduction in urinary copper levels should be seen.
3. Since zinc, unlike chelating agents, does not promote cupruresis, an increase in urinary copper is not found.
4. The 24-hour urinary excretion of zinc should be 2.5 mg or higher in a compliant patient taking 150 mg of zinc per day.

Ammonium Tetrathiomolybdate

Therapy with ammonium tetrathiomolybdate requires a laboratory that can measure both free copper and molybdenum levels. The ratio of serum molybdenum to free copper should be measured. A ratio of 1:1 indicates that all the free serum copper has been complexed with molybdenum and albumin.

Penicillamine and Trientine

Patients receiving penicillamine should have urinalysis, complete blood cell count with platelets, and skin examination once a month and should measure their temperature daily.

All patients receiving chelation therapy require regular follow-up.

1. Slit-lamp examination is used to monitor KF rings. The examination is done weekly during the first month of therapy, monthly during the first year, and yearly thereafter.
2. Urinary excretion of copper in 24 hours is measured after the first 2 weeks of therapy and then monthly during the first year. The goal is to eliminate 2 to 5 mg of copper through the urine daily during the initial phase of therapy.
3. After the first year, excretion of copper in the urine diminishes as the copper stores are gradually normalized in the body. Urinary copper concentration is determined annually thereafter. If urinary copper excretion increases after the first several years of therapy, noncompliance should be suspected.

SUMMARY

1. WD is treatable but is uniformly progressive and fatal if left untreated.
2. A high degree of awareness is the most important aspect of diagnosis.
3. WD is largely a disease of children and young adults. It does not occur before age 3 years or after age 60.
4. The clinical manifestations, whether hepatic, neurologic, or psychiatric, are protean.
5. Parents and siblings of affected persons should be evaluated for the disease.
6. Serum ceruloplasmin determination, 24-hour urinary copper measurement, and slit-lamp examination for KF rings are the most useful tests in a patient with neurologic manifestations. Liver biopsy is the next step if the likelihood of disease is high.

SELECTED READING

Biglan AW, Taylor SL. Kayser-Fleischer ring in a patient with Wilson's disease. *N Engl J Med* 1993; 328: 1820.

Bowcock AM, Farrer LA, Cavalli-Sforza LL, Hebert JM, Kidd KK, Frydman M, Bonne-Tamir B. Mapping the Wilson disease locus to a cluster of linked polymorphic markers on chromosome 13. *Am J Hum Genet* 1987; 41: 27-35.

Brewer GJ. Practical recommendations and new therapies for Wilson's disease. *Drugs* 1995; 50: 240-249.

Brewer GJ, Yuzbasiyan-Gurkan V. Wilson disease. *Medicine (Baltimore)* 1992; 71: 139-164.

Cartwright GE. Diagnosis of treatable Wilson's disease. *N Engl J Med* 1978; 298: 1347-1350.

Farrer LA, Bowcock AM, Hebert JM, Bonne-Tamir B, Sternlieb I, Giagheddu M, St George-Hyslop P, Frydman M, Lossner J, Demelia L, Carcassi C, Lee R, Beker R, Bale AE, Donis-Keller H, Scheinberg IH, Cavalli-Sforza LL. Predictive testing for Wilson's disease using tightly linked and flanking DNA markers. *Neurology* 1991; 41: 992-999.

Marsden CD. Wilson's disease. *Q J Med* 1987; 65: 959-966.

Rathbun JK. Neuropsychological aspects of Wilson's disease. *Int J Neurosci* 1996; 85: 221-229.

Saatci I, Topcu M, Baltaoglu FF, Kose G, Yalaz K, Renda Y, Besim A. Cranial MR findings in Wilson's disease. *Acta Radiol* 1997; 38: 250-258.

Shah AB, Chernov I, Zhang HT, Ross BM, Das K, Lutsenko S, Parano E, Pavone L, Evgrafov O, Ivanova-Smolenskaya IA, Anneren G, Westermark K, Urrutia FH, Penchaszadeh GK, Sternlieb I, Scheinberg IH, Gilliam TC, Petrukhin K. Identification and analysis of mutations in the Wilson disease gene (ATP7B): population frequencies, genotype-phenotype correlation, and functional analyses. *Am J Hum Genet* 1997; 61: 317-328.

Starosta-Rubinstein S. Treatment of Wilson's disease, in *Treatment of Movement Disorders* (Kurlan R ed), JB Lippincott, Philadelphia, 1995, pp. 115-151.

Strickland GT, Beckner WM, Leu ML. Absorption of copper in homozygotes and heterozygotes for Wilson's disease and controls: isotope tracer studies with ^{67}Cu and ^{64}Cu. *Clin Sci* 1972; 43: 617-625.

Thomas GR, Forbes JR, Roberts EA, Walshe JM, Cox DW. The Wilson disease gene: spectrum of mutations and their consequences. *Nat Genet* 1995; 9: 210-217.

Wilson SAK. Progressive lenticular degeneration: a familial nervous disease associated with cirrhosis of the liver. *Lancet* 1912; 1: 1115-1119.

Yarze JC, Martin P, Munoz SJ, Friedman LS. Wilson's disease: current status. *Am J Med* 1992; 92: 643-654.

33

Gait Disorders: Recognition of Classic Types

Frank A. Rubino, MD

CONTENTS

OVERVIEW
ANATOMY AND PHYSIOLOGY OF LOCOMOTION
INITIAL EVALUATION
EFFECTS OF AGE ON GAIT AND BALANCE
FALLS AMONG ELDERLY PERSONS
GAIT DISORDERS RELATED TO NEUROLOGIC ABNORMALITIES
SUMMARY
SELECTED READING

OVERVIEW

Walking is a highly refined, remarkable, and automatic skill of humans we all take for granted. The basic reflex for walking, which is probably located in the spinal cord, is present at birth. Parents, relatives, and friends are all very pleased, excited, and proud when an infant takes the first steps. At the other end of the time spectrum, abnormalities of gait and falling tend to be problems of the elderly. Disorders of gait and mobility are second only to impaired mental function as the most frequent neurologic effects of aging.

Normal gait, stance, and balance require precise input from proprioceptive (position sense), vestibular (inner ear mechanisms and their connections within the brain stem), and visual pathways as well as auditory and tactile information. Two of the three major afferent systems (proprioceptive, vestibular, and visual) must be intact to maintain balance. Afferent data must be integrated in the brain stem and brain through motor (pyramidal and extrapyramidal) and cerebellar pathways, which then serve as the efferent arc of the important skill of walking. Dysfunction in the afferent or efferent systems or in the central integrating centers can lead to gait problems. Gait disorders in the elderly are frequently heterogeneous and often multifactorial in origin.

Most gait disorders are neurologic in origin, but certainly the musculoskeletal and cardiovascular systems contribute importantly. Arthritis is a common cause of gait disturbance in the elderly; it usually leads to an antalgic (painful) gait due to pain in the low back, hips, knees, or feet. Cardiovascular causes of abnormal gait are usually easy to identify and include orthostatic hypotension, heart failure, edema of the legs, various

From: *Parkinson's Disease and Movement Disorders:*
Diagnosis and Treatment Guidelines for the Practicing Physician
Edited by: C. H. Adler and J. E. Ahlskog © Mayo Foundation
for Medical Education and Research, Rochester, MN

arrhythmias, and intermittent claudication. Syncope accounts for 4% to 8% of falls and should direct the clinical evaluation toward a cardiovascular origin.

Drug therapy, especially polypharmacy, can adversely affect gait and cause falls. The benzodiazepines are a common culprit in this situation. Medications affect gait by many mechanisms. A number of drugs lead to orthostatic hypotension, others cause confusion and lethargy, and still others cause problems in the central integrating system of the brain and brain stem.

Two important and common mechanisms for gait abnormalities, especially in the elderly, need to be emphasized. Acute deterioration in gait and balance is a common finding in elderly persons with acute medical illness. Gait, balance, and postural adjustments are impaired. Acute confusional states induced by non-neurologic illnesses, such as any organ failure, dehydration, electrolyte imbalance, medications, and systemic infections (especially urinary tract infection and pneumonia), are the common causes of this problem. The cause is often multifactorial, and the gait disturbance, which resembles that of so-called lower-half parkinsonism, is potentially reversible. Some of these confused patients also have obvious asterixis ("flapping") of the hands and feet, which may add to the balance problems. In the acute confusional state, patients are inattentive, disoriented, confused, often hallucinating, and unable to maintain a coherent stream of thought. Difficulty in choosing the right word to say and difficulty with writing and drawing are also common. Therapy is aimed, of course, at finding and correcting the cause or causes.

The other common gait problem of the elderly is the so-called benign disequilibrium of aging (multiple sensory deficit syndrome). This syndrome, which unfortunately is not widely recognized, causes vaguely described dizziness. The main problem is that of disequilibrium, or so-called dizziness in the feet, but often the patient also complains of feeling lightheaded or vaguely dizzy when standing, walking, and turning. Holding on to a solid object, such as a wall or a person, is very helpful, and the patient is not dizzy when sitting and lying. Patients usually have a combination of impairment of position and tactile sense, vision, and hearing. In addition, they often have impairment of vestibular and baroreceptor functions. These patients can be helped with gait training and the use of an assistive device, such as a cane, and should *not* be given medication (e.g., meclizine hydrochloride).

ANATOMY AND PHYSIOLOGY OF LOCOMOTION

The characteristic adult pattern of walking does not occur until about age 7 to 9 years. Equilibrium and locomotion are the two main essentials for walking. Equilibrium is the ability to assume the upright posture and to maintain balance. Righting reflexes allow one to get from sitting or lying to the vertical position to walk. Afferent inputs important in this mechanism are vestibular, proprioceptive, tactile, and visual. Upright position is maintained by contraction of antigravity muscles (supporting reflex). The line of gravity for normal erect posture is 3 to 8 cm anterior to the ankles and fluctuates within narrow limits. For postural control, one must maintain the body center of mass over the base of support. Normal persons exhibit little increase in postural sway with eyes closed (Romberg's test). Body sway increases with age, particularly in women.

Locomotion is the ability to initiate and to maintain rhythmic stepping. One needs to initiate, or ignite, gait before rhythmic alternating movements of the legs take place.

Stepping next occurs by alternating, coordinated movement of the legs and trunk. Locomotion is influenced by the bones and joints of the legs and trunk, the strength of muscles, and the integrating central nervous system network. The walking cycle consists of a stance, or support, phase and a swing phase. During the stance phase, a given foot is in contact with the ground, and during the swing phase, the foot is in the air. The heel strikes the ground first, followed by the sole and then the toes. As soon as the weight is transferred to the supporting leg, the opposite leg begins its movement. The body remains erect, with the head straight and the arms hung loosely at the sides. The arms move rhythmically forward with the opposite leg and swing symmetrically. The legs are parallel and form a base about 6 inches between the feet.

So far as the anatomy is concerned, the spinal cord interneurons appear to function as "local motor generators." These generators are probably important in producing locomotion in humans but are insufficient by themselves to allow walking. The brain stem generates postural responses. The medial pathways in the brain stem affect proximal limb and trunk muscle synergies. In the basal ganglia, the dopaminergic system appears to be important in gait ignition and postural responses. The cerebellum is important for balance and equilibrium, but its role in postural responses is unclear. The cerebral cortex appears to be important for precise foot placement for highly skilled tasks, such as walking on a narrow beam.

Thus, in review, for the maintenance of gait and station, one needs afferent information from proprioceptive, tactile, vestibular, visual, and hearing pathways. This afferent information is integrated in the brain stem, cerebellum, basal ganglia, and cerebral cortex, with efferent data from this central network passing into the spinal cord, where the locomotor generators appear to be located. This information is then transmitted via the peripheral nervous system to the muscles. This complex and intricate mechanism is supported by the cardiovascular system, including the baroreceptors in the neck and heart, and the musculoskeletal system, including the bones and the joints. These mechanisms are all reviewed in Table 33-1.

INITIAL EVALUATION

As in any medical disorder, a carefully obtained history is extremely important in the initial evaluation of gait disorders. This includes a review of the patient's general medical history and neurologic history. It is extremely important to review medication and drug history, especially exposure to alcohol, benzodiazepines, neuroleptic drugs, and vasodilators.

The physical examination should focus on the musculoskeletal, cardiovascular, visual, and neurologic systems. Supine, sitting, and standing blood pressures and pulses must be recorded. Observation of gait and mobility is important, and many neurologists consider examination of the gait to be the single best test of neurologic function. Watching a patient sit, stand, and walk, including regular walking and toe, heel, and tandem walking, followed by standing with eyes opened and closed gives the examiner much information about proximal and distal strength, proprioception, and coordination. One can also look for abnormal movements, such as tremors, dyskinesia, dystonia, myoclonus, and asterixis. Quite often, the cause of a gait disturbance can be detected by the company it keeps. Difficulty getting out of a chair and standing might be due to proximal weakness of the legs, a painful syndrome, or the bradykinesia of parkinsonism.

Table 33-1
Basic Requirements for Walking and Basic Abnormalities of Gait

Neural control mechanisms	Physiologic mechanisms	Abnormalities
Equilibrium		
Arising to erect posture support upright position	Righting reactions	Inability to rise
	Supporting reactions	Inability to stand
Correct perturbations and adapt to circumstances	Anticipatory postural reactions	Inability to protect upright stance
	Reactive postural responses	
	Rescue reactions	
	Protective reactions	
Locomotion		
Initiate steps	Shift center of gravity	Gait ignition failure
	Start stepping	
Stepping	Locomotion	Alterations in pattern stepping
Adapting stepping to circumstances	Voluntary	Inability to perform dexterous stepping
Non-neurologic factors		
The mechanical support system	Bones, joints	Limp
General health (cardiorespiratory)	Exercise tolerance	Slowness

From Nutt JG, Marsden CD, Thompson PD. Neurology 1993; 43: 269. By permission of Lippincott-Williams and Wilkins.

In doing a neurologic examination, one always looks for symmetries on both sides of the body. Thus, with walking, one looks at symmetries of the shoulders and hips and in the movements of the arms and legs. The rest tremor of Parkinson's disease is well seen when the patient walks with arms hanging at the sides. An external rotation of the leg might indicate hip disease or an upper motor neuron type of weakness. Likewise, an arm that does not swing well, is flexed at the elbow and wrist, and is adducted at the shoulder might also indicate an upper motor neuron type of weakness. An arm held rigidly that does not swing well might indicate parkinsonism. A wide-based, unsteady gait usually indicates cerebellar dysfunction. Losing balance while turning, turning rigidly or with small steps, or losing balance easily when pushed or pulled gently might indicate parkinsonism. An inability to maintain station with eyes closed that is corrected with eyes open indicates a proprioceptive impairment. Difficulty walking on heels and toes might indicate a peripheral nerve problem, whereas a waddling gait usually indicates a muscle problem involving the proximal muscles. The examiner can also get an idea of muscle tone by asking the patient to relax and go limp while gently shaking the patient by grasping onto the shoulders. In this way, one can evaluate axial and appendicular tone and any asymmetries of appendicular tone (e.g., unilateral or asymmetrical arm rigidity in Parkinson's disease) (Table 33-2).

Table 33-2
Evaluation of Balance and Gait

Abnormal maneuver	Causes
Difficulty arising from a chair	Proximal muscle weakness
	Arthritides
	Parkinson's syndrome
	Hemiparesis or paraparesis
	Deconditioning
Instability standing with eyes open	Parkinson's syndrome
	Postural hypotension
	Cerebellar disease
	Multisensory deficits
	Foot or back pain
	Deconditioning
Instability standing with eyes closed	Multisensory deficits
	Peripheral neuropathy (e.g., B_{12} deficiency, diabetes)
Instability on turning or sitting	Cerebellar disease
	Visual deficits
	Hemiparesis, ataxia
	Proximal myopathies
Decreased step height and length	Parkinson's syndrome
	Pseudobulbar palsy
	Frontal lobe gait
	"Normal pressure" hydrocephalus (symptomatic hydrocephalic)
	Spastic gait
	Fear of falling
	Habit

Modified from Tinetti ME. *J Am Geriatr Soc* 1986; 34: 124. By permission of Lippincott-Williams and Wilkins.

EFFECTS OF AGE ON GAIT AND BALANCE

There are a variety of normal and abnormal gaits in the elderly. Many gait disturbances are multifactorial. Normal aging alone overlaps with many disease states, so that it may be difficult to tell the normal aging state from the pathologic state. These overlaps can occur with dementing diseases, such as Alzheimer's disease, Parkinson's disease, cerebellar ataxias, motor neuron disease, and peripheral neuropathies, as shown in Figure 33-1.

With normal aging, one can see atrophy of the cerebral cortex, especially in the frontal lobes; atrophy of the cerebral white matter; enlargement of the ventricular system, especially the frontal horns; and atrophy of the anterior vermis of the cerebellum. Neuronal loss is common and also occurs in the basal ganglia, Purkinje's cells of the cerebellum, and motor neurons of the spinal cord. In addition, the density and number of nerve fibers in peripheral nerves decrease, and skeletal muscle loses fibers and gradually weight. Thus, with normal aging, changes occur in both the central and the peripheral nervous systems.

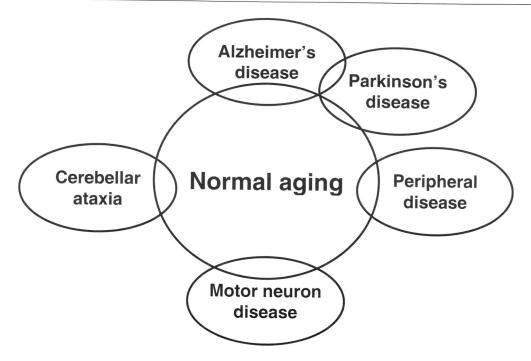

Fig. 33-1. Overlap of normal aging with disease states.

In healthy elderly persons of both sexes, one sees a moderate decline in both stride length and walking speed. In balance testing, elderly subjects demonstrate greater sway and tendency to fall when visual and somatosensory clues are decreased. The line of gravity tends to shift posteriorly in the elderly, especially with the development of abnormal spinal curvature and pelvic rotation, resulting in a tendency to fall backward. There is also a decline in striatal dopamine in normal elderly persons, so that many have parkinsonian features. They are often bradykinetic, walk with small steps and forward flexion, and may have difficulty with postural reflexes, especially with turning. However, they do not have the other features of Parkinson's disease and should not be treated with medication.

In general, older men tend to walk with a flexed posture at both the head and torso, with increased flexion at both the elbows and knees and with diminished arm swings. Elderly women tend to have a more narrow-based gait with a waddling quality. However, there is so much variation that one cannot describe a characteristic "senile" gait. It is the ability to walk without serious limitations or without falling that differentiates a normal aged gait from a dysfunctional gait.

White matter abnormalities, especially on magnetic resonance imaging (MRI), and ventriculomegaly are common in both demented and cognitively normal elderly patients. White matter changes are increased by hypertension, coronary artery disease, and diabetes mellitus. Significant white matter changes, especially in the frontal lobes, and significant ventriculomegaly, especially in the frontal horns, can lead to abnormalities of gait and equilibrium in the elderly.

It is important to remember that conditioning, even in very elderly persons, can improve strength and balance. It is also important to remember that active elderly persons are likely to have more functional motor reserves than their sedentary counterparts and thereby may be less susceptible to the effects of both age and disuse.

Finally, decreased vision, decreased hearing, vestibular hypofunction, decreased tactile sensation, baroreceptor hypofunction, and decreased ability to cardioaccelerate to compensate for the hypotensive effect of many medications are all part of normal aging and add to the impairment of gait and balance. From the musculoskeletal standpoint, the normal aging patient has degeneration of mechanoreceptors in large joints, tends to activate proximal muscles (hip) before more distal muscles (ankle) rather than distal before proximal in balance testing, and tends to have cocontraction of agonist and antagonist muscles, resulting in joint stiffening rather than compensatory movements during postural perturbations. Despite all these factors, however, many elderly persons do not experience dizziness or imbalance; thus, gait imbalance is not an inevitable consequence of aging but probably the result of specific nervous system impairments. In most large series, a treatable disorder is found in about 25% of patients with gait abnormality.

FALLS AMONG ELDERLY PERSONS

It is estimated that one-fourth to one-third of community dwelling people older than 65 and up to one-half of institutionalized elderly persons fall at least once a year. An estimated 8% of persons older than 75 suffer a serious fall each year. Fear of falling is common among older people who have or have not fallen. Some elderly persons restrict their activities because of fear of falling. Risk factors for falling include older age, female sex, medication usage, specific chronic diseases (such as arthritis, Parkinson's disease, and stroke), and impairments in muscle strength, balance, gait, vision, hearing, and cognition. Independent risk factors for serious falls include older age, white race, decreased bone and mineral density, decreased body mass index, cognitive impairment, abnormal neuromuscular findings (such as decreased reaction time or balance disturbance), poor visual acuity, history of previous falls and fall injuries, and the presence of specific chronic diseases, such as diabetes mellitus and stroke. The environment and housing situation also pose risk factors for falling. Common hazards are excessive furniture, door stops, slippery floors, throw rugs, poor lighting, and unsafe stairways.

GAIT DISORDERS RELATED TO NEUROLOGIC ABNORMALITIES

Neurologic gait disorders have been classified as lowest-level gait disorder, middle-level gait disorder, and highest-level gait disorder (Table 33-3).

The lowest-level gait disturbances are due to peripheral musculoskeletal or peripheral sensory problems. These disorders include arthritic, myopathic, peripheral neuropathic, sensory ataxic, vestibular ataxic, and visual ataxic gaits. Dysfunction at these low levels is generally well compensated for by an intact central nervous system. Midlevel gait disturbances include disorders of pyramidal, cerebellar, and basal ganglia motor systems. The highest-level gait disturbances are the most complex and the least understood. There is often an overlap or mixture of these gait disorders.

Frontal Gait Disorder

This disorder has been considered an apraxia of gait. It is caused by bilateral frontal lobe dysfunction, although occasionally it has been reported with unilateral frontal lobe

Table 33-3
Classification of Gait Syndromes

Lowest-level gait disorders
   Peripheral skeletomuscle problems
      Arthritic gait
      Myopathic gait
      Peripheral neuropathic gait
   Peripheral sensory problems
      Sensory ataxic gait
      Vestibular ataxic gait
      Visual ataxic gait
Middle-level gait disorders
   Hemiplegic gait
   Paraplegic gait
   Cerebellar ataxic gait
   Parkinsonian gait
   Choreic gait
   Dystonic gait
Highest-level gait disorders
   Cautious gait
   Subcortical disequilibrium
   Frontal disequilibrium
   Isolated gait ignition failure
   Frontal gait disorder

From Nutt JG, Marsden CD, Thompson PD. *Neurology* 1993; 43: 271. By permission of Lippincott-Williams and Wilkins.

dysfunction in the dominant hemisphere. Balance, initiation, and locomotion are disturbed during walking. With the initial attempt at walking, the foot may appear to be glued to the floor. The gait is slow and shuffling, with hesitation in starting, difficulty with turns, and poor standing balance. After a few small steps, the gait slowly improves and steps lengthen. During turns, the pivotal foot may stay in a single spot with the other making a series of small steps. In severe occurrences, walking may not be possible at all. During walking, trunk and leg posture is upright and arm swings are preserved. Often, start hesitation, shuffling, and freezing spells or motor blocks mimic the gait of Parkinson's disease, but the upright trunk and arm swings differentiate it.

The lesions in the frontal lobes for this disorder can be infarctions, hemorrhages, neoplasms, hydrocephalus, or even a degenerative process. The patient may have other bifrontal symptoms, including mild dementia, with slowness and paucity of thought but correct answers; emotional lability or flat affect; urinary frequency, urgency, and incontinence; palmar and plantar grasping reflexes ("frontal lobe release signs"); increased tone, especially in the legs (paratonia or gegenhalten); and Babinski's signs. This pattern of gait can be seen with severe white matter disease or symptomatic hydrocephalus. In the latter, absorption of spinal fluid is decreased, so that the typical radiographic picture is that of ventriculomegaly, especially in the frontal horns, with comparatively normal cerebral sulcal patterns.

No other diagnostic tests determine whether a shunting procedure (usually ventriculoperitoneal shunt) will be of any benefit. Although many authors claim that the use of a cerebrospinal fluid drain or removal of 50 to 100 mL of fluid to see whether the patient's gait improves is a reasonable diagnostic study to predict success of shunting, I believe that these diagnostic procedures are potentially dangerous and may lead to infection or subdural hematoma. If the patient has the clinical and radiographic criteria for symptomatic hydrocephalus, shunting should be considered. If there are any doubts, especially concerning the clinical criteria, the patient's gait should be videotaped and the patient should be reexamined in 4 to 6 months to see whether gait or cognitive function has further deteriorated.

Cautious Gait

Cautious gait is the most common gait disorder of the elderly. The cautious gait pattern is an appropriate response to real or perceived disequilibrium. This is a nonspecific primary disorder seen in elderly patients and characterized by a mildly wide-based gait with a shortened stride, slowness in walking, and turning en bloc (the whole body turns as one unit). The neurologic signs and symptoms are mild and do not form a consistent pattern. The primary problem can be very difficult to differentiate from the compensatory mechanisms. This gait pattern is likened to that of a person walking on ice. The best approach to this problem is a good physical therapy program with gait training and conditioning.

Primary Progressive Freezing Gait Disorder

Progressive freezing is considered a primary gait disorder because the anatomic location and neurochemical abnormality are unknown. This disorder occurs mainly in older men and is restricted to the legs. The gait disturbance is characterized by start hesitation, motor blocks or freezing spells, and recurrent falls that appear only during walking. The patient walks as if the feet were glued to the floor. The upper part of the body is normally mobile. Progression of this disorder may lead to total inability to walk and severe functional disability. It has also been called "lower body parkinsonism," "motor blocks," "apraxia of gait," and "freezing in movement." Symptoms of this disorder are *not* relieved by dopaminergic drugs. This disorder most likely is the same as or at least similar to the primary gait disorder termed "isolated gait ignition failure."

Spastic Gaits

SPASTIC HEMIPARETIC GAIT

A spastic hemiparetic gait is caused by impairment of the corticospinal tract in either the opposite cerebrum or the brain stem or even ipsilaterally in the high cervical spinal cord. Characteristically, the patient drags and circumducts the stiff affected spastic leg, with the foot everted and plantar-flexed. The paretic arm is usually held flexed and immobile across the chest or abdomen as the patient walks.

SPASTIC PARAPARESIS

The gait is stiff-legged, with reduced toe clearance and a tendency toward circumduction. A degree of plantar flexion in the feet superficially mimics the footdrop often seen with lesions of the peripheral nerve or nerve roots, but it is probably better termed "spastic footdrop." The thighs may hyperadduct and nearly cross during walking ("scissors gait"). Bilateral spastic weakness of the legs without involvement of the arms can occur with lesions as high as the parasagittal area or with lesions of the thoracic spinal cord. Common entities are parasagittal meningioma, primary or metastatic spinal cord tumor, and subacute combined degeneration of the spinal cord, as seen in vitamin B_{12} deficiency syndromes. When the arms are also involved, one must consider a compressive lesion in the upper cervical region by a neoplasm or spondylosis. Noncompressive intramedullary spinal cord lesions, such as motor neuron disease and multiple sclerosis, can cause similar manifestations.

The most common cause of cervical myelopathy is degenerative arthritis of the cervical spine (cervical spondylosis). Typically, spasticity and hyperreflexia in the legs can be asymmetrical and at times unilateral, with occasional posterior column signs and urinary urgency. Neck discomfort and radicular pain in the upper limbs are helpful in making the diagnosis of a cervical spinal cord problem, but they may be absent, even with cervical spondylosis (only half the patients have neck pain, and signs of radiculopathy are commonly absent). Often, however, patients complain of numb, clumsy hands. The diagnosis is confirmed by computed tomographic myelography or MRI scanning. It is important to remember, however, that most persons older than 60 have radiographic evidence of cervical spondylosis on plain films and on MRI and computed tomography scans of the cervical spine, but most do *not* have clinical symptoms; the same is true for radiographic studies of the lumbar spine.

Parkinsonian Gait

The function of the extrapyramidal system is to modulate posture, righting reactions, and associated movements. The parkinsonian gait is characterized by a flexed posture, diminished arm swing, and rigid, small-stepped, shuffling gait. Arising from a sitting position may be slow or impossible. Patients often have difficulty with initiation of movement and turns. Disturbances of balance are often present (impairment of postural reflexes). The legs are stiff and bent at the knee and hips. As the patient walks, the upper part of the body gets ahead of the lower part, and the steps become smaller and more rapid (festination). Turning is accomplished with multiple unsteady steps, with the body turning as a single unit (en bloc). Many patients who have a bradykinetic or rigid syndrome, especially with early falls and impairment of postural reflexes, are found to have something other than idiopathic Parkinson's disease. These diseases include progressive supranuclear palsy, Shy-Drager syndrome (multiple system atrophy with primary orthostatic hypotension), olivopontocerebellar degeneration, striatonigral degeneration, diffuse Lewy body disease, and corticobasal ganglionic degeneration. These diagnoses should be considered in patients with postural instability and especially in those without response to levodopa. Drug-induced parkinsonism is a common disorder, particularly in nursing home patients who are receiving neuroleptic drugs.

Cerebellar Disorders

When the vermis of the cerebellum is predominantly involved by disease, the result is generally one of so-called truncal ataxia. The patients walk unsteadily on a broad base with the feet wide apart, staggering from one side to the other. The patients are unable to walk heel-to-toe (tandem gait) and have difficulty stopping and turning. There may also be a rhythmic swaying of the trunk or head or both (titubation). Symptoms often are reduced with support, such as using a cane or holding on to another person's arm.

The vermis and anterior lobe of the cerebellum are responsible for the coordination of proprioceptive, vestibular, and visual information. Unilateral lesions that do not involve the vermis cause ataxia of the ipsilateral arm and leg and have a greater effect on limb coordination than on balance. One form of cerebellar degeneration that leads to atrophy of the anterior vermis is associated with chronic alcoholism. Familial and sporadic forms of cerebellar degeneration of late onset have also been described. Olivopontocerebellar atrophy is the most commonly recognized type, but other types have been described. The cerebellum, of course, can also be affected directly by primary and secondary neoplasms. Other causes of cerebellar dysfunction are toxins, such as alcohol and phenytoin; vitamin E deficiency; hypothyroidism; and paraneoplastic syndromes. Diagnostic studies include MRI of the head, determination of thyrotropin and vitamin E levels, paraneoplastic antibody panel, and genetic studies for the spinocerebellar ataxias.

Gaits Associated With Abnormal Movements

CHOREIC

Choreic movements are brief, abrupt, unpredictable jerking motions of the limbs, face, and trunk. Thus, the patient may sway and jerk while standing, and choreic gait may mimic the gait of a drunken person. Chorea most commonly occurs in Huntington's disease but can also be seen in Sydenham's chorea and other forms of chorea. Dyskinetic movements resembling chorea are often drug-induced and may appear as tardive phenomena related to neuroleptic drugs. Dyskinetic movements are also commonly seen in patients with Parkinson's disease treated with dopaminergic drugs.

DYSTONIC

Dystonia is characterized by sustained muscle contractions. Often, movements of agonist and antagonist muscles cause abnormal postures and positions. The dystonic gait is characterized by slow, twisting, repetitive movements of the trunk and often proximal muscles in any direction. The gait is often bizarre and may be mistaken for that of a psychogenic disorder. Dystonia of the foot can cause in-turning or mimic footdrop.

Peripheral Neuromuscular System

The motor unit is made up of the anterior horn cell, its axon and terminal twigs, the myoneural junction, and the muscle fibers supplied by that axon. Lesions in these areas may lead to weakness, atrophy, fasciculations (especially with anterior horn cell disease), decreased motor tone (flaccidity), and decreased muscle stretch reflexes. Electromyography and nerve conduction velocity studies are really an extension of the neurologic examination for this group of disorders. These studies are essential in evaluating diseases of the motor unit.

MOTOR NEURON DISEASE (AMYOTROPHIC LATERAL SCLEROSIS)

Patients present with both upper and lower motor neuron signs and thus may have a combination of spastic gait with atrophy and fasciculations of muscle.

NEUROPATHY

Neuropathies may be mainly motor, mainly sensory, or a combination. Usually, patients have distal weakness, although with neuropathies such as the Guillain-Barré syndrome, they have significant proximal weakness as well. Patients often have bilateral footdrop and are unable to walk on their heels or toes. A flaccid footdrop may accompany lesions of the common peroneal nerve (usually as it winds laterally around the neck of

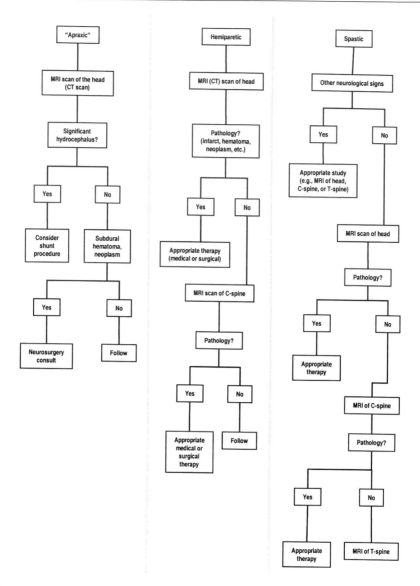

Fig. 33-2. Algorithm for management of abnormal gaits. B_{12}, vitamin B_{12}; CT, computed tomography; C-spine, cervical spine; EMG, electromyographic; ETOH, ethanol; L-spine, lumbar spine; MRI, magnetic resonance imaging; RPR, rapid plasma reagin; T-spine, thoracic spine.

the fibula), which are typically painless, or lesions of the L5-S1 nerve root, which are typically painful. Patients with a lumbosacral radiculopathy usually lean away from the involved side, and when weight is put on the painful side, a limp is produced. Patients with symptomatic lumbar spinal stenosis, essentially a compression of multiple lumbosacral nerve roots, may present with neurogenic claudication (pseudoclaudication).

ABNORMALITIES OF PROPRIOCEPTIVE SENSATION (SENSORY ATAXIA)

If position and joint sense in both legs and feet are impaired, the patient tends to compensate by lifting the feet higher than usual and striking the ground heavily with each step (steppage gait). The patient lifts each foot as high as possible, causing an exaggerated

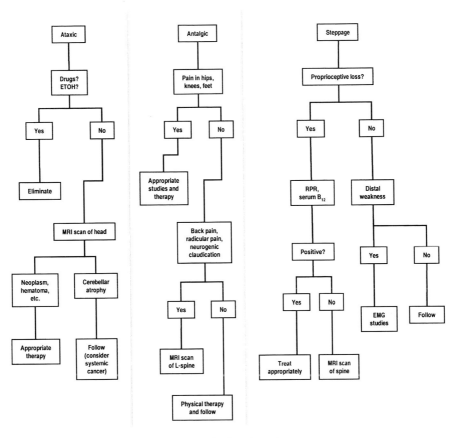

Fig. 33-2. Continued.

flexion at the hip and knee. Then the toe comes down hard on the floor before the heel or ball of the foot strikes the ground, producing a characteristic slapping sound. Vision helps compensate for this proprioceptive loss. Walking in the dark or with the eyes closed is difficult or impossible. A peripheral neuropathy with involvement of large sensory fibers (proprioceptive fibers) can lead to this disorder. This gait is also seen typically in patients with tabes dorsalis due to neurosyphilis, in subacute combined degeneration of the spinal cord due to vitamin B_{12} deficiency, and in any disorder causing dysfunction of the posterior columns of the spinal cord or large sensory fibers of the peripheral nerves.

DISORDERS OF THE VOLUNTARY MUSCLES

Dysfunction of the myoneural junction, as in myasthenia gravis, rarely causes a gait disorder. If the disease is severe and generalized, however, a gait disorder may occur, but the diagnosis depends on the typical findings of fatigability and involvement of ocular, bulbar, and upper limb muscles. The diagnosis of myopathy is suggested by symmetrical proximal weakness of the limbs, especially the legs. Patients with proximal weakness of the legs have difficulty rising from a chair and climbing stairs. They usually stand and walk with a pronounced lordosis and have a broad-based gait with exaggerated pelvic rotation. The characteristic waddling gait results because the hips are thrown side-to-side with each step to help shift the weight of the body. The inherited myopathies, or muscular dystrophies, and the acquired myopathies, such as inflammatory myopathies (polymyositis), are in this category. Diagnosis is usually confirmed by increased serum muscle enzyme levels and findings on electromyography and muscle biopsy.

Psychogenic Gait Disorders

Recognition of a psychogenic gait disorder is often not easy. Depressed patients may have psychomotor retardation and walk with a slow lifting motion, not propelling themselves forcefully. Patients with fear of falling can have a phobic disorder that is probably a variant of cautious gait. The patients are timid, often move with their arms abducted as if walking on a slippery surface, and cling to the walls or furniture for support.

The hysterical gait is a difficult one to categorize. Quite often, the gait disorder has an underlying organic cause, but it is accompanied by tremendous psychologic overlay or response. The diagnosis of hysteria must never be made by exclusion only but must be based on positive evidence of emotional disease. Hysteria often involves an attempt to escape from an exceptionally stressful situation in a patient with an underlying histrionic personality. There are no abnormal neurologic signs, but the gait is often bizarre and histrionic, calculated to impress, and much worse when the patient is being observed. The patient may stagger widely, dip and sway, or walk with gross flexion or hyperextension of the trunk or in a curiously syncopated manner. The patient rarely falls but has a consistent tendency to fall toward or away from the examiner. The term "astasia-abasia" has been used for this abnormality of stance and gait. Level of impairment often fluctuates moment to moment, extreme slowness resembles slow motion, and uneconomical postures waste muscular energy. Performance can be improved by distraction.

A psychogenic gait is also suggested by scissoring of the legs during walking in a patient who nevertheless has a wide-based gait.

SUMMARY

Normal gaits in the elderly population vary widely. Characteristics of normal and abnormal gaits often overlap, making diagnosis of ambulation disorders difficult in some patients. Abnormal gaits are usually caused by musculoskeletal, cardiovascular, or neurologic abnormalities, which often can be readily identified clinically. Obviously, treatment depends on cause, and a treatable cause can be identified in about 25% of patients with gait disorders. An algorithm suggesting a workup for abnormal gaits is given in Figure 33-2.

For all patients, physical therapy with gait training is recommended, and assistive devices, such as canes, walkers, and footdrop braces, can be helpful. Conditioning, even in the elderly, can improve strength and balance. Footwear is important; lightweight shoes with good heel stability and traction are recommended. In addition, one should not forget to assess the home environment.

SELECTED READING

Achiron A, Ziv I, Goren M, Goldberg H, Zoldan Y, Sroka H, Melamed E. Primary progressive freezing gait. *Mov Disord* 1993; 8: 293-297.

Fife TD, Baloh RW. Disequilibrium of unknown cause in older people. *Ann Neurol* 1993; 34: 694-702.

Lipsitz LA. An 85-year-old woman with a history of falls. *JAMA* 1996; 276: 59-66.

Marsden CD, Thompson P. Toward a nosology of gait disorders: descriptive classification, in *Gait Disorders of Aging: Falls and Therapeutic Strategies* (Masdeu JC, Sudarsky L, Wolfson L eds), Lippincott-Raven Publishers, New York, 1997, pp. 135-146.

Nutt JG, Marsden CD, Thompson PD. Human walking and higher-level gait disorders, particularly in the elderly. *Neurology* 1993; 43: 268-279.

O'Keeffe ST, Kazeem H, Philpott RM, Playfer JR, Gosney M, Lye M. Gait disturbance in Alzheimer's disease: a clinical study. *Age Ageing* 1996; 25: 313-316.

Rubino FA. Gait disorders in the elderly. Distinguishing between normal and dysfunctional gaits. *Postgrad Med* 1993; 93: 185-190.

Steiger MJ, Berman P. Gait disturbances in the acute medically ill elderly. *Postgrad Med J* 1993; 69: 141-146.

Sudarsky L. Geriatrics: gait disorders in the elderly. *N Engl J Med* 1990; 322: 1441-1446.

Tell GS, Lefkowitz DS, Diehr P, Elster AD. Relationship between balance and abnormalities in cerebral magnetic resonance imaging in older adults. *Arch Neurol* 1998; 55: 73-79.

Thomann KH, Dul MW. Abnormal gait in neurologic disease. *Optom Clin* 1996; 5: 181-192.

Tinetti ME, Speechley M. Prevention of falls among the elderly. *N Engl J Med* 1989; 320: 1055-1059.

Wolfson L. Falls and gait, in *Principles of Geriatric Neurology* (Katzman R, Rowe JW eds), F. A. Davis Company, Philadelphia, 1992, pp. 281-299.

34 Post-traumatic Movement Disorders

Sotirios A. Parashos, MD, PhD

CONTENTS

The relationship between trauma and movement disorders was recognized as early as the 19th century. Different types of trauma (e.g., mechanical injury, electrical injury, anoxic injury, repetitive injury) incurred on different parts of the nervous system (brain, spinal cord, peripheral nerves) in various ways (single or repetitive injury) have produced a great variety of movement disorders (parkinsonism, dystonia, tremors, hyperkinesis). Trauma can cause movement disorders directly, it can increase the risk for certain diseases affecting the extrapyramidal system, or it can act as a catalyst for the manifestation of an underlying subclinical condition. The significance of risk factors (such as family history of movement disorders and previous use of antidopaminergic agents) for the development of post-traumatic movement disorders remains controversial. The underlying pathophysiology is poorly understood, but it appears that regardless of the location of the trauma, both peripheral and central nervous system mechanisms may be involved.

Although it is customary in the related literature to refer to "central" or "peripheral" trauma, it is often difficult to ascertain the locus of the lesion in the nervous system in most patients. Usually, the head and neck or the neck and an extremity are involved, and a specific lesion cannot be found in the brain, spinal cord, nerve roots, or peripheral nerves. In this chapter, we focus on movement disorders resulting directly from trauma to the head, spine (neck and back), or extremity.

The differential diagnosis of post-traumatic movement disorders may be difficult. Thus, some post-traumatic movement disorders have delayed onset, so that the temporal proximity of cause and effect may not always be of help. Severe motor dysfunction may follow a mild injury, and such a mismatch does not necessarily imply a nonorganic cause. Malingering or conversion disorder may further complicate the diagnosis, particularly when continuing litigation and secondary gain are factors. In such instances, the criteria discussed in Chapter 35 for the differentiation between organic and functional movement

From: *Parkinson's Disease and Movement Disorders:*
Diagnosis and Treatment Guidelines for the Practicing Physician
Edited by: C. H. Adler and J. E. Ahlskog © Mayo Foundation
for Medical Education and Research, Rochester, MN

disorders can be applied. Knowledge of the specific clinical pictures typically resulting from trauma, as described in the literature, may enhance diagnostic accuracy.

HEAD TRAUMA

Krauss and associates (1996) reported that movement disorders developed in 50 of 221 survivors of severe head trauma (22.6%). Half were transient, and the other half persisted. Although this finding refers to severe trauma, mild head injuries may be associated with similar clinical phenomena.

The pathophysiology of such movement disorders is diverse and often obscure. As a rule, specific types of motor dysfunction correlate with the cerebral localization of the inciting lesion. Thus, a traumatic subthalamic hemorrhage is associated with hemiballism because of the direct involvement of the subthalamic nucleus. In most instances, however, no apparent anatomical lesion is found to account for the clinical picture. Several hypotheses have implicated diverse underlying mechanisms, including apoptosis, synaptic reorganization, free radicals, and delayed inflammatory changes, and it is possible that multiple mechanisms are involved.

Because brain injuries affect multiple cortical and subcortical areas and pathways, their extrapyramidal sequelae are usually accompanied by cognitive, cranial nerve, pyramidal, sensory, or paroxysmal symptoms and signs. As for the particular types of movement disorders resulting from brain trauma, the authors quoted above found tremors in 42 of the 50 patients, dystonia in 9, parkinsonism in 2, and myoclonus, hyperekplexia, and orofacial, hand, and paroxysmal hypnogenic dyskinesias in 1 each.

Tremor

A great variety of post-traumatic, nonparkinsonian tremors have been described in the literature. A large number of these reports appear in the neurosurgical literature because of the relative resistance of such tremors to pharmaceutical treatment and the favorable response to stereotactic thalamotomy. Both early (1 to 4 weeks after trauma) and delayed (6 weeks to 2 years after trauma) onset of tremors have been described. The tremor may involve one or both upper extremities or the head. Unusual tremors, such as orthostatic tremor or tremor of the tongue, have also been described. Upper extremity tremors may be kinetic, postural, or both. Depending on the anatomical location of the lesion, cerebellar or pyramidal signs and myoclonus may be present. Lesions associated with kinetic tremor usually involve any of the components or pathways of the dentatorubrothalamic circuit. Post-traumatic kinetic tremor is usually of low frequency, 2.5 to 4 Hz.

Biary and associates (1989) described seven patients with head and neck trauma who had head or upper extremity tremor, or both, associated with myoclonus but had no other signs of cerebellar dysfunction. Results of brain imaging studies (computed tomography [CT] in all patients and magnetic resonance imaging [MRI] in three) were normal. In a different report (Louis and co-workers, 1996), three patients had kinetic tremors as part of a delayed-onset post-traumatic cerebellar syndrome. Early neuroimaging studies (brain CT and MRI) revealed thalamic and midbrain lesions in two, whereas only cortical atrophy was found in the third patient 5 years after the injury.

Treatment of post-traumatic tremors is similar to that of nontraumatic tremors. Most post-traumatic tremors involve the cerebellar circuits in some way, and, as in nontraumatic cerebellar tremors, pharmacologic treatment may not be effective. Primidone, beta-blockers, and benzodiazepines may be tried. Valproate may be helpful in the control of superimposed myoclonus but may aggravate the tremor. In cases of disabling, treatment-resistant tremor, ventrolateral thalamotomy or long-term thalamic stimulation is reportedly effective (see Chapter 22 for more details).

Dystonia

Both severe and mild head injuries may be associated with dystonia. Brain trauma was the cause in 3 of 22 patients with hemidystonia (13.6%) described by Pettigrew and Jankovic (1985). Chan and colleagues (1991), in a review of 266 patients with idiopathic torticollis, reported that 4 patients had recent head trauma without loss of consciousness. Krauss and associates (1996) reported that dystonia occurred in 9 of 221 survivors of severe head injury (4.1%). Goetz and Pappert (1992) pointed out the need to recognize pseudodystonia, a compensatory, nondystonic posturing of the head after trochlear nerve or occipital lobe lesions.

Lee and associates (1994) described 10 patients with dystonia (hemidystonia in 1 and focal appendicular dystonia in 9) after head injury. In these patients, the delay between trauma and onset of dystonia varied between 14 days and 9 years (median, 1 year). In eight patients, the dystonia was progressive. Additional movement disorders were present in five patients, and additional neurologic findings were present in nine. Early CT scan of the head in eight patients revealed thalamic or basal ganglia hemorrhages (three), extra-dural or subdural hematoma (three), and contusions or lobar hemorrhage (two). Brain imaging (CT or MRI) after the onset of dystonia revealed unilateral thalamic or basal ganglia lesions in 6 of the 10 patients. The same authors reviewed an additional 19 cases described in the literature, 17 with basal ganglia or thalamic lesions. The authors observed that dystonia due to mild head injury may occur earlier than dystonia due to severe head injury and that progressive dystonia may be associated with younger age at the time of the trauma.

The treatment of post-traumatic dystonia may be quite frustrating. The methods involved are similar to those for nontraumatic dystonias, such as administration of medications (e.g., anticholinergics, benzodiazepines, baclofen), botulinum toxin injections, intrathecal delivery of baclofen, and stereotactic thalamotomy (see Chapter 23 for more details). The effectiveness of treatment has been reported to be none, slight, or, at best, transient.

Parkinsonism

The relationship between head trauma and parkinsonism is rather complex. Acute head trauma has been implicated as a direct cause of parkinsonism and investigated as a potential risk factor for Parkinson's disease. Repetitive head trauma can cause a syndrome with prominent parkinsonian features.

Acute severe head trauma may cause parkinsonism in rare instances, particularly in association with intracranial hemorrhage. Although no convincing evidence exists to support parkinsonism as a result of mild head trauma, a parkinsonian syndrome may follow severe brain injury. In addition to parkinsonism, this syndrome usually includes cognitive changes, cranial nerve and pyramidal tract dysfunction, and pain. Parkinsonian tremor is usually absent.

In the previously quoted series of Krauss and associates (1996), only 2 of 221 survivors of severe head injury manifested nonprogressive parkinsonian features, including hypokinesia, bradykinesia, hypomimia, hypophonia, and postural instability. Both patients also had nonparkinsonian movement disorders. Williams and colleagues (1991) reported the neurologic sequelae of brain trauma in a population-based cohort. Among 821 subjects (859 brain injuries), parkinsonian syndrome developed in 9 and Parkinson's disease in 7. The prevalence of parkinsonism or Parkinson's disease in this cohort was no higher than that expected for the general population in the same area; therefore, the authors concluded that brain trauma is an unlikely risk factor for parkinsonism or Parkinson's disease. This conclusion is corroborated by multiple epidemiologic studies that failed to confirm a high frequency of head trauma among patients with Parkinson's disease.

Cumulative head trauma (such as that in professional boxers) has been recognized as a cause of a degenerative disease of the central nervous system. This syndrome, referred to as the "punch-drunk syndrome" in the early literature, includes a variable degree of progressive cognitive decline and is usually unresponsive to treatment with dopaminergic agents. Positron emission tomography may differentiate this disorder from Parkinson's disease.

Post-traumatic parkinsonism is generally resistant to treatment with levodopa, dopamine agonists, or other antiparkinsonian medications.

Other Movement Disorders

Krauss and associates (1996) found only rare instances of other movement disorders in their group of 221 survivors of severe brain trauma. They described two patients with hyperkinetic disorders, one patient with myoclonus, one with paroxysmal nocturnal dyskinesia, and one with hyperekplexia. The authors concluded that such movement disorders are only a rare consequence of severe head injury. Additionally, they pointed out that violent kinetic tremors may often be misdiagnosed as hemiballism. Myoclonus, sometimes also misdiagnosed as hemiballism, may often accompany post-traumatic tremor in the form of action myoclonus, particularly after cortical trauma, or may occur as palatal myoclonus after brain stem trauma.

SPINAL TRAUMA

It is often difficult to attribute a movement disorder to spinal cord trauma. With the exception of disorders following injuries that produced a specific anatomical lesion in the spinal cord (e.g., hematoma, dissection, syrinx), concomitant cerebral (in head and neck injuries) or peripheral (in neck and shoulder injuries) involvement cannot be excluded.

Dystonia

The predominant form of dystonia after neck trauma is spasmodic torticollis (cervical dystonia). This type of trauma may involve injury to peripheral nerves, muscles, and bones but may often affect the spinal cord as well. Chan and associates (1991), in a review of 266 patients with idiopathic cervical dystonia, reported that 24 (9%) had a history of trauma in the 3 months preceding onset. In 13 patients, the trauma was to the neck or back or was a whiplash injury. Truong and colleagues (1991) analyzed six instances of torticollis following neck trauma. Mild head injury was also present in two patients. Goldman and Ahlskog (1993) supplied five additional patients with cervical dystonia after mild neck injury. In both studies, onset was 1 to 4 days after the injury. The diagnosis was confirmed by muscle hypertrophy on clinical examination or by co-contraction of agonist and antagonist muscles on electromyographic examination. Post-traumatic torticollis

Table 34-1
Clinical Characteristics
in Post-traumatic and Idiopathic Torticollis

	Post-traumatic	Idiopathic
Range of motion	↓	↓ or normal
Persistence during sleep	Yes	No
Improvement after night's sleep	No	Yes
Sensory trick	No	Yes

differed somewhat from idiopathic torticollis in its clinical presentation (Table 34-1). Range of neck motion was uniformly decreased, and the symptoms persisted through sleep, did not abate with rest, and did not worsen with activity. In most patients, a morning period of improvement was absent, and patients could not alleviate the dystonia by using "sensory tricks" or "antagonistic gestures" (maneuvers used by persons with idiopathic torticollis to mitigate the abnormal movement and posture). Brain and neck imaging did not reveal specific abnormalities in these patients. Risk factors reported in previous series of post-traumatic dystonia were not present in these patients.

All patients lacked a response to oral medications. Botulinum toxin, when extent of involvement allowed its use, was of mild to moderate benefit at best. Physical therapy and physical means of treatment, such as transcutaneous electrical nerve stimulation, have occasionally been successful.

Tremor

Cardoso and Jankovic (1995) described 39 patients with post-traumatic tremor, 7 of whom sustained trauma to the neck and 2 to the lower back. None of those with neck trauma had parkinsonian features. Parkinsonian tremor developed in one of the patients with lower back trauma; dopaminergic agents did not produce a response. Although dystonic head tremor may often accompany idiopathic spasmodic torticollis, the patients in the series mentioned above (Truong and associates, 1991; Goldman and Ahlskog, 1993) did not have tremor.

Myoclonus

Segmental myoclonus, a rare consequence of spinal cord injuries, may involve muscles at or below the level of the lesion. Often, onset is delayed, particularly when myoclonus is the manifestation of post-traumatic syringomyelia. Sensory stimulation may trigger the disorder (proprioceptive myoclonus). Without a lesion amenable to surgery, myoclonus may be quite resistant to oral medications. Clonazepam and valproate may offer limited control.

EXTREMITY TRAUMA

The mechanism by which minor or major injury to a limb may precipitate a movement disorder is unclear. Motor dysfunction after such trauma is usually associated with pain, although the pain may behave differently from the movement disorder. Animal studies have demonstrated spinal and cerebral synaptic reorganization after either single or repetitive peripheral trauma. Similar alterations occur in instances of chronic pain or sympathetic dysfunction and may conceivably trigger dystonias, tremors, and other "unique" dyskinesias. Repetitive peripheral trauma may also contribute to the development of task-specific or occupational dystonias, such as writer's cramp. Finally, electrical injuries may result in post-traumatic movement disorders.

Dystonia

Schott (1985) described four cases of appendicular dystonia and dyskinesia after minor extremity trauma without concomitant head or neck injury. All patients had severe pain. Onset of the movement disorder was insidious, following the injury by weeks or months. The condition continued to progress over time. The author attributed the movement disorder to aberrant sensory input reorganization and proposed that pain may be a factor in such a process. Indeed, the role of pain in the development of movement disorders after peripheral injury is reinforced by the frequent association of movement disorders with reflex sympathetic dystrophy. In addition to minor trauma, repetitive injury may be associated with the development of task-specific dystonias (e.g., writer's cramp). Laboratory investigations have demonstrated the involvement of central mechanisms in the development of such peripherally triggered disorders.

Tremor and Parkinsonism

Although tremor is recognized as a sequela of extremity trauma and often accompanies reflex sympathetic dystrophy, the relationship between limb injuries and parkinsonism is more controversial. Factor and associates (1988) published an extensive review of the literature and concluded that there was no evidence supporting the hypothesis that peripheral trauma can cause Parkinson's disease. Nevertheless, such associations are occasionally encountered in the literature.

In the study previously mentioned, Cardoso and Jankovic (1995) reported that of 18 patients with upper extremity trauma, 10 also had parkinsonian features. The authors reported that 6 of the 10 patients with parkinsonian features had a good response to dopaminergic agents and that 2 patients without parkinsonian features who received botulinum toxin had a good response to this treatment. Stereotactic thalamotomy was performed with good results in one patient with and one without parkinsonism. Notably, a large proportion of the patients in this study (32%) had a family history of movement disorders.

Specific Syndromes After Peripheral Injury

MOVEMENT DISORDER OF REFLEX SYMPATHETIC DYSTROPHY

Reflex sympathetic dystrophy (a syndrome consisting of extremity pain, tenderness, edema, and vasomotor disturbances) is usually associated with a specific motor disturbance of the affected limb. This disturbance includes weakness, spasms, hyperreflexia, tremor, and dystonia. The tremor resembles enhanced physiologic tremor, responds to sympathetic block, and is attributed to sensitization of the muscle spindles. The dystonia is more difficult to explain, and although it may initially be responsive to sympathetic block, it tends to become resistant to such treatment with time. Moreover, dystonia may spread ipsilaterally or develop in a mirror fashion in the contralateral extremity. Such spread implies centralization of the process. Spinal cord neuropeptide and brain stem norepinephrine systems have been implicated. As mentioned above, sympathetic blocks may be helpful early on, but when resistance develops, conventional dystonia treatments may also be applied, usually with limited results. Although the syndrome of the movement disorder of reflex sympathetic dystrophy has been well described in the literature, psychogenic movement disorders often coexist.

PAINFUL LEGS AND MOVING TOES

This syndrome consists of abnormal, involuntary, athetotic movements of the toes or the entire foot and is associated with deep, crushing, or burning pain that involves the foot or the entire leg. It may involve one or both lower extremities and occurs months or years after minor trauma of the extremity due to injury or surgery. Additional symptoms of reflex sympathetic dystrophy are generally not present. Laboratory investigations, including electromyography and imaging studies, are usually unrevealing. Treatment is usually unrewarding, and medications or modalities that alleviate the pain may or may not affect the movement disorder. Combinations of benzodiazepines, anticonvulsants, tricyclic antidepressants, and baclofen have very limited success. In one series, lumbar sympathetic blockade offered transient relief in about half the patients, and a very few had spontaneous remission or improvement. A similar syndrome for the upper extremities has been described as "painful hand and moving fingers."

JUMPING STUMP

Sudden painful, clonic movements of the stump of an amputated limb characterize this syndrome. Such movements are also associated with the patient's perception of painful cramping of the phantom limb. The syndrome was first described during the Civil War. As with other movement disorders after peripheral trauma, sensory and sympathetic mechanisms have been implicated as the basis for the underlying pathophysiology

BELLY DANCER DYSKINESIA

Several dyskinesias that involve the muscles of the abdominal wall may occur after abdominal surgery. They are usually associated with pain or abdominal discomfort, although, as with the painful legs and moving toes syndrome, the dyskinesia and the pain tend to behave independently. Treatment is usually unrewarding, although occasional improvement has been reported with the usual antidystonia medications.

SELECTED READING

Biary N, Cleeves L, Findley L, Koller W. Post-traumatic tremor. *Neurology* 1989; 39: 103-106.

Cardoso F, Jankovic J. Peripherally induced tremor and parkinsonism. *Arch Neurol* 1995; 52: 263-270.

Chan J, Brin MF, Fahn S. Idiopathic cervical dystonia: clinical characteristics. *Mov Disord* 1991; 6: 119-126.

Dressler D, Thompson PD, Gledhill RF, Marsden CD. The syndrome of painful legs and moving toes. *Mov Disord* 1994; 9: 13-21.

Factor SA, Sanchez-Ramos J, Weiner WJ. Trauma as an etiology of parkinsonism: a historical review of the concept. *Mov Disord* 1988; 3: 30-36.

Goetz CG, Pappert EJ. Trauma and movement disorders. *Neurol Clin* 1992; 10: 907-919.

Goldman MS, Kelly PJ. Symptomatic and functional outcome of stereotactic ventralis lateralis thalamotomy for intention tremor. *J Neurosurg* 1992; 77: 223-229.

Goldman S, Ahlskog JE. Posttraumatic cervical dystonia. *Mayo Clin Proc* 1993; 68: 443-448.

Krauss JK, Trankle R, Kopp KH. Post-traumatic movement disorders in survivors of severe head injury. *Neurology* 1996; 47: 1488-1492.

Lee MS, Rinne JO, Ceballos-Baumann A, Thompson PD, Marsden CD. Dystonia after head trauma. *Neurology* 1994; 44: 1374-1378.

Louis ED, Lynch T, Ford B, Greene P, Bressman SB, Fahn S. Delayed-onset cerebellar syndrome. *Arch Neurol* 1996; 53: 450-454.

Pettigrew LC, Jankovic J. Hemidystonia: a report of 22 patients and a review of the literature. *J Neurol Neurosurg Psychiatry* 1985; 48: 650-657.

Ribbers G, Geurts AC, Mulder T. The reflex sympathetic dystrophy syndrome: a review with special reference to chronic pain and motor impairments. *Int J Rehabil Res* 1995; 18: 277-295.

Schott GD. The relationship of peripheral trauma and pain to dystonia. *J Neurol Neurosurg Psychiatry* 1985; 48: 698-701.

Truong DD, Dubinsky R, Hermanowicz N, Olson WL, Silverman B, Koller WC. Posttraumatic torticollis. *Arch Neurol* 1991; 48: 221-223.

Turjanski N, Lees AJ, Brooks DJ. Dopaminergic function in patients with posttraumatic parkinsonism: an ^{18}F-dopa PET study. *Neurology* 1997; 49: 183-189.

Williams DB, Annegers JF, Kokmen E, O'Brien PC, Kurland LT. Brain injury and neurologic sequelae: a cohort study of dementia, parkinsonism, and amyotrophic lateral sclerosis. *Neurology* 1991; 41: 1554-1557.

35 Psychogenic Movement Disorders: Theoretical and Clinical Considerations

David S. Glosser, ScD
and Matthew B. Stern, MD

Although the nomenclature surrounding "psychogenic movement disorders" has undergone many changes over the past century, the clinical significance of these behavioral phenomena is no less important now than when they were described by the pioneers of modern neurology and psychiatry. Patients with an undiagnosed psychogenic movement disorder (PMD) are at risk for iatrogenic complications, such as continuing unnecessary exposure to drugs, emergency room misadventures, and needless operations. Patients with PMD often have very long "illness careers" during which the probability of productive employment is greatly reduced, and substantial medical and social welfare resources may be fruitlessly expended.

Although classification and diagnosis of the organic movement disorders for which these behaviors may be mistaken have greatly improved, a more orderly understanding of the PMD behaviors themselves has been more elusive. The most likely reasons are that the PMDs are wildly heterogeneous but are typically classified either as symptoms of a

From: *Parkinson's Disease and Movement Disorders:*
Diagnosis and Treatment Guidelines for the Practicing Physician
Edited by: C. H. Adler and J. E. Ahlskog © Mayo Foundation
for Medical Education and Research, Rochester, MN

presumed underlying psychiatric disorder or in terms of their superficial resemblance to the various neurologic diseases for which they are mistaken. Neither strategy is productive.

DEFINITION

The most parsimonious and satisfying definition of non-neurologic movement disorders is also the simplest and makes the least reference to a presumed cause. A nonorganic movement disorder is a repeated or persistent behavioral event that has been interpreted to be a neurogenic movement disorder when, in fact, it is not. The neurogenic movement disorders that may be falsely diagnosed include dystonia, tremors, myoclonus, parkinsonism, tics, and assorted gait disturbances. Non-neurogenic movement disorders may arise from either normal or abnormal organic or "psychogenic" processes and may occur among persons with or without neurologic illness. Among the organic processes that may be mistaken for neurogenic movement disorders, and that a responsible workup must consider, are several orthopedic and rheumatologic conditions, epilepsy, and some other neurologic diseases. The various organic illnesses that may be mistaken for neurogenic movement disorder are well described in the preceding chapters and are excluded by definition from this discussion.

Although PMD behaviors are real physical events, clinically significant, and presumably subject to scientific analysis, there is no reason to believe that a class of phenomena, construed mainly in reference to unrelated diseases, are homogeneous or have a similar cause. Nonetheless, the practicing clinician has ample reason to become familiar with PMDs and to appreciate the various factors that can contribute to their emergence.

EPIDEMIOLOGY

Precise estimates of the incidence and prevalence of PMD are difficult for several reasons. No population-based surveys of the phenomenon have been attempted, and virtually every estimate has been derived either from an opportunity sample of tertiary care movement disorders centers or from psychiatric populations. In addition, referral patterns to various sites are likely to be biased, the patients seen in movement disorders centers tend to be the most refractory cases, and diagnostic methods and criteria vary somewhat from program to program. Moreover, owing to the heterogeneity of the behavior constituting the phenomenon, patients with similar or identical symptoms (sometimes the identical patients) are as likely to be seen for various other pseudo-neurologic symptoms in an epilepsy center, rheumatology clinic, or chronic pain program as in a movement disorders center and would thus be excluded from epidemiologic estimates.

Nonetheless, various published reports of clinical opportunity samples are instructive and generally support the notion that the problem is not rare. Reports of sequential neurology admissions at different hospitals in England and Denmark noted predominantly psychogenic disorders in up to 60% of patients with movement disorders. In a contrasting German report of 4,470 consecutive neurology admissions, 9% were for psychogenic disorders; 1.5% of the total had psychogenic disorders of stance and gait alone (Lempert and associates, 1991). Most series have revealed that a great majority of patients with PMD are female, similar to the finding reported for other pseudo-neurologic syndromes, such as psychogenic nonepileptic seizures, in which the prevalence of women is 75%.

In a retrospective analysis of 842 consecutive patients with movement disorders seen at an academic center, Factor and associates (1995) identified 28 with PMD. In this group, the most common psychogenic symptoms were tremor (50%), dystonia (14%), myoclonus (14%), and parkinsonism (7%). Other unusual behaviors were observed that did not fit into any of these disease-referenced classifications.

IMPORTANCE OF ACCURATE DIAGNOSIS

Two major classes of diagnostic error and their logical consequences can be described. First, PMD can be erroneously diagnosed, although the symptoms are actually neurogenic. This is most likely if the neurologic symptoms are atypical or in an early stage of development or if the patient has significant histrionic personality traits. Naturally, the experience and acumen of the examining physician are important in this situation. This is a high-risk error because mislabeling a neurogenic movement disorder as psychogenic may delay or prevent correct diagnosis and treatment until unequivocal symptoms are eventually recognized. Moreover, it may expose the patient to prejudicial reactions from health insurers, employers, and family members. The error may occasion the administration of psychiatric medications, which at best will be benign but which may worsen the symptoms or, with typical neuroleptic agents, produce new organic movement disorders. Some medications, such as tricyclic antidepressants and benzodiazepines, may coincidently suppress movement disorder symptoms and thereby add to the confusion by seeming to confirm a psychogenic origin for the symptoms.

Sometimes an astute psychiatrist catches the misdiagnosis by noting the incongruence of the psychiatric and motor symptoms, but most psychiatric care is provided by primary care physicians and nonphysician subdoctoral level counselors with no experience of movement disorders at all. Finally, many movement disorders, such as Parkinson's disease, Huntington's disease, and tic disorders, have prominent neuropsychiatric symptoms that may emerge before the motor signs become obvious.

A related subclass of error is to diagnose all the symptoms as psychogenic when some of them are of neurologic origin. This error may be even more difficult to detect and correct, since the comorbid PMD symptoms may continue and the observation of neurologically nonsensical behaviors on subsequent occasions may falsely reconfirm the misdiagnosis of exclusive PMD. Among patients insured through managed care schemes, it may be difficult to obtain neurologic specialty referral, and psychiatric care is likely to be obtained through a capitated clinic with minimal physician contact. The detailed record review, repeated observations, laboratory studies, imaging, and multidisciplinary neurologic, psychologic, and psychiatric assessments needed to arrive at a firm diagnosis in obscure cases can be difficult to obtain in the best of circumstances, especially in areas physically or administratively remote from a tertiary care center.

The second major class of error occurs when a neurogenic movement disorder is diagnosed although all the events are psychogenic. In this situation, the risks involve the initiation or continuation of ineffective neurologic, orthopedic, and neurosurgical therapies. Moreover, patients with undiagnosed PMD are diverted from proper psychiatric and behavioral treatments, and official confirmation of the organicity of their symptoms inevitably "hardens" them and imposes additional disability. The longer the PMD behavior continues, the lower the probability that psychosocial morbidity will lessen even if the correct diagnosis is later made. If PMD behaviors even temporarily decline as a placebo response to neurologic therapy, the belief in the diagnosis can be so strong that subsequent physicians are dissuaded from reconsidering the case or discontinuing therapy.

Finally, all the symptoms may be diagnosed as neurogenic although the patient actually has mixed organic and psychogenic complaints. This situation usually occurs after a neurogenic disorder has been correctly diagnosed. The patient may derive substantial subjective benefit from the experience of illness: being cared for, becoming the subject of professional and family concern, being excused from noxious responsibilities, receiving disability compensation, or obtaining dependence-producing drugs. Among some patients, the illness role may come to displace more "normal" behaviors and sources of satisfaction in life. Even if the neurogenic symptoms remit in response to therapy, abnormal illness behavior can emerge to secure the benefits of the sick role. Having earlier been certified as neurologically ill, the patient can assume that the original diagnosis accounts for any subsequent symptoms that may appear.

Some clinicians may be inclined to render the "benefit of the doubt" in the direction of a false-positive diagnosis of neurogenic movement disorder. Although this response bias minimizes false-negative diagnostic errors and is often quite palatable to psychogenic patients, it is not without risk. In the authors' experience and in many published reports, it is not especially rare for patients with predominantly or exclusively psychogenic symptoms to be bound to wheelchair or bed, to have undergone thalamotomy or tendon release, or to have experienced polypharmacy, frank narcotic dependence, fixed deformities, total disability, and a host of other severe consequences so varied and bizarre as to beggar the imagination.

Less widely appreciated is the possibility that the psychogenic patient's slavish devotion to the duped doctor can rapidly turn to rage, blame, and litigation if useless invasive therapies are undertaken and the patient is later "decertified" by the diagnosis of PMD. "I would never have had the surgery, or injections, or device if my doctor hadn't told me I had multiple sclerosis or Parkinson's disease or dystonia, etc. Now they tell me it was all because of stress in the first place; and meanwhile I have had continual abdominal spasms and chest pain since the operation. I understand that I was mostly depressed before, but now I am really injured." Accordingly, there are many good reasons to understand PMDs and to adopt a rigorous approach toward their diagnosis.

ASSESSMENT

Assessment of patients thought to be at risk of PMD is a two-dimensional process which can best be conducted by a multidisciplinary team that minimizes the risks of both false-positive and false-negative conclusions. Ideally, the assessment should be carried out by a movement disorders neurologist in concert with a psychologist or psychiatrist with specialty training in the behavioral aspects of both neurologic disorders and psychogenic disorders. The first stage of the process is the establishment or rejection of the diagnosis of a neurogenic movement disorder. This is done on the basis of the congruence of the objectively gathered medical evidence and the presenting signs and symptoms. It is occasionally necessary to admit the patient to a unit in which continuous video monitoring can be accomplished. Although inconvenient and expensive, this step may provide the evidence needed to establish a diagnosis in complicated or obscure cases. The second stage consists of assessment of the psychologic and environmental factors that have shaped the PMD behavior, that maintain its expression in the patient's social environment, and that may presumably be targets of treatment efforts to modify the behavior.

Since earlier chapters have detailed the characteristics and diagnosis of the neurogenic movement disorders, this discussion focuses primarily on the characteristics and diagnostic procedures most particular to PMD. As in other illnesses, it is critically important to obtain a thorough history of the problem. In PMD and other psychogenic disorders, however, the reported history may be so colored by interim events and conflicting motives that special care must be taken. Obtaining records from previous physicians and hospitalizations is very important. As a prelude to the collection of the history, it is vital to obtain a list of all the doctors previously and currently involved in the care of the movement problem as well as other health issues. As later described, psychogenic patients often have very elaborate health histories, replete with many diagnoses that are difficult to definitively confirm: chronic fatigue syndrome, environmental toxic exposure, fibromyalgia, rare deficiency disorders, undocumented hypoglycemia, and the like.

The contemporaneous descriptions of the original symptoms and diagnostic impressions included in previous records may significantly vary from the patient's retrospective report. Today's right-handed tremor may initially have appeared as drop attacks, vertigo, or whole body pain or may reveal that symptom laterality has shifted. Patients with PMD have often reported to us that they have already been examined by a psychologist or psychiatrist and been "cleared," although the hospital discharge summary reveals the opposite. Hesitance to allow the release of previous records is commonly justified by the incompetence of the preceding doctors or the desire to avoid "prejudice" of the current examination. Although these concerns may seem superficially legitimate, such patients have often already seized upon a diagnosis and are shopping for confirmation rather than diagnostic opinion. By the same token, patients who are reluctant to allow the examining physician to communicate the findings to other physicians are preserving their option to maintain their identity as organically ill persons if the new diagnosis is not what is hoped for.

The patient's report that the symptoms are unrelated to any particular stress or life conflict is often unreliable. Patients with organic movement disorders often first notice their symptoms under conditions of life stress and frequently express hope that their illness is a reaction to stress. By contrast, psychogenic patients often deny any emotional distress other than their reaction to physical symptoms and are actively hostile to psychologic interpretations. For this reason, histories should be obtained separately from both the patient and family members to place the symptoms within the context of the patient's prior health history and of the family and social environment.

CLINICAL CHARACTERISTICS

Although the behaviors constituting PMD are infinitely variable, certain characteristics are frequently observed and are illustrated here by case material.

1. *PMD symptoms often arise suddenly with a high degree of severity, unlike the gradual evolution characteristic of most organic movement disorders.* For example, a 64-year-old woman with a long history of polysomatic complaints, narcotic analgesic dependency, and accident litigation complained of bilateral hand flapping in which she rhythmically slapped the tops of her thighs. The symptoms emerged several weeks after an airline stewardess accidentally dropped an object from an overhead bin, causing jaw and dental injuries. After her final dental treatment, she had sudden onset of a "weird feeling" followed by shaking of all extremities and then wild flailing of her arms without change of consciousness. The disorder did not begin as tremor or gradually evolve, but new components were subsequently added and substituted.

2. *The movements are not stereotypical and change in amplitude, frequency, and participation of different body parts.* The hand flapping and slapping movements occurred in episodes lasting from 1 minute to 90 minutes. Some episodes featured synchronous slapping, and others were asynchronous or had mixed features. Frequency and amplitude of the movements were related, varying from low amplitude at two per second to high amplitude at one per 2-second frequency. Sometimes the head rotated left and right; at other times it moved up and down in a nodding motion.

3. *The symptoms can be alternately provoked and stopped by presentation of nonphysiologic placebo stimuli.* In this case, the flapping always continued until the patient's spouse was summoned to place a lorazepam tablet in her mouth and hold a cup to her lips. The flapping generally stopped within 1 minute of drug acquisition, regardless of proximity to the patient's last meal and long before the drug could have exerted a direct action.

4. *The movements are said to be either constant or unpredictable but remit when there is no attention or regularly occur only in specific social contexts.* The patient's flapping never arose from sleep and occurred only when her spouse was with her or easily summoned.

5. *The movements remit in response to distraction.* While the lorazepam and water were being given, the patient's flapping stopped, and her head became steady long enough for her husband to administer the pill.

6. *The neurologic examination demonstrates associated false signs, such as give-way weakness, non-neurologic sensory findings, and preserved function in one but not another posture or setting.* In this case, the patient had bilateral give-way weakness and denied the sensation of cranial vibration delivered to the left of midline. Her events occurred only while she was seated.

7. *The symptoms do not conform to those of any known organic movement disorder (extremely bizarre gait is the most frequent example).* Careful review of her videotaped signs by three movement disorders specialists yielded a unanimous judgment of non-neurogenic origin.

DEGREES OF DIAGNOSTIC CERTAINTY

Fahn and Williams originally developed categories of diagnostic certainty for psychogenic dystonia, but their schema is considered useful for diagnosis of all other PMD syndromes as well. In *documented* PMDs, the movements are consistently relieved by placebo, suggestion, or psychotherapy or the patient is observed to be asymptomatic when presumably left alone or unobserved. In the *clinically established* category, the movements are not typical of known movement disorder types and are inconsistent over time. A sustained posture might resemble dystonia, but if it is absolutely resistant to passive movement while functioning normally at other times, the disorder is most likely to be psychogenic. Accompanying features in this category are false weakness and sensory loss with or without multiple somatizations. In *probable* PMDs, movements are atypical yet no other features suggesting psychogenic origin are present. Further, the movement might appear typical, but other neurologic signs are clearly psychogenic or there are multiple accompanying somatizations. Finally, a PMD is *possibly* present when the movement is indistinguishable from a classic movement disorder in a patient with obvious psychiatric illness.

PSYCHOLOGIC ASSESSMENT

The exclusion of neurogenic movement disorder is based on the neurologic evidence rather than on any observed psychiatric morbidity. However, assuming that the abnormal movements have been judged to be non-neurologic and that other organic causes have been reasonably excluded, the diagnostic effort shifts to the identification of the variables that do control the behavior. Theories of causation of PMD typically rely on classification of the patient's problem into a psychiatric diagnostic disorder, and the movements are thought to represent symptoms of the disorder. Hysteria, conversion disorder, and various subtypes of somatization disorder are the most commonly encountered conditions, and the first two diagnoses imply a specific theory of etiology: reduction of repressed unconscious emotional conflict or unconscious expression of a repressed psychologic need. However, few data support this notion, and the reasoning tends to be circular. It assumes that the patient is motivated to emit the behaviors but cannot tell us what the motivation is, therefore proving the "repressed" conflict. If the motivation is insufficiently obscure to qualify as unconscious, the patient is accused of malingering. In either event, reliance on a mental disease explanation of complex behavior such as PMD is beset by many problems.

The first issue is homogeneity. Although there are many different neurogenic movement disorders, the symptomatic presentation and history of each are remarkably similar regardless of the social circumstances or personal history of the person with the diagnosis. That the same behavior is common to people of different circumstance defines its usefulness as a symptom of underlying neurologic disease. Neurogenic movement disorders do not rely on learning and can be regarded as a stereotyped response to unconditioned neuropathologic stimuli. By contrast, PMD behaviors are almost infinitely heterogeneous and differ widely from person to person. Even if the PMD behaviors of very different people are morphologically similar, clinical experience reveals that the causes may be quite divergent.

Categorization of groups of patients with PMD into *Diagnostic and Statistical Manual of Mental Disorders, Revised Third Edition (DSM-III-R)* or *International Classification of Diseases, Ninth Revision* diagnoses has been done, but specificity has not been demonstrated. In the better researched area of psychogenic nonepileptic seizures, lifetime Axis-1 and Axis-2 *DSM-III-R* diagnoses have not been found to discriminate between patients with epilepsy only and those with exclusively psychogenic seizures. There is a high base rate of various psychiatric disorders among patients with epilepsy, and this may further serve to reduce the discriminative utility of psychiatric diagnosis. A high base rate of psychiatric disorder prevails among patients with neurogenic movement disorders as well. Thus, while it is clinically useful to identify treatable psychiatric disorders among patients with PMD, the same is true of patients with neurogenic movement disorders.

Although many patients with PMD may have psychiatric disorders, with a preponderance of them classifiable as somatoform, personality, and anxiety disorders, no data exist on the prevalence of PMD among the general population of psychiatric patients in those categories. Just because a person with PMD behavior has a psychiatric disorder does not necessarily mean that the disorder is etiologically relevant to the PMD behavior, that the PMD is a symptom of the disorder, or that the PMD is caused by the disorder. The psychiatric disorder may or may not be a predisposing risk factor for the development of PMD, but it almost certainly does not determine it.

A SOCIAL LEARNING INTERPRETATION

Given the heterogeneity and complex nature of PMD behavior, it is most parsimonious to regard it as normally acquired behavior. It is normal in the sense that it is probably learned, shaped, maintained, cued, and presumably extinguished in the same way that both culturally approved and deviant behaviors are learned. The PMD has value to the patient and cannot be expected to remit until the conditions giving rise to it change or the patient learns alternative behaviors that can provide the same or better levels of perceived value.

A number of risk factors thought to be relevant to the development of PMD have been observed and reported in the literature (Table 35-1). They substantially overlap risk factors associated with psychogenic nonepileptic seizures. Psychogenic nonepileptic seizures are a usefully analogous group to PMD since the behaviors overlap, and an accepted definitive diagnostic standard in epilepsy is available. Using a very similar list of risk factors in a study of successive admissions to an epilepsy monitoring unit, Glosser and associates (1998) found that the presence of any four risk factors correctly predicted exclusively psychogenic rather than epileptic seizures with 88% accuracy. The presence of 5 or more risk factors accurately classified better than 95% of patients as having psychogenic nonepileptic seizures. The authors' clinical experience suggests that a similar degree of predictive accuracy may prevail in PMD.

OPERANT CONDITIONING AND SOCIAL LEARNING VARIABLES

If it is assumed that PMD behavior is learned, regardless of membership in a particular psychiatric diagnostic category, all the risk factors cited above can be reasonably understood as correlates of operant conditioning and social learning variables that control the acquisition of complex behavior. The three main learning variables are modeling, positive reinforcement, and negative reinforcement (escape from aversive stimulus).

Modeling

The acquisition of complex motor behavior in humans is greatly facilitated by exposure to a model of the behavior. In PMD, the model can be direct observation of others, exposure to television images of neurogenic movement disorders, or self-observation in patients with comorbid PMD and neurogenic movement disorder. Membership in support groups, frequent diagnostic interviews, patient information brochures, and physicians' waiting room exposure to other patients can all provide the information that can influence the shape of the symptoms. However, modeling should not be expected to produce persistent symptoms unless the behavior is subsequently reinforced.

Positive Reinforcement

As a general rule, a positive reinforcer is any stimulus event that a person will work to obtain. Positive reinforcement of PMD behavior may occur in obvious ways, such as drug acquisition and monetary awards, or more subtly through potent social reinforcers made contingent on the emission of the PMD behavior. In some cases, the PMD behavior may become so frequent that it crowds out more normal elements of the patient's behavioral repertoire. Since most cultures demand a response to illness behavior, it is a more reliable source of social contact for some people than normal social exchanges. This process may be understood to underlie Lipowski's clinical description of somatization, the "tendency to experience and communicate somatic distress and symptoms unaccounted for by pathologic findings, to attribute them to physical illness, and to seek medical help for them." In other literature, the same process has been described as

Table 35-1
Risk Factors Associated With Psychogenic Movement Disorder

1. Symptom variability (more than 3 distinctly different types of movements)
2. No response to conventional medical therapy or "made worse"
3. History of psychiatric treatment
4. History of suicide attempt
5. History of self-injurious behavior
6. History of drug abuse
7. History of alcohol abuse
8. Reports history of sexual abuse or assault
9. Reports history of physical abuse, assault, or domestic violence
10. History of abnormal illness behavior or somatization
11. Symptoms atypical for organic disorder
12. Exposure to a model of abnormal movements (e.g., friend or relative with neurogenic movement disorder)
13. Previous or current medicolegal conflict (e.g., workers' compensation claim, personal injury suit, disability application pending, lawyer referred)
14. Impaired ability to instrumentally solve problems encountered in the environment (poor skills or cognitive impairment)

"abnormal illness behavior" and betokens the emission of behavioral signs of illness far in excess of what is culturally normative for a given severity of disease. When positive reinforcement is delivered on an intermittent schedule, the abnormal illness behavior's frequency and resistance to spontaneous extinction can be astonishing.

Attempts to withdraw the positive reinforcement that had been contingent on symptom display often result in an escalation of the frequency and intensity of the PMD behavior, called an "extinction burst." Thus, the first attempts to tell the patient that there is nothing wrong typically produce more extreme symptoms, which may in turn elicit a new round of doctors, examinations, and tests that reinforce the newly intensified PMD behavior. Even if the symptom display can be extinguished, inadvertent reinforcement of even a single response can reinitiate the behavior at the same high rate that prevailed before extinction.

Assessment of the role of positive reinforcement is done by identifying the immediate consequences to the patient of emitting the behavior in his or her natural environment. In whose presence does the behavior usually occur? How does the audience react? What exactly do the observers do when the signs of the PMD are imminent? It is helpful to separately interview the patient and each of the family members. Have them provide descriptions and evidence about specific instances rather than their conclusions about causation. Elicit the affective impact of the PMD behavior on each of the family members, as this provides an index of the power of the behavior over other members of the immediate social environment.

When the PMD behavior is primarily supported by positive reinforcers, the main treatment implications are for differential reinforcement of other incompatible behaviors. Often, the patient's family becomes exhausted by the care and worry burden of the PMD and obtains time off from these responsibilities by avoiding the patient when help is not needed (when behavior is normal). In other situations, the PMD behavior becomes the dominant medium for exchange of other emotions, such as love and hate, and has coercive power over others. In these situations, the patient may quickly learn that when behaving normally, no one pays attention. If the situation is chronic, other sources of positive reinforcement have very likely faded out, and loss of the social contact contingent on illness is intolerable. Although cessation of positive reinforcement for PMD behavior is a sensible and appealing strategy, control over reinforcement contingencies in family, institutional, and legal settings may be practically very difficult to arrange.

The treatment implications are to provide alternative normal motor behaviors for the other family members to coach and encourage in lieu of responding to the PMD behavior. In this way, the patient can still command the attention of family members while gradually resuming a more normal repertoire. This reduces the burden of care, makes it more rewarding to interact with the patient, and sets the stage for social exchanges not based on the coercion of illness display. Thus, even though attention is withdrawn from the PMD behavior, the total amount of social reinforcement need not be put in peril by the patient.

Physical medicine and rehabilitation interventions can be useful because of the specific instructions and training they usually entail. If the emphasis is on "relearning normal movement," the patient can obtain improved physical conditioning, positive reinforcement for successive approximations to normal motor behavior, and a face-saving way to escape from the role of invalid without having to defend charges of willful misbehavior. Teasell and Shapiro (1994) described an interesting variation on this theme in which patients with PMD were informed that failure to recover normal function proved that the movement disorder was of psychiatric origin. Thus, recovery of function yielded the additional benefit of certification of blamelessness. Like all strategies based on deception, this approach can be challenged on ethical and practical grounds, but it demonstrates the importance of not punishing recovery by requiring the acceptance of stigma or blame. Interventions relying on passive modalities, such as heat and massage, may have some placebo value but are generally contraindicated, since they tend to certify the existence of organic disease and are delivered contingent on illness behavior rather than improved motor function.

Negative Reinforcement

Negative reinforcement refers to the increased probability of a behavior's occurrence if it has been followed by escape from a noxious stimulus or situation. People will work to escape, and subsequently to avoid, any number of situations. PMD behaviors can be reinforced by escape from hated work, school, military, or penal situations. When the escaped situation is obvious and avoidance is socially proscribed, the behavior is termed malingering. More subtle aversive stimuli that can be escaped or avoided may include relief from performance demands exceeding the patient's cognitive or skill level, social criticism, or exposure to a wide range of external or internal anxiety-provoking situations. Events preceding the withdrawal of positive reinforcement discriminate for extinction and acquire the properties of an aversive stimulus as well. Thus, major losses, such as the death of a close family member, are obviously aversive, and avoidance of situations that resemble the event may motivate PMD behavior.

Sexual and physical trauma are the aversive stimuli most often cited as causal events in both psychogenic nonepileptic seizures and PMD and have been the most extensively studied of the proposed features underlying both dissociative events and post-traumatic stress symptoms. Reported sexual abuse history is not specific for PMD and is commonly found among those with chronic nonmalignant pain syndrome as well as psychogenic nonepileptic seizures.

The relevance of sexual and physical assault to PMD probably has to do with the intensity of the aversive stimuli, the degree of generalization to nondangerous stimuli that resemble the original aversive situation, and the impressive resistance of avoidance behavior to extinction. Typical clinical accounts of those with reported history of abuse include fears of repeated abandonment or mistreatment when approaching emotional or sexual intimacy in later social relations. More commonly, the escape or avoidance value of the PMD centers on the desire to avoid more mundane situations, such as school, work, or overwhelming family responsibilities.

Treatment implications for PMD principally motivated by avoidance of aversive situations hinge on discovery of the relevant current or past situations that are being avoided. These are typically not evident, and the patient ordinarily makes little connection between the current PMD and escape from other stressors in life. This is not to say that the conflicts are "forgotten" or otherwise "unconscious." To the contrary, usually the stressors are very bothersome and quite accessible, though the patient may have despaired of finding instrumental means of actually solving the problem or coping with it.

For example, a 55-year-old construction worker has intermittent gait instability, bilateral hand tremor, and weakness that arose after occupational exposure to carbon monoxide (CO). Although he did have subtle symptoms of CO poisoning, he was promptly removed without loss of consciousness, and his symptoms not only were greatly in excess of expectation but also became clinically significant at a time remote from the acute exposure. His exposure occurred during renovation of an indoor shopping mall, and he has now become phobic of all similar environments and avoids them. He knows that CO is not detectable by odor or color and is now fearful of any situation in which fresh air ventilation is not assured. He complains of nightmares about the incident and has become clinically depressed. During his initial psychologic evaluation, it was discovered that 20 years ago, his older brother died of CO asphyxiation while repairing a gasoline-powered chain saw in his garage. The patient discovered him unconscious and summoned emergency help but had to watch helplessly as he died. As he tells the story, he becomes obviously very upset. Until he is desensitized to his fear of poisoning, his anxiety, depression, and movement disorder are likely to continue.

In actual clinical practice, modeling and positive and negative reinforcement exert their effects in combination. For example, the job held by a dissatisfied, underpaid production worker requires exposure to potentially dangerous high voltages. After a minor electrical shock, the worker learns that a co-worker recently suffered a serious burn after an accidental contact, and this knowledge generates understandable fear. His co-worker was compensated for his injury and sick time, and disability retirement left him financially no worse off than when working. The safe performance of the detested job requires steady hands. On returning to the work site after medical clearance, the affected worker is scared and shaky; a psychogenic tremor develops, and unpredictable episodes of weakness emerge. His symptoms get him out of the factory without having to admit to fear, produce a compensation check, elicit the interest of doctors, and generate the nurture of family members. Moreover, it later emerges that his adult daughter has recently moved back into his home to escape physical abuse and death threats by her sociopathic husband. The patient is enraged at the errant son-in-law, turned stalker, but is afraid of him as well. A gun fancier and collector, the patient has the means to avenge his daughter's mistreatment but is afraid both of botching it up and of succeeding and ending in prison. With constantly shaking hands, he cannot be expected to physically avenge her.

PROVOCATION TESTS

Since definitive laboratory tests are not typically available to diagnose movement disorders, clinical observation occupies a vital role in making the discrimination between psychogenic and organic causation. Provocation testing refers to the presentation of a placebo stimulus to either start or stop the movements. To the extent that either ongoing abnormal movements can be stopped or quiescent movements can be initiated immediately after presentation of the placebo stimulus, strong evidence that the events are learned behavior is obtained. The type of placebo stimulus is of little importance and can range from injection of normal saline to mock hypnosis or pressure on "trigger points." However, a number of threats exist to the reliability of the obtained data.

First, the provoked movements must be absolutely typical of the patient's clinical complaint. To assure that they are, it is important to obtain a fully detailed description of the movements from both the patient and observing family members. The provocation should be videotaped for review by the patient and family after successful provocation so that the test results can be confirmed. From time to time, patients who want to comply with a doctor's directives and conform to the perceived demands of the situation produce a false-positive result. If the resulting behavior is not typical of the clinical complaint and has not also been a co-occurring feature of the movement disorder, it lacks clinical relevance and must be discarded. If the patient has both a neurogenic movement disorder and a PMD, the provocation test can be useful in discriminating between the two elements.

A negative provocation test result provides no evidence that the movement disorder is not psychogenic, particularly in instances of well-informed intentional deception or among intelligent and medically sophisticated patients. Moreover, among those who experience their symptoms in response to anxiety or situations reminiscent of past traumas, the provocation stimulus may simply be irrelevant and ineffective. In such cases, exposure to stimuli closely resembling the actual inciting trauma is often necessary.

Some have voiced vigorous ethical objections to the provocation of abnormal movements, largely because it can be interpreted as a deceitful maneuver that exposes fragile people to ridicule, may motivate flight from treatment, or may simply be unreliable. However, in patients with psychogenic nonepileptic seizures, worse clinical outcomes or reduced psychiatric follow-up participation after provocation has not been demonstrated. Since provocation methods can occasion false-positive results, they should not be undertaken unless the treatment center is competent both to produce acceptably low diagnostic error rates and to manage any complications.

DISCRIMINATION
OF SPECIFIC NEUROGENIC DISORDERS FROM PMD

Idiopathic Torsion Dystonia

Most patients suffer gradual onset of this condition from childhood and can experience twisting, cramping, sustained muscle contraction, and distorted posture in any part of the body. Psychogenic dystonia is thought to be rare, and there has historically been a greater risk of falsely diagnosing psychiatric disorder than of missing a PMD. Nonetheless, Fahn and Williams (1988) reported that in a series of 1,185 patients with dystonia, 21 received a diagnosis of psychogenic dystonia. The condition has been reported elsewhere as well. Although the general risk factors for PMD are relevant, Fahn and Williams suggested the following additional specific clues to psychogenic dystonia:

1. Excessive pain reported on light touch (low pain threshold)
2. Onset with at-rest symptoms rather than action dystonia
3. Bizarre extraneous movements not ordinarily seen in dystonia
4. False weakness
5. Excessive slowness of voluntary movement
6. Polysomatic complaints
7. Inconsistency of motor symptoms (variable body site, distractible)
8. Paroxysmal symptoms

Neurogenic dystonia rarely remits spontaneously, with the exception of torticollis, which then often relapses. In psychogenic dystonia, remission often occurs or the symptoms are expressed most specifically in response to a specific situation. Care must be taken not to confound the waxing and waning diurnal pattern of symptom severity experienced by some neurogenic patients as situation-specific psychogenic symptoms. Similarly, certain motor tasks can provoke dystonia—as in writer's cramp, which is not psychogenic. Typically, the provoking task is an aversive event the patient seeks to avoid or a situation in which there is expectation of reinforcement contingent on symptom display, as in other PMDs.

Gait Disorders

Lempert and associates (1991) noted six main characteristics of psychogenic disorders of stance and gait in 37 patients, only 1 of whom had a mild comorbid neurologic disorder. With use of these criteria (listed below), 90% of the patients could be readily classified. Obviously, however, the accuracy of the classification depends on the examiner's skill in eliciting and discriminating the symptoms.

1. Fluctuation of symptoms (reduction on distraction or encouragement)
2. Excessive slowness (generally due to simultaneous innervation of antagonist muscles)
3. Gait hesitancy not overcome after the initial step; feet stuck to ground after each step
4. Psychogenic Romberg's sign
   a. Consistent falls toward (or away from) the observer independent of position; fall usually prevented by clutching examiner
   b. Large-amplitude body sway with gradual buildup after latency of a few seconds
   c. Improvement of balance with distraction
   d. "Walking on ice" gait; broad-based stiff ankle shuffling with arms extended and innervation of antagonist muscles
5. Uneconomical postures in which the distorted gait requires a higher degree of energy and skill than normal locomotion despite objectively normal strength and orthopedic status; knees and hips often flexed
6. Sudden partial buckling of knees with last-moment self-rescue through activation of antigravity muscles; no other symptoms of cataplexy

Many of the patients in this series had features overlapping those of other PMDs. The psychologic aspects and historical risk factors in psychogenic gait disorders are no different from those in other PMDs.

Psychogenic Parkinsonism

Lang and colleagues (1995) cited several features of history and symptom presentation that betoken a psychogenic origin of parkinsonism:

1. Abrupt onset without a history of toxic exposure or other relevant cerebral insult
2. Static course; maximum disability occurs shortly after symptom onset
3. Dominant side most affected
4. Resting, postural, and action tremors tend to decrease with distraction and increase with attention
5. On rigidity testing, resistance increases in proportion to the examiner's effort; no cogwheeling
6. Atypical bradykinesia; no true fatiguing, exaggerated slowness, bizarre features
7. Atypical gait that may be antalgic, or affected arm held stiffly at side
8. Exaggerated or bizarre response to minimal displacement on postural stability testing

Again, the other historical risk factors for PMD are relevant, and the behaviors of these patients are best understood as socially learned and maintained.

Pediatric Psychogenic Movement Disorder

Pediatric PMD virtually always requires family counseling as well as intensive instruction of parents and caregivers in the methods of behavior therapy. Often, parental conflicts are enacted through the sick child, and in rare but terribly sad cases, a parent feels needed and important only when demonstrating devotion to the sick child's needs. In extreme cases, the parent may intentionally cause injury to keep the child sick and protect the role of devoted and admirable caretaker. This behavior has been termed "Munchausen by proxy syndrome" and requires involvement of civil authorities to protect the child from abuse.

More commonly, a child with PMD has observed a sibling or cousin with a hereditary neurogenic movement disorder and copied the behavior, which is then inadvertently reinforced by parental and medical attention. Sometimes, the neurologically ill sibling has required so much care that the well child feels deprived and mimics the motor symptoms to compete for a fair share of parental attention.

PRESENTING THE DIAGNOSIS

The most sophisticated diagnostic skills and efficient workup are of little practical value if the disclosure of findings to the patient is avoided or bungled. The disclosure session sets the stage for treatment planning and heavily conditions its likelihood of success. Particularly in chronic PMD, patients have predicated a great portion of their identity and social behavior on the premise of needing help because of a physical illness over which they have no control. At the disclosure of PMD, the patient is being asked not only to jettison access to the reinforcing aspects of illness behavior but also to incur the risk of being regarded as a mentally ill person or a willful manipulator of other people's sentiments.

Accordingly, the goals of disclosure are to present a truthful and understandable explanation of how the symptoms developed and to map a route to recovery of a more normal behavioral repertoire without requiring adoption of a stigmatizing identity. The diagnosis must be presented constructively so that treatment is not impeded by the patient's attempts to prove that "I really am sick." It is best for the neurologist and psychologist to jointly review the test results, videotape, and psychologic findings together with the patient and, when applicable, the family.

Patients who have researched their condition often attempt to debate the technical accuracy of the diagnosis and to offer alternative organic interpretations. Their arguments are often peppered with medical jargon, which is only superficially understood, and extended attempts by the physician to offer scientific arguments in support of the diagnostic opinion are seldom helpful. Psychogenic patients are more interested in determining whether the physician will offer care and help than in having their theories demolished. It is best to simply explain that the best assessment efforts, using accepted methods, have led to the diagnosis of PMD. It is also reasonable to acknowledge that 100% certainty is seldom achieved in medicine but that a confident diagnosis has been made. It should also be explained that it may be reasonable to seek a second opinion, but that the opinion should be sought from a recognized movement disorders specialist or center. It is pointless to engage in a debate with a patient determined to defend the organic source of the problem, but a confident and clear explanation of the findings should leave no room for doubt that a neurologic origin has been excluded.

Some neurologists are very hesitant to tell the patient that the symptoms are of psychogenic origin, because of the frequently encountered hostile reaction that may ensue. Failure to report the findings represents an unacceptable level of care in this condition. Imagine the analogous situation of an oncologist approached by a somatically anxious patient with a persistent benign cough who delusionally insists that it is caused by cancer. It would be inadequate not to make an effort to educate the patient and clearly explain one's best diagnostic opinion. It would be ridiculous and unethical to initiate radiotherapy merely to avoid offending the patient.

Some physicians are hostile to psychogenic patients and regard them as time wasters, complainers, or something worse. Patients are very sensitive to this sentiment, and in turn use the physician's scorn as a basis of their rejection of the psychogenic findings. Frequently, this motivates patients to seek out a practitioner who is willing to endorse their own organic theories of causation and render compassionate, if misguided, care. Again, the standard of adequate care for a potentially disabling condition such as PMD requires that the patient not be simply dismissed from treatment. Relevant therapy, referral, and follow-up should always be offered.

Fortunately, in a proper disclosure, much of the resistance to the findings can be avoided by compassionate explanation that the patient has a common and often disabling disorder that has reasonably well understood causes and for which a rational treatment can be devised. When the PMD is of recent onset, it is often sufficient to explain that it is a self-limiting emotional reaction to life stressors, perhaps occurring after a physical illness, affective disorder, or family crisis that served to catalyze the movements. Patients can be told that the movements will remit now that there is no more need to worry about serious neurologic illness. The importance of rapidly resuming normal life responsibilities to prevent deconditioning and loss of self-confidence can be stressed as well. Referral for active physical therapy modalities often is helpful, since it offers a positive ceremony by which recovery can occur. The therapy protocol should be arranged in collaboration with the diagnosing neurologist and psychologist.

An often overlooked problem that can serve as a barrier to acceptance of the PMD diagnosis is the patient's quandry about how to explain the new diagnosis to friends and employers. In chronic severe cases, this is less of a problem because employment has long since ceased to be an issue and most nonobliged family and friends have dropped out of contact. In more acute cases, it helps the patient if the diagnostic team offers an honest nonstigmatizing explanation. At some level, all behavior, both pathologic and normal, is based on neurologic and neurochemical substrates, but this can be hard to explain to unsophisticated laypersons.

The authors have found it useful to explain brain function and PMD by using the following computer metaphor. A computer program may fail to run properly because of an electronic or "hardware" problem, analogous to a classic neurologic disease or injury to the brain. Computer hardware problems are solved by repair or replacement of the damaged components. Neurogenic movement disorders are solved by manipulating brain function with drugs or surgery.

However, computer programs most often fail to run properly because of flawed instructions to the computer—"software" problems. The software problem makes the computer act as though it is broken and often is difficult to differentiate from a hardware failure. These software problems are analogous to the behavioral abnormalities observed in PMD. Software problems are fixed by correcting the ineffective operating instructions contained in the program. Human "software" problems are corrected by relearning or replacing the dysfunctional learning that is causing the body to behave as though it were neurologically ill.

If a comorbid psychiatric condition has been diagnosed, treatment of it should be offered and started at the disclosure session. The psychologist or psychiatrist should explain the findings and describe recommendations to more instrumentally meet the patient's needs. Generic referral for outside psychiatric help or counseling does not typically result in patient follow-up and may simply reignite the process of somatizations for a new audience. If coordinated neurologic and psychologic care is not available within the same institution, it is very important to select knowledgeable outside colleagues.

Even in the case of exclusively psychogenic motor symptoms, the neurologist should schedule the patient for follow-up at prearranged intervals after diagnosis. This communicates a commitment to offer help, reassures the somatically anxious patient that any neglected organic disease will be discerned, and offers the opportunity to reinforce progress. If follow-up visits are conditioned on the emergence of worse symptoms rather than at intervals, new symptoms are certain to occur to justify further care.

Finally, there is the problem of "dueling doctors." Sometimes the diagnosing physician sees a patient who asserts that "none of my other doctors agree with your diagnosis" or some variation on that theme. Although it can be tempting to contrast one's own superior diagnostic acumen to the bumbling errors of previous doctors, this response is almost always unwise, to say nothing of unkind. It is generally a mistake to respond to this observation for a number of reasons. First, the patient's report of the other doctors' opinions may be dead wrong. Even in the event that the patient's report is correct, condemnation of prior practitioners serves to foster the patient's rationalization that since the doctors disagree, none of them know anything. Moreover, the patient was probably content to accept the previous organic attribution and valued both the incorrect diagnosis and the doctor who bestowed it. If previous doctors are discredited, the patient is being told that the valued diagnosis must be abandoned and that both the patient and the doctors are fools. The best course of action is to acknowledge the difference of opinion, explain that the new diagnosis was made on the basis of information and test results not available to previous doctors, and to obtain consent to directly communicate this diagnosis to the other practitioners.

If the PMD is chronic, severely disabling, and deeply entrenched, it is often necessary to arrange admission to a physical medicine and rehabilitation unit with strong psychiatric and psychologic support. In such an environment, the contingencies of reinforcement can be completely altered to support the emission of more normal behavior. The social environment of a good physical medicine and rehabilitation unit favors reinforcement of effort to regain function. However, it should not be expected that progress automatically generalizes to the home environment after dismissal unless the conditions favoring establishment of the PMD have been changed as well. Again, planned follow-up after dismissal is needed to prevent relapse. It is necessary to adopt realistic expectations and therapeutic goals among patients whose adaptive function has never been adequate or who are embedded in inescapable and aversive social conditions. For example, an abused dependent person who can escape from the environment only through illness behavior needs to be helped through direct relief of the inciting conditions rather than through modification of the PMD behavior.

SELECTED READING

Arnold LM, Privitera MD. Psychopathology and trauma in epileptic and psychogenic seizure patients. *Psychosomatics* 1996; 37: 438-443.

Bandura A. Influence of model's reinforcement contingencies on the acquisition of imitative responses. *J Personality Soc Psychol* 1965; 1: 589-595.

Betts T, Boden S. Diagnosis, management and prognosis of a group of 128 patients with non-epileptic attack disorder. Part I. *Seizure* 1992; 1: 19-26.

Bowman ES. Etiology and clinical course of pseudoseizures. Relationship to trauma, depression, and dissociation. *Psychosomatics* 1993; 34: 333-342.

Factor SA, Podskalny GD, Molho ES. Psychogenic movement disorders: frequency, clinical profile, and characteristics. *J Neurol Neurosurg Psychiatry* 1995; 59: 406-412.

Fahn S, Williams DT. Psychogenic dystonia. *Adv Neurol* 1988; 50: 431-455.

Glosser DS, Nei M, Bruno D, Gross K, Sirven J. A brief structured interview to accurately identify psychogenic non-epileptic seizure (PNES) patients. Presented at the American Epilepsy Society annual conference, San Diego, California, December 1998.

Henry TR, Drury I. Non-epileptic seizures in temporal lobectomy candidates with medically refractory seizures. *Neurology* 1997; 48: 1374-1382.

Keane JR. Hysterical gait disorders: 60 cases. *Neurology* 1989; 39: 586-589.

Koller WC, Findley LJ. Psychogenic tremors. *Adv Neurol* 1990; 53: 271-275.

Lang AE, Koller WC, Fahn S. Psychogenic parkinsonism. *Arch Neurol* 1995; 52: 802-810.

Lempert T, Brandt T, Dieterich M, Huppert D. How to identify psychogenic disorders of stance and gait. A video study in 37 patients. *J Neurol* 1991; 238: 140-146.

Linton SJ. A population-based study of the relationship between sexual abuse and back pain: establishing a link. *Pain* 1997; 73: 47-53.

Lipowski ZJ. Somatization and depression. *Psychosomatics* 1990; 31: 13-21.

Ranawaya R, Riley D, Lang A. Psychogenic dyskinesias in patients with organic movement disorders. *Mov Disord* 1990; 5: 127-133.

Slater JD, Brown MC, Jacobs W, Ramsay RE. Induction of pseudoseizures with intravenous saline placebo. *Epilepsia* 1995; 36: 580-585.

Teasell RW, Shapiro AP. Strategic-behavioral intervention in the treatment of chronic nonorganic motor disorders. *Am J Phys Med Rehabil* 1994; 73: 44-50.

Walczak TS, Papacostas S, Williams DT, Scheuer ML, Lebowitz N, Notarfrancesco A. Outcome after diagnosis of psychogenic nonepileptic seizures. *Epilepsia* 1995; 36: 1131-1137.

Appendix

FOR MORE INFORMATION

We have compiled a listing of national organizations in the United States that specifically deal with the various movement disorders discussed. These organizations may provide further information for both patients and health care providers. Some of the organizations send out periodic newsletters, and some provide other educational materials. Additionally, some of these organizations provide research funding through grants to researchers and academic institutions.

ATAXIA

National Ataxia Foundation
2600 Fernbrook Lane, Suite 119
Minneapolis, MN 55447
Tel: (612) 553-0020
Fax: (612) 553-0167
E-mail address: naf@mr.net
Web site: http://www.ataxia.org

DYSTONIA

Benign Essential Blepharospasm Research Foundation, Inc.
P.O. Box 12468
Beaumont, TX 77726-2468
Tel: (409) 832-0788
Fax: (409) 832-0890
E-mail address: bebrf@ih2000.net
Web site: http://www.blepharospasm.org/ bebrf/

Dystonia Medical Research Foundation
One East Wacker Drive, Suite 2430
Chicago, IL 60601-1905
Tel: (312) 755-0198; (800) 377-3978
Fax: (312) 803-0138
E-mail address: dystonia@dystonia-foundation.org
Web site: http://www.dystonia-foundation.org/

From: *Parkinson's Disease and Movement Disorders:*
Diagnosis and Treatment Guidelines for the Practicing Physician
Edited by: C. H. Adler and J. E. Ahlskog © Mayo Foundation
for Medical Education and Research, Rochester, MN

National Spasmodic Dysphonia Association
One East Wacker Drive, Suite 2430
Chicago, IL 60601-1905
Tel: (312) 755-0198; (800) 795-6732
Fax: (312) 803-0138
E-mail address: nsda@aol.com
Web site: http://www.dystonia-foundation.org/spasdysp.html

National Spasmodic Torticollis Association
9920 Talbert Avenue, Suite 233
Fountain Valley, CA 92708
Tel: (714) 378-7838; (800) 487-8385
Fax: (714) 378-7830
E-mail address: NSTAmail@aol.com
Web site: http://www.torticollis.org

Spasmodic Torticollis/Dystonia, Inc.
P.O. Box 64
Waupaca, WI 54981
Tel: (888) 445-4588
Fax: (715) 258-8757
E-mail address: info@spasmodictorticollis.org
Web site: http://www.spasmodictorticollis.org

HUNTINGTON'S DISEASE

Huntington's Disease Society of America
158 West 29th Street, 7th Floor
New York, NY 10001-5300
Tel: (212) 242-1968
Fax: (212) 239-3430
E-mail address: curehd@idt.net
Web site: http://www.hdsa.org/

MYOCLONUS

Myoclonus Research Foundation
200 Old Palisade Road, Suite 17D
Fort Lee, NJ 07024
Tel: (201) 585-0770
Fax: (201) 585-8114
E-mail address: research@myoclonus.com
Web site: http://www.myoclonus.com

PARKINSON'S DISEASE

The American Parkinson Disease Association, Inc.
1250 Hylan Boulevard, Suite 4B
Staten Island, NY 10305-1946
Tel: (718) 981-8001; (800) 223-2732
Fax: (718) 981-4399
E-mail address: apda@admin.con2.com
Web site: http://www.apdaparkinson.com

National Parkinson Foundation, Inc.
Bob Hope Parkinson Research Center
1501 N.W. 9th Avenue/Bob Hope Road
Miami, FL 33136-1494
Tel: (305) 547-6666; (800) 327-4545
Fax: (305) 243-4403
E-mail address: mailbox@npf.med.miami.edu
Web site: http://www.parkinson.org/

Parkinson's Disease Foundation
William Black Medical Building
Columbia-Presbyterian Medical Center
710 West 168th Street, 3rd Floor
New York, NY 10032-9982
Tel: (212) 923-4700; (800) 457-6676
Fax: (212) 923-4778
E-mail address: info@pdf.org
Web site: http://www.parkinsons-foundation.org/

Parkinson's Disease Foundation Midwest Regional Office
833 West Washington Boulevard
Chicago, IL 60607
Tel: (312) 733-1893
Fax: (312) 733-1896
E-mail address: pdfchgo@enteract.com
Web site: http://www.pdf.org

PROGRESSIVE SUPRANUCLEAR PALSY

Society for Progressive Supranuclear Palsy, Inc.
Woodholme Medical Building
1838 Greene Tree Road, Suite 515
Baltimore, MD 21208
Tel: (800) 457-4777
Fax: (410) 486-4283
E-mail address: spsp@erols.com
Web site: http://www.psp.org

RESTLESS LEGS SYNDROME

Restless Legs Syndrome Foundation
819 Second Street S.W.
Rochester, MN 55902-2985
Tel: (507) 287-6465
Fax: (507) 287-6312
E-mail address: RLSFoundation@rls.org
Web site: http://www.rls.org

TARDIVE DYSKINESIA, TARDIVE DYSTONIA

Tardive Dyskinesia/Tardive Dystonia National Association
4424 University Way N.E.
P.O. Box 45732
Seattle, WA 98145-0732
Tel: (206) 522-3166
Fax: (206) 522-3166
E-mail address: skjaer@halcyon.com

TOURETTE'S SYNDROME

Tourette Syndrome Association
42-40 Bell Boulevard
Bayside, NY 11361-2820
Tel: (718) 224-2999; (800) 237-0717
Fax: (718) 279-9596
E-mail address: tourette@ix.netcom.com
Web site: http://TSA.mgh.harvard.edu

TREMOR

International Tremor Foundation
7046 West 105th Street
Overland Park, KS 66212-1803
Tel: (913) 341-3880
Fax: (913) 341-1296
E-mail address: IntTremorFnd@worldnet.att.net
Web site: http://www.essentialtremor.org

WILSON'S DISEASE

Wilson's Disease Association
4 Navaho Drive
Brookfield, CT 06810
Tel: (203) 775-9666; (800) 399-0266
Fax: (203) 743-6196
E-mail address: hasellner@worldnet.att.net
Web site: http://www.wilsonsdisease.org/

WORLDWIDE EDUCATION AND AWARENESS FOR MOVEMENT DISORDERS (WE MOVE)

WE MOVE
204 West 84th Street
New York, NY 10024
Tel: (212) 241-8567; (800) 437-6682
Fax: (212) 987-7363
E-mail address: wemove@wemove.org
Web site: http://www.wemove.org

INDEX